Papers presented at the
International Conference on

Fan Design & Applications

Guildford, England
September 1982

Organised and sponsored by
BHRA Fluid Engineering
England

Editors: H.S. Stephens
Mrs. G.B. Warren

NOTES

The Organisers are not responsible for statements or opinions made in the papers.

The papers have been reproduced by offset printing from the authors' original typescript to minimise delay.

When citing papers from this volume, the following references should be used:-

Title, Author(s), Paper No., Pages, International Conference on Fan Design and Applications, Guildford, U.K. BHRA Fluid Engineering, Cranfield, Bedford, U.K. 7-9 September, 1982.

British Library Cataloguing in Publication Data

International Conference on Fan Design and
Applications *(1982 : Guildford)*
Papers presented at the International Conference
on Fan Design & Applications.
1. Fans (Machinery)—Congresses
I. Title II. BHRA Fluid Engineering
621.6'1 TH7683.F3

ISBN 0-906085-72-1

Printed and published by
BHRA Fluid Engineering
Cranfield, Bedford MK43 0AJ, UK

ISBN 0-906085-72-1
ISSN 0263-421X

ACKNOWLEDGEMENTS

The valuable assistance of the Organising Committee and Panel of referees is gratefully acknowledged.

ORGANISING COMMITTEE

M.P. Hunt (Chairman)	Standard and Pochin Ltd.
P. Baker	BHRA Fluid Engineering
J. Davidson	Davidson & Company Ltd
J. Marples	Aerex Ltd
C.L.B. Noon	Woods of Colchester Ltd
Dr. B.R. Pursall	University of Sheffield
A. Smith	Grubb Parsons
H.S. Stephens	BHRA Fluid Engineering
Mrs. G.B. Warren	BHRA Fluid Engineering

International Conference on
FAN DESIGN AND APPLICATIONS
Guildford, England: 7-9th September, 1982

LIST OF PAPERS PRESENTED

Paper		Page

Paper		Page

SESSION K: FLOW CONTROL AND ENERGY SAVING

International Conference on

Fan Design & Applications

Guildford, England: September 7-9, 1982

PAPER A1

IMPLICATIONS TO THE INDUSTRY OF THE NEW AND PROPOSED BS 848 FAN PERFORMANCE STANDARDS

W.M. Deeprose

Woods of Colchester Limited, U.K.

Summary

The main differences between the new BS 848 fan performance standard and its predecessor of 20 years standing are described particularly with reference to the four standardised airways installations. The resultant changes to the fan aerodynamic performance are shown to be minor for the case of guide vane fans but significant for non-guide vane axial fans when operated in a ducted outlet configuration. It is recognised that this latter difference confirms existing effects rather than suggest that these fan types have different and lower performance. The basic proposals for a new fan noise standard are highlighted in relation to the existing standard and the need to determine in-duct sound power levels from measurements made in anechoically terminated ducts. Typical effects of different duct configurations on fan inlet and outlet sound power levels are shown. Since up to four sets of performance data and up to eight sets of noise data may be needed to characterise the performance of a fan a sample presentation of data is shown to stimulate user discussion of the need and usefulness of such information.

Organised and sponsored by
BHRA Fluid Engineering, Cranfield, Bedford MK43 0AJ, England.

0263 - 421X/82/01 00 - 0001 $5.00
The entire volume can be purchased from
BHRA Fluid Engineering for $82.00

NOMENCLATURE

M = Mach number at duct exit

k = Wave number

a = Duct exit radius

p_{da} = Dynamic pressure in the fan annulus

p_{df} = Dynamic pressure in the fan duct

p_{dd} = Dynamic pressure in the diffuser outlet duct

η_T = Fan total pressure in accordance with BS 848 Part 1

Code or type A fan with GV, 20° or 32°, refers to a free inlet and free outlet guide vane fan whose impeller is set at 20° or 32° to the tangent at the blade tip. Other configurations are defined similarly with the duct connections as stated in the text.

1. INTRODUCTION

The new fan aerodynamic performance standard BS 848 Part 1 published in late 1980 has superseded its predecessor of some 20 years standing. This standard embodies technical resolutions agreed by the International Standards Organisation Technical Committee responsible for fan performance. It is known that some organisations are already using the new standard for specifications. Moreover, because of its similarity to other actual and proposed National Standards, it is likely to be readily accepted by international specifiers of fan equipment for determining satisfactory levels of performance.

The new BS 848 Part 1 has now been available for about 15 to 18 months and in that time many manufacturers will have begun to assess the implications of the standard, not only in terms of test facilities, but in terms of the presentation of the test information. The complication of presentation of performance confronts both manufacturer and specifier alike. On the one hand the information must be reliably and accurately provided by the manufacturer and on the other the specifier must provide greater detail of the fan application to allow a proper fan selection to be made. This is likely to cause problems.

A further major part of the requirements for a new standard is the up-dating of the Part 2 of BS 848. This up-dating could well appear within the next 15 months, and with those two new standards the fan manufacturers in the UK will have available standards embodying the most up to date technical thinking in fan aerodynamics and acoustics.

It is proposed to give a brief indication of where the main technical differences are between the new Part 1 and the old Part 1, and where the proposed Part 2 may differ from the old Part 2. An indication of the main difficulty of implementation of these standards will be discussed as well as an indication of the likely differences which the user might see in the performance of fans. A possible presentation of the various data is included to stimulate interest.

2. USE OF STANDARDISED AIRWAYS

2.1 Duct connections

Fan performance is a function of the ducts connected to the fan at its inlet and outlet. It is well known that a right angled bend adjacent to the fan inlet can have a dramatic effect on the fan performance generation due to

(i) skewing of the meridional velocity profile

(ii) generation of circumferential components of velocity if successive bends are in same clockwise direction.

Straight ducts also have a major effect, this time on velocity pressure recovery, when the extremes of no duct and a long duct are considered at the outlet section.

Clearly, a very large number of combinations may exist for duct connections. However, these fall conveniently into two categories, namely, fans with at least one connection to a long uniform duct of closely similar area to the fan, and fans where at least one connection is free from any influence inhibiting the free in-flow or out-flow of the fan mass flow. These two categories yield four standardised combinations, which are termed installation types, and are as follows.

A Free inlet and free outlet fan, usually a wall or diaphragm mounted exhaust fan.

B Free inlet ducted outlet, usually a supply fan.

C Ducted inlet free outlet, usually a return or exhaust fan.

D Ducted inlet and ducted outlet, usually a fully ducted supply fan.

2.2 Types of test facility

Although duct installations from BS 848 Part 1 1963 may still usefully be part of a new facility, a plenum rig, if chosen, is both versatile and capable of providing test information on, not only the four types of installation, but also on roof units. A fan manufacturer must firstly decide what the scope of his product is to be in relation to installation types, and then provide test facilities according to those needs. Usually the four types of installation data will be required at minimum, and therefore a plenum rig of some description is likely to be necessary. This may be seen by examining Fig. 1, where types B, C and D tests can be conducted from existing ducts, excepting the straightener, although a type A test requires a facility to enable flow to be measured on the upstream side and to allow accurate simulation of a free inlet. This may be achieved using a plenum whose area is 6 times that of the inlet to the fan under test. Such a facility is shown schematically in Fig. 2.

2.3 Use of the common part

The main problem in reliably determining fan static pressure rise in the presence of a downstream duct such as in a B or D installation is the probability that measurement is in the presence of skewed meridional flow and spinning or swirling flow. This is resolved by firstly allowing time for the meridional flow to become uniform, which is a fairly rapid process, and secondly, by incorporating a standardised flow straightener in the discharge to remove swirl only. This length of duct and straightener is termed a 'common part' since it may be common irrespective of method of determining flow rate (Fig. 3) and common from country to country. If a similar loss allowance is given for the common part in other industrialised countries then an important step forward for international standardisation can take place. The use of this common part will ensure that fan static pressure in B and D type installations will be the same internationally.

It is the use of the common part which leads to the most significant problem in the use of the new BS 848 Part 1, especially when a non-guide vane axial fan is considered. This type of fan contains at its outlet a proportion of energy in the form of swirl energy and when tested with a short discharge duct (2D) some recovery of that swirl energy and the always present excess axial energy may occur. However, the intermediate ducted outlet case is not one of the recognised standard installation types and therefore this fan, when tested without 2D discharge, will appear to have a lower pressure rise. Moreover, when tested with a fully ducted discharge, including the common part, the extra recovery which is obtained through using a ducted discharge may be, and in many cases will certainly be, more than compensated by the higher actual losses across the standard straightener due to the presence of swirl. These high losses cannot, of course, be credited to the fan since the loss permitted to be credited is based only on uniform swirl free flow. The actual losses of the straightener vary considerably depending on the degree of swirl and axial energy present and its radial distribution, and it has not been possible to produce losses which give a better measure of the true loss in the presence of swirling flow.

3. TYPICAL FAN PERFORMANCE

The fan facility shown schematically in Fig. 2 permits fans up to 1.4 metre diameter to be tested in accordance with the new standard in each of the 4 standardised conditions. This facility permits all data to be gathered electronically, sampled and analysed by on-line computer. In order that the salient points of difference between the old and new standards may be illustrated a 1 metre diameter axial flow fan was tested with and without downstream guide vanes at an intermediate and high pitch setting. Additionally, tests were conducted with both a previously conventional 2D discharge and a conical diffuser.

3.1 Guide vane axial fan

If care is taken to ensure good in-flow conditions, namely fully developed flow, when a free inlet exists, then the difference in performance to be expected is small for the same outlet configuration, irrespective of the inlet. Therefore, a type B and D fan installation should provide similar results, and a type A and C installation should provide similar results, but different to B and D types. Examination of tests on a guide vane axial fan at both pitch angle settings indicates the systematic difference found. These differences in both fan static pressure and total efficiency shown in Fig. 4, whilst varying in detail, show the A and C types to be identical but lower than the B and D type results which are themselves identical. This difference in principle has been long recognised since fan performance with a 2D discharge is usually very much better than an unducted discharge, due to the recovery of both the axial energy and the residual swirl energy at the fan outlet plane. A further example of this is the use of conical or annular diffusers to improve fan performance by the recovery of otherwise unused but available energy. The amount of recovery which may be obtained in any of these cases is obviously related to the proportion of the total pressure rise contributed by the fan annular velocity pressure, and in a complex manner to the amount of residual swirl discharging into the downstream duct.

3.2 Non-guide vane axial fan

In this case since the B and D type tests yield the same performance and A and C types yield indentical results but different to B and D, the difference between C and D types only are shown for clarity in Figs. 5 and 6. In order to emphasise the difference in a C and D type performance Fig. 5 shows the fan in its C configuration when fitted with a diffuser to be the equivalent of 2° in pitch angle better than the performance of the same fan in a D type installation. This fan pitch angle was set at 20°. When the pitch angle was set at 32° the influence of the test configuration on fan performance is seen dramatically. At peak efficiency flow rate the change in fan static pressure is around 15% in comparing type C to type D. This may be seen in context by examining the hitherto conventional performance presentation of a non-guide vane fan with a 2D discharge duct. The performance with a 2D discharge is a further 10 to 15% above that of the C type installation. The addition of a simple conical diffuser to the fan in a C type test significantly improves the fan performance - but only for a C type installation. Here the diffuser is transforming the swirl energy and excess axial energy to useful pressure rise. This emphasises that it is not good practice to install a non-guide vane fan with a long outlet duct, since the swirl present will cause larger friction losses in the system. The system losses are calculated on fully developed flow profile and incoming swirl will cause these losses to be different and higher. Moreover the utility of diffusers is clearly shown in return fan systems in terms of energy saving since in Fig. 6 an extra 10% of volume on a system line is achieved at this maximum pitch angle setting by use of a diffuser. The apparently poor performance of the fan in a type D installation shows that the extra recovery in a ducted outlet is more than compensated for by the extra losses of swirling flow over the straightener.

Therefore, a non-guide vane fan with relatively short ducts will still allow the extra recovery noticed on conventional tests with 2D discharge to be achieved, since the extra recovery more than compensates for the extra loss. This should be so for ducts up to say 5 or 6D long.

4. ACOUSTIC PERFORMANCE

4.1 The proposed BS 848 Part 2

Both installation categories previously discussed will yield systematic differences in aerodynamic performance, especially where outlet swirl exists. Acoustically a standard is being prepared where the four installation types are also used as standardised airways. The existing BS 848 Part 2 recognises the possibility that inlet and outlet fan noise may be different and the proposed standard continues to recognise this. As an example, fan inlet noise for a type B and D installation can be different due to the presence or absence of a duct at the fan inlet. The inlet noise for these cases must be assessed in the inlet duct, if it is present, or in a reverberant or free field, if the inlet duct is absent. At the same time, and for the same two installations, the noise levels measured in the ducted outlet are themselves liable to be different. This allows the possibility of up to 8 different noise levels being necessary to characterise a fan noise level. Like the existing BS 848 Part 2 the proposed document permits presentation of a very much reduced amount of information, provided that the ducted or unducted levels at inlet or outlet are within 2 dB of each other. Moreover, and in a similar way, if the inlet and outlet noise levels are within 2 dB of each other then these may be assumed to be equal to the higher value. The proposal currently indicates that simultaneous acoustic and aerodynamic measurements are not possible when there is a ducted discharge due to the need for the common part to be present for aerodynamic measurements. This arises because the proposal requires a duct, wherever present, to be terminated anechoically. If this is not done a non-standard acoustic impedance will exist and the noise output from the fan will become dependent on the system properties of the manufacturer's test duct arrangement.

In contrast to this view AMCA are working to retain the determination of in-duct sound power levels by use of measurements made in a reverberant field. These measurements will be corrected by the well known end reflection factors to enable in-duct levels to be estimated and further corrected to allow for flow effects. The flow correction given by Bechert (1) to predict the attentuation for sound power transmission is

$$R = 10 \log \frac{L_W \text{ radiated}}{L_W \text{ transmitted}} = 10 \log \left(\frac{1 + (4/3)M^2}{(4M)} (ka)^2 \right)$$

when $4M \gg (ka)^2$ and for small ka.

This Mach number correction does give good agreement[(2)] when comparing sound power levels likely to be present in a duct with those measured in a reverberant room. However, this approach does not recognise that the fan itself may (and in many cases will) generate a different level of noise as a function of frequency depending upon the acoustic impedance presented to the fan inlet and outlet connections. Thus the end reflection and flow correction of this end reflection will more reliably predict levels in-duct from reverberation room measurements, they will not provide any information on the effect of an unterminated duct on fan noise generation.

Lastly the proposed standard recognises the problem in fan duct systems of turbulence. Turbulence, when acting upon a microphone, creates pressure fluctuations which, so far as the microphone is concerned, appear as if they were real noise. In fact they are pseudo noise. The proposal advocates the use of either a turbulence screen or a similar attachment to minimise these turbulent pressure fluctuations at the microphone, and thereby allow measurement of mainly real noise. The measured noise levels, as a result, are likely to be lower than measurements using earlier standards. An example of this is shown in Fig. 7 where the important low frequency measurements are lower through use of certain windshields which reduce any turbulent pressure fluctuation at the microphone.

These are the main differences between the proposed standard and the existing standard, and will result in a test facility which can be significantly longer than currently required, with a maximum length for a type D test of 20D.[(3)] This is primarily due to the need for anechoic terminations. However recent work at the National Engineering Laboratories has shown that short, inexpensive and satisfactory terminators can be designed which will make the test rig more manageable in size and cost.

4.2 Acoustic performance in standardised airways

A series of fan types representative of each specific speed type were evaluated at the National Engineering Laboratories in order to gain an appreciation of the likely effects of duct configurations on fan noise generation. Table 1 and Figs. 8 and 9 summarise these results. Accepting that type A tests are absent the results nevertheless demonstrate that the duct impedances presented to the fan influence the fan noise generation in a very variable manner. The most dramatic effect in the data presented are from a mixed flow fan test. At both inlet and outlet the noise levels in the non-plane wave region are encompassed within a range of ± 2 dB although the level is different between inlet and outlet. The most significant effect is in the low frequency or plane wave region. The presence or absence of a terminated duct on the non-measurement side affects the inlet or outlet by 7 to 10 dB. The presence or absence of a duct on the measured side is equally significant. Whilst this effect may be qualitatively considered to be caused by the total impedance effects of both inlet and outlet ducts as the load impedance on the source, for a detailed explanation knowledge of the source impedance is necessary. The proposed standard does not suggest measurement of this quantity but clearly it is possible for fan noise generation to be greatly affected by both total and constituent load impedances. The data in Table 1 show much smaller effects on a range of fans with little systematic effect evident. However, these data are related only to the difference in ducted sound power levels between a type C and type B test. This evidence shows the outlet noise to be greatest with the medium specific speed mixed flow fan but lowest with the in-line radial and the bifurcated axial. These differences in inlet and outlet noise on the ducted side extend over the entire frequency range.

5. IMPLICATIONS FOR IMPLEMENTATION OF THE NEW STANDARDS

5.1 Aerodynamic performance

When a basic set of data for all types of installation has been established, a series of charts or factors must be developed to convert information from a type A or C test to suitable data for types B and D installations, or vice versa. Furthermore, if additional outlet side connections are to be supplied, such as diffusers, which may be conical or annular, and short or long, then a further series of correction charts will be necessary for each additional outlet side connection to be used.

With such a comprehensive amount of information it is readily seen that there could be complications in the selection and specification of fans. If a user does not fully specify the installation requirements in terms of one of these standard types, great care is necessary when the manufacturer makes a selection, since the performance of the fan is clearly dependent on the duct system to which it is applied. Having said that, the fans themselves, of course, are no different to those previously offered, and, therefore, the performances in reality, in the systems selected, should be the same. What we are talking about is the recognition that there is a different level of performance for fans in different systems. Once again, since the standard installation types are not wholly universal, a series of installation factors may well be required which will place a greater load on manufacturers' laboratories and ingenuity in attempting to synthesise what will ultimately become a considerable amount of data into a manageable and reliable form. For this reason the Fan Manufacturers' Association have agreed that the full implementation of this standard will be achieved only over the next 4 to 5 year period.

5.2 Acoustic performance

Once again, with the acoustic proposal, more comprehensive facilities are needed, and due to the volume of work necessary, many manufacturers will find it essential to use installations which have both computerised data reduction equipment and at least automated data acquisition equipment. For production of product noise levels in catalogues, if both inlet and outlet levels for all installation types are not sufficiently similar, then ingenuity will be required by the manufacturer to present this information in a simple form but conveying the fan noise level differences clearly. Since the main differences in acoustic performance with the new proposal are likely to be in the low frequency region, it is readily recognised why it is important to obtain the most reliable data. This frequency region controls silencer installation costs, and consequently the importance of the proposal is evident.

As the effects of impedance are not readily calculable or constant from one fan to another, it will be necessary to evaluate these effects on a fan by fan basis, and again provide reliable but accurate factors to enable the system designer, specifier or user to more certainly predict the resultant noise levels. For interest Fig. 10 has been constructed as a possible method of presenting test information in future fan catalogues. It is necessary to find a method whereby

(i) the standard motor absorbed power is shown;

(ii) the fan static pressure rise performance is shown for a stated installation type;

(iii) this static pressure rise is corrected to each of the other 3 installation types;

(iv) corrections are provided for use of say an annular diffuser, conical diffuser, a 2D discharge connection, a restricted plenum arrangement such as in air handling units;

(v) the total sound power levels for either inlet or outlet in a stated installation configuration;

(vi) correction factors for inlet to outlet or vice versa for each other condition;

(vii) possible corrections to the frequency spectrum.

It will be noted that no emphasis is placed on fan efficiency. The reason for this is twofold. Firstly it should be considered as an index of quality, since the numerical efficiency value is a matter of convention rather than a definition of the aerodynamic effectiveness of an impeller and/or guide vane. Secondly, impeller power is here unambiguous and use of this term, together with static pressure rise and the fan velocity and annular pressure gives sufficient information to enable sensible calculations of fan performance in a system to be made.

6. CONCLUSION

With the possibility of this comprehensive information becoming available, the user, specifier and designer must consider how much of it is needed and to what use this information will be put. It is easy to see the source of confusion where lack of information to the supplier on installation type can lead to the manufacturer interpreting the customer requirement in different ways. For example, it may be assumed that the user wants a fan for a type D static pressure requirement and the supplier offers a non-guide vane fan tested in a C type installation which in a type D installation will be unlikely to provide the designed performance of the system. The non-guide vane fan is obviously cheaper than a guide vane fan or fan with diffuser. Therefore much careful consideration of the implications of these new standards must be made by all parties to the contract for purchase of a heating and ventilating system. The new BS 848 Part 1 and the proposed 848 Part 2, whilst more complex, will allow greater confidence from the user that the system performance will be a further step closer to that which the designer expects. The new installation types, whilst not suitable for all practical cases, do recognise acoustically and aerodynamically, the systematic differences which may be found in fan performance. Furthermore, whilst manufacturers must create more comprehensive facilities, those manufacturers exporting a significant proportion of their products may take confidence that a broad measure of international agreement exists at the base of these new fan standards.

7. ACKNOWLEDGEMENTS

The presentation of information contained in Figs. 8 and 9 and in Table 1 is made with the permission of the Director, National Engineering Laboratories, East Kilbride, Glasgow.

8. REFERENCE

1. Bechert, D.W. "Sound Absorption caused by Vorticity Shedding demonstrated with a Jet Flow". AIAA 5th Aeroacoustics Conference Seattle Washington March 1979 Paper 79-0575.

2. Sandbakkan, M., Parde, L. and Crocker, M.J. "Investigation of End Reflection Coefficient Accuracy Problems with AMCA Standard 300-67". Purdue University Report HL 81-16 April 1981.

3. Bolton, A.N. and Margetts E.J. "Anechoic terminations for in-duct fan noise measurement". National Engineering Laboratory, UK. To be presented at this Conference.

Fan Type

Frequency Hz	Mixed Flow	Mixed Flow	In Line Radial	Bifurcated Axial	Axial	Backward Curved Centrifugal
	Duct Configuration					
	Type D	Type B -Type C	Type B -Type C	Type B -Type C	Type B -Type C	Type B - Type C
	dBW	dBW	dBW	dBW	dBW	dBW
50	5.0	17.0	-1.9	7.1	-0.4	-4.0
63	8.5	9.9	-2.1	1.9	1.8	0.8
80	6.0	12.3	-2.9	-5.3	-1.1	2.1
100	8.0	5.2	-7.8	-5.2	-1.7	4.4
125	9.5	8.0	-10.7	-9.3	-11.3	4.7
160	9.0	16.3	-1.6	-16.2	-10.7	4.1
200	5.5	5.3	-4.8	-4.0	-3.4	-2.7
250	4.0	4.8	-4.2	0.7	-2.0	0.5
315	6.5	6.6	-4.2	-6.1	-0.5	3.9
400	7.0	9.6	0.1	-5.0	-0.2	3.0
500	7.0	8.4	3.5	-3.3	0	5.1
630	5.5	3.1	4.8	-2.6	-2.1	5.0
800	5.0	1.5	1.5	-1.8	-2.2	4.9
1000	2.5	-0.7	-0.7	-3.2	-1.8	5.3
1250	-2.5	-3.7	-0.6	-4.3	-2.3	3.6
1600	1.0	2.4	-0.8	-3.3	-2.5	3.3
2000	2.0	4.7	-1.7	-3.7	-2.7	3.0
2500	0	3.0	-2.6	-3.7	-3.0	1.8
3150	-4.5	1.2	-3.3	-4.7	-3.7	2.2
4000	-2.5	3.2	-3.6	-5.7	-3.4	2.5
TOTAL	4.6	6.8	-2.1	-2.9	-3.7	-0.9

TABLE 1: DIFFERENCE BETWEEN OUTLET AND INLET INDUCT SOUND POWER LEVELS FOR VARIOUS FAN TYPES EACH AT THEIR DESIGN FLOWRATE EXPRESSED AS (OUTLET - INLET) IN SOUND POWER LEVEL RE 10^{-12} WATTS

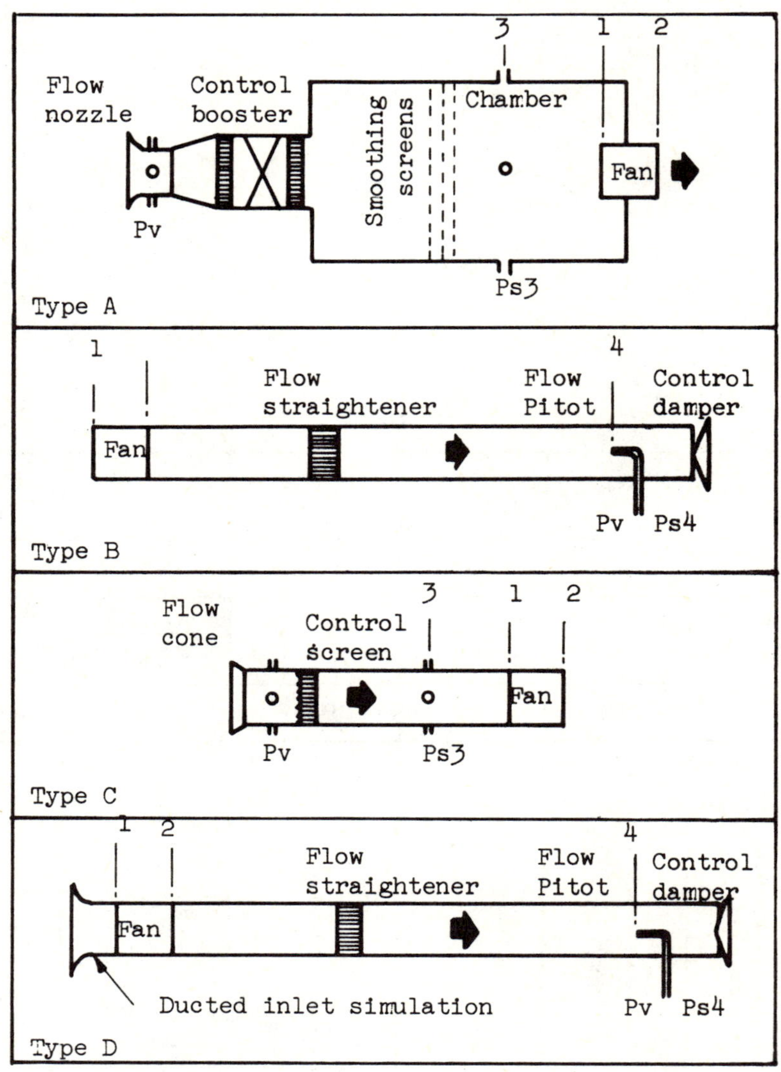

Fig.1 EXAMPLES OF STANDARDISED AIRWAYS

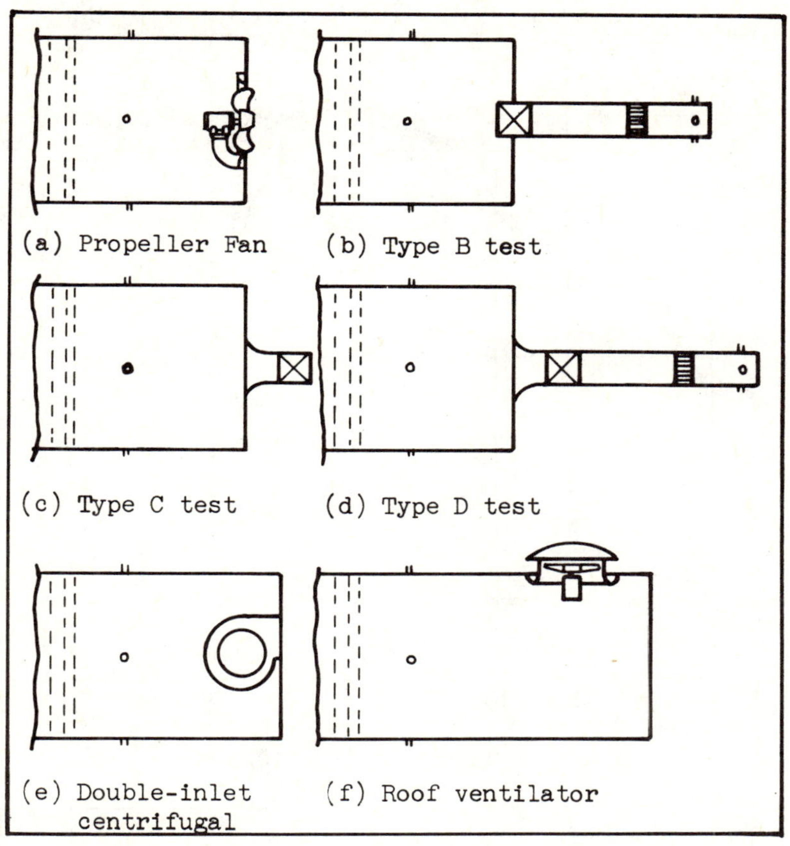

Fig. 2 STANDARDISED AIRWAY TESTS WITH AN INLET SIDE CHAMBER

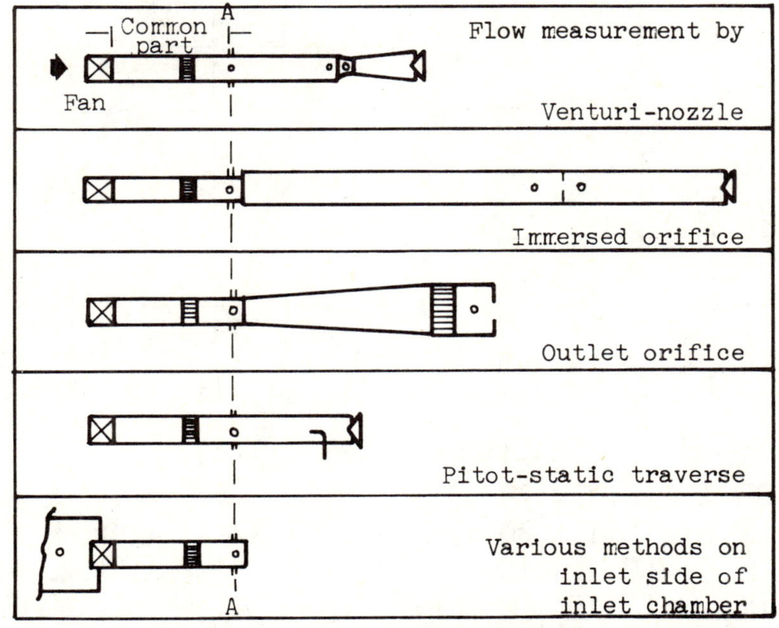

Fig. 3 PRINCIPLE OF THE COMMON PART APPLIED TO TYPE B TEST AIRWAYS

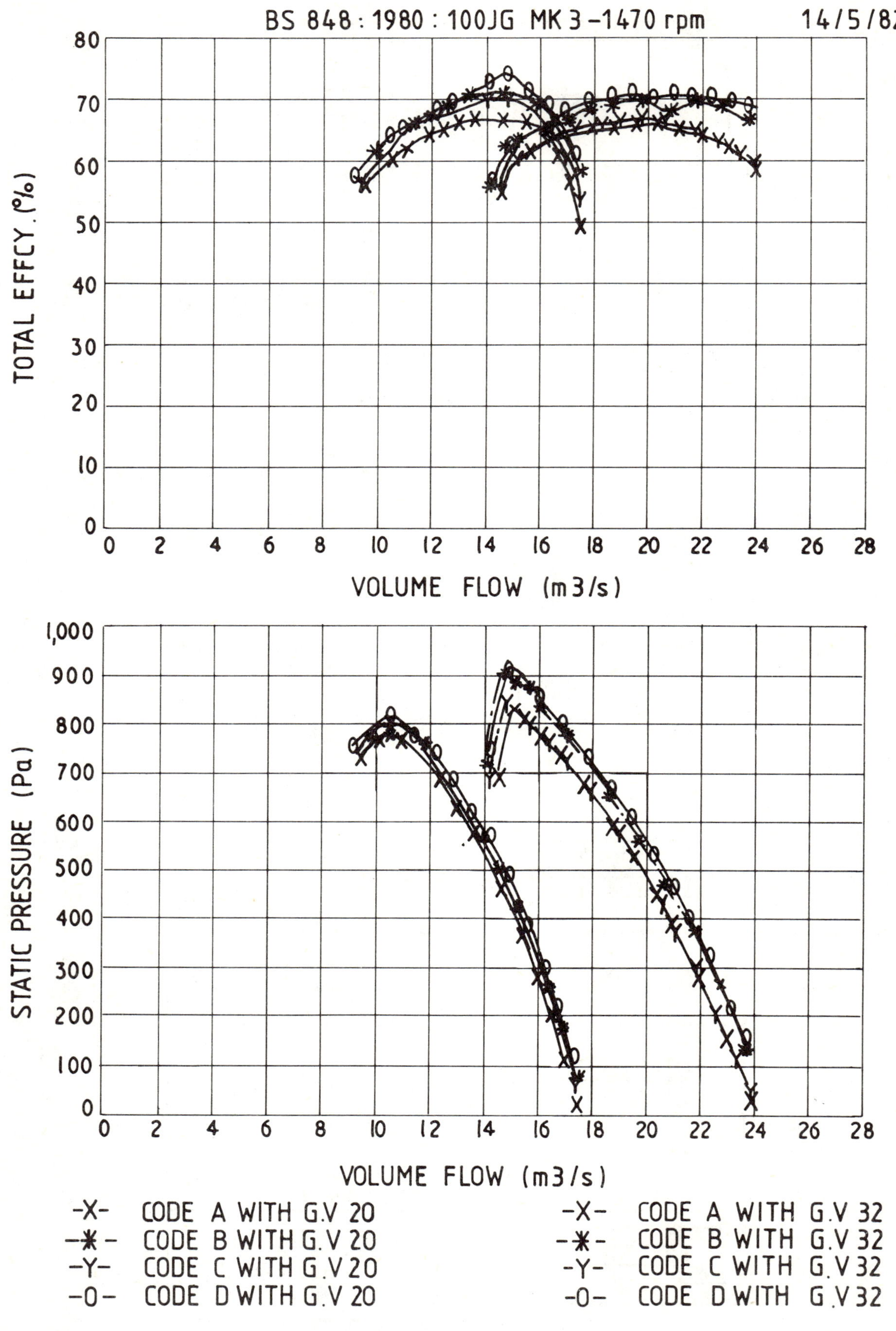

Fig.4 EFFECT OF TEST INSTALLATION ON GUIDE VANE FAN PERFORMANCE

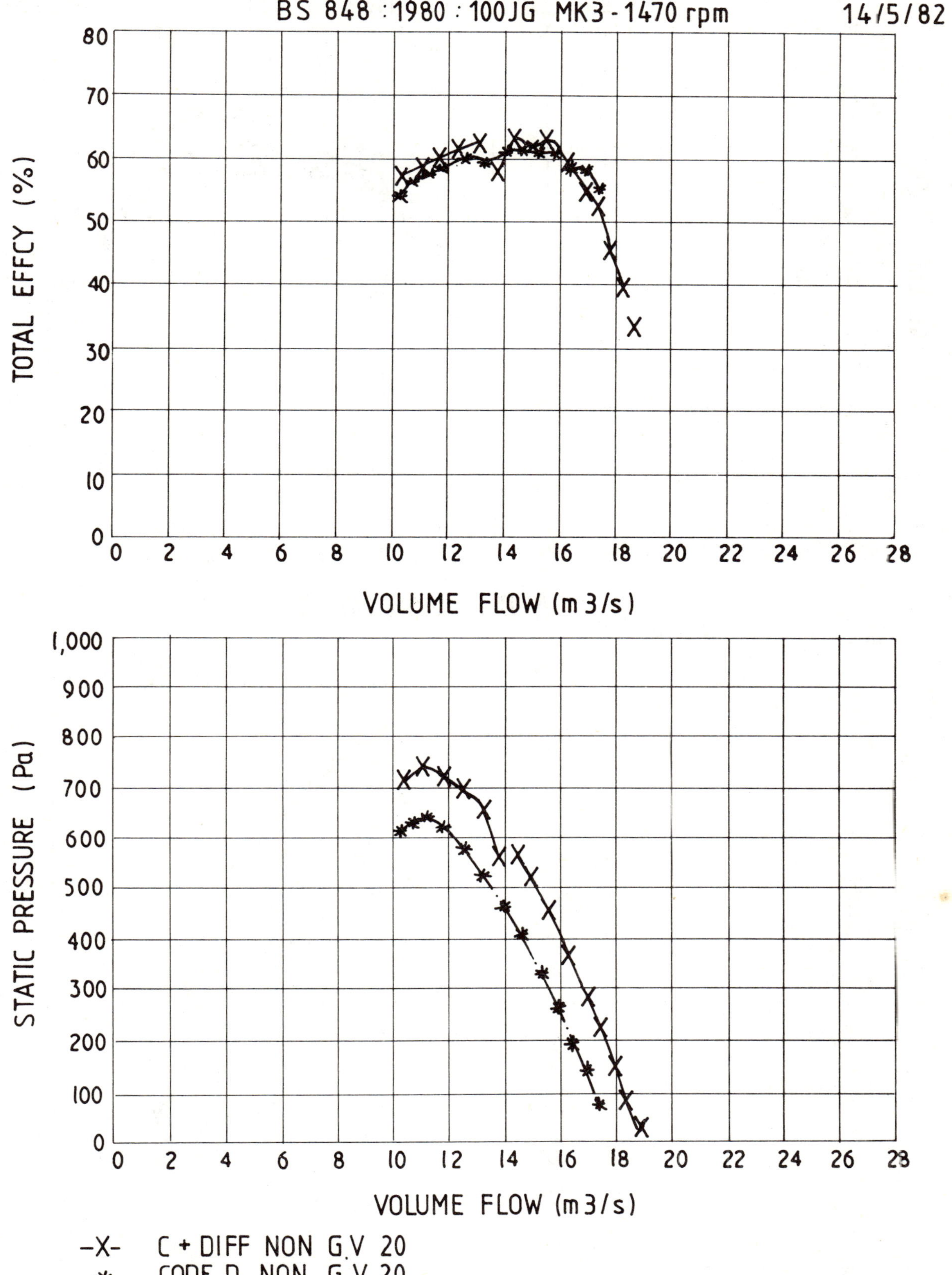

Fig.5 EFFECT OF OUTLET CONNECTIONS ON LOW PITCH ANGLE PERFORMANCE

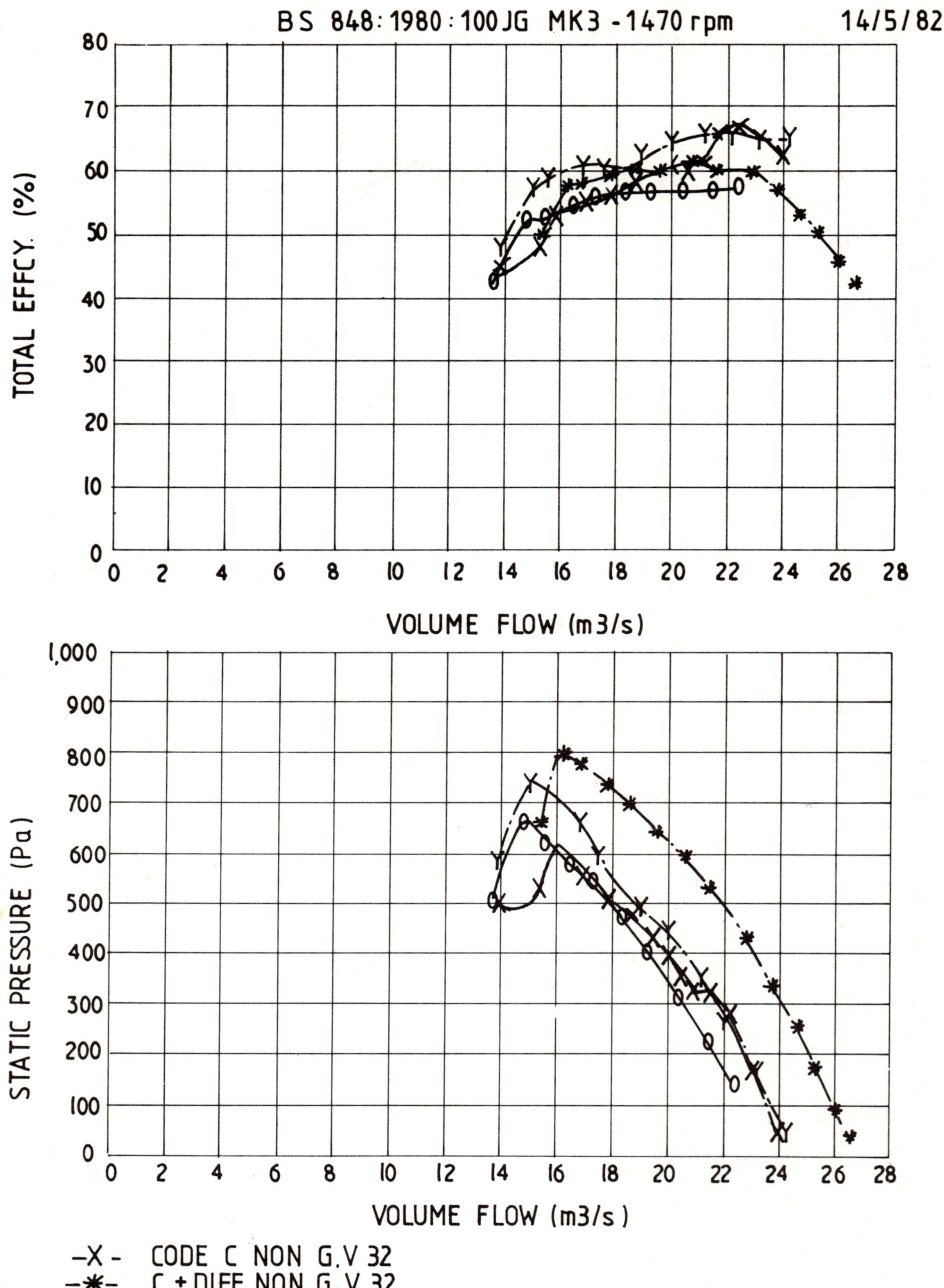

Fig.6 EFFECT OF OUTLET CONNECTIONS ON HIGH PITCH ANGLE PERFORMANCE

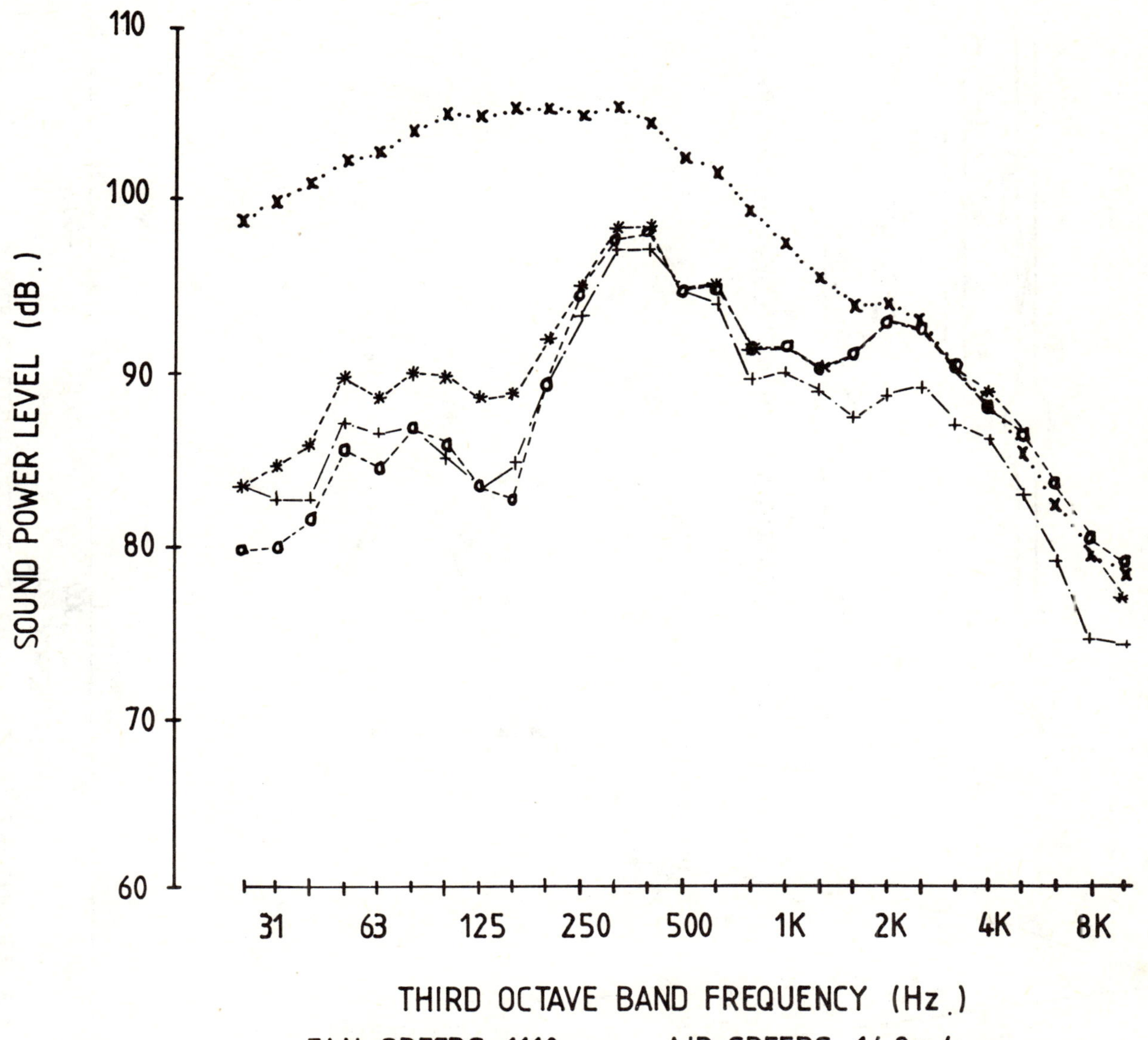

FAN SPEEDS. 1110 rpm. AIR SPEEDS. 14·9 m/sec.

x········x GRID ONLY
a- - - - - - -a FOAM BALL
✱- - - - - - -✱ NOSECONE
+——·——·+ TURBULENCE SCREEN

Fig. 7 IN-DUCT TESTS ON B.S. 848
OUTLET TESTS – HIGH SPEED
MIXED FLOW FAN

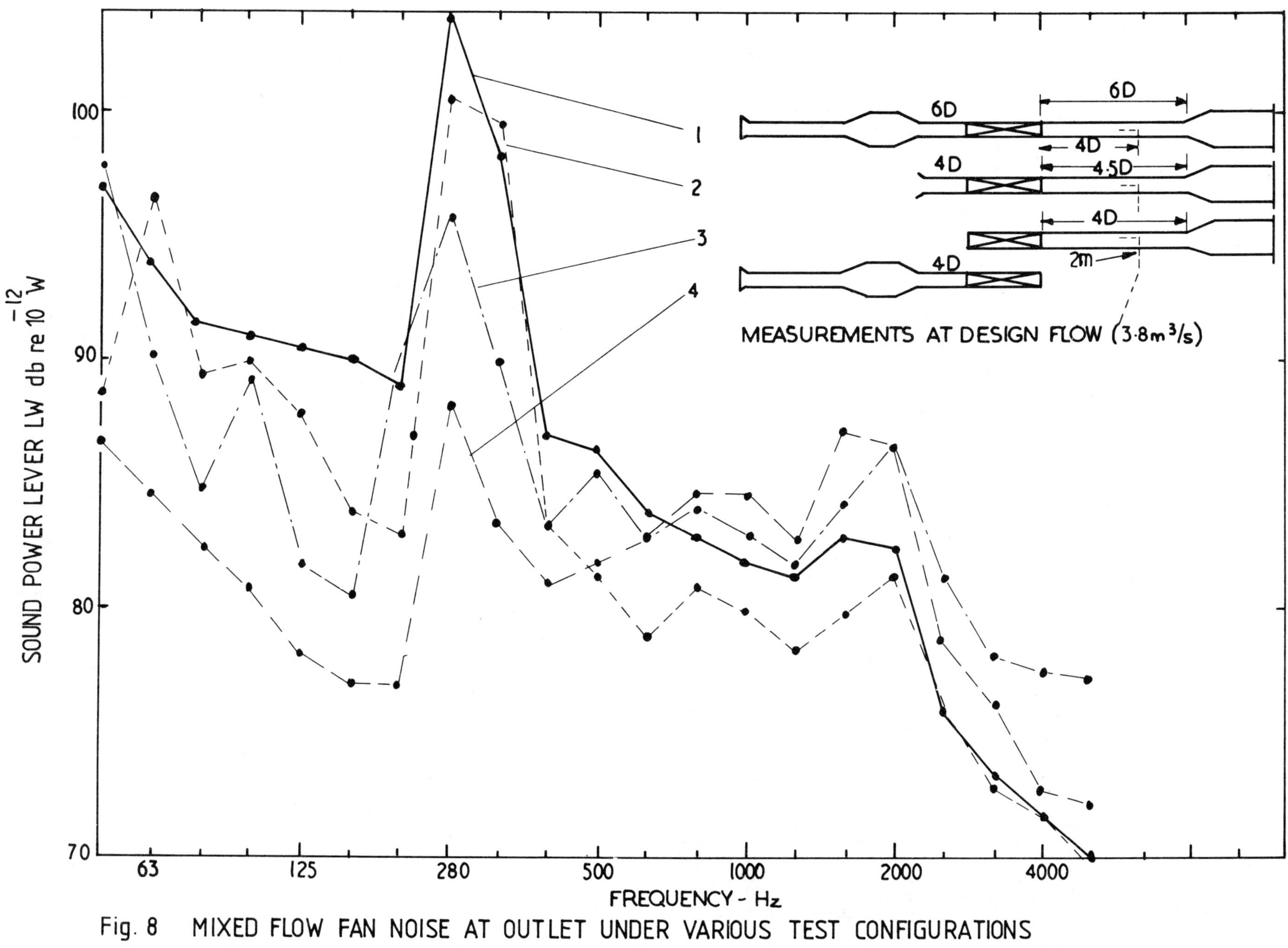

Fig. 8 MIXED FLOW FAN NOISE AT OUTLET UNDER VARIOUS TEST CONFIGURATIONS

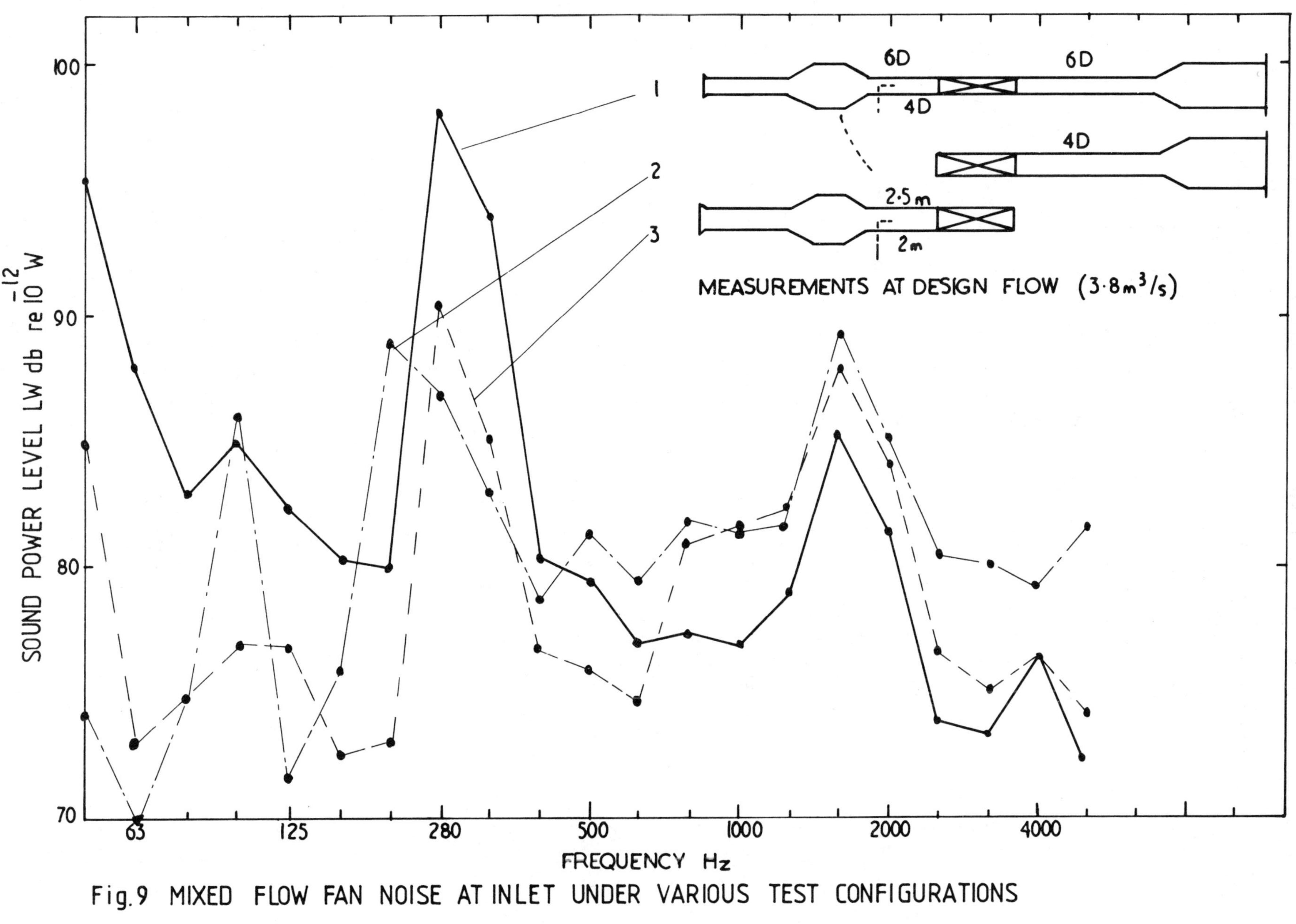

Fig.9 MIXED FLOW FAN NOISE AT INLET UNDER VARIOUS TEST CONFIGURATIONS

FAN CODE: 100 JG 40A-4-9 REV/MIN: 1470 Hz: 50

CORRECTION TO D TYPE OUTLET TOTAL SOUND POWER LEVEL db

f	TOTAL	63	125	250	500	1k	2k	4k	8k
A_0	-6	-10	-17	-9	-18	-18	-20	-22	-25
B_0	-4	-8	-16	-9	-17	-19	-17	-24	-31
C_0	-10	-17	-23	-16	-19	-18	-16	-27	-33
D_0	0	-7	-12	-2	-15	-20	-20	-30	-35
A_I	-8	-12	-19	-10	-20	-20	-18	-23	-29
B_I	-10	-28	-20	-15	-22	-20	-15	-21	-24
C_I	-11	-21	-26	-15	-26	-20	-17	-26	-30
D_I	-5	-10	-19	-7	-12	-13	-19	-27	-33

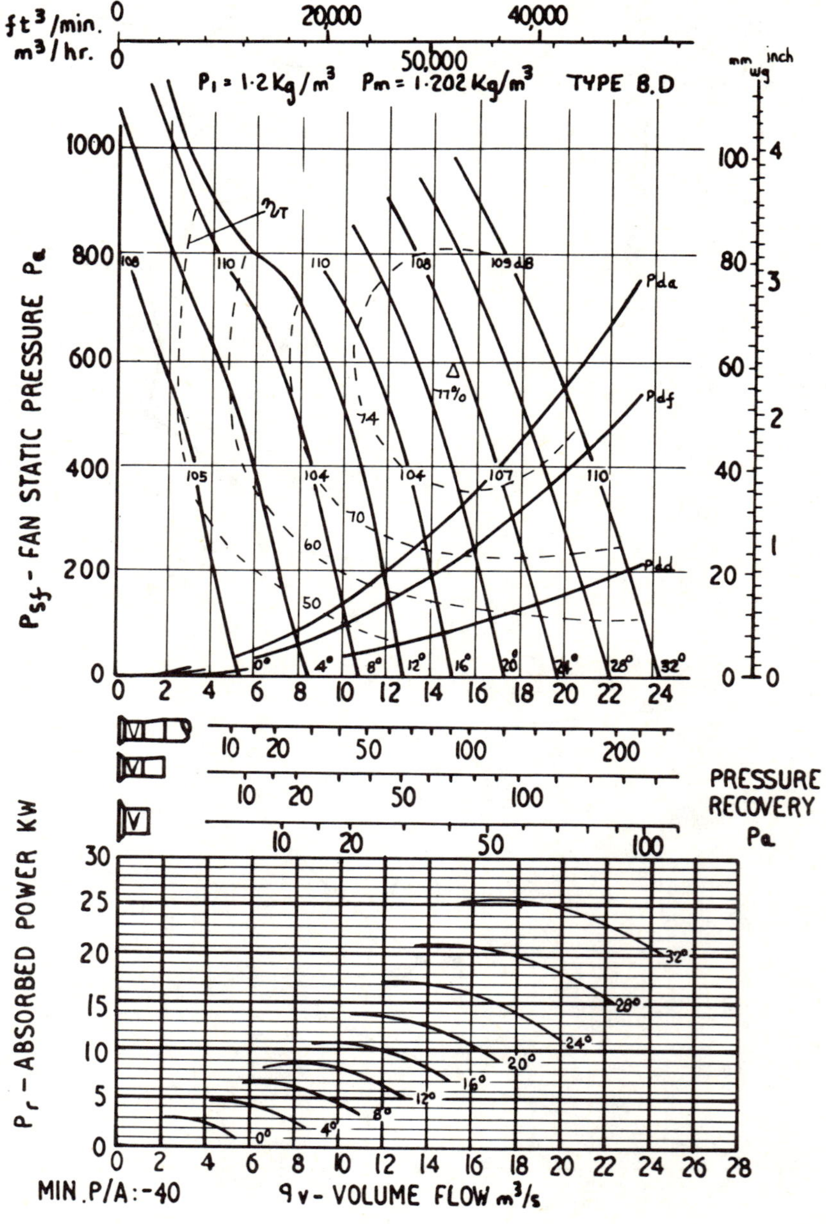

Fig. 10 : POSSIBLE DATA PRESENTATION OF FAN PERFORMANCE IN STANDARD AIRWAYS

International Conference on

Fan Design & Applications

Guildford, England: September 7-9, 1982

PAPER A2

FAN PERFORMANCE TESTING AND THE EFFECTS OF THE SYSTEM

W.T.W. Cory

Keith Blackman Limited, U.K.

Summary

BS 848 1980 shows that fans may be tested a number of ways. These give similar results for the same category but there are differences according to the conditions at fan inlet and/or outlet. Four installation types are recognised.

The writer's company have tested their centrifugal fans as standard to Category D (ducted inlet and outlet) but comparisons for other installation types have been made. Generally the greater variations are with high specific speed fans. Performance curves for various categories are presented and the differences explained.

The design of inlet and outlet connections has an important effect on the fan performance. Air/gas should preferably be delivered to the fan inlet as a straight line flow. Spin (pre or contra) alters the pressure developed. An asymmetrical velocity profile has a similar effect. Sufficient straight on the outlet ensures adequate conversion of velocity pressure at the fan throat to static pressure within the duct. Despite the need for good connections there are many installations where site geometry precludes their use. Test results are presented to help assess the effects of less than ideal ducting.

System designers have access to information giving pressure losses as some fraction of the local velocity pressure, but these assume a developed symmetrical velocity profile. Where fittings are adjacent to the fan, an additional effect may be anticipated as the profile is far from this ideal. The additional 'loss' for a number of different duct configurations is given. These, being virtually impossible to measure directly, are best calculated. The method described is to test both with and without the fitting interposed at appropriate points.

Organised and sponsored by
BHRA Fluid Engineering, Cranfield, Bedford MK43 0AJ, England.

0263 - 421X/82/01 00 - 0001 $5.00
The entire volume can be purchased from
BHRA Fluid Engineering for $82.00

NOMENCLATURE.

Symbol	Term	Unit
p_s	Static pressure at a point	Pa
p_{sx}	Average static pressure at an airway section 'x'	Pa
p_d	Dynamic pressure at a point	Pa
p_{dx}	Effective dynamic pressure at an airway section 'x'	Pa
p_t	Total pressure at a point	Pa
p_{tx}	Effective total pressure at an airway section	Pa
p_{tF}	Fan total pressure	Pa
p_{dF}	Fan dynamic pressure	Pa
p_{sF}	Fan static pressure	Pa
Δp	Differential pressure across a flow measurement device	Pa
Δp_j	Differential pressure at point j in a pitot-static traverse	Pa
Δp_{eff}	Effective differential pressure at the section traversed	Pa
p_u	Absolute pressure upstream of a flow measurement device	Pa
p_{fxy}	Loss of pressure due to friction between airway sections 'x' and 'y'	Pa
p_a	Atmospheric pressure in test enclosure	Pa
t	Temperature	°C
t_a	Dry-bulb temperature in test enclosure	°C
t_x	Average temperature at an airway section 'x'	°C
ρ_a	Density of air in test enclosure	kg/m^3
ρ_x	Average density at an airway section 'x'	kg/m^3
ρ_i	Inlet density	kg/m^3
q_m	Mass flow	kg/s
q_v	Inlet volume flow	m^3/s
Re_d Re_D	Reynolds number $\frac{vd\rho}{\mu}$ or $\frac{vD\rho}{\mu}$	-
η_t	Fan total efficiency	-
η_s	Fan static efficiency	-
v	Velocity of air at a point	m/s
v_x	Average velocity of air at an airway section 'x'	m/s
n	Rotational speed (in revolutions per second)	r/s
D	Diameter of airway	m
D_R	Impeller tip diameter	m
d	Diameter of orifice or nozzle throat	m
L	Length of airway	m
L_{T2}	Length of transitional airway	m
L_{s1}	Length of transitional section	m
l	Length of cell of square or hexagonal cross section	m
w	Width of cell of square or hexagonal cross section	m
$b \times h$	Dimensions or a rectangular airway	m

Symbol	Term	Unit
A_x	Cross-sectional area of airway at a section 'x'	m^2
μ	Dynamic viscosity	Pa s
ν	Kinematic viscosity = $\frac{\mu}{\rho}$	m^2/s
k_p	Compressibility coefficient	-
α	Flow coefficient	-
ϵ	Expansibility factor	-
λ	Darcy friction factor	-
ζ	Conventional friction loss coefficient	-
D_h	Hydraulic diameter = $\frac{\text{4 x area of cross section}}{\text{perimeter of cross section}}$	m
m	Number of straight traverse lines	-
k	Number of measuring points	-
g	Acceleration due to gravity	m/s^2

Subscripts. The following subscripts identify the locations specified for particular measurements, air conditions and calculated quantities.

Subscript	Location
a	The ambient atmosphere in the test enclosure
1	The test fan inlet
2	The test fan outlet
3	The pressure measurement section in an inlet side airway
4	The pressure measurement section in an outlet side airway
xy	The airway length from plane x to plane y

Multiples And Sub-Multiples. The following are the only multiples and sub-multiples of the SI units used.

Pressure	kilopascal	1kPa = 10^3 Pa
Barometric pressure	millibar	1mbar = 100 Pa
Volume flow	litre per second	1 litre/s = 10^{-3} m^3/s
Power	kilowatt	1 kW = 10^3 W
Efficiency	per cent	% = 100 x ratio
Rotational speed	revolutions per minute	60 r/min = 1 r/s
Dimensions	millimetre	1 mm = 10^{-3} m

1. INTRODUCTION.

Perhaps the most important event in the fan world over the last few years has been the publication of the long awaited standard for performance testing - B.S. 848 : Part 1 : 1980 (Ref. 1). The need to revise the 1963 edition had been evident for sometime. Whilst the original intention was to await the completion by I.S.O. of a new international standard, it was seen that this was still some way off. The decision to produce a new national standard was therefore understandable especially as it incorporated all the decisions to date made at the international level.

It is an unfortunate fact that the test methods of different countries give different results. These differences can be as much as 13% on efficiency for axial flow fans without guide vanes tested under free inlet, ducted outlet conditions (Ref. 2). Recent tests on centrifugal fans with forward curved multivane impellers conducted by the author's company suggest that the same fan tested according to the requirements of different national codes and for different installation categories may result in calculated efficiency differences of up to 10%. Needless to say, B.S. 848 : 1980 tests and calculations gave the lowest results, which was to be expected. The differences depend chiefly on the allowable flow patterns at inlet and outlet and to a lesser extent on the values assumed for flow coefficients of the various primary devices such as orifice plates, inlet cones and venturi nozzles. Agreement within I.S.O. backed by an experimental programme has resulted in the specification and design of common parts adjacent to the fan sufficient to ensure consistent determination of fan pressure without the 'benefit' of any rotational energy. To date however only B.S. 848 : 1980 incorporates these recommendations. It is to be hoped that an international standard will not be long delayed for it is only when this exists that meaningful comparisons will be possible between the performance claims of different nations' fans. Unfortunately, whilst units of a given design may be manufactured in quantity the systems to which they are attached are invariably unique. Ductwork designers therefore usually add a safety margin to their calculated system resistance and often being of the order of 10% to 20% it is sufficient to mask such differences. In an energy starved world however this cannot continue for increased system resistance means increased power consumption.

2. INSTALLATION CATEGORY.

Fans may be tested in a number of ways. Whilst these should give similar results for the same category, it is known that there will be differences according to the conditions at fan inlet and/or outlet. Both I.S.O. and B.S.I. recognise the following four installation arrangements :-

Type A : free inlet and free outlet
Type B : free inlet and ducted outlet
Type C : ducted inlet and free outlet
Type D : ducted inlet and ducted outlet

In this classification the terms have the following meanings :-

Free inlet or outlet signifies that the air enters or leaves the fan directly from or to the unobstructed free atmosphere except that in installations of type A a partition in which the fan is mounted may support a pressure difference between the inlet and outlet sides.

Ducted inlet or outlet signifies that the air enters or leaves the fan through a long, straight duct directly connected to, and of the same cross-sectional area as, the fan inlet or outlet respectively. It is important for the supplier to state the installation types for which it is intended whilst the user should select from the installation types available, the one which matches or is closest to his application. As a general rule it is found that the greatest difference in performance curves will exist with the higher specific speed fans in both the axial and centrifugal categories. For axial flow fans these differences will be greatest where there are no guide vanes to reclaim rotational energy and where flows are sufficiently high to require minimum inlet losses. For centrifugal fans the throat area is generally a smaller percentage of the fan outlet area, the velocity pressure is a greater proportion of the total, and the fan inlet is larger relative to the impeller diameter. Thus on the discharge side further conversion of velocity pressure to useful static pressure takes place in a straight outlet duct. On the inlet side the larger area can result in variable velocity distributions. Again, the greater the volume flow the greater will be these effects.

3. INSTALLATION EFFECTS.

Despite the need to establish good inlet and outlet conditions it has to be recognised that there will be many installations where lack of space or the site geometry will preclude the fitting of ideal duct connections. Normally the system designer will have access to information (Refs. 3 and 4) giving the pressure loss in fittings expressed as some fraction of the local velocity pressure. These loss calculations invariably assume a fully developed and symmetrical velocity profile upstream of the fitting and sufficient straight downstream for these conditions to be regained. Where fittings are adjacent to the fan inlet and/or outlet an additional effect may be anticipated as the profile may be far from ideal. Whilst it is possible to look on this as an additional loss it should be realised that the mechanisms involved are far more complex.

Elementary fan theory (Refs. 5 and 6) shows that the design pressure development of a centrifugal fan will only be achieved with a fully developed symmetrical and straight line flow into the inlet. Indeed many flow control devices such as variable radial inlet vanes make use of this fact to induce a controlled pre-swirl. Inlet bends not only result in asymmetrical flow but according to position, attitude and number can produce pre or contra swirl.

The velocity profile at the outlet of a centrifugal fan is far from uniform (Fig. 1). Straight ducting, as used in the normal laboratory tests, controls the diffusion process making the profile more uniform before discharging to atmosphere or into a plenum chamber. Alternatively where there is a ducting system on the fan outlet a length of a straight ducting is desirable to ensure that the pressure losses in bends etc., will be kept to a minimum.

4. TESTING PROGRAMME.

4.1 General.

Taking into account all the above, it seemed apparent that there was merit in arranging a test programme to ascertain :-

a) the differences in performance according to installation category

b) the effects of inlet and outlet bends

c) if the inlet and outlet connections specified in BS 848 are adequate for their purpose and/or whether they can be modified or relaxed in any way

4.2 Design Of 'Common' Parts.

At the present time sufficient progress has been made internationally on the design of common inlet and outlet duct parts for them to have been included in the methods of test with standardized airways. It has been felt that the use in the 1963 edition of an outlet duct simulation section two diameters in length is not permissible for type B or D standardized airway tests. The simulation section should reproduce the common part incorporating a standardized flow straightener.

Whilst these comments are particularly relevant to the testing of tubular axial flow fans with appreciable swirl at the fan outlet, the writer felt that they might be less applicable to the testing of centrifugal fans where appreciably straighter flow exists.

Some relaxation in the design of the common part on the inlet side is allowed where a fan tested for use with a free inlet may also be used with a ducted inlet. Conversion from the former to the latter is made by attaching an inlet duct simulation section to its inlet. This normally consists of a cylindrical airway of the same diameter as the fan inlet to which it is attached. A bell mouth entry is also fitted. The length of this duct is defined as one diameter and should provide a true ducted inlet fan characteristic over the normal duty range. In certain cases the standard allows a longer duct "to enable the fan to develop its full ducted inlet pressure at or near zero volume flow". It was felt that this length might be important even in the region of peak fan efficiencies and tests at various lengths could be of value.

4.3 Design Of Straighteners.

One of the important elements in the standardized test ducts is the flow straightener. It is convenient at this point to consider the two designs which are recommended within BS 848 : 1980 : Part 1. Their use is also recommended in a new draft international standard for fluid flow measurement in ducts (Ref. 7).

a) The A.M.C.A. straightener (Fig. 2) is used only to prevent the growth of swirl in a normally axial flow. It does not improve asymmetric velocity distributions. The device consists of a nest of equal cells of square cross-section. It has a very low pressure loss and is typically used either side of an auxiliary booster fan where this is necessary to overcome the resistance of the airway when a complete fan characteristic is required.

b) The Etoile straightener (Fig. 3) is again designed to eliminate swirl and is of no use in the equalization of asymmetric velocity distributions. The eight radial vanes should be of sufficient thickness to provide adequate strength but should not exceed 0.007 D_4 for pressure loss considerations. This straightener has a similar pressure drop to the A.M.C.A. straightener i.e. approximately 0.25 times the approach velocity pressure, but it is also easier to manufacture. In addition it allows the static pressure to equalize radially as the air flows through it. This is not the case with the A.M.C.A. straightener which can produce variations in the static pressure across the duct downstream. The étoile straightener is preferred in the common duct on the fan outlet.

5. THE TEST METHOD.

5.1 Choice Of A Suitable Standardized Airway.

Bearing in mind the requirements of the test programme, it was decided that an inlet side chamber would be employed. Flow rate determination would be by a conical inlet and the diameter of this was chosen to give measurable pressures. In order to keep pressure losses through the rig to a minimum, the chamber was made very large - its cross-sectional measurements being 2m x 2m with a chamber length of 3m. By this means an auxiliary fan was deemed unnecessary nor were A.M.C.A. straighteners included as it was only desired to produce fan characteristic curves over the working range of the unit under test. The general arrangement was as shown in Fig. 4.

5.2 Inlet And Outlet Duct Simulation.

Such a test arrangement is recommended for category A installations (free inlet and outlet). BS 848 does however indicate how such a chamber may be used to test a fan for use on type B, C or D installations. The fan was therefore fitted with inlet and/or outlet duct simulation sections as described later. Where a bellmouth entry was fitted, this was sunk within the end wall of the chamber. The outlet end of all outlet duct simulation sections was left open to atmosphere.

5.3 Additional Duct Effects.

It was also possible with such a rig to interpose additional lengths of straight ducting or bends at appropriate points. Their effects on fan performance when positioned adjacent to the fan inlet and/or outlet could thus be ascertained.

6. THE FAN AND ANCILLARIES.

6.1 Choice Of A Suitable Fan.

Previous programmes elsewhere (Ref. 8) had indicated the difficulties encountered in measuring system effects, particularly where the flowrates were small. It was therefore decided that the initial range of experiments would be conducted on a 630mm diameter straight backward inclined bladed centrifugal fan having an inlet 630mm diameter, outlet 630mm x 450mm and operating at speeds around 1500 rev/min.

This unit is of fairly high specific speed but low Fan Reynolds Number. The fan is of the single width pattern with open inlet and overhung impeller. This impeller is mounted on a shaft running in ball and roller bearings contained in a rigid housing. Drive is by means of a TEFC squirrel cage motor having an output of 30kW at 1455 rev/min on a 415 volt 3 phase 50 Hz AC supply. Interposed between the motor and fan is a torquemeter having a rating of up to 0.046 kW/rev/min at up to 12000 rev/min, the drive being through flexible couplings. An electrical output from the torquemeter is taken to a control panel giving direct digital readout n of rotational speed rev/min and P power absorbed kW. The general arrangement of the fan is shown in Fig. 5.

6.2 Choice Of Ancillary Ducting.

A range of duct pieces for the inlet and outlet of the fan was manufactured, the dimensions and identification marks being as shown in Fig. 6. All the pieces for the inlet side of the fan have an internal diameter of 630mm to exactly match the fan. According to how they were arranged it was possible to simulate the conditions of open or ducted inlet. The lobster backed bend could be positioned as necessary. Outlet ducting is either 630mm x 450mm inside to match the fan outlet or 600mm inside diameter which is within0.3% less of the same cross-sectional area. It therefore meets the requirement of the standard which allows for differences up to 7% greater or 5% less. Again, according to disposition it was possible to simulate open or fully ducted outlet conditions together with some intermediate situations. 'Easy' and 'hard' bends could be interposed as appropriate.

7. THE TEST PROGRAMME.

7.1 General.

By running the fan attached to the inlet chamber and gradually closing the multi-leaf opposed blade damper, it was possible to obtain pressure, temperature and power readings as specified in BS 848. By this means characteristic curves were generated for the various duct configurations specified below. An overall impression of the complete ducting and fan rig can be obtained by reference to the photographs (Figs. 7 and 8). Barometric pressure and ambient temperature were also monitored. The input of datum offset (running the rig without an impeller fitted) and torquemeter shaft stiffness enabled direct readings of P_R impeller power kW to be obtained.

Throughout the test programme the importance of an unchanged fan was realised. The impeller and shaped inlet were therefore fixed in position both axially and radially and no changes allowed.

7.2 Tests For Installation Category.

A series of tests were conducted with various additions and subtractions to the duct connections to simulate the four installation categories. Certain other duct connections were tried which were not in accordance with the standard but which were relevant to the normal usage of centrifugal fans. Details of this part of the programme are detailed in Table 1 with the writer's comments. Allowances for any friction losses were in accordance with those prescribed in the standard for the particular test method, but these were shown separately and fan performance calculated both with and without their value included.

7.3 Tests For The Effect Of Additional Inlet Duct Pieces.

A second series of tests was conducted with additional inlet connecting pieces interposed between the standard inlet simulation piece item m and the fan inlet flange. Throughout this series the outlet ducting remained unchanged and comprised the pieces specified for fans used on installation category D i.e. duct pieces items T, U, V, and W. The inlet pieces comprised either additional lengths of straight or a lobster backed 90° bend positioned at various distances from the fan inlet flange. It was anticipated that these tests would indicate whether the standard inlet simulation piece was sufficient for fully ducted performance figures. The inclusion of the bend would enable its effect on fan performance to be assessed according to placement and this compared with standard pressure loss figures.

Where a bend is adjacent to the fan inlet an additional effect may be anticipated. Whilst not strictly true, it may be considered as an additional loss.

Table 1.

Test No.	Inlet Duct Pieces	Outlet Duct Pieces	Comments
IV01	None	None	Category A
IV02	Y	Z	Short inlet and outlet connections as normally recommended for Category A installations
IV03	Y	T, U, V & W	Category B as normally recommended
IV04	M	None	Category C
IV05	M	Z	Short outlet connection as normally recommended for Category C installations.
IV06	M	T, U, V & W	Category D
IV06a	M	Z, T, U, V & W	Category D but short straight to outlet dimension before common part

Note: The order of the duct items is listed sequentially from inlet chamber to open outlet.

It should be realised that this additional loss or 'system effect factor' (Ref. 9) would be extremely difficult or impossible to measure directly. It is best obtained by calculation as follows :-

Standard fan performance - fan performance with inlet bend

= System effect factor + standard inlet bend loss

It should be noted that in all cases this bend was positioned such that its inlet and outlet and the fan outlet were all in the same horizontal plane whilst it was also 'with' the fan such that the airflow described 180° (see Fig. 9).

Allowance for the friction loss of additional straight ducting was in accordance with Appendix D of the standard. The standard pressure loss for the right angle bend was calculated as 0.4 x velocity pressure. Results were calculated with and without this loss included. Details and comments for this part of the programme are in Table 2.

During test number IV09, pitot traverses were additionally carried out either side of the lobster backed bend. Readings of velocity and static pressure were taken on three diameters. Eight points per diameter were read, these being spaced according to the log-linear rule. No attempt was made to find the true flow direction but it was hoped that the traverses might indicate the amount of swirl or asymmetry which was present.

Table 2.

Test No.	Inlet Duct Pieces	Outlet Duct Pieces	Comments
IV07	M and N	T, U, V & W	2.4 diameters approx. of additional straight on fan inlet
IV08	M, N and P	T, U, V & W	4.8 diameters approx. of additional straight on fan inlet
IV09	M, N, Q and P	T, U, V & W	Right angle bend interposed between straight lengths
IV10	M, N, P and Q	T, U, V & W	Long straight then right angle bend to fan inlet
IV10a	M, N, P, Q and X	T, U, V & W	As previous test but with 1 diameter straight interposed before fan inlet
IV11	M, Q and X	T, U, V & W	Right angle bend and 1 diameter of straight on fan inlet

7.4 Tests For The Effect Of Additional Outlet Duct Pieces.

A third series of tests was conducted with additional outlet duct pieces interposed between the short outlet duct and the normal items specified for use on the outlet of a Category D fan i.e. duct pieces T, U, V and W. These additional pieces comprised 'easy' and 'hard' bends. Again the inclusion of these bends would enable their effect on fan performance to be assessed and this compared with standard pressure loss figures.

Where a bend is adjacent to the fan outlet the bend loss may differ from standard figures as the approach velocity profile will be far from the ideal of a fully developed symmetrical pattern. Again it should be noted the bends were generally 'with' the fan and the outside of the bend took the higher velocities.

The standard pressure losses for these right angle bends were taken as 0.23 x velocity pressure ('easy' bend - item R) and 0.27 x velocity pressure ('hard' bend - item S). Results were again calculated with and without these losses included. Details and comments for this part of the programme are given in Table 3.

During test IV12 and IV13, pitot traverses were additionally carried out either side of the rectangular bends. Readings of velocity and static pressure were taken by a 30 point (6 x 5) traverse according to a Log-Tchebycheff distribution as specified in BS 848. Again no attempt was made to find the true flow direction, but the readings are of interest in showing the deviation from ideal conditions.

Table 3.

Test No.	Inlet Duct Pieces	Outlet Duct Pieces	Comments
IV12	M	Z, R, T, U, V, W	Short straight and easy bend before standard Category D outlet
IV13	M	Z, S, T, U, V, W	Short straight and hard bend before standard Category D outlet
IV14	M, N, Q & P	Z, S, T, U, V, W	Bends on inlet and outlet but 'protected' by straights

8. CALCULATIONS AND ASSUMPTIONS.

8.1 General Comments.

The rotational speed n rev/min and the impeller power P_R W were taken directly from the torquemeter panel. Air conditions in the test laboratory p_a and t_a were also measured and the atmospheric density ρ_a determined. The differential pressure across the conical inlet was recorded for each damper closure position together with the negative pressure p_{s3} and temperature t_3 within the inlet chamber. In all cases the conditions immediately upstream of the conical inlet were assumed to be ambient. Rather than calculate the temperature t_2 at the fan outlet, an attempt was made to measure this although the differences between t_a, t_2 and t_3 were all 1°k or less. As all measured pressures were less than 2500 p_a it was assumed that $k_p = 1$. Furthermore, in view of its large cross-sectional area, the dynamic pressure p_{d3} within the chamber could be ignored and $p_{t3} = p_{s3}$. The relevant flow coefficient for the conical inlet was taken from clause 21 figure 16 of the standard.

8.2 Formulae.

$$\rho_3 = \rho_a \left(\frac{p_a + p_{s3}}{p_a}\right)\left(\frac{273 + t_a}{273 + t_3}\right) \quad ;$$

$$q_m = \alpha \varepsilon \frac{\pi d^2}{4} \times \sqrt{2 \rho_a \Delta p} \quad ; \qquad q_v = \frac{q_m}{\rho_i}$$

$$p_{dF} = \frac{q_m^2}{2 \rho_2 A_2^2} \qquad \eta_s = \frac{q_v \, p_{sF}}{P_R}$$

$$p_{tF} = p_{sF} + p_{dF} \qquad \eta_T = \frac{q_v \, p_{tF}}{P_R}$$

$$\rho_2 = \rho_a \left(\frac{273 + t_a}{273 + t_2}\right)$$

In all cases results were converted to standard conditions of n' = 25 rev/sec (1500 rev/min) and inlet air density = 1.2 kg/m³ by use of the following rules:-

$$\frac{q'_v}{q_v} = \frac{n'}{n} \qquad \frac{p_{tF}'}{p_{tF}} = \frac{p_{dF}'}{p_{dF}} = \frac{p_{sF}'}{p_{sF}} = \left(\frac{n'}{n}\right)^2 \frac{\rho_i'}{\rho_i}$$

$$\frac{P_R'}{P_R} = \left(\frac{n'}{n}\right)^3 \frac{\rho_i'}{\rho_i} \; ; \quad \eta_T' = \eta_T \; ; \quad \eta_s' = \eta_s$$

In the above formulae the symbols q_v' etc. with the prime indicate the performance quantities after conversion from the test values to the specified conditions.

8.3 Fan Performance - Category A Installations.

The conditions for tests IV01 and IV02 could be considered under this category and the following expressions applied :-

fan static pressure $p_{sF} = -p_{s3}$

inlet volume flowrate $q_v = \frac{q_m}{\rho_3}$ (as $\rho_i = \rho_3$

No allowance was made for friction in test IV02 and the short outlet duct was considered part of the fan.

8.4 Fan Performance - Category B Installations.

The conditions for test IV03 could be considered under this category although a very short inlet connection $0.16D_i$ in length was used to facilitate assembly. This duct was flush with the end of the chamber.

Measurements were taken at various damper openings of p_{s3} and t_3, which was negative. It was noted that over the range of flows encountered during the experiments p_{d3} and p_{s4} essentially remained at zero. As $A_2 = A_4$ within 0.3% then $p_{d4} = p_{d2}$. In accordance with the standard it was assumed that:

$$\rho_i = \rho_1 = \rho_3 \; : \; \rho_4 = \rho_2 \; : \; p_{t1} = p_{t3}$$

then fan total pressure $p_{tF} = p_{t2} - p_{t1}$

$$= (1 + \zeta_{24}) \; p_{d2} - p_{s3}$$

fan static pressure $p_{sF} = p_{tF} - p_{dF}$

$$= p_{t2} - p_{t1} - p_{dF}$$

$$= (1 + \zeta_{24}) \; p_{dF} - p_{t3} - p_{dF}$$

$$= \zeta_{24} p_{dF} - p_{s3}$$

and $q_v = \frac{q_m}{\rho_3}$

In this instance calculations were made of fan total pressure/efficiency and fan static pressure/efficiency with and without the friction increment $\zeta_{24} p_{dF}'$ included.

8.5 Fan Performance - Category C Installations.

The conditions for tests IV04 and IV05 could be considered in this category. Proceeding as before, measurements were taken of p_{s3} and t_3. In accordance with the standard, it was assumed that $\rho_i = \rho_1 = \rho_3$ and $p_{s2} = 0$.

No allowance was made in test IV05 for the friction in the short outlet duct which was considered to be part of the fan.

Then $p_{d1} = \dfrac{q_m^2}{2\rho_3 A_1^2}$; $p_{t1} = p_{s3} - \zeta_{31}\, p_{d1}$

$p_{sF} = -p_{t1}$; $q_v = \dfrac{q_m}{\rho_3}$

Again, calculations were made of the fan pressures and efficiencies with and without the friction increment $\zeta_{31}\, p_{d1}'$ included.

8.6 Fan Performance - Category D Installations.

Tests IV06 to IV14 could all be considered as fully ducted on inlet and outlet, thus conforming to Category D. Again p_{s4} and p_{d3} remained substantially zero and $p_{d4} = p_{d2}$. In accordance with the standard $\rho_i = \rho_1 = \rho_3$ whilst $\rho_4 = \rho_2$

Then $p_{d1} = \dfrac{q_m^2}{2\rho_3 A_1^2}$; $p_{t1} = p_{s3} - \zeta_{31}\, p_{d1}$

$p_{d2} = \dfrac{q_m^2}{2\rho_2 A_2^3} = p_{d4}$; $p_{t2} = (1 + \zeta_{24})\, p_{d4}$

$p_{tF} = p_{t2} - p_{t1}$: $q_v = \dfrac{q_m}{\rho_3}$

8.7 Additional Pressure Loss Formulae.

For tests IV07 to IV14 there were further lengths of straight which were positioned on the inlet or outlet as previously described whilst tests IV09 to IV14 incorporated bends. The frictional resistance of these items was calculated as follows :-

additional inlet ducting $p_{f1} = \lambda_1 \dfrac{L_1}{D_1} \dfrac{1}{2}\rho_1 v_1^2 = \dfrac{L_1}{D_1} p_{d1} \lambda_1$

additional outlet ducting $p_{f2} = \lambda_2 \dfrac{L_2}{D_2} \dfrac{1}{2}\rho_2 v_2^2 = \dfrac{L_2}{D_2} p_{d2} \lambda_2$

[λ_1 and λ_2 were obtained from Fig. 37 of BS 848]

inlet bend loss $p_{fb1}' = 0.4\, p_{d1}'$

item R : 'easy' outlet bend loss $p_{fb2R}' = 0.23\, p_{d2}'$

item S : 'hard' outlet bend loss $p_{fb2S}' = 0.27\, p_{d2}'$

9. READINGS AND RESULTS.

The formulae given above were programmed for a minicomputer so that results could be rapidly produced. These are appended in Tables 4 to 19 inclusive.

For further evaluation of the results graphs have been plotted (Figs 9 to 12) and these immediately illustrate the differences in performance according to category and ducting arrangement. Where applicable, the system effect could be obtained by subtraction, although only approximate at this stage of the programme.

10. CONCLUSIONS.

10.1 General.

This series of experiments was conducted on a particular fan during its development. Whilst the results are an indication of likely effects, their exact magnitude remains to be determined. The fan, being of the backward inclined type, suffered from a definite stall point at low flows. It was of interest to note that the onset of stall appeared to be affected by the duct arrangement.

The tests depended on the accurate determination of fan pressures, as it was the performance difference between various duct/fan arrangements in which we were interested. Analysing the previous work conducted in the U.S.A. (Ref. 8) it was felt that they may have had difficulties due to the relatively small size of fan whose impeller was only 345mm diameter. This resulted in a low performance, typically up to 1.7m³/s flowrate at free air, pressures up to 400Pa and absorbed powers up to 1400W. In this series of tests, therefore, a larger unit was purposely selected. By using a 630mm diameter fan running at 1500rev/min, flowrates up to 4.5m³/s have been possible with pressures up to 1500Pa and powers to 4700W. Even so, the performance differences have been particularly small, and future tests will be conducted with even larger fans at higher speeds to achieve the desired accuracy.

Nevertheless these experiments have been of value in confirming previously held "beliefs" and also in indicating certain ways forward, both in determining future test programmes and also in providing information for site test engineers and users. Whilst many more such tests are desirable, the writer's immediate reaction is that previously published "system effect factors" need to be viewed with caution, and that their values should be as much a function of the particular fan design and duty point as they are of the duct arrangement. Certainly this part of the programme gave low results for the effect.

10.2 Performance Differences according to Category.

It was noted that the best performance for the fan in terms of the maximum possible flowrate was achieved when working in the Category D mode ie with ducting on inlet and outlet. At the best efficiency point, however, Category B was very close to Category D and indeed the efficiency was slightly higher. Presumably any inlet losses were negligible due to the low inlet velocity of 8.3m/s. The results for Categories C and A were somewhat lower as the absence of the common duct parts on the fan outlet resulted in little diffusion. There was zero recovery of the high velocity pressure at the fan throat area to the somewhat lower velocity which would have existed in a ducted outlet.

It was of interest to note that by the addition of only 1080mm of rectangular ducting (the equivalent of 2 diameters) to the fan outlet, a reasonable degree of velocity pressure regain took place. Category A results then approached Category B whilst Category C results were as good as Category D.

10.3 Effects of Increased Lengths of Inlet Duct.

Whilst the differences are small, and the results thus have a low level of accuracy, it was noted that a higher maximum efficiency was achieved with an increased length to the inlet simulation piece. Figures improved with each additional piece, even when ignoring friction losses, from $1D_i$ through $3.4D_i$ and $5.8D_i$. The performance towards free air appeared to reduce in like manner. There is therefore some doubt with this fan at least as to whether the length of inlet simulation piece as specified in BS848 is adequate.

10.4 Effect of Inlet Bends.

Whilst the inclusion of a bend directly on the fan inlet seriously effected fan performance, the addition of only $1D_i$ of straight between this bend and the fan recovered some of this loss.

Bends with $2.4D_i$ of straight either side resulted in a loss which could be allowed for by the normal pressure drop calculation.

10.5 Straight Outlet Ducting.

It was noted that the design of the common part was not ideal for centrifugal fans. These usually have the impeller offset in the casing and nearer to the side opposite the inlet. The velocity profile is distorted in like manner. As the transformation piece is symmetrical to the casing outlet, the diffusion is too rapid towards the inlet side. By the inclusion of a rectangular piece 1080mm (equivalent to 2 diameters) before the common parts an improved performance was obtained. It is suggested that this should either be allowed within the standard or that manufacturers should specify the need for this piece in their data.

10.5 Outlet Bends.

These have less effect than those on the inlet, but bend losses are higher than calculated due to the higher air velocities on the scroll side of the outlet. Again, the inclusion of 2 diameters of rectangular straight before any bend is to be recommended to allow some diffusion to take place.

10.6 System Effect Factors.

Previous work in this field has suggested two possible approaches. The first has been to see the effect as an additional pressure drop over and above the normal calculations and expressed as some fraction of the inlet/outlet velocity pressure. The second has been to calculate $\frac{\text{Impeller tip speed}}{\text{Inlet/outlet velocity}}$

and from this non-dimensional figure again express the result as a fraction of velocity pressure.

It is believed that these tests have given little encouragement for either approach. The system effect increases towards free air but not in anyway as the square of flowrate. It is inevitably related to a particular design of fan - its impeller inlet, radial and actual air velocities. Outlet effects will be a function of velocity profile which again will vary from design to design. As an example forward curved fans may be expected to differ from other designs and in many cases these have a more symmetrically positioned impeller. The ratio of throat area to outlet duct area will also vary so that regain of velocity pressure will also be different.

10.7 Pitôt Traverses

During certain of the tests pitot traverses were taken on the inlet and outlet of bends. These were essentially to emphasize the distortion in the velocity profiles. Calculations have however been made to obtain the flowrate in accordance with normal methods. The divergence from the true flow as obtained from the inlet cone measurements is an indication of the amount of swirl and yaw which was present. These results are shown in Figs 13 to 15.

It was of interest to note the divergences and that inlet calculations were closer than those for outlet traverses.

10.8 Future Work Programme.

This paper describes the "nuts and bolts" approach which has been employed to date. There is undoubtedly a need for more experiments. As an indication of the enormity of the problems involved, please refer to Fig. 16 which shows the performance of a narrow forward curved bladed centrifugal fan. As might be expected, because much of the static pressure development is through conversion from velocity to static within the casing, and less is developed by the impeller, differences are larger between categories. Power absorbed remained unchanged. The effects of inlet and outlet bends still have to be determined.

There is also a need to test both lower and higher specific speed fans together with other impeller blade types. Axial flow fans could also benefit from a similar programme. Bends have to date been working with the fan eg., the axis of inlet bends has been parallel to the direction of the outlet ducting and the air has been turned through 180°. Outlet bends have been with the impeller rotation. Other attitudes need to be compared.

10.9 Some Final Thoughts.

For too long there has been a need for a meeting of minds between users and manufacturers. The former will continue to instal fans in less than ideal ducting systems. There is no mileage in the manufacturer just giving characteristic curves obtained under laboratory conditions. Is there not a case for compromise? Could additional standard tests be devised for inlet/ outlet bend performance to give the system designer an indication of the likely effects of less than perfect ducting?

ACKNOWLEDGEMENTS.

The writer wishes to acknowledge the very valuable assistance given by Dr. David J. Allen, Research Manager, and Mr. Paul Hunt, Research Engineer, at Keith Blackman Limited in carrying out the experiments described in this paper.

The writer also wishes to place on record the considerable help given by Miss Janice Lewin and Mrs. Eileen Buse in the preparation of the paper.

REFERENCES.

1.	B.S. 848 : Part 1 : 1980 Fans For General Purposes - Method Of Testing Performance		B.S.I.
2.	H & V News Vol 25 No. 20 Moss Slams False Fan Claims	B. Moss	Maclaren Publishers Limited
3.	Woods Practical Guide To Fan Engineering	B. Daly	Woods Of Colchester Limited
4.	Internal Flow Systems	D.S. Miller	BHRA Fluid Engineering
5.	Fans	W.C. Osborne	Pergamon
6.	Centrifugal Fan Guide	W.T.W. CORY	Keith Blackman Limited
7.	DIS 7194 - Measurement Of Fluid Flow In Closed Conduits - Velocity-area Methods Of Flow Measurement In Swirling Or Asymmetric Flow Conditions In Circular Ducts By Means Of Current-meters Or Pitot Static Tubes		I.S.O.
8.	The Effects Of System Connections On Fan Performance - No. 2506 RP 139	M.S. Clarke J.T. Barnhart F.J. Bubsey & E. Neitzel	American Association Of Heating, Refrigeration And Airconditioning Engineers.
9.	Publication 201 - Fans And Systems		Air Movement And Control Association (U.S.A.)

TEST No. 1V01 FRICTION LOSSES ZERO ON INLET AND ZERO ON OUTLET
BAROMETER 998·3 mbar
INLET DUCTING NOTHING × 630 mm dia
OUTLET DUCTING NOTHING × (630 mm × 450 mm) AND NOTHING × 600 mm dia

CATEGORY 'A'

										Excluding Friction Losses					Including Friction Losses			
Damper Setting	Δp Pa	ps3 Pa	t_a °C	t_3 °C	t_2 °C	n rev/min	P_R W	q_V' m³/s	P_R' W	p_{sF}' Pa	p_{tF}' Pa	η_s %	η_t %	Friction Losses Pa	p_{sF}' Pa	p_{tF}' Pa	η_s %	η_t %
1	375	200	20	20	20	1498	3930	4·21	4020	205	337	21·4	35·3					
2	350	310	20	20	20	1497	4180	4·08	4290	318	442	30·3	42·0					
3	300	565	20	20	20	1497	4450	3·78	4570	583	689	48·2	57·0					
4	250	660	20	20	20	1497	4210	3·46	4330	682	771	54·4	61·5					
5	220	780	20	20	20	1497	4200	3·25	4330	808	886	60·6	66·5					
6	180	910	20	20	20	1496	4165	2·94	4310	947	1010	64·7	69·1					
7	140	1060	20	20	20	1496	4060	2·60	4200	1110	1160	68·4	71·5					
8	100	1180	20	20	20	1496	3850	2·20	3990	1230	1270	68·0	70·0					
9	70	1260	20	20	20	1496	3620	1·84	3760	1320	1350	64·7	66·0					
10	60	1210	20	20	20	1496	3260	1·70	3380	1270	1290	63·9	64·9					
11	5	1360	20	20	20	1497	1960	0·487	2030	1430	1430	34·2	34·2					
12	0	1300	20	20	20	1497	1500	0	1550	1360	1360	0	0					

TABLE No 4.

TEST No. 1V02 FRICTION LOSSES NEGLIGIBLE ON INLET AND NEGLIGIBLE ON OUTLET
BAROMETER 994·6 mbar
INLET DUCTING 'Y'– 100 mm × 630 mm dia.
OUTLET DUCTING 'Z'– 1080 mm × (630 mm × 450 mm) AND NOTHING × 600 mm dia.

CATEGORY 'A' (NORMAL RECOMMENDATIONS)

										Excluding Friction Losses					Including Friction Losses			
Damper Setting	Δp Pa	ps3 Pa	t_a °C	t_3 °C	t_2 °C	n rev/min	P_R W	q_V' m³/s	P_R' W	p_{sF}' Pa	p_{tF}' Pa	η_s %	η_t %	Friction Losses Pa	p_{sF}' Pa	p_{tF}' Pa	η_s %	η_t %
1	400	215	16	16	16	1498	3980	4·32	4030	217	356	23·3	38·2					
2	350	444	16	16	16	1499	4290	4·05	4340	450	572	41·9	53·3					
3	300	655	17	17	17	1500	4510	3·77	4580	670	775	55·7	63·7					
4	260	740	17	17	17	1498	4360	3·51	4450	760	852	60·0	61·2					
5	220	835	17	17	17	1498	4240	3·24	4330	860	937	64·2	70·0					
6	180	960	17	17	17	1498	4180	2·93	4280	991	1050	67·9	72·2					
7	140	1110	17	17	17	1498	4070	2·59	4170	1150	1200	71·3	74·4					
8	100	1210	17	17	17	1499	3860	2·19	3950	1250	1290	69·4	71·4					
9	70	1300	18	18	18	1500	3610	1·84	3710	1360	1380	67·3	68·5					
10	60	1220	18	18	18	1500	3240	1·70	3320	1270	1290	65·0	66·1					
11	5	1360	18	18	18	1502	1915	0·485	1960	1420	1420	35·1	35·1					
12	0	1250	18	18	18	1502	1310	0	1340	1300	1300	0	0					

TABLE No 5.

TEST No. 1V03 FRICTION LOSSES NEGLIGIBLE ON INLET AND $\zeta_{24}\frac{1}{2}\rho_2 v_2^2$ ON OUTLET
BAROMETER 994·6 mbar
INLET DUCTING Y – 100 mm × 630 mm dia.
OUTLET DUCTING NOTHING × (630 mm × 450 mm) AND T, U, V, W, × 600 mm dia
CATEGORY 'B'

										Excluding Friction Losses					Including Friction Losses			
Damper Setting	Δp Pa	p_{s3} Pa	t_a °C	t_3 °C	t_2 °C	n rev/min	P_R W	q_V' m³/s	P_R' W	p_{sF}' Pa	p_{tF}' Pa	η_s %	η_t %	Friction' Losses Pa	p_{sF}' Pa	p_{tF}' Pa	η_s %	η_t %
1	390	220	16	16	16	1496	3970	4·27	4030	223	359	23·6	38·1	44	267	403	28·3	42·8
2	350	400	16	16	16	1495	4260	4·06	4350	407	530	38·0	49·5	40	447	570	41·7	53·2
3	300	620	16	16	17	1496	4480	3·76	4570	633	739	52·1	60·8	34	667	783	54·9	63·6
4	260	695	16	16	17	1496	4440	3·51	4490	710	802	55·4	62·6	30	740	832	57·7	64·9
5	220	805	16	16	17	1496	4240	3·23	4330	825	902	61·4	67·2	26	851	928	63·3	69·1
6	180	950	16	16	17	1495	4180	2·93	4290	977	1040	66·7	71·0	22	999	1062	68·2	72·5
7	140	1095	18	18	18	1495	4000	2·60	4140	1150	1200	72·0	75·2	17	1167	1217	73·1	76·3
8	100	1180	18	18	18	1494	3810	2·20	3950	1240	1270	69·0	71·0	13	1253	1283	69·7	71·7
9	66	1290	18	18	18	1496	3590	1·79	3710	1350	1380	65·2	66·3	9	1359	1389	65·6	66·7
10	55	1210	18	18	18	1496	3160	1·63	3270	1270	1290	63·3	64·3	7	1277	1297	63·6	64·6
11	5	1350	18	18	18	1498	2010	0·486	2070	1410	1420	33·2	33·2	1	1411	1421	33·2	33·2
12	0	1250	18	18	18	1499	1300	0	1340	1300	1300	0	0	0	1300	1300	0	0

TABLE No. 6.

TEST No. 1V04 FRICTION LOSSES $\zeta_{31}\frac{1}{2}\rho_1 v_1^2$ ON INLET AND ZERO ON OUTLET
BAROMETER 998·3 mbar
INLET DUCTING M – 630 mm × 630 mm dia + BELLMOUTH
OUTLET DUCTING NOTHING × (630 mm × 450 mm) AND NOTHING × 600 mm dia
CATEGORY 'C'

										Excluding Friction Losses					Including Friction Losses			
Damper Setting	Δp Pa	p_{s3} Pa	t_a °C	t_3 °C	t_2 °C	n rev/min	P_R W	q_V' m³/s	P_R' W	p_{sF}' Pa	p_{tF}' Pa	η_s %	η_t %	Friction' Losses Pa	p_{sF}' Pa	p_{tF}' Pa	η_s %	η_t %
1	380	215	19	19	19	1498	4050	4·23	4130	219	353	22·5	36·2	5	224	358	23·0	36·7
2	350	365	19	19	19	1498	4300	4·07	4390	373	496	34·6	46·0	4	377	500	34·9	46·4
3	300	615	19	19	19	1497	4550	3·78	4660	633	739	51·3	59·9	4	637	743	51·6	60·2
4	260	680	19	19	19	1497	4350	3·52	4460	701	793	55·3	62·5	3	704	796	55·5	62·7
5	220	810	19	19	19	1497	4290	3·24	4410	837	915	61·6	67·3	3	840	918	61·8	67·5
6	180	940	19	19	19	1497	4190	2·94	4310	974	1040	66·4	70·7	2	976	1042	66·5	70·8
7	140	1070	19	19	19	1496	4050	2·59	4180	1110	1160	69·1	72·2	2	1112	1162	69·2	72·3
8	100	1190	19	19	19	1497	3830	2·19	3950	1240	1270	68·9	70·8	1	1241	1271	69·0	70·9
9	70	1270	19	19	19	1497	3590	1·84	3700	1320	1350	65·7	66·9	1	1321	1351	65·7	66·9
10	60	1200	19	19	19	1497	3160	1·70	3260	1250	1270	65·2	66·3	1	1251	1271	65·2	66·3
11	5	1340	19	19	19	1497	1870	0·486	1930	1400	1400	35·2	35·3	0	1400	1400	35·3	35·3
12	0	1230	19	19	19	1498	1400	0	1440	1280	1280	0	0	0	1280	1280	0	0

TABLE No. 7

TEST No. 1V05 FRICTION LOSSES $\zeta_{31} \frac{1}{2} \rho_1 v_1^2$ ON INLET AND NEGLIGIBLE ON OUTLET
BAROMETER 997·3 mbar
INLET DUCTING M - 630 mm × 630 mm dia + BELLMOUTH
OUTLET DUCTING Z - 1080 × (630 mm × 450 mm) AND NOTHING × 600 mm dia
CATEGORY 'C' (NORMAL RECOMMENDATIONS)

										Excluding Friction Losses					Including Friction Losses			
Damper Setting	Δp Pa	ps3 Pa	t_a °C	t_3 °C	t_2 °C	n rev/min	P_R W	q_v' m³/s	P_R' W	p_{sF}' Pa	p_{tF}' Pa	η_s %	η_t %	Friction' Losses Pa	p_{sF}' Pa	p_{tF}' Pa	η_s %	η_t %
1	400	240	17	17	17	1500	3990	4·33	4030	243	382	26·1	41·0	5	248	387	26·6	41·5
2	350	460	17	17	17	1500	4410	4·05	4460	467	590	42·5	53·6	4	471	594	42·9	54·0
3	300	686	17	17	17	1500	4600	3·76	4660	700	805	56·5	65·0	4	704	809	56·8	65·3
4	260	740	17	17	17	1500	4350	3·50	4410	756	847	60·1	67·3	3	759	850	60·3	67·5
5	220	860	17	17	17	1499	4270	3·23	4340	882	959	65·6	71·3	3	885	962	65·8	71·5
6	180	990	17	17	17	1499	4220	2·93	4300	1020	1080	69·3	73·6	2	1022	1082	69·4	73·7
7	140	1130	17	17	17	1498	4100	2·59	4190	1170	1220	72·0	75·0	2	1172	1222	72·1	75·1
8	100	1230	17	17	17	1498	3880	2·19	3970	1270	1310	70·1	72·1	1	1271	1311	70·2	72·2
9	70	1300	17	17	17	1498	3630	1·83	3720	1350	1370	66·4	67·6	1	1351	1371	66·4	67·6
10	60	1230	17	17	17	1498	3195	1·69	3270	1270	1290	66·0	67·0	1	1271	1291	66·0	67·0
11	5	1360	17	17	17	1498	1810	0·484	1850	1410	1410	36·8	36·9	0	1410	1410	36·8	36·9
12	0	1270	17	17	17	1477	1380	0	1420	1320	1320	0	0	0	1320	1320	0	0

TABLE No. 8

TEST No. 1V06 FRICTION LOSSES $\zeta_{31} \frac{1}{2} \rho_1 v_1^2$ ON INLET AND $\zeta_{24} \frac{1}{2} \rho_2 v_2^2$ ON OUTLET
BAROMETER 1004·1 mbar
INLET DUCTING M - 630 mm × 630 mm dia + BELLMOUTH
OUTLET DUCTING NOTHING × (630 mm × 450 mm) AND T, U, V, W, × 600 mm dia
CATEGORY 'D'

										Excluding Friction Losses					Including Friction Losses			
Damper Setting	Δp Pa	ps3 Pa	t_a °C	t_3 °C	t_2 °C	n rev/min	P_R W	q_v' m³/s	P_R' W	p_{sF}' Pa	p_{tF}' Pa	η_s %	η_t %	Friction' Losses Pa	p_{sF}' Pa	p_{tF}' Pa	η_s %	η_t %
1	400	220	16	16	17	1494	4020	4·32	4060	222	362	23·6	38·5	50	272	412	28·9	43·6
2	350	440	16	16	17	1495	4330	4·05	43 80	445	568	41·2	52·5	44	489	612	45·3	56·6
3	300	670	16	16	17	1496	4680	3·75	4730	680	786	53·9	62·3	38	718	804	56·9	63·7
4	260	760	16	16	17	1495	4350	3·50	4410	774	866	61·4	68·7	33	807	899	64·0	71·3
5	220	840	16	16	17	1495	4330	3·22	4390	857	934	62·8	68·5	29	886	963	64·9	70·6
6	180	980	16	16	17	1494	4240	2·92	4320	1000	1070	67·9	72·2	24	1024	1094	69·5	73·8
7	140	1100	16	16	17	1494	4100	2·58	4180	1130	1180	69·7	72·7	19	1149	1199	70·9	73·9
8	100	1210	16	16	17	1494	3890	2·18	3970	1250	1280	68·4	70·3	14	1264	1294	69·2	71·1
9	70	1290	16	16	17	1495	3650	1·83	3720	1330	1350	65·1	66·3	10	1340	1360	65·6	66·8
10	60	1230	16	16	17	1496	3260	1·69	3320	1260	1280	64·3	65·4	8	1268	1288	64·7	65·8
11	5	1360	16	16	17	1496	1900	0·482	1930	1400	1400	34·9	34·9	1	1401	1401	34·9	34·9
12	0	1300	16	16	17	1497	1410	0	1430	1330	1330	0	0	0	1330	1330	0	0

TABLE No. 9

TEST No. 1V06a FRICTION LOSSES $\zeta_{31}\frac{1}{2}\rho_1 v_1^2$ ON INLET AND $\zeta_{24}\frac{1}{2}\rho_2 v_2^2$ ON OUTLET

BAROMETER 994·6 mbar

INLET DUCTING 'M' - 630 mm × 630 mm dia. + BELLMOUTH

OUTLET DUCTING 'Z' - 1080 mm × (630 mm × 450 mm) AND T, U, V, W, × 600 mm dia

CATEGORY 'D' (INCLUDING TWO DIAMETERS OF STRAIGHT BEFORE TRANSFORMATION - NO ADDITIONAL LOSS FOR FRICTION ALLOWED.)

										Excluding Friction Losses					Including Friction Losses			
Damper Setting	Δp Pa	p_{s3} Pa	t_a °C	t_3 °C	t_2 °C	n rev/min	P_R W	q_v' m³/s	P_R' W	p_{SF}' Pa	p_{tF}' Pa	η_s %	η_t %	Friction' Losses Pa	p_{SF}' Pa	p_{tF}' Pa	η_s %	η_t %
1	395	200	19	19	19	1496	4010	4·34	4120	206	347	21·7	36·5	50	256	397	27·0	41·8
2	350	420	19	19	19	1497	4380	4·09	4500	434	559	39·4	50·8	44	478	603	43·4	54·8
3	300	640	18	18	18	1498	4550	3·78	4660	658	765	53·4	62·0	38	696	803	56·5	65·1
4	260	720	18	18	18	1500	4400	3·51	4490	740	832	57·9	65·1	33	773	865	60·5	67·7
5	220	834	18	18	18	1500	4320	3·24	4410	859	937	63·0	68·7	29	888	966	65·1	70·8
6	180	970	18	18	18	1501	4230	2·93	4320	1000	1060	67·9	72·2	24	1024	1084	69·5	73·8
7	140	1125	19	19	19	1502	4060	2·60	4160	1170	1220	73·1	76·2	19	1189	1239	74·3	77·4
8	100	1220	19	19	19	1502	3830	2·20	3930	1270	1310	71·1	73·1	14	1284	1324	71·9	73·9
9	70	1290	19	19	19	1498	3580	1·84	3700	1350	1380	67·4	68·7	10	1360	1390	67·9	69·2
10	60	1220	19	19	19	1498	3170	1·71	3280	1280	1300	66·6	67·7	8	1288	1308	67·0	68·1
11	5	1330	19	19	19	1499	1715	0·487	1770	1400	1400	38·4	38·4	1	1401	1401	38·4	38·4
12	0	1230	19	19	19	1499	1300	0	1340	1290	1290	0	0	0	1290	1290	0	0

TABLE No. 10.

TEST No. 1V07 FRICTION LOSSES $\zeta_{3N1}\frac{1}{2}\rho_1 v_1^2$ ON INLET AND $\zeta_{24}\frac{1}{2}\rho_2 v_2^2$ ON OUTLET

BAROMETER 1004·1 mbar

INLET DUCTING M+N = 2130 mm × 630 mm dia. + BELLMOUTH

OUTLET DUCTING NOTHING × (630 mm × 450 mm) AND T, U, V, W, × 600 mm dia

CATEGORY 'D' PLUS ADDITIONAL STRAIGHT DUCTING ON INLET

										Excluding Friction Losses					Including Friction Losses			
Damper Setting	Δp Pa	p_{s3} Pa	t_a °C	t_3 °C	t_2 °C	n rev/min	P_R W	q_v' m³/s	P_R' W	p_{SF}' Pa	p_{tF}' Pa	η_s %	η_t %	Friction' Losses Pa	p_{SF}' Pa	p_{tF}' Pa	η_s %	η_t %
1	395	200	17	17	17	1500	4040	4·29	4050	201	339	21·3	36·0	52	253	391	26·8	41·5
2	350	410	16	16	17	1499	4380	4·03	4390	412	535	37·9	49·1	47	457	582	42·2	53·4
3	300	625	16	16	17	1498	4580	3·75	4610	632	737	51·4	59·9	41	673	778	54·7	63·2
4	260	705	16	16	17	1498	4400	3·49	4430	714	806	56·2	63·4	35	749	841	59·0	66·2
5	220	820	16	16	17	1499	4320	3·21	4350	832	909	61·4	67·1	31	863	940	63·7	69·4
6	180	955	16	16	17	1500	4320	2·91	4350	970	1030	64·9	69·1	26	996	1056	66·6	70·8
7	140	1110	16	16	17	1501	4120	2·57	4140	1130	1180	70·0	73·0	20	1150	1200	71·2	74·2
8	100	1215	16	16	17	1502	3920	2·17	3940	1240	1270	68·2	70·1	15	1255	1285	69·0	70·9
9	70	1290	16	16	17	1502	3670	1·82	3690	1320	1340	64·8	66·0	11	1331	1351	65·3	66·5
10	60	1230	16	16	17	1502	3340	1·68	3360	1250	1270	62·8	63·8	9	1259	1279	63·3	64·3
11	5	1360	16	16	17	1501	1890	0·481	1910	1390	1390	35·1	35·1	1	1391	1391	35·1	35·1
12	0	1290	16	16	17	1501	1400	0	1410	1320	1320	0	0	0	1320	1320	0	0

TABLE No. 11

TEST No. 1V08 FRICTION LOSSES $\zeta_{3NP1}\ \tfrac{1}{2}\rho_1 v_1^2$ ON INLET AND $\zeta_{24}\ \tfrac{1}{2}\rho_2 v_2^2$ ON OUTLET

BAROMETER 1005·8 mbar

INLET DUCTING M+N+P=3630mm × 630 mm dia. + BELLMOUTH

OUTLET DUCTING NOTHING × (630 mm × 450 mm) AND T,U,V,W, × 600 mm dia.

CATEGORY 'D' PLUS ADDITIONAL STRAIGHT DUCTING ON INLET

										Excluding Friction Losses					Including Friction Losses			
Damper Setting	Δp Pa	p_{s3} Pa	t_a °C	t_3 °C	t_2 °C	n rev/min	P_R W	q_v' m^3/s	P_R' W	p_{sF}' Pa	p_{tF}' Pa	η_s %	η_t %	Friction' Losses Pa	p_{sF}' Pa	p_{tF}' Pa	η_s %	η_t %
1	390	210	19	19	19	1499	4000	4·27	4040	212	349	22·4	36·9	55	267	404	28·2	42·7
2	350	400	19	19	19	1499	4300	4·05	4350	406	529	37·8	49·3	50	456	579	42·5	54·0
3	300	640	19	19	19	1501	4550	3·76	4590	651	756	53·2	61·8	44	695	800	56·8	65·4
4	260	695	19	19	19	1501	4350	3·50	4390	707	799	56·3	63·6	37	744	836	59·2	66·5
5	220	840	19	19	19	1502	4330	3·22	4370	856	933	63·1	68·8	33	889	966	65·5	71·2
6	180	980	19	19	19	1502	4260	2·92	4310	1000	1070	67·8	72·1	28	1028	1098	69·7	74·0
7	140	1125	19	19	19	1504	4130	2·57	4160	1150	1200	71·1	74·1	21	1171	1221	72·4	75·4
8	100	1220	19	19	19	1504	3930	2·18	3970	1250	1280	68·6	70·5	16	1266	1296	69·5	71·4
9	70	1300	19	19	19	1503	3650	1·82	3690	1340	1360	65·9	67·1	12	1352	1372	66·5	67·7
10	55	1240	19	19	19	1503	3220	1·62	3260	1270	1290	63·1	64·1	10	1280	1300	63·6	64·6
11	5	1380	19	19	19	1504	1920	0·482	1940	1420	1420	35·2	35·2	1	1421	1421	35·2	35·2
12	0	1330	19	19	19	1505	1390	0	1400	1360	1360	0	0	0	1360	1360	0	0

TABLE No. 12.

TEST No. 1V09 FRICTION LOSSES $\left(0{\cdot}4+\zeta_{3NP1}\right)\tfrac{1}{2}\rho_1 v_1^2$ ON INLET AND $\zeta_{24}\ \tfrac{1}{2}\rho_2 v_2^2$ ON OUTLET

BAROMETER 1009·8 mbar

INLET DUCTING M+N+Q+P=3630mm+BEND × 630 mm dia. + BELLMOUTH

OUTLET DUCTING NOTHING × (630 mm × 450 mm) AND T,U,V,W, × 600 mm dia.

CATEGORY D PLUS STRAIGHT EITHER SIDE OF BEND ON INLET.

* PITÔT TRAVERSE BEFORE AND AFTER BEND TAKEN FOR THIS DAMPER SETTING.

										Excluding Friction Losses					Including Friction Losses			
Damper Setting	Δp Pa	p_{s3} Pa	t_a °C	t_3 °C	t_2 °C	n rev/min	P_R W	q_v' m^3/s	P_R' W	p_{sF}' Pa	p_{tF}' Pa	η_s %	η_t %	Friction' Losses Pa	p_{sF}' Pa	p_{tF}' Pa	η_s %	η_t %
* 1	395	250	18	18	19	1498	4280	4·23	4290	251	339	25·0	38·8	100	351	439	35·0	50·2
2	351	440	18	18	19	1499	440	4·04	4490	443	565	39·8	50·8	91	534	656	48·0	59·0
3	300	610	18	18	19	1498	4630	3·74	4660	617	722	49·6	58·0	79	696	801	56·0	64·3
4	250	810	18	18	19	1497	4580	3·43	4630	823	911	61·0	67·5	67	890	978	66·0	72·5
5	220	850	18	18	19	1497	4380	3·22	4430	865	942	62·8	68·4	59	924	1001	67·1	72·7
6	180	920	18	18	19	1497	4180	2·91	4230	937	1000	64·5	68·9	49	986	1049	67·9	72·3
7	140	1060	18	18	19	1497	4030	2·57	4080	1080	1130	68·2	71·3	38	1118	1168	70·6	73·7
· 8	100	1180	18	18	19	1497	3810	2·18	3860	1210	1240	68·1	70·0	28	1238	1268	69·7	71·6
9	60	1290	18	18	19	1499	3460	1·69	3500	1320	1340	63·6	64·6	20	1340	1360	64·6	65·6
10	52	1200	18	18	19	1501	3000	1·56	3020	1220	1240	63·4	64·3	17	1237	1257	64·3	65·2
11	5	1310	18	18	19	1501	1770	0·48	1780	1340	1340	36·0	36·1	2	1342	1342	36·1	36·2
12	0	1230	18	18	19	1502	1390	0	1400	1250	1250	0	0	0	1250	1250	0	0

TABLE No 13

TEST No. 1V10 FRICTION LOSSES $\left(0{\cdot}4+\zeta_{3NP1}\right)\frac{1}{2}\rho_1 v_1^2$ ON INLET AND $\zeta_{24}\frac{1}{2}\rho_2 v_2^2$ ON OUTLET

BAROMETER 1010·8 mbar

INLET DUCTING M+N+P+Q=3630mm+BEND × 630 mm dia. + BELLMOUTH

OUTLET DUCTING NOTHING × (630 mm × 450 mm) AND T,U,V,W, × 600 mm dia.

CATEGORY 'D' PLUS LONG STRAIGHT AND THEN BEND DIRECTLY ON INLET

										Excluding Friction Losses					Including Friction Losses			
Damper Setting	Δp Pa	pss Pa	t_a °C	t_3 °C	t_2 °C	n rev/min	P_R W	q_v' m³/s	P_R' W	p_{sF}' Pa	p_{tF}' Pa	η_s %	η_t %	Friction' Losses Pa	p_{sF}' Pa	p_{tF}' Pa	η_s %	η_t %
1	380	170	21	21	22	1492	3990	4·24	4090	174	309	18·0	32·0	100	274	409	28·3	42·4
2	340	350	21	21	22	1492	4260	4·01	4370	359	480	32·9	44·0	91	450	571	41·2	52·3
3	300	515	21	21	22	1493	4370	3·77	4490	529	636	44·5	53·5	79	608	715	51·1	60·1
4	255	700	21	21	22	1496	4480	3·48	4580	719	809	54·6	61·5	67	786	876	59·7	66·6
5	215	800	21	21	22	1500	4370	3·19	4440	818	894	58·8	64·3	59	877	953	63·0	68·5
6	180	900	21·5	21·5	22·5	1501	4230	2·92	4300	924	988	62·9	67·3	49	973	1037	66·2	70·6
7	150	960	21·5	21·5	22·5	1499	4060	2·67	4140	990	1040	63·9	67·3	41	1031	1081	66·5	70·0
8	120	1060	21·5	21·5	22·5	1500	3950	2·39	4030	1090	1140	65·0	67·5	33	1122	1173	66·9	69·5
9	90	1140	21·5	21·5	22·5	1500	3790	2·07	3870	1180	1210	63·2	64·9	28	1208	1238	64·7	66·4
10	60	1240	21·5	21·5	22·5	1500	3530	1·69	3600	1280	1310	60·4	61·4	20	1300	1330	61·3	62·3
11	30	1210	21·5	21·5	22·5	1500	2730	1·19	2790	1250	1260	53·7	54·1	10	1260	1270	54·1	54·5
12	0	1270	21·5	21·5	22·5	1500	1610	0	1640	1320	1320	0	0	0	1320	1320	0	0

TABLE No. 14

TEST No. 1V10 a FRICTION LOSSES $\left(0{\cdot}4+\zeta_{3NPX1}\right)\frac{1}{2}\rho_1 v_1^2$ ON INLET AND $\zeta_{24}\frac{1}{2}\rho_2 v_2^2$ ON OUTLET

BAROMETER 1010·8 mbar

INLET DUCTING M+N+P+Q+X × 630 mm dia. + BELLMOUTH

OUTLET DUCTING NOTHING × (630 mm × 450 mm) AND T, U,V,W, × 600 mm dia.

CATEGORY D PLUS LONG STRAIGHT THEN BEND AND ONE DIAMETER OF STRAIGHT ON INLET. (ie. AS 1V10 PLUS SHORT STRAIGHT ON INLET)

										Excluding Friction Losses					Including Friction Losses			
Damper Setting	Δp Pa	pss Pa	t_a °C	t_3 °C	t_2 °C	n rev/min	P_R W	q_v' m³/s	P_R' W	p_{sF}' Pa	p_{tF}' Pa	η_s %	η_t %	Friction' Losses Pa	p_{sF}' Pa	p_{tF}' Pa	η_s %	η_t %
1	395	170	24	24	25	1498	3940	4·32	4030	174	315	18·7	33·7	105	279	420	30·0	44·9
2	365	340	24	24	25	1499	4230	4·16	4330	349	479	33·5	46·0	99	448	578	43·0	55·5
3	325	485	24	24	25	1498	4350	3·93	4460	499	615	44·0	54·2	80	579	695	51·1	61·3
4	280	640	24	24	25	1499	4450	3·65	4560	660	760	52·9	60·9	73	733	833	58·8	66·8
5	235	740	24	24	25	1501	4220	3·35	4320	763	847	59·2	65·7	64	827	911	64·2	70·7
6	200	830	24	24	25	1500	4150	3·09	4260	858	930	62·4	67·5	50	908	980	66·0	71·1
7	160	950	24·5	24·5	25·5	1501	4010	2·77	4120	987	1040	66·4	70·3	45	1032	1085	69·4	73·3
8	120	1060	24·5	24·5	25·5	1499	3870	2·41	3990	1110	1150	66·7	69·3	33	1143	1183	68·7	71·3
9	80	1180	24·5	24·5	25·5	1498	3640	1·97	3770	1240	1270	64·8	66·1	28	1268	1298	66·3	67·6
10	40	1170	24·5	24·5	25·5	1498	2790	1·39	2890	1230	1240	59·0	59·7	10	1240	1250	59·5	60·2
11	0	1270	24·5	24·5	25·5	1499	1540	0	1590	1330	1330	0	0	0	1330	1330	0	0
12																		

TABLE No. 15

TEST No. 1V11 FRICTION LOSSES $(0{\cdot}4 + \zeta_{31})\frac{1}{2}\rho_1 v_1^2$ ON INLET AND $\zeta_{24}\frac{1}{2}\rho_2 v_2^2$ ON OUTLET
BAROMETER 1010·8 mbar
INLET DUCTING M-630mm + Q × 630 mm dia. + BELLMOUTH
OUTLET DUCTING NOTHING × (630 mm × 450 mm) AND T, U, V, W, × 600 mm dia

CATEGORY 'D' BUT BEND ON INLET

										Excluding Friction Losses					Including Friction Losses			
Damper Setting	Δp Pa	ps3 Pa	t_a °C	t_3 °C	t_2 °C	n rev/min	P_R W	q_v' m³/s	P_R' W	p_{SF}' Pa	p_{tF}' Pa	η_s %	η_t %	Friction' Losses Pa	p_{SF}' Pa	p_{tF}' Pa	η_s %	η_t %
1	400	190	26	26	27	1501	3930	4·35	4020	194	337	21·0	36·4	54	248	391	26·8	42·2
2	365	360	26·5	26·5	27·5	1500	4270	4·17	4400	371	502	35·2	47·6	48	419	550	39·8	52·2
3	320	520	26·5	26·5	27·5	1499	4440	3·92	4590	539	654	46·0	55·8	42	581	696	49·6	59·4
4	280	670	26·5	26·5	27·5	1499	4450	3·67	4600	696	797	55·5	63·5	36	732	833	58·4	66·4
5	235	780	26·5	26·5	27·5	1499	4160	3·36	4310	812	897	63·4	70·0	32	844	929	65·9	72·5
6	200	860	26·5	26·5	27·5	1499	4230	3·11	4380	897	969	63·5	68·6	27	924	996	65·4	70·5
7	155	965	27	27	28	1499	3960	2·74	4120	1010	1070	67·4	71·1	22	1032	1092	68·8	72·2
8	120	1060	27	27	28	1498	3840	2·42	4000	1120	1160	67·3	69·9	17	1137	1177	68·3	70·9
9	80	1190	27	27	28	1498	3570	1·98	3730	1260	1280	66·5	68·1	12	1272	1292	67·1	68·7
10	40	1160	27	27	28	1498	2760	1·39	2380	1220	1240	59·2	59·9	5	1225	1245	59·4	60·1
11	0	1200	27	27	28	1500	1410	0	1470	1260	1260	0	0	0	1260	1260	0	0
12																		

TABLE No. 16

TEST No. 1V12 FRICTION LOSSES $\zeta_{31}\frac{1}{2}\rho_1 v_1^2$ ON INLET AND $(0{\cdot}23 + \zeta_{24})\frac{1}{2}\rho_2 v_2^2$ ON OUTLET
BAROMETER 1010·2 mbar
INLET DUCTING M-630mm × 630 mm dia. + BELLMOUTH
OUTLET DUCTING Z-1080mm and R-90° BEND. × (630 mm × 450 mm) AND T, U, V, W. × 600 mm dia.
CATEGORY 'D' INCLUDING TWO DIAMETERS OF RECTANGULAR STRAIGHT (NO ADDITIONAL LOSS FOR FRICTION) THEN TIGHT RECTANGULAR (EASY) BEND BEFORE COMMON PARTS.
* PITÔT TRAVERSE BEFORE AND AFTER BEND TAKEN FOR THIS DAMPER SETTING

										Excluding Friction Losses					Including Friction Losses			
Damper Setting	Δp Pa	ps3 Pa	t_a °C	t_3 °C	t_2 °C	n rev/min	P_R W	q_v' m³/s	P_R' W	p_{SF}' Pa	p_{tF}' Pa	η_s %	η_t %	Friction' Losses Pa	p_{SF}' Pa	p_{tF}' Pa	η_s %	η_t %
* 1	400	200	19	19	19	1500	4000	4·30	4010	200	339	21·5	36·4	82	282	421	30·2	45·1
2	350	420	19	19	20	1500	4380	4·03	4400	422	544	38·7	49·9	72	494	616	45·2	56·4
3	300	640	19	19	20	1500	4500	3·74	4530	646	751	53·4	62·1	63	709	814	58·5	67·2
4	260	680	19	19	20	1498	4340	3·49	4390	689	780	54·8	62·0	54	743	834	59·1	66·3
5	220	810	20	20	20	1496	4280	3·23	4370	831	909	61·5	67·2	47	878	956	64·9	70·7
6	180	970	20	20	20	1495	4210	2·93	4310	1000	1060	67·9	72·2	39	1039	1099	70·6	74·7
7	140	1110	20	20	20	1494	4080	2·59	4190	1150	1200	70·9	74·0	31	1181	1231	73·0	76·1
8	100	1225	20	20	20	1494	3850	2·19	3960	1270	1310	70·3	72·2	22	1292	1332	71·5	73·7
9	72	1290	20	20	20	1196	3610	1·86	3700	1340	1360	67·0	68·3	16	1356	1376	68·2	69·2
10	60	1225	20	20	20	1496	3240	1·69	3320	1270	1290	64·7	65·7	12	1282	1302	65·3	66·3
11	5	1390	20	20	20	1497	1880	0·484	1930	1440	1440	36·2	36·2	1	1441	1441	36·2	36·2
12	0	1350	20	20	20	1498	1450	0	1480	1400	1400	0	0	0	1400	1400	0	0

TABLE No. 17

TEST No. 1V13 FRICTION LOSSES $\zeta_{31}\frac{1}{2}\rho_1 v_1^2$ ON INLET AND $(0{\cdot}27+\zeta_{24})\frac{1}{2}\rho_2 v_2^2$ ON OUTLET

BAROMETER 1012·6 mbar

INLET DUCTING M - 630 mm × 630 mm dia + BELLMOUTH

OUTLET DUCTING Z-1080mm+S-90° BEND × (630 mm × 450 mm) AND T, U, V, W, × 600 mm dia.

CATEGORY 'D' INCLUDING TWO DIAMETERS OF RECTANGULAR STRAIGHT (NO ADDITIONAL LOSS FOR FRICTION) THEN LONG RECTANGULAR (HARD) BEND BEFORE COMMON PARTS.

* PITÔT TRAVERSE BEFORE AND AFTER BEND TAKEN FOR THIS DAMPER SETTING.

										Excluding Friction Losses					Including Friction Losses			
Damper Setting	Δp Pa	ps3 Pa	t_a °C	t_3 °C	t_2 °C	n rev/min	P_R W	q_v' m³/s	P_R' W	p_{sF}' Pa	p_{tF}' Pa	η_s %	η_t %	Friction' Losses Pa	p_{sF}' Pa	p_{tF}' Pa	η_s %	η_t %
* 1	390	200	22	22	22	1498	4000	4·28	4060	203	340	21·4	35·9	87	290	427	30·6	45·0
2	350	390	22	22	22	1496	4300	4·07	4390	399	522	36·9	48·4	77	476	599	44·1	55·5
3	300	600	22	22	22	1497	4500	3·77	4590	615	721	50·5	59·2	67	682	788	56·0	64·7
4	260	660	22	22	22	1496	4270	3·51	4370	678	770	54·5	61·9	58	736	828	59·1	66·5
5	220	800	22	22	22	1498	4270	3·23	4360	822	900	61·0	66·7	50	872	950	64·6	70·4
6	180	955	22	22	22	1497	4260	2·93	4360	986	1050	66·2	70·5	42	1028	1092	69·1	73·4
7	140	1110	22	22	22	1496	4120	2·59	4240	1150	1200	70·4	73·4	33	1183	1233	72·3	75·3
8	100	1215	22	22	22	1497	3870	2·19	3970	1260	1300	69·5	71·4	23	1283	1323	70·8	73·0
9	70	1300	22	22	22	1498	3650	1·83	3740	1350	1370	66·0	67·3	17	1367	1387	66·9	67·9
10	60	1220	22	22	22	1498	3280	1·70	3360	1260	1290	63·7	64·8	13	1273	1303	64·4	65·9
11	5	1380	22	22	22	1498	1880	0·485	1930	1430	1440	35·9	36·1	1	1431	1441	36·0	36·2
12	0	1320	22	22	22	1499	1440	0	1470	1370	1370	0	0	0	1370	1370	0	0

TABLE No. 18

TEST No. 1V14 FRICTION LOSSES $(0{\cdot}4+\zeta_{3NP1})\frac{1}{2}\rho_1 v_1^2$ ON INLET AND $(0{\cdot}27+\zeta_{24})\frac{1}{2}\rho_2 v_2^2$ ON OUTLET

BAROMETER 1010·8 mbar

INLET DUCTING M+N+Q+P=3630mm + BEND × 630 mm dia.

OUTLET DUCTING Z-1080mm & S-90° BEND × (630 mm × 450 mm) AND T, U, V, W, × 600 mm dia.

CATEGORY 'D' PLUS STRAIGHT EITHER SIDE OF BEND ON INLET AND TWO DIAMETERS OF RECTANGULAR (NO ADDITIONAL LOSS FOR FRICTION) THEN LONG RECTANGULAR (HARD) BEND BEFORE COMMON PARTS.

										Excluding Friction Losses					Including Friction Losses			
Damper Setting	Δp Pa	ps3 Pa	t_a °C	t_3 °C	t_2 °C	n rev/min	P_R W	q_v' m³/s	P_R' W	p_{sF}' Pa	p_{tF}' Pa	η_s %	η_t %	Friction' Losses Pa	p_{sF}' Pa	p_{tF}' Pa	η_s %	η_t %
1	395	200	21	21	22	1499	4070	4·3	4110	202	341	21·2	35·7	137	339	478	35·5	50·1
2	360	350	21	21	22	1498	4180	4·11	4240	356	483	34·5	46·8	124	480	607	46·5	58·8
3	325	500	21	21	22	1498	4350	3·91	4420	510	625	45·2	55·3	116	626	741	55·4	65·6
4	280	710	21·5	21·5	22·5	1498	4650	3·65	4740	729	829	56·1	63·8	102	831	931	64·0	71·7
5	240	800	21·5	21·5	22·5	1498	4470	3·38	4560	823	909	61·0	67·3	88	911	997	67·5	73·9
6	195	880	21·5	21·5	22·5	1497	4230	3·05	4330	908	978	64·0	68·9	72	980	1050	69·0	73·9
7	165	970	21·5	21·5	22·5	1497	4070	2·81	4170	1000	1060	67·6	71·5	61	1061	1121	71·5	75·5
8	135	1050	21·5	21·5	22·5	1497	3940	2·54	4040	1090	1140	68·5	71·5	52	1142	1192	71·8	74·9
9	95	1160	21·5	21·5	22·5	1496	3700	2·14	3810	1210	1240	67·7	69·6	37	1247	1277	70·0	71·7
10	60	1260	21·5	21·5	22·5	1496	3420	1·70	3520	1310	1330	63·4	64·4	26	1336	1356	64·5	65·5
11	30	1230	21·5	21·5	22·5	1496	2590	1·20	2670	1280	1290	57·5	58·0	14	1294	1304	58·2	58·6
12	0	1200	21·5	21·5	22·5	1495	1390	0	1430	1250	1250	0	0	0	1250	1250	0	0

TABLE No. 19.

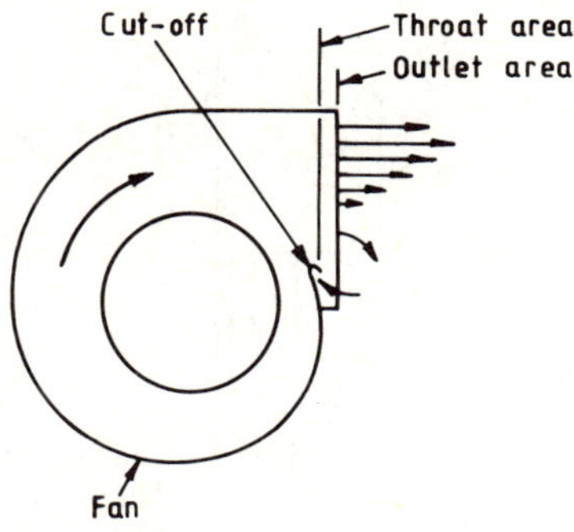

Non uniform velocity profile when discharging into atmosphere

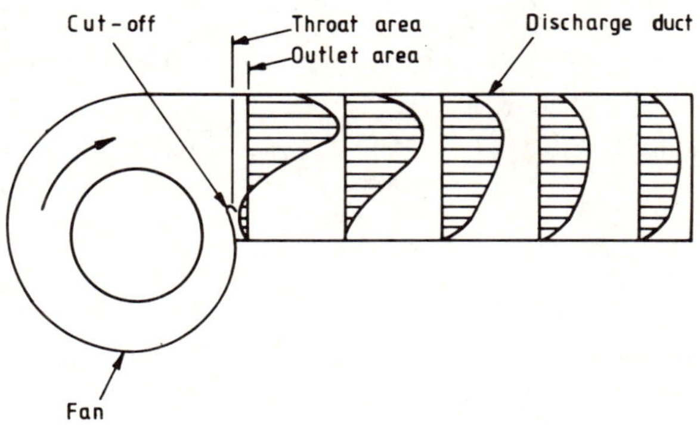

Controlled diffusion and establishment of uniform velocity profile in straight duct.

Figure 1

Velocity profiles at the outlet of a centrifugal fan.

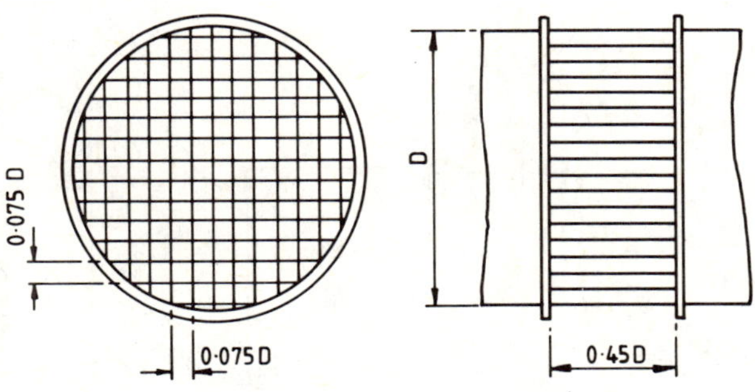

Figure 2

AMCA straightener

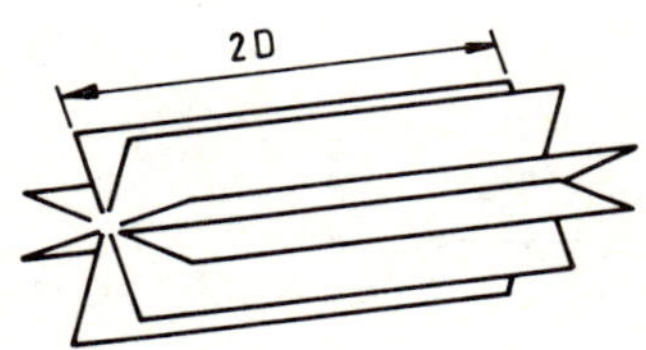

Figure 3

Etoile straightener

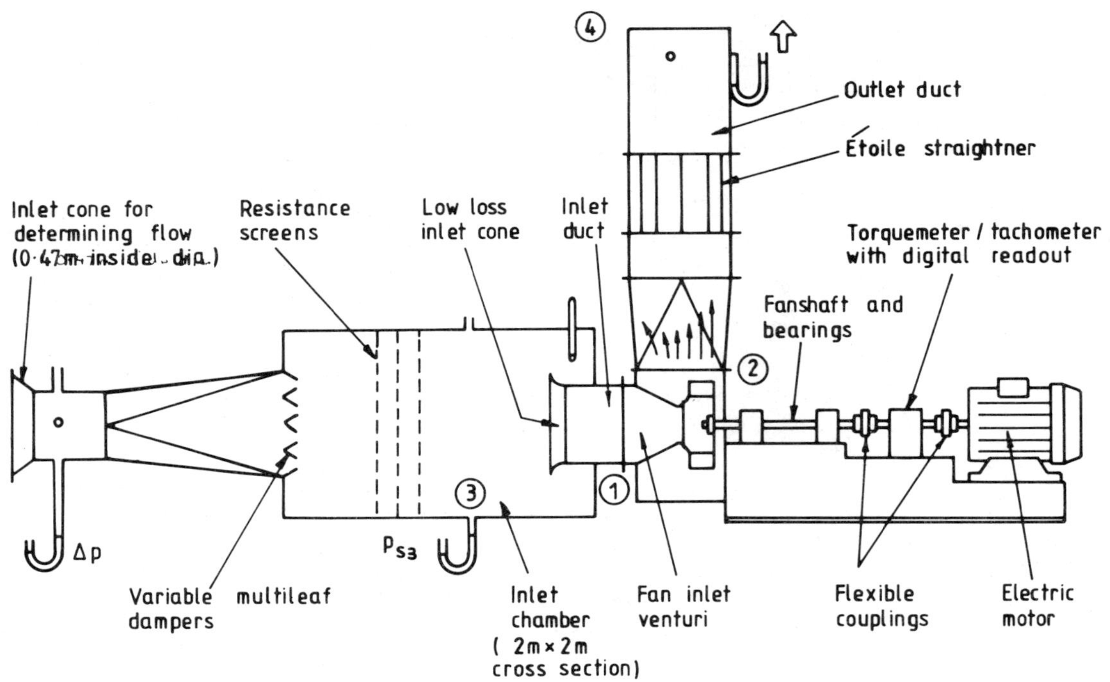

Figure 4

Testing procedure for centrifugal fans as determined by BS848:1980 for installation type D (diagrammatic only)

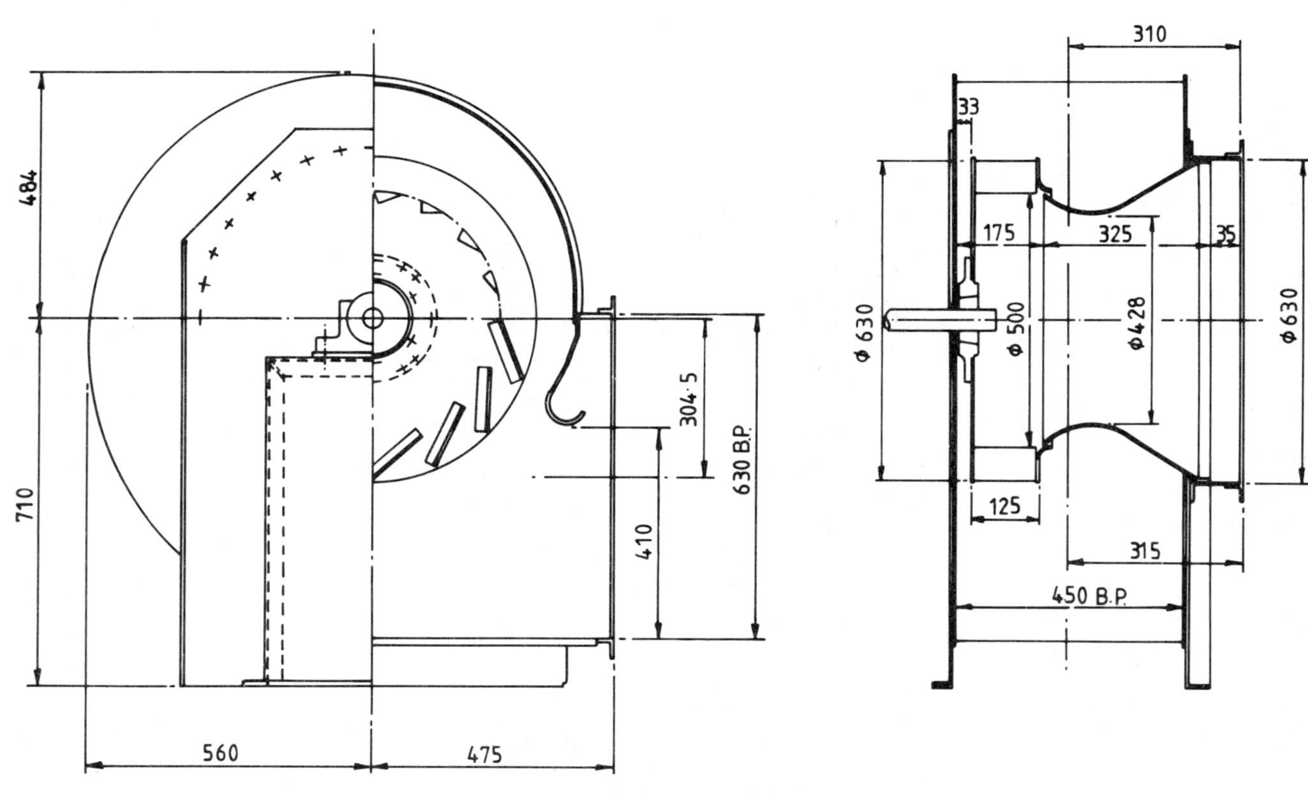

Figure 5

General arrangement of centrifugal fan used in test programme

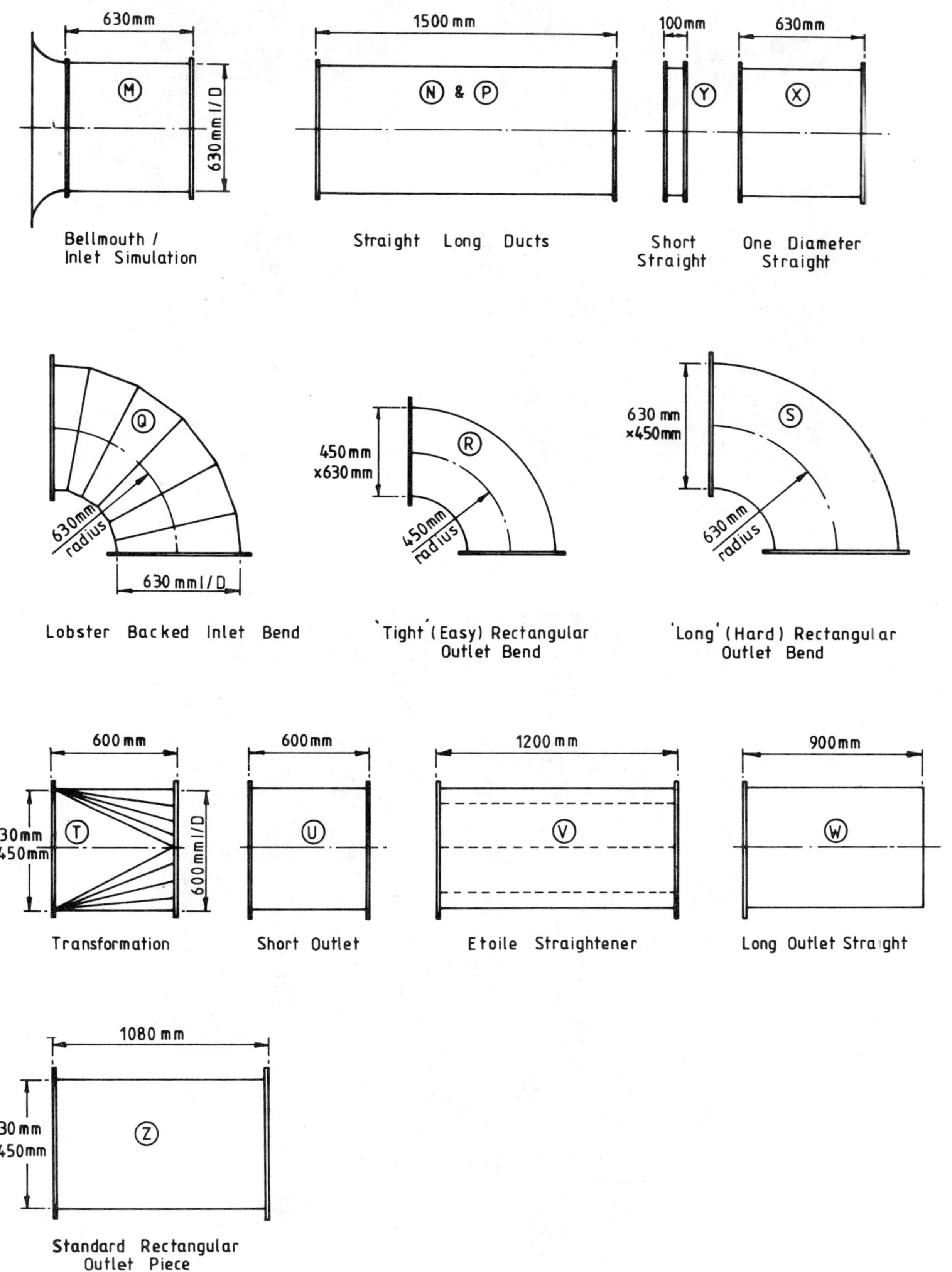

Figure 6

Range of inlet and outlet ducting

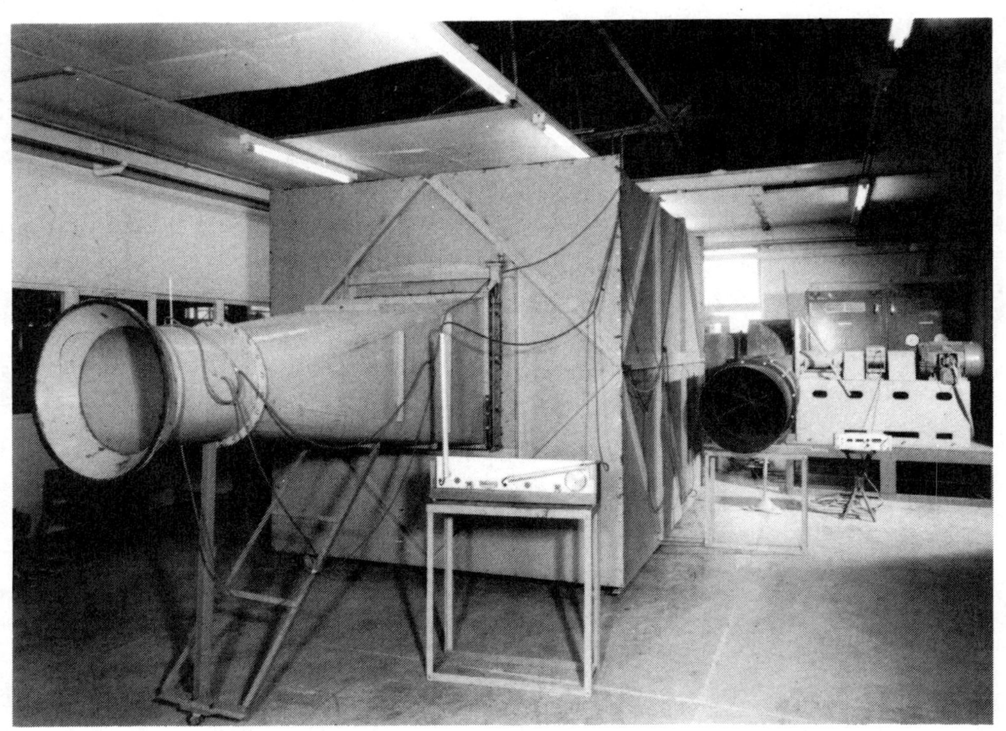

Figure 7

Photograph of test chamber and inlet cone etc.

Figure 8

Photograph showing close-up of fan and torque meter

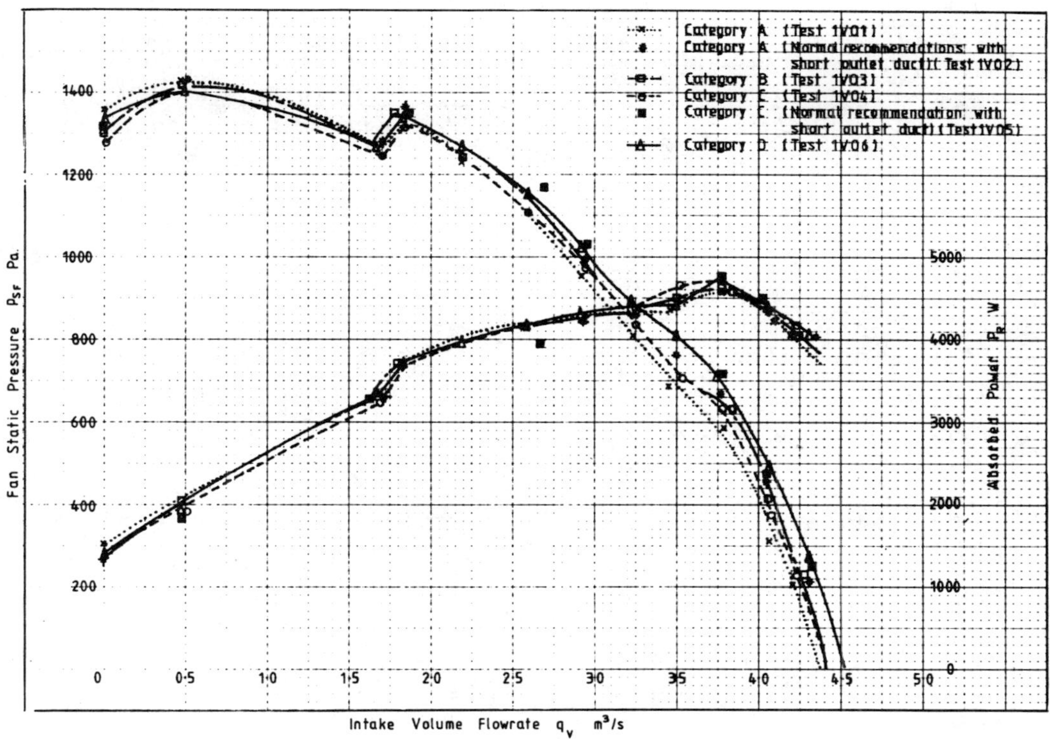

Figure 9

Performance comparisons for 630mm diameter centrifugal fan at 1,500 rev/min. Handling air 1.2kg/m³ and operating under different installation categories.

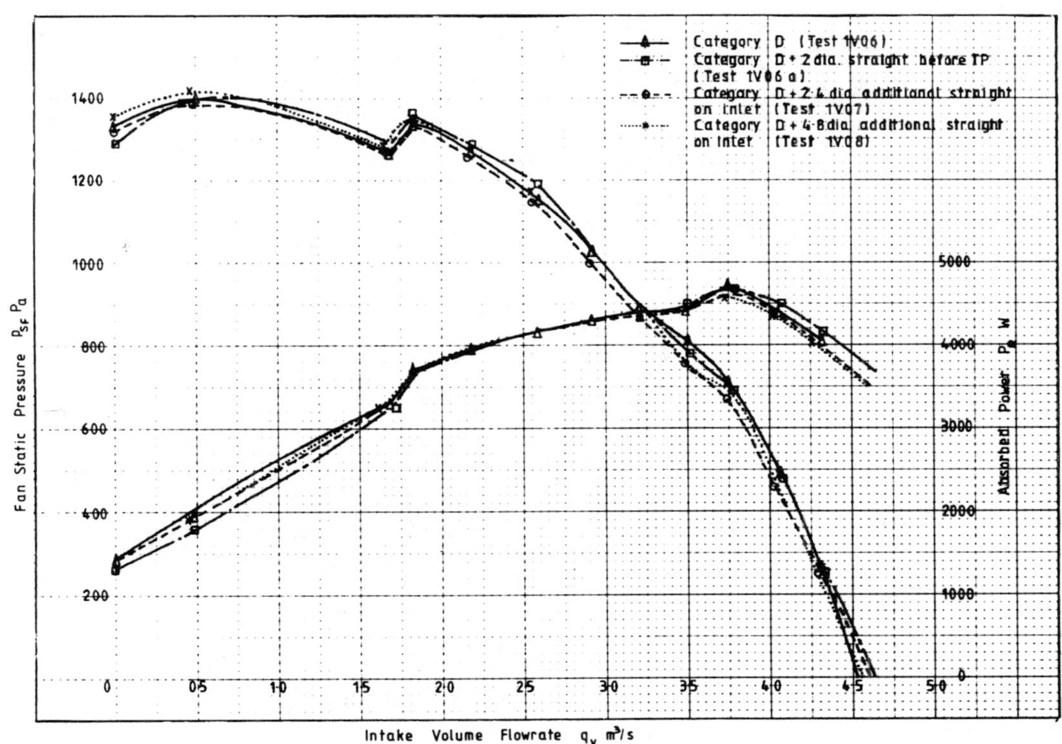

Figure 10

Performance comparisons for 630mm diameter centrifugal fan at 1,500 rev/min. Handling air 1.2kg/m³ and effects of additional inlet/outlet straight on category D fan installations

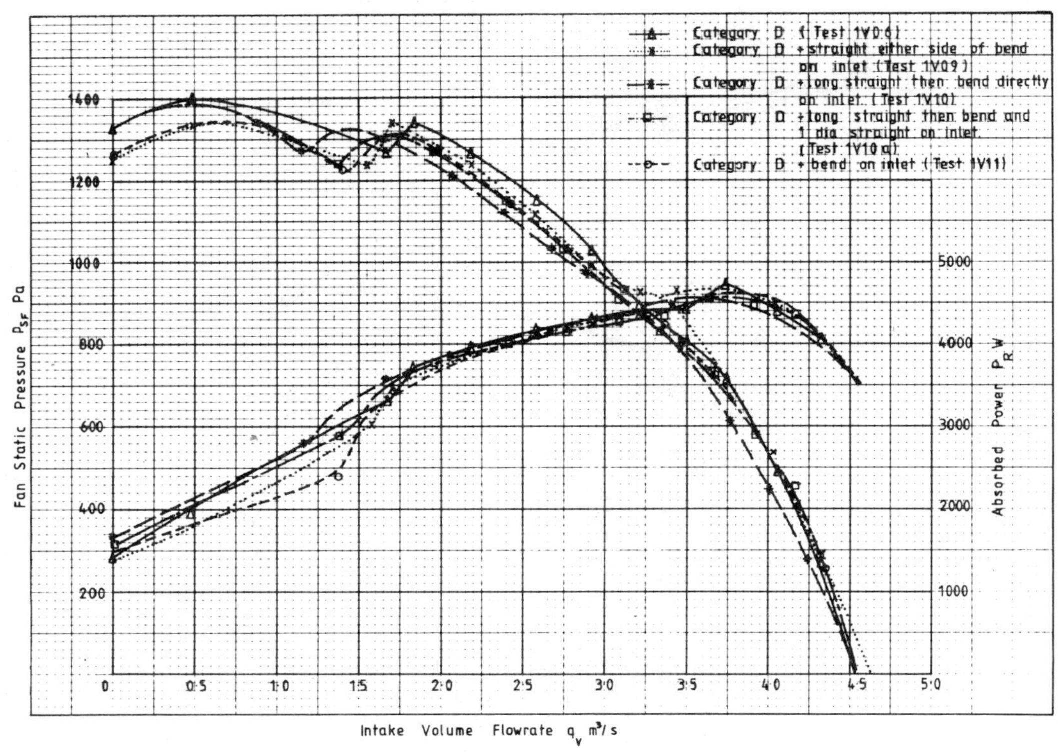

Figure 11

Performance comparisons for 630mm diameter centrifugal fan at 1,500 rev/min. Handling air 1.2kg/m³ and effects of bends on inlet side

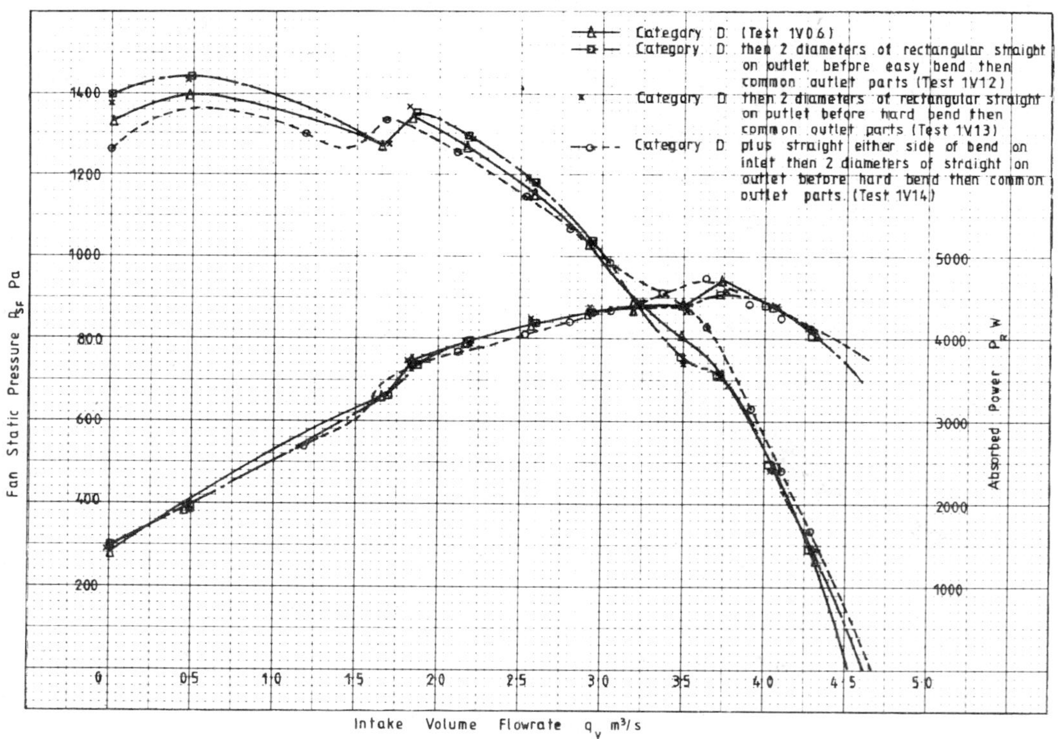

Figure 12

Performance comparisons for 630mm diameter centrifugal fan at 1,500 rev/min. Handling air 1.2kg/m³ and effects of bends on outlet side

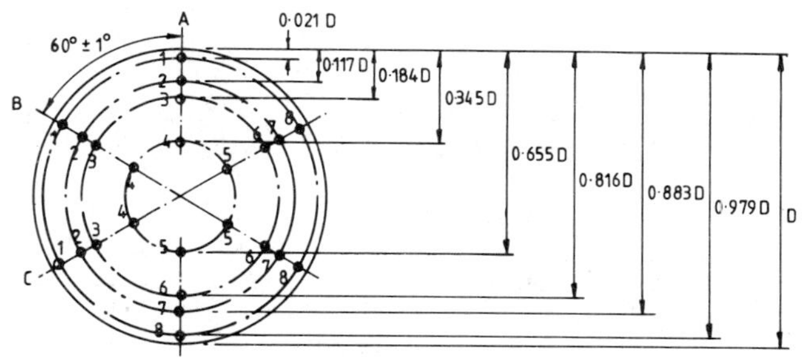

Velocity pressures

Posn	Before bend			After bend		
	A	B	C	A	B	C
1	115	115	120	140	60	120
2	132	128	130	145	100	160
3	130	135	135	145	135	160
4	125	143	140	135	145	150
5	120	125	130	135	120	125
6	125	100	115	135	100	120
7	115	85	115	140	100	115
8	90	55	90	123	110	130

Before bend:

$av\sqrt{p_v} = 10 \cdot 8$

$\rho = 1 \cdot 205 \text{ kg/m}^3$

$v = 13 \cdot 91 \text{ m/s}$

$q_v = 4 \cdot 34 \text{ m}^3/\text{s}$

= 102·6% of flow measured at conical inlet.

After bend:

$av\sqrt{p_v} = 11 \cdot 22$

$v = 14 \cdot 65 \text{ m/s}$

$q_v = 4 \cdot 54 \text{ m}^3/\text{s}$

= 107·3% of flow measured at conical inlet.

Figure 13

Pitot traverses taken during test number IV09

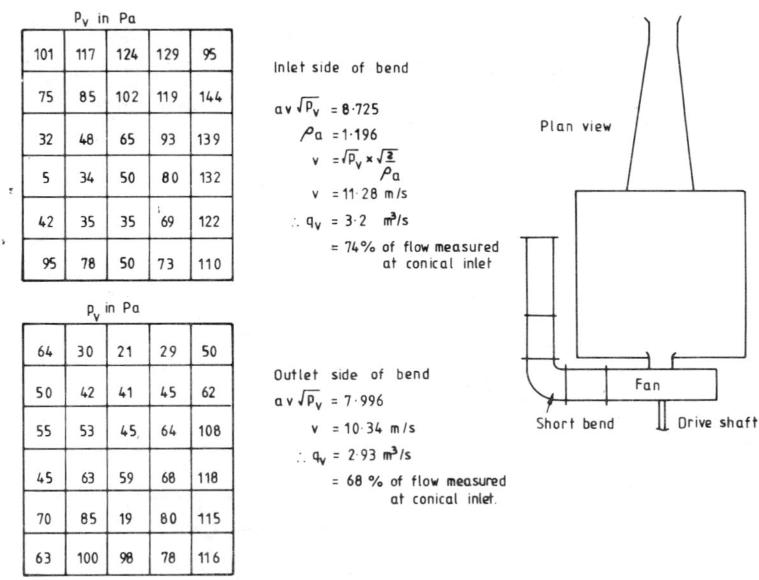

p_v in Pa

101	117	124	129	95
75	85	102	119	144
32	48	65	93	139
5	34	50	80	132
42	35	35	69	122
95	78	50	73	110

Inlet side of bend

$av\sqrt{p_v} = 8 \cdot 725$

$\rho a = 1 \cdot 196$

$v = \sqrt{p_v} \times \sqrt{\frac{2}{\rho a}}$

$v = 11 \cdot 28 \text{ m/s}$

$\therefore q_v = 3 \cdot 2 \text{ m}^3/\text{s}$

= 74% of flow measured at conical inlet

p_v in Pa

64	30	21	29	50
50	42	41	45	62
55	53	45	64	108
45	63	59	68	118
70	85	19	80	115
63	100	98	78	116

Outlet side of bend

$av\sqrt{p_v} = 7 \cdot 996$

$v = 10 \cdot 34 \text{ m/s}$

$\therefore q_v = 2 \cdot 93 \text{ m}^3/\text{s}$

= 68 % of flow measured at conical inlet.

Figure 14

Pitot traverses taken during test number IV12

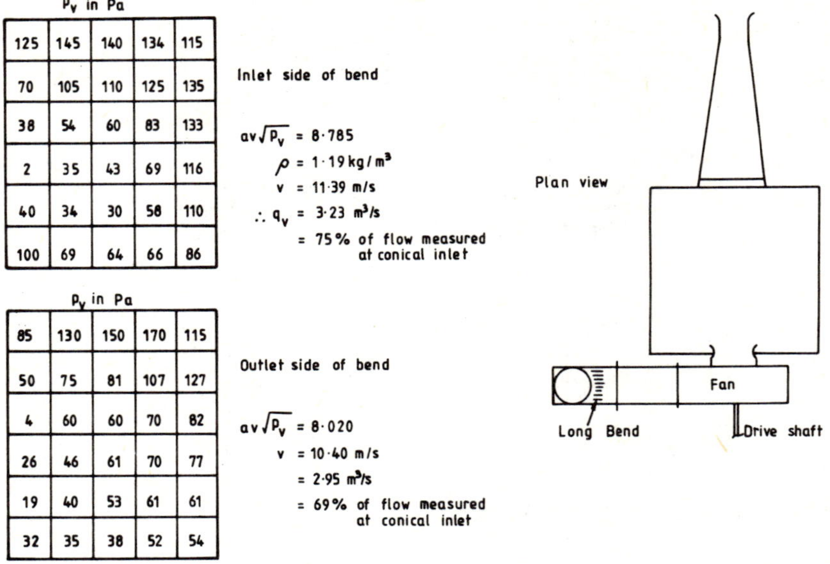

p_v in Pa

125	145	140	134	115
70	105	110	125	135
38	54	60	83	133
2	35	43	69	116
40	34	30	58	110
100	69	64	66	86

p_v in Pa

85	130	150	170	115
50	75	81	107	127
4	60	60	70	82
26	46	61	70	77
19	40	53	61	61
32	35	38	52	54

Figure 15

Pitot traverses taken during test number IV13

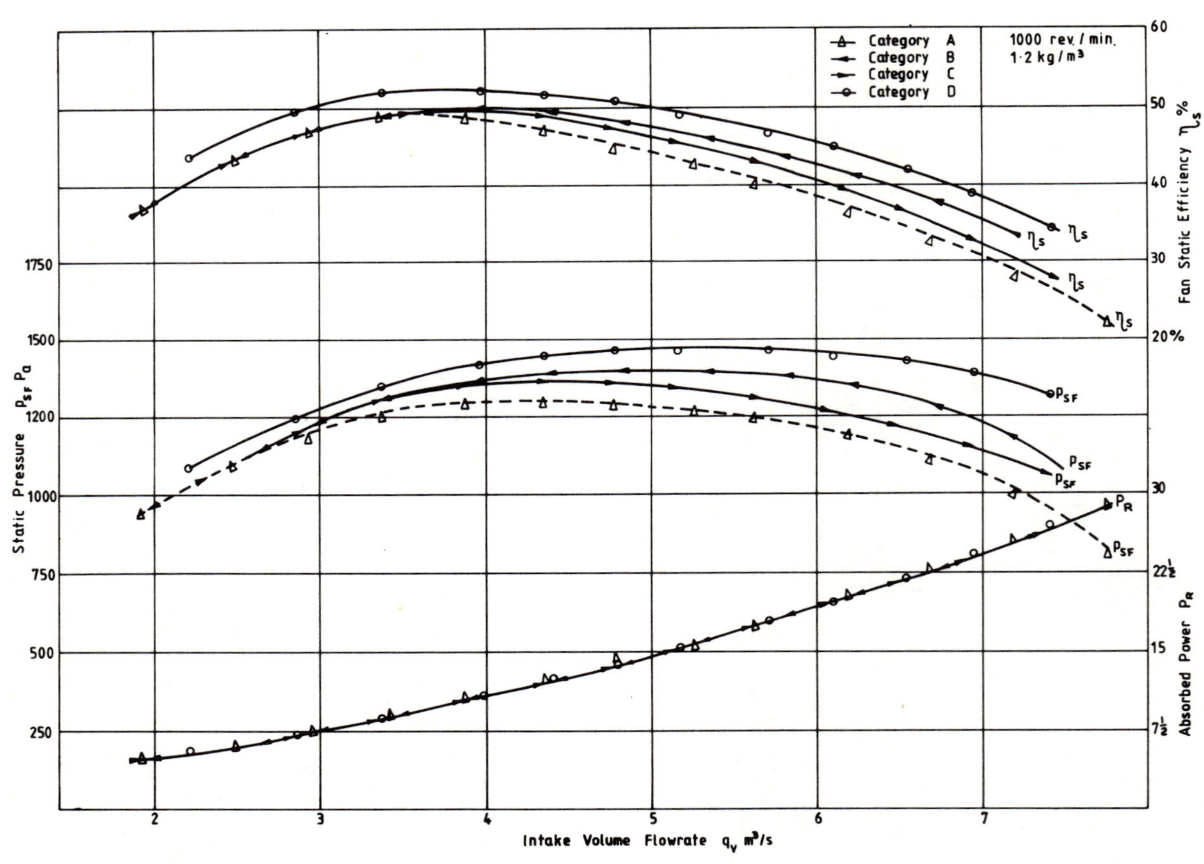

Figure 16

630mm diameter 48 bladed narrow forward curved fan

International Conference on

Fan Design & Applications

Guildford, England: September 7-9, 1982

PAPER B1

LOSSES IN THE INLET AND THE IMPELLER OF A RADIAL FAN

J. Vandevenne

Rijksuniversiteit Gent, Belgium

Summary

In this paper the losses occurring in the inlet and the impeller of a radial fan are measured and compared with empirical formulas available in the literature. Especially for fans with a small number of blades the one-dimensional calculation of the flow rate for shock-less entry seems to be inadequate.

Organised and sponsored by
BHRA Fluid Engineering, Cranfield, Bedford MK43 0AJ, England.

0263 - 421X/82/01 00 - 0001 $5.00
The entire volume can be purchased from
BHRA Fluid Engineering for $82.00

NOMENCLATURE

A = area of channel

b = width of impeller

D = diameter

f = friction coefficient

H = $b(r).r$ (this product is a constant for the impellers R6, R10, R14, S10)

N = blade number

p = static pressure

p_T = total pressure

Δp_L = total pressure loss

Q = flow rate through channel

r = radius

R = radius of the position of the probe

T = torque

u = peripheral velocity

v = absolute velocity

v_{sh} = shock component of velocity

w = relative velocity

$\bar{w}$ = $(w_1 + w_2)/2$

z = distance from shroud

β = blade angle

ν = velocity ratio : w_2/w_1

ρ = specific mass

Indices

0 = inlet of impeller

1 = blade inlet

2 = blade outlet

C = calculated

M = measured

r = radial component

θ = tangential component

Superscript

$\sim$ = mass averaged value

1. INTRODUCTION

The ultimate goal of many fan designers is probably the development of an analytic procedure, that makes it possible to determine in some optimal way the dimensions of a fan for given input values of the flow rate, the total pressure rise, etc. Therefore, however, it is necessary that the losses occurring in the different parts of the fan can be expressed as functions of the geometrical parameters.

In this paper we shall limit ourselves to the study of inlet and impeller losses. Formulas available in the literature will be compared with experimental data.

2. THEORETICAL CONSIDERATIONS

2.1. Inlet losses.

According to Eck (Ref. 1) inlet losses (Δp_{LI}) can be calculated as :

$$\Delta p_{LI} = \zeta_o \frac{\rho}{2} v_o^2$$

with ζ_o = 0.15 ... 0.25 and v_o the highest velocity in the inlet. We shall assume that the flow enters the impeller without shock, so that :

$$v_{r1} = u_1 \tan\beta_1$$

For fans with a high value for the ratio inlet width to inlet diameter ($b_1/D_1 > 0.25$) the highest velocity, calculated in a one-dimensional way, will occur in the suction eye, so that :

$$\frac{\Delta p_{LI}}{\rho u_2^2} = 8\ \zeta_o \left(\frac{b_1}{D_1}\right)^2 \left(\frac{D_1}{D_2}\right)^2 \text{cotan}^2 \beta_1$$

If b_1/D_1 takes a smaller value, the velocity is the highest in front of the blades, so that in this case :

$$\frac{\Delta p_{LI}}{\rho u_2^2} = \frac{\zeta_o}{2} \left(\frac{D_1}{D_2}\right)^2 \text{cotan}^2 \beta_1$$

The losses in the inlet are due completely to the presence of the boundary layers. On the hub the flow exhibits a stagnation point in the center of the axis. The velocity increases in the radial direction, so that generally the boundary layer will be very thin. On the shroud, however, the flow decelerates in the second half of the bend, so that boundary layer separation may be present. Even when there is no separation, the flow deceleration results in thicker boundary layers. So, the pressure distribution on the shroud wall will be a factor of major influence on the losses.

It can be argued against the formulas presented earlier, that they do not sufficiently take into account the geometrical parameters influencing that pressure distribution. This of course is one of the reasons for the rather large spread of the coefficient ζ_o. As the aim of an optimisation in design is primarily the determination of the principal dimensions, it is not necessary, however, to take into account all possible parameters.

Concerning the principal dimensions especially the ratio b_1/D_1 determines the form of the inlet. So, it is sufficient to have a relationship $\zeta_o(b_1/D_1)$, where ζ_o is the minimal loss coefficient, attainable for a certain value of b_1/D_1.

A first point to keep the inlet losses into limits is to prevent boundary layer separation. Even for impellers with small b_1/D_1 values, the round-off radius of the shroud should be sufficiently high. For the wider impellers, however, this is not sufficient. Here some boundary layer control has to be applied. Mostly the boundary layer is regenerated by means of the leakage flow entering through the gap.

The shape of the inlet does not just influence the losses. The boundary layer thickness in front of the blades is responsible for a change in shockless flow rate. The importance of that influence can be seen in (Ref. 2).

2.2. Impeller losses.

The losses in the impeller (Δp_{LR}) can be subdivided into diffusion (Δp_{LRD}) and friction (Δp_{LRF}) losses.

According to (Ref. 1) :

$$\Delta p_{LRD} = \zeta' \frac{\rho}{2}(w_1^2 - w_2^2) \qquad \text{with} \qquad \zeta' = 0.1 \dots 0.2$$

or

$$\frac{\Delta p_{LRD}}{\rho u_2^2} = \frac{\zeta'}{2}(1 - \nu^2)\left(\frac{D_1}{D_2}\right)^2 \frac{1}{\sin^2\beta_1}$$

and

$$\Delta p_{LRF} = f \frac{1}{Q} A \frac{\rho}{2} \bar{w}^2 \qquad \text{with} \qquad f = 0.004 \dots 0.0045$$

or :

$$\Delta p_{LRF} = \frac{f}{4}\left(\frac{1+\nu}{2}\right)^3 \left(1 - \frac{D_1}{D_2}\right) \frac{\pi \frac{D_1}{b_1}\left(1 + \frac{D_1}{D_2}\right) + \frac{N}{\cos\frac{\beta_1+\beta_2}{2}} \frac{D_1}{D_2}\left(1 + \frac{b_2}{b_1}\right)}{\pi \cos\beta_1 \sin^2\beta_1}$$

The formula for the friction loss has been derived in the assumption that the flow on the blades, the hub and the shroud can be calculated in the same way as the flow over a flat plate. This assumption is far from reality and the really occurring friction losses will be higher than the calculated ones. The most important points of difference between a boundary layer on a flat plate and the one in an impeller channel is on one hand the adverse pressure gradient, on the other the centrifugal and coriolis forces.

Centrifugal forces on the particles in the neighbourhood of the wall make it possible for the boundary layers to overcome higher pressure gradients. The charging part for the boundary layer is just the one due to diffusion.

Centrifugal forces, due to blade curvature, and the coriolis forces cause secondary flows in the impeller channels, comparable to the ones in a bend. Depending on absolute value and direction of those forces, low energy fluid can be collected on the pressure or the suction side of the blades.

The correct calculation of the different loss mechanisms in the impeller is still a rather problematic issue. Therefore we decided to do some experiments to see how far the fore-mentioned formulas can be used.

3. MEASUREMENTS

The impellers were tested in axisymmetric inlet and outlet conditions (Ref. 3). At the outlet the spiral casing had been replaced by two parallel flat plates perpendicular to the impeller axis. So, even when the fan is not operating in design conditions, the boundary conditions remain axisymmetric.

For every impeller the torque T and the inlet pressure were measured as a function of the sucked flow rate for a constant angular velocity. For a number of flow rates, static and total pressures and the absolute values and the directions of the velocities were determined at the rotor outlet with the help of a probe.

The pressures were measured at 3 cm from the impeller outlet (radius R). This distance could not be made smaller due to constructional difficulties with the probe. However, it is not recommended to measure very close to the outlet due to the blade wakes.

All the tested impellers have a diameter ratio $D_1/D_2 = 0.7$, an outlet diameter D_2 = 0.5 m and circular arc blades. The other characteristics of the impellers are presented in table 1.

Table 1. Impeller characteristics

Impeller	β_1	β_2	N	shroud	Rounding-off radius of the shroud
R6	15	30	6	$H = 0.01771\ m^2$	46 mm
R10	15	30	10	$H = 0.01771\ m^2$	46 mm
R14	15	30	14	$H = 0.01771\ m^2$	46 mm
S10	26.75	30	10	$H = 0.01771\ m^2$	46 mm
T6	19.8	22.4	6	$b = 0.149\ m$	52 mm
T10	19.8	22.4	10	$b = 0.149\ m$	52 mm

From the total pressure and the tangential components of the velocities at impeller outlet, the sum of inlet and impeller losses can be calculated :

$$\left(\frac{\Delta p_L}{\rho u_2^2}\right)_M = \left(\frac{\Delta p_{LI} + \Delta p_{LR}}{\rho u_2^2}\right)_M = \frac{\tilde{v}_{\theta 2}}{u_2} - \frac{\tilde{p}_{T2} - p_{T0}}{\rho u_2^2}$$

where :

$$\frac{\tilde{v}_{\theta 2}}{u_2} = \frac{1}{u_2} \frac{\int_0^b v_r v_\theta dz}{\int_0^b v_r dz} \frac{R}{r_2}$$

and

$$\tilde{p}_{T2} = \frac{\int_0^b p_T v_r dz}{\int_0^b v_r dz}$$

In the Fig. 1 to 6 the dimensionless losses are represented as a function of the dimensionless flow rate $\Phi = v_{r2}/u_2$.

The marks indicate the flow rates for shockless entry according to one-dimensional theory (Φ_1^*) and according to potential flow calculations (Φ_2^*) (Ref. 4, 6).

It is quite clear that especially for impellers with a small blade number the losses for $\Phi = \Phi_1^*$ are much larger than for $\Phi = \Phi_2^*$.

If in a design optimisation an impeller with a small number of blades is found, then it will surely be necessary to use a blade inlet angle smaller than the one calculated with one-dimensional theory, to take into account the relative vortex.

In table 2 the calculated and the measured losses are presented. For the last three impellers the flow rate for shockless entry falls outside the measured range. There we took the measured losses for the highest Φ-values, for which measurements were present. The losses were calculated with the smallest loss coefficients. So we can conclude that for $\Phi = \Phi_2^*$ the formulas can give realistic results. For R10 and R14 the results do not agree as well.

Table 2

Impeller	$(\frac{\Delta p_L}{\rho u_2^2})_M$	$\frac{\Delta p_{LI}}{\rho u_2^2}$	$\frac{\Delta p_{LRD}}{\rho u_2^2}$	$\frac{\Delta p_{LRF}}{\rho u_2^2}$	$(\frac{\Delta p_L}{\rho u_2^2})_C$
R6	0.020	0.010	-	0.013	0.023
R10	0.031	0.007	0.002	0.014	0.023
R14	0.038	0.006	0.006	0.016	0.028
S10	0.032	0.017	-	0.015	0.032
T6	0.042	0.029	-	0.011	0.040
T10	0.042	0.023	0.002	0.013	0.038

A boundary layer calculation on the basis of (Ref. 5), however, shows a small boundary layer separation on the suction side of the blades.

On both sides of the point for shockless entry the losses rise very rapidly. For an infinite number of blades the shock losses are given by $\rho/2\ v_{sh}^2$. For a finite number this expression is multiplied by a factor μ $(0.5 < \mu < 0.9)$.
However, if we assume the flow enters the impeller without shock for $\Phi \neq \Phi_1^*$, it is no longer possible to write the shock losses as $\rho/2\ v_{sh}^2$.

4. CONCLUSIONS

If boundary layer separation is avoided on the blades, the formulas agree with the measured values in the point for shockless entry. Especially for impellers with a small number of blades it is important to use the flow rate following from a two-dimensional calculation.

5. REFERENCES

1. Eck, B. : "Ventilatoren". Berlin - Heidelberg - New York, Springer Verlag, 1972.

2. Toyokara, T., Akaike, S. and Hosaka, H. : "Absolute flow near the impeller inlet of a centrifugal turbomachine". Bulletin of the JSME, 15, 85, 1972.

3. Somerling, H. and Vandevenne, J. : "Studie van de stroming door een radiale ventilator met open stator". M-tijdschrift, 24, 4, 1978.

4. Vandevenne, J. : "Potential flow calculation in the rotor of a radial fan with variable width using a singularity method".M-tijdschrift, 28, 1, 1982.

5. Herring, H.J. and Mellor, J.L. : "A computer program to calculate incompressible laminar and turbulent boundary layer development". NASA CR-1564, march 1970.

6. Vandevenne, J. : "Die theoretische Kennlinie eines Radialventilators", accepted for publication in Zeitschrift für Heizung, Lüftung, Klimatechnik, Haustechnik.

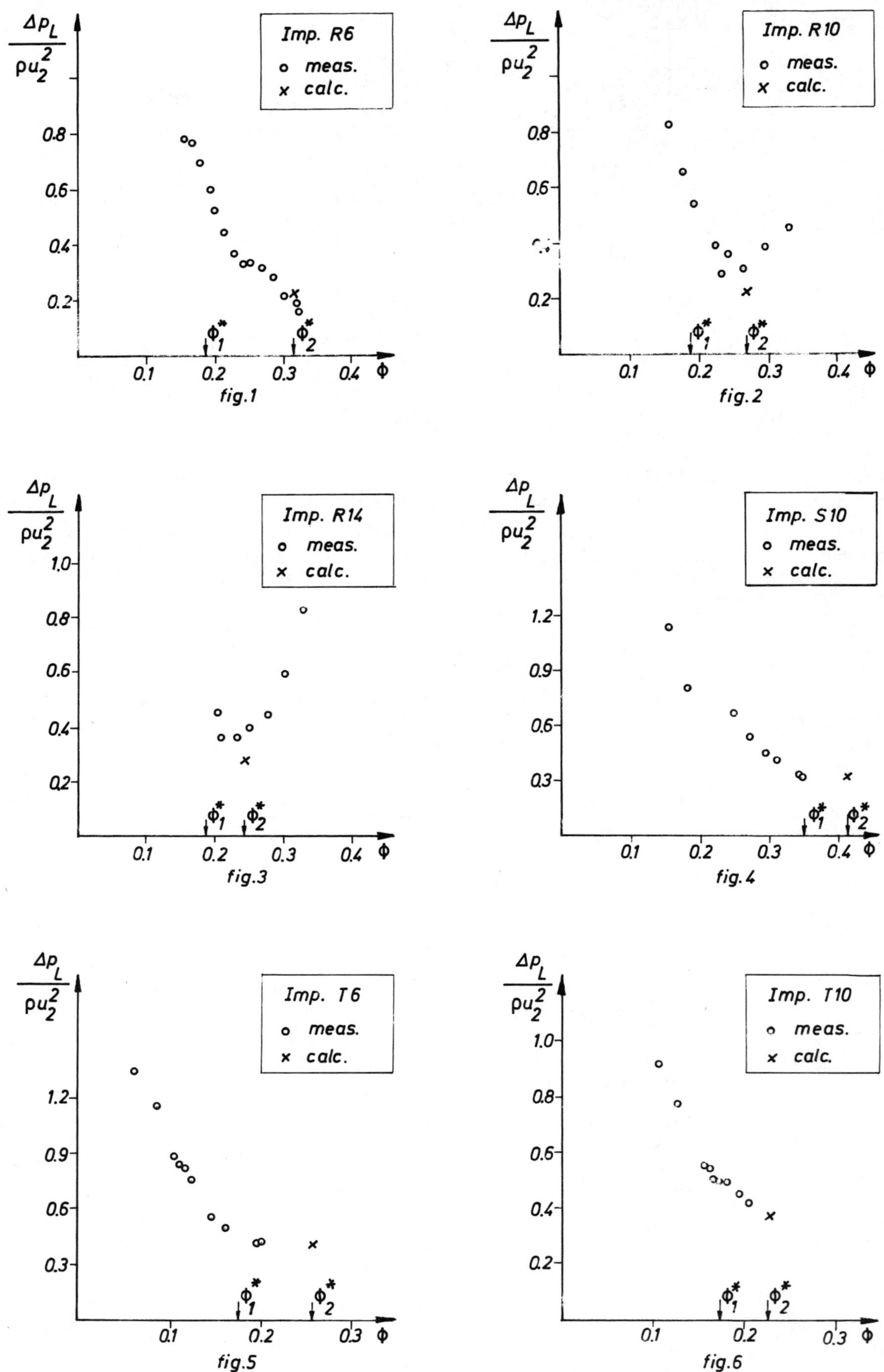

fig.1 fig.2 fig.3 fig.4 fig.5 fig.6

International Conference on

Fan Design & Applications

Guildford, England: September 7-9, 1982

PAPER B2

THE ADVERSE EFFECTS OF POOR DIFFUSION OF FAN PERFORMANCE - A CASE HISTORY

J. Marples

Aerex Limited, U.K.

Summary

The correct design of a fan outlet diffuser has always been accepted as making an essential contribution towards the achievement of high static efficiencies from the majority of fans. However it is less readily appreciated in practice that an outlet diffuser of inadequate proportions can result in an unacceptable reduction in total efficiency and overall fan performance.

This paper illustrates, by detailing a case history, how the performance of an axial flow fan of proven high efficiency design was severely affected by the dimensional constraints placed upon the outlet diffuser due to site restrictions, and how some improvement was achieved by relatively simple internal modifications.

Organised and sponsored by
BHRA Fluid Engineering, Cranfield, Bedford MK43 0AJ, England.

0263 - 421X/82/01 00 - 0001 $5.00
The entire volume can be purchased from
BHRA Fluid Engineering for $82.00

1. INTRODUCTION

An arrangement of 4, 1.8m diameter axial flow fans was required to work in parallel for the purpose of supplying air to a heater/cooler battery nominally 6m sq., and thence via a nozzle to an environmental test chamber. A recirculation system of ducting directed air from the test chamber back to the inlets of the fans. Such was the space available for the fans that it was necessary for each unit to have a radial air entry to the impeller, and to provide a rapidly changing section from the 1.8m diameter circular discharge downstream of the guide vanes to a 3m square section covering one quarter of the area of the heater/cooler battery. Fig. 1 shows the arrangement of the four fans and Fig. 2 a larger representation of a single unit.

The performance specification required an air quantity of $234m^3/s$ to be passed by all four fans when working in parallel, against a circuit resistance measured between the radial inlet and rectangular outlet of each fan of 2,600 Pa.

It was agreed that one fan would have a full performance characteristic test carried out in the works to establish that the specified performance could be obtained and to demonstrate that the characteristic was of such a shape to ensure that satisfactory parallel operation could be achieved.

It was also agreed that having proved the fan performance capability thus far, no further obligation to guarantee the performance in situ would be required as the design parameters of the fan chamber had been established by others and no alterations were possible.

2. GENERAL DESIGN OBJECTIVES

It has been adequately established elsewhere that with proven blade and guide vane geometry, high fan efficiencies can only be maintained if inlet and outlet losses are kept to a minimum. Hence good aerodynamic profiles at the inlet and outlet are essential parameters of an axial flow fan rotor/guide vane unit. It will be clear from Figs. 1 & 2 that the usual proportions associated with fan inlet and outlet geometry could not be achieved in any way. Accordingly means of adapting these to the particular circumstances had to be established and a re-assessment of the increased internal losses associated with them, investigated.

Clearly if an axial flow fan blade is to achieve optimum lift or pressure development, then the pattern of the airflow approaching it must be uniform, with no areas of turbulence of undecayed wakes. Therefore great attention must be paid to the design of the fan inlet. Fig. 2 illustrates the proportions of the inlet so designed and having thus defined an arrangement for the inlet, impeller and guide vane section, the space then remaining had to accommodate an expansion section. It will be readily apparent that the resulting proportions giving expansion angles at each side and in the corners of approximately 20° and 40° respectively far exceeded the design parameters used in normal practice. Evidence readily available indicated that such a configuration would result in low conversion efficiencies, but it was believed that the presence of considerable back pressure due to the heater/cooler battery would compensate in some measure for this and a value of 50% conversion efficiency was assessed accordingly. Fig. 3 shows the expected fan performance curve based on this configuration and shown also is the fan stage total pressure which was expected, the difference between the two being the calculated diffuser and expansion loss.

3. FAN TEST ARRANGEMENT

The test arrangement was not in accordance with BS.848 1963 Part 1, Test Method No. 1, the standard for axial flow fans at the time the units were manufactured, because of its impracticability for the particular fan in question. A blowing test similar to that described in Test Method No. 3, was agreed upon but having a square section plenum and settling chamber incorporated at the termination of the diffuser. This was followed by a reducing nozzle connected to a more appropriately sized test duct in which the fan air quantity would be established using the recommended method of pitot traverse. Fig. 4 and plate 1 illustrate the test arrangement adopted. The rectangular plenum will appear unusually short, but it was considered that this would more closely represent the site arrangement. This was an important consideration, for the purchaser was interested in the velocity distribution which would be established against the heater/cooling coils as these had been selected by others based on the assumption that generally uniform flow would occur across the whole surface area. The test fan was driven by a variable speed thyristor controlled test motor very similar to that which was to be employed in practice.

4. INITIAL PERFORMANCE DATA

Preliminary tests, measuring fan pressures in the plenum chamber and velocity pressures in the smaller test duct, and with the fan running at normal operating speed, indicated that there was a considerable shortfall in performance. Examination of the results showed that:

(a) The fanshaft power absorbed was very close to that forecast.

(b) An unmeasured resistance or loss of pressure at constant volume was being experienced.

Fig. 5 illustrates the difference in performance between that forecast and that measured. As the fan power absorbed was substantially correct then this was taken as indication that the profile of the inlet air velocity approaching the blade element was fully developed. Comprehensive smoke tests carried out over the inlet area confirmed this, illustrating the total decay of all wakes generated by the supporting structure. Notwithstanding this it was decided to take a total head traverse across the annulus just downstream of the guide vane discharge to determine the fan stage total head characteristic. The velocity pressure for air quantity computation was again measured in the smaller test duct. The results of this test are compared with the forecast data in Fig. 5 and it is readily apparent from this comparison that the stage performance was generally as anticipated over the useful range of fan performance so adequately demonstrating that the fan inlet configuration was not responsible for the measured performance deficiency.

Accordingly it was clear that the severe loss in pressure being experienced could be attributed to the inadequate proportions of the diffusing section of the fan and notwithstanding the presence of a resistance of some 2,000 to 3,000 Pa at the end of the test duct the flow profile in the plenum chamber was not becoming fully developed resulting in excessive expansion losses. Calculation of these losses from the test data shown on Fig. 5 illustrates that they were more than twice the value originally estimated.

5. FURTHER INVESTIGATIONS AND CORRECTIVE MEASURES

A pitot traverse was taken at the downstream end of the plenum chamber. Fig. 6 shows the velocity distribution determined from this test data. This diagram shows the presence of a high velocity core but indicates unacceptably low velocities at the sides, top and bottom and particularly in the corners where the steeply diverging surfaces had clearly encouraged extensive flow separation despite the presence of the high back pressure in the small test duct. It has been stated that the fans were to supply air to a heater/cooler battery to which the diffuser outlets were commonly connected. It is well known that when a region of high resistance is fitted to a diffuser exit, it assists the diffuser to run "full". Such a feature has been used in Wind Tunnel design. Notwithstanding the fact that the high test duct back pressure had produced seemingly little benefit, it was decided to incorporate such a resistance. This was to simulate the heater/ cooler battery and took the form of a fine wire mesh screen of similar resistance positioned toward the end of the plenum chamber and in a place relative to the fan outlet approximating as close as possible that which the actual battery would take up in the site installation.

Fig. 7 shows the small increase in performance achieved and whilst suggesting that some improvement in velocity distribution might have been obtained it was quite apparent that the improvement was unsatisfactory. Further means of improving the diffuser operation had to be investigated therefore. To achieve this and at the same time to obtain the necessary improvement in velocity distribution at the downstream end of the plenum, within the existing dimensional constraints of the diffuser it was clear that the provision of internal flow control devices had to be considered. It would be necessary for these to direct air from the annulus into the starved sides and particularly corner areas, thereby lowering the peak velocities and substantially reducing the degree of outer surface separation which was occurring. Fig. 8 illustrates the air deflection devices incorporated and Fig. 9 the improved velocity distribution in the plenum box. The curve on Fig. 7 illustrates the further improvement in performance.

A useful increase in pressure was now achieved together with a much improved outlet velocity distribution. However whilst the air quantity was now well within the limits allowed by British Standard tolerances, the peak operating pressure was not being achieved and the stall characteristic showed a lower "dip" pressure than that originally forecast. As is well known this area of a fan characteristic has considerable bearing on the ability of a fan to operate in parallel with other fans of like design. Fig. 10 shows the combined characteristics which the four fans would achieve when operating in parallel and the difficulty of safe parallel operation in this particular instance is easily recognisable. Further improvement in the fan characteristic was still essential therefore but time was no longer available to continue detailed investigations utilising more sophisticated flow control devices within the diffuser. Accordingly the blade angle of the fan was reduced and the speed increased to compensate for the difference. The result was a pressure/volume characteristic nearly identical to that forecast although an increase in peak power was now necessary to achieve this. Figs. 11 and 12 illustrate the final characteristic and the combined four fan system characteristic respectively.

6. OBSERVATIONS AND CONCLUSIONS

The reduction in expansion losses which was achieved in this instance by relatively simple modification, illustrates the degree to which diffusion efficiency could be improved where the incorporation of more sophisticated flow control devices could be made a fundamental part of the diffuser design.

Fig. 13 shows a direct comparison between the estimated expansion losses and those actually calculated from the test data for differing outlet conditions.

It is apparent that the diffusion efficiency is not constant over the range of air volume flows measured, a rapid fall-off being indicated as the flow reduces. This is despite the presence of an increasing back pressure due to the greater outlet resistance. It was a fact that the Reynolds number at the guide vane outlet reduced from approximately 6.5×10^6 to 3.5×10^6 and to one fifth of these values at the diffuser outlet. However airflow conditions at the inlet of a diffuser are recognised as having a major influence on diffuser performance and the presence of adverse pressure gradients with uneven velocity patterns as stall approached was probably a major contributing factor to the reducing efficiency. Time did not permit this to be investigated in more detail.

Many technical references compare conversion efficiency merely with diffuser profile and expansion angle. Whilst this is no doubt satisfactory for in-duct diffusers having uniform inlet velocity distribution, the case history illustrates clearly the need for considerable caution where fan terminations having larger than normal expansion angles are concerned.

7. ACKNOWLEDGMENTS

The author wishes to record his appreciation to his Co-Directors of Aerex Limited for their permission to prepare, present and publish this paper.

8. REFERENCES

WALLIS R.A.: "Axial Flow Fans", London, George Newnes, 1961.

ECK, BRUNO: "Fans" First English Edition. Oxford, Pergamon Press, 1973.

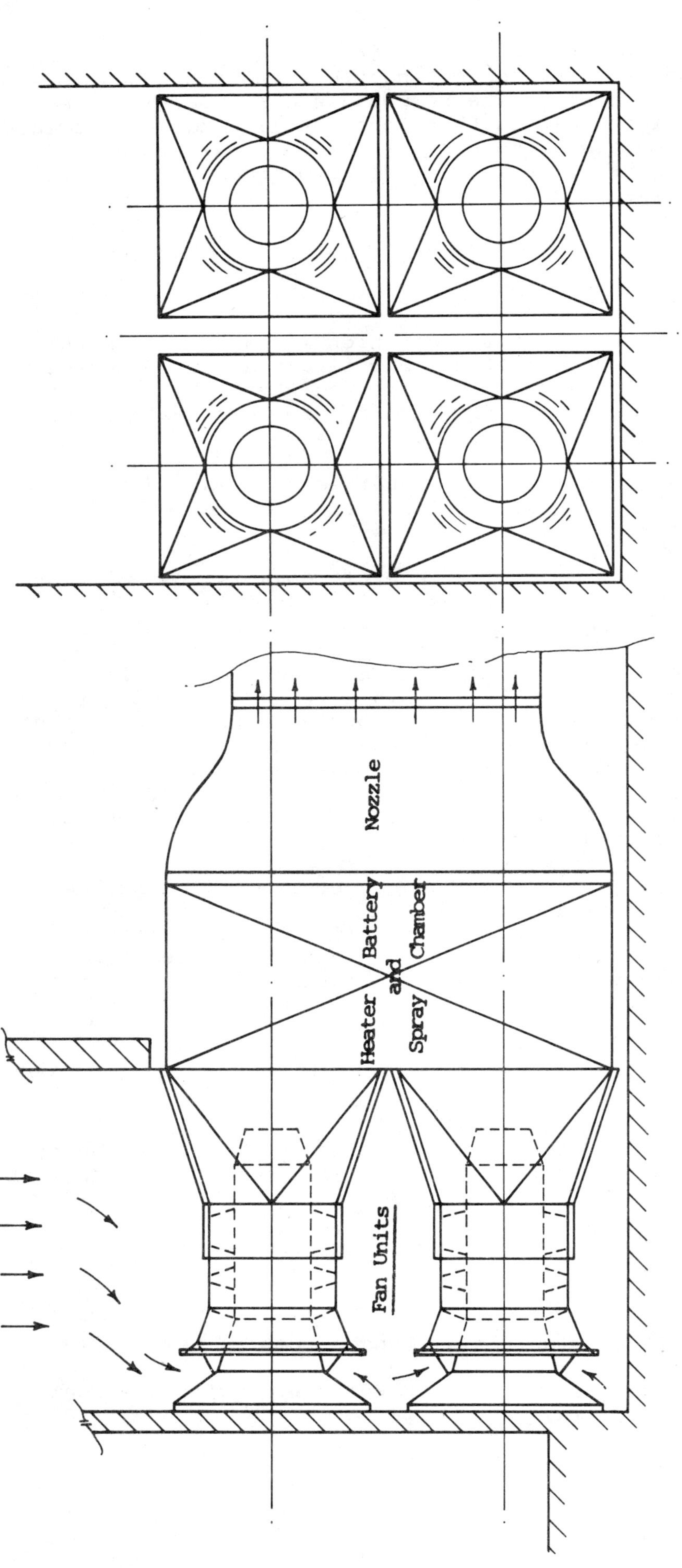

Figure 1.

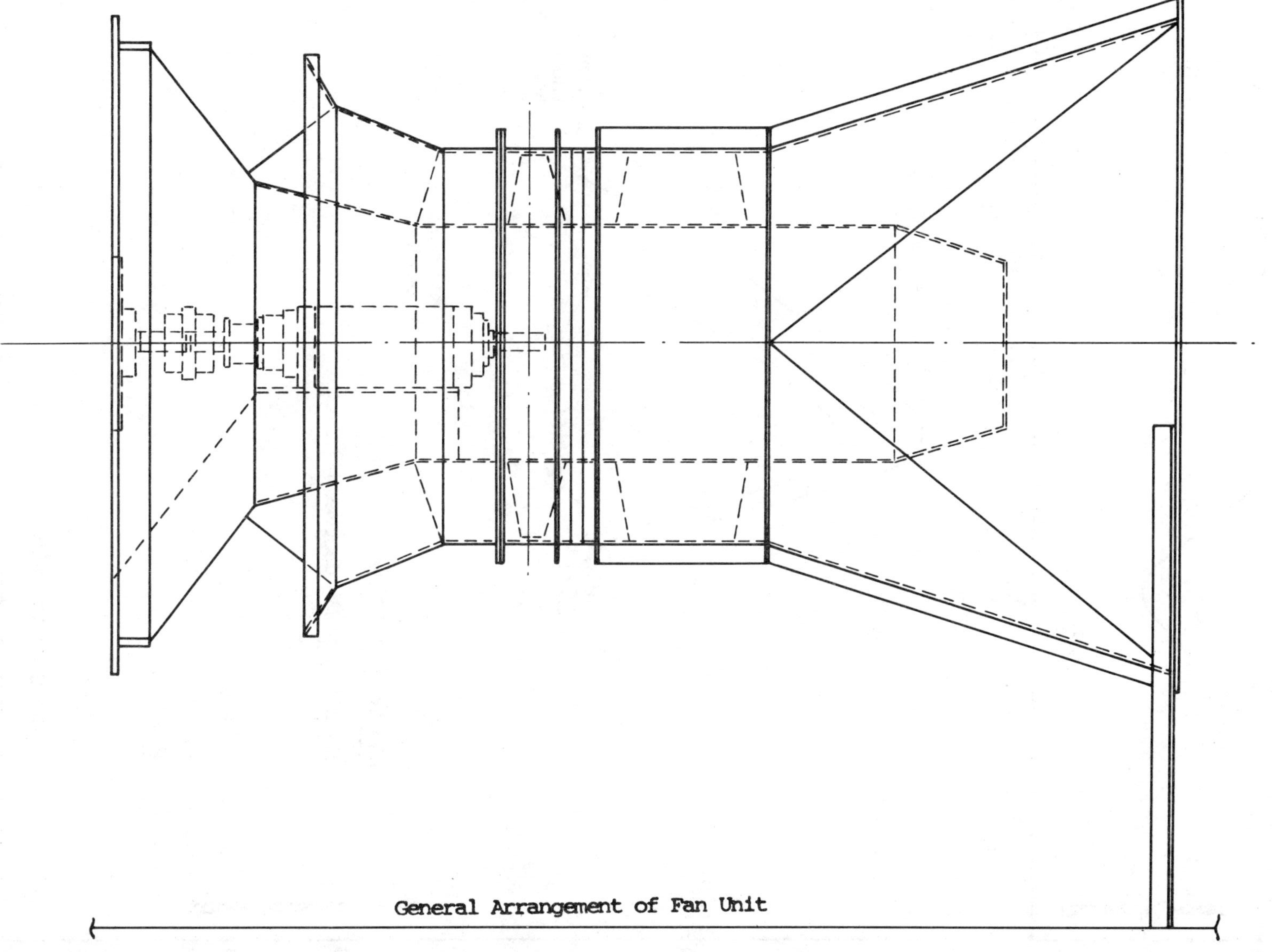

General Arrangement of Fan Unit

Figure 2.

250
200
150
100
Fanshaft Power - kW
204 kW
10
20
30
40
50
60
70
80
4000
3500
3000
2500
2000
1500
1000
500
Total Pressure - Pa
2600 Pa
Specified Duty
Stage Total Pressure
Fan Total Pressure
Fan Speed 1200 rpm
Blade Angle 23½ degrees
58.5 m3/s
Air Quantity m3/s

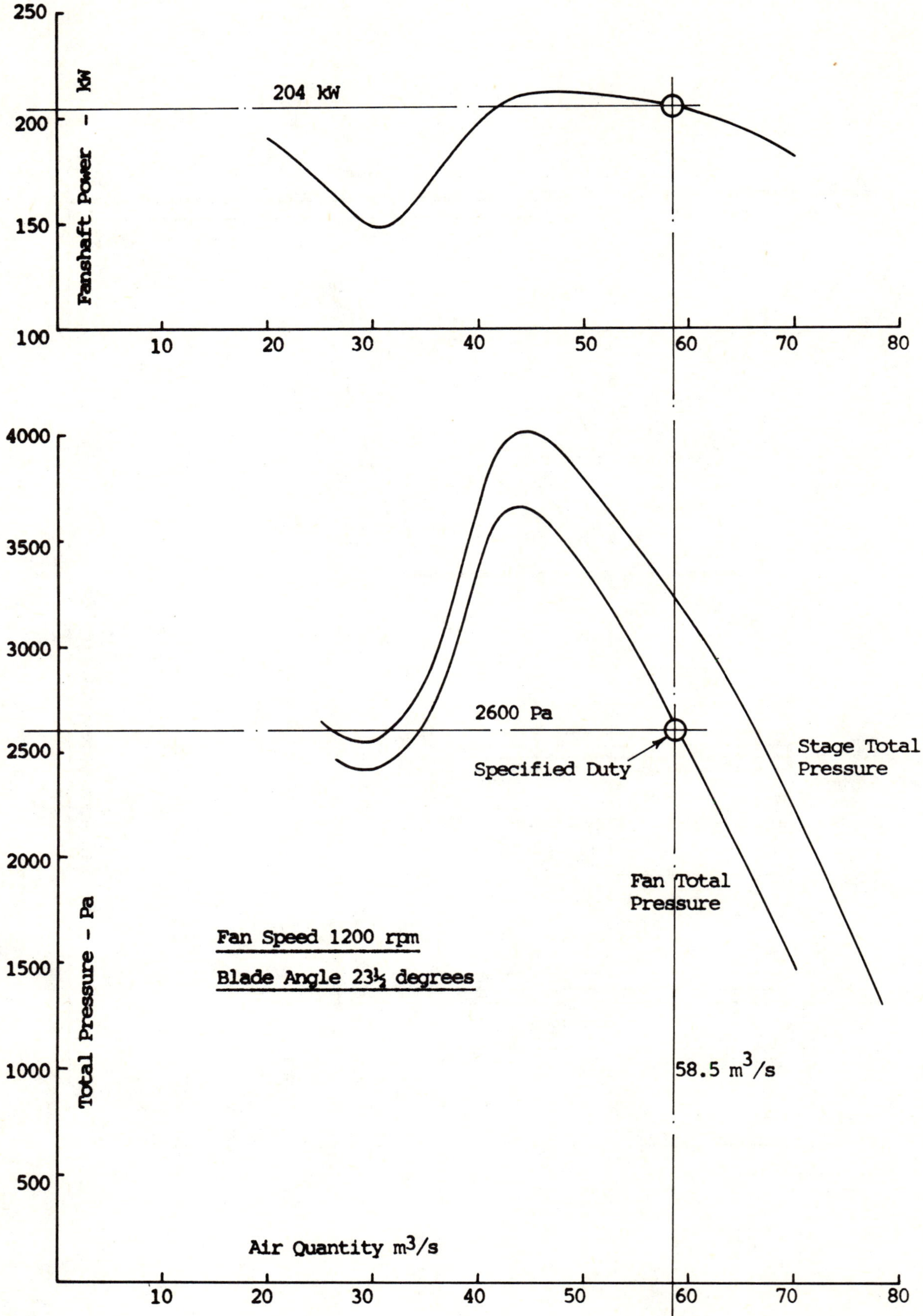

Figure 3.

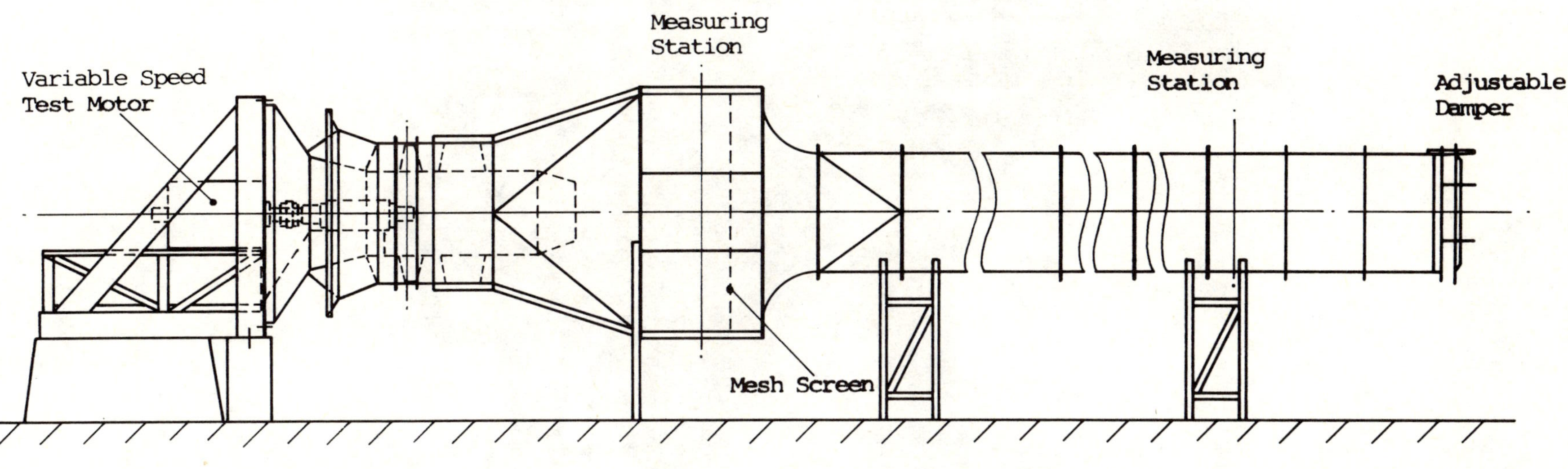

Test Arrangement

Figure 4.

Plate 1

250
200
150
100
Fanshaft Power kW
10
20
30
40
50
60
70
80
4000
3500
3000
2500
2000
1500
1000
500
0
Total Pressure - Pa
Measured Stage Pressure
Measured Fan Total Pressure
+ Denotes Test Points
Fan Speed 1200 rpm
Blade Angle 23½ degrees
Air Quantity - m3/s

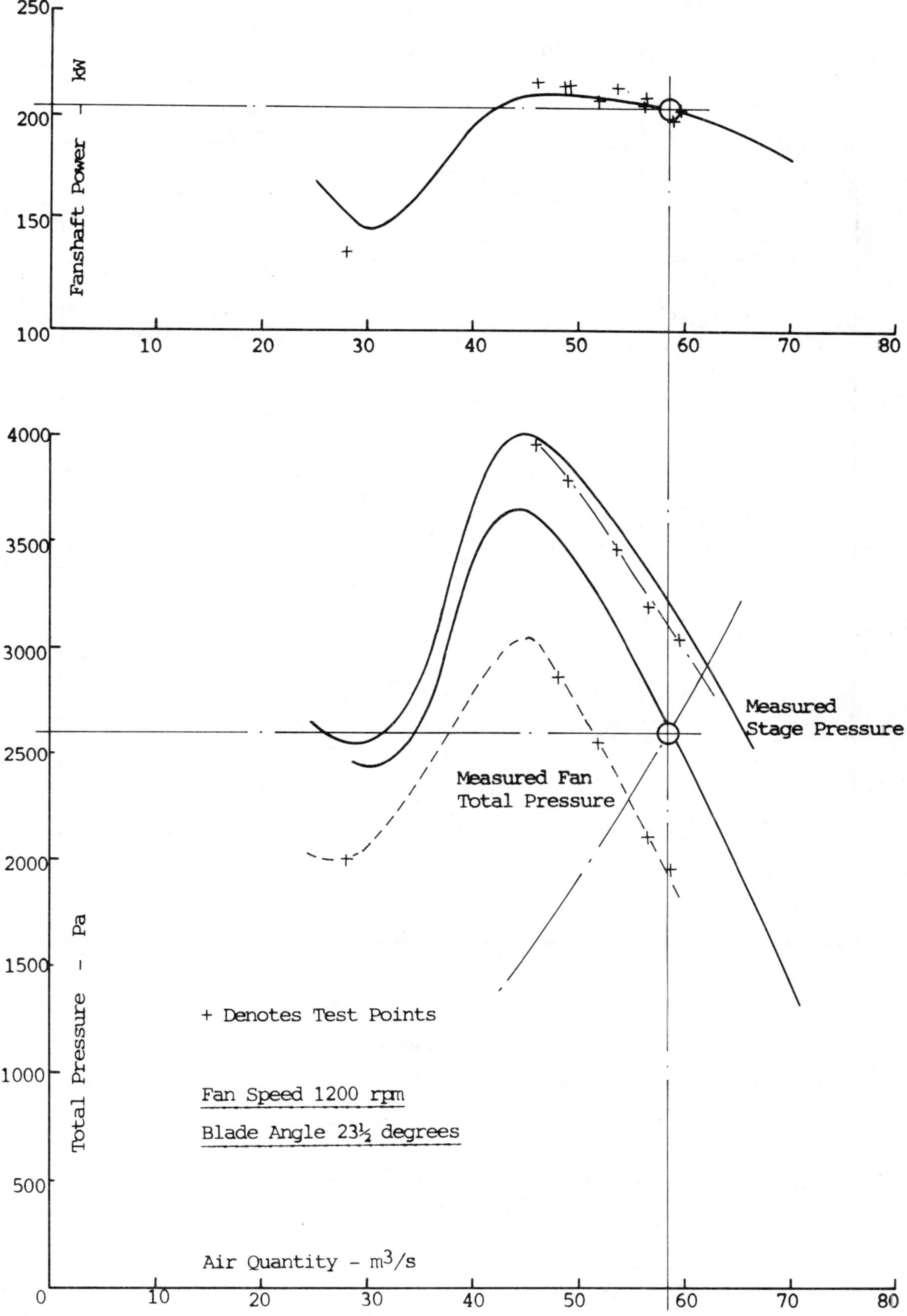

Figure 5.

Air Quantity - 59 m^3/s

Mean Calculated Velocity - 7.22 m/s

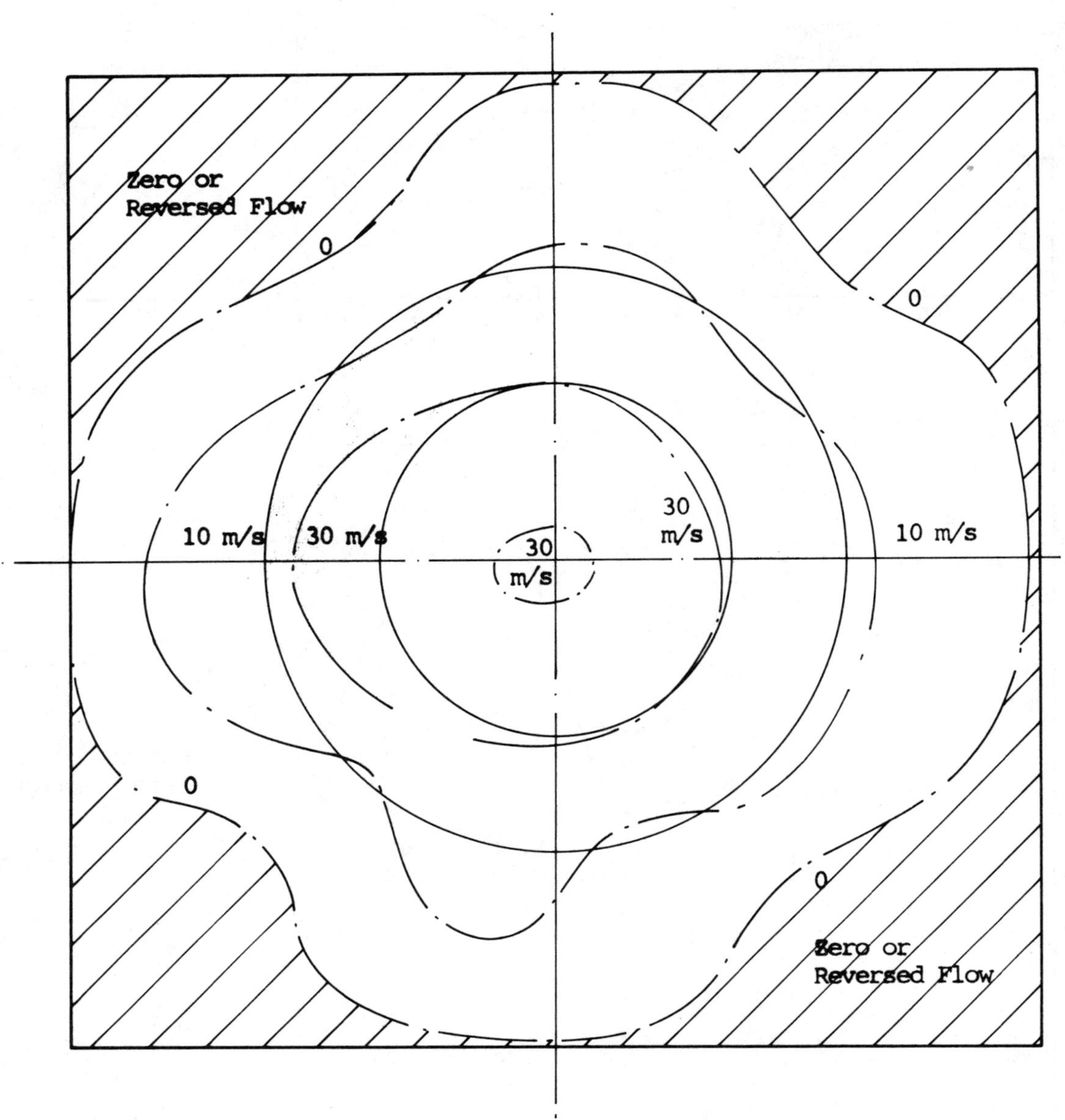

No Screen in Plenum

No Diffuser Deflectors

Velocity Distribution in Plenum Chamber

(Looking towards Fan)

Figure 6.

Efficiency/Load
Axial Flow Fan with
Inlet Vane Control
Variable Pitch
Axial Flow Fan
100
80
60
Eff'y
40
20
10 20 30 40 50 60 70 80 90 100 110 120
% Design Flow

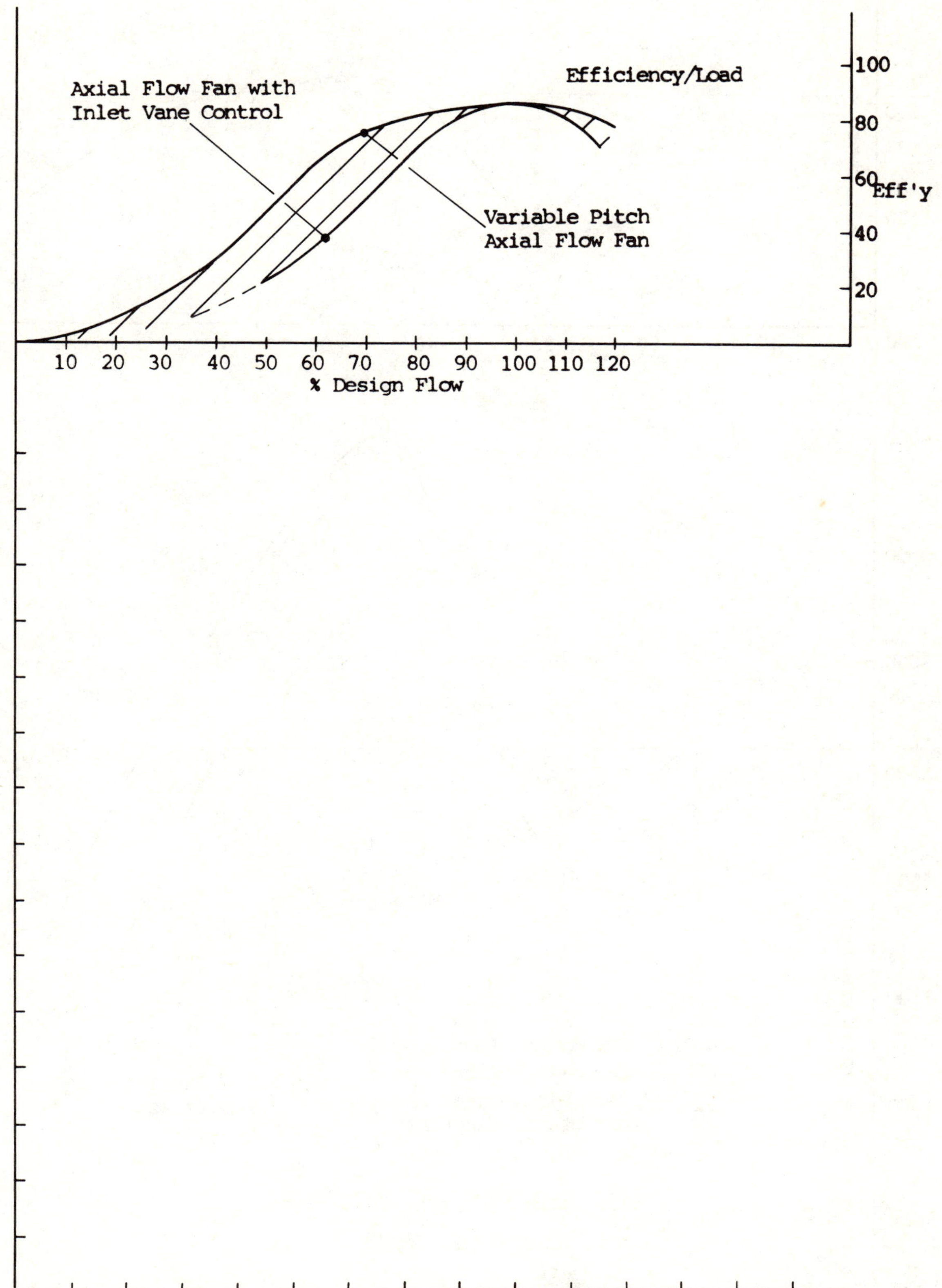

Figure 6A.

250
200
150
100
Fanshaft Power - kW
10
20
30
40
50
60
70
80
4000
3500
3000
2500
2000
1500
1000
500
Total Pressure - Pa
Stage Total Pressure
Fan Total Pressure
First test from Fig. 5.
test with Plenum Screen
test with Screen and diffuser deflectors
Fan Speed 1200 rpm
Blade Angle 23½ degrees
Air Quantity m3/s

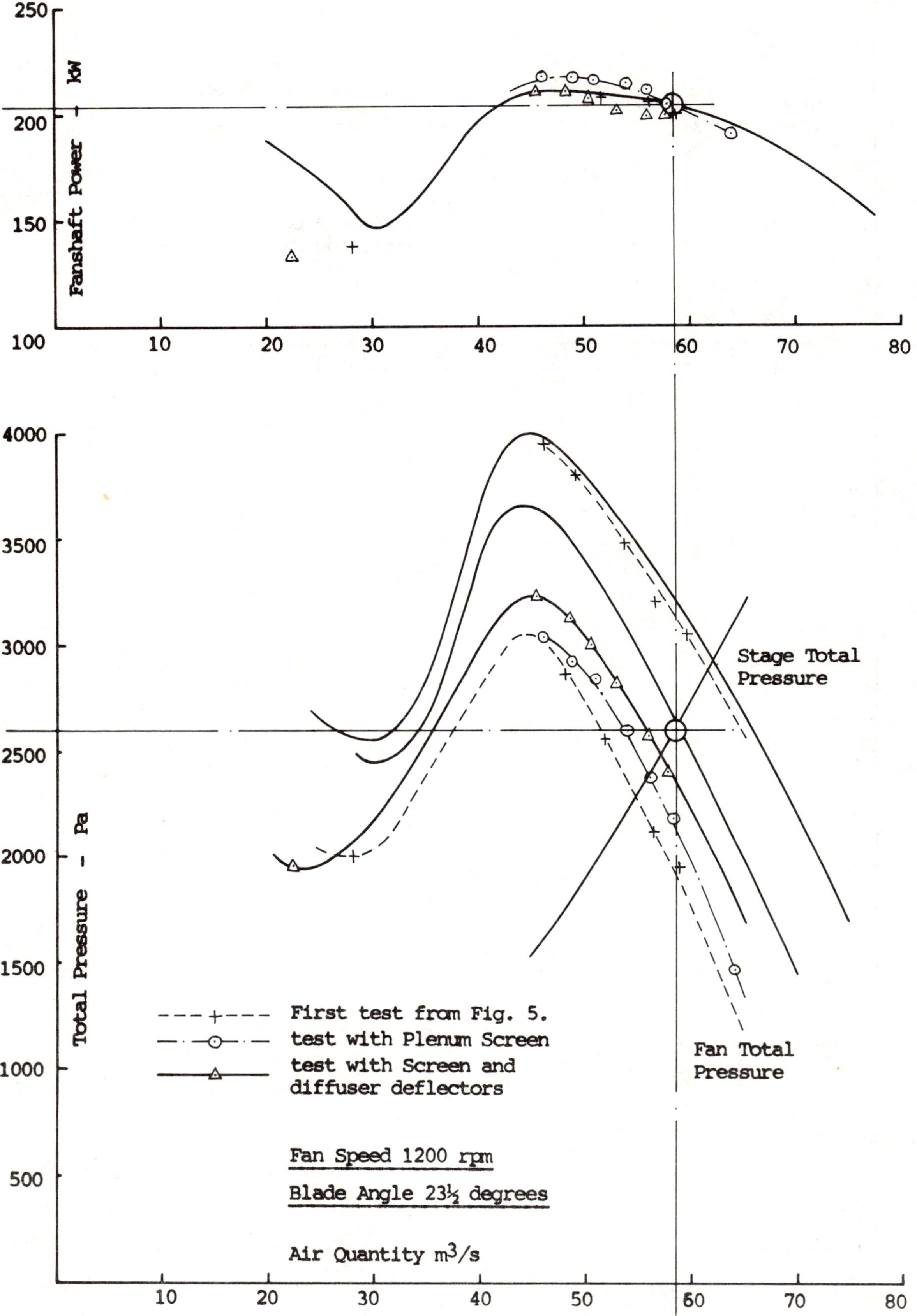

Figure 7.

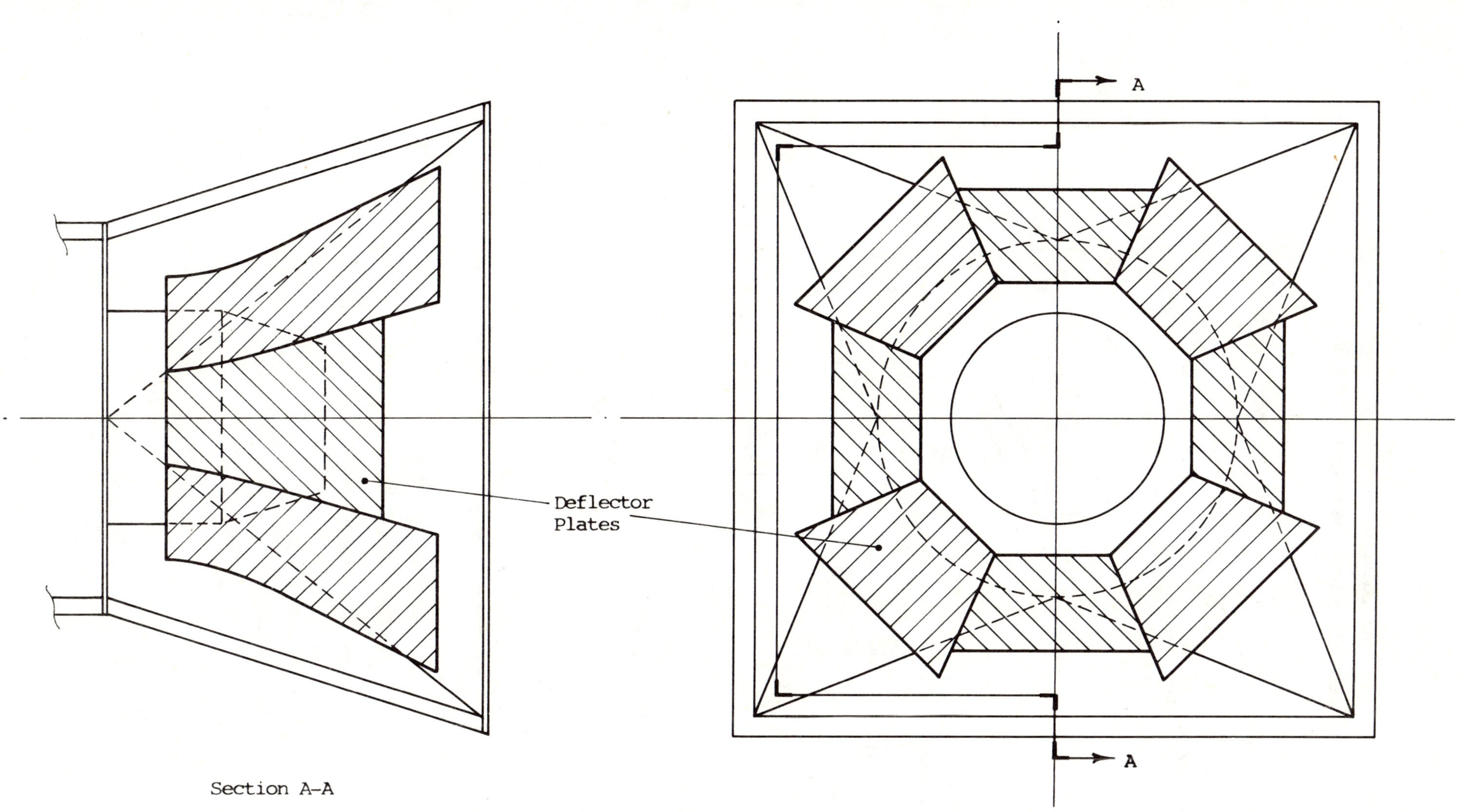

Figure 8.

Air Quantity 60 m^3/s

Mean Calculated Velocity 7.34 m/s

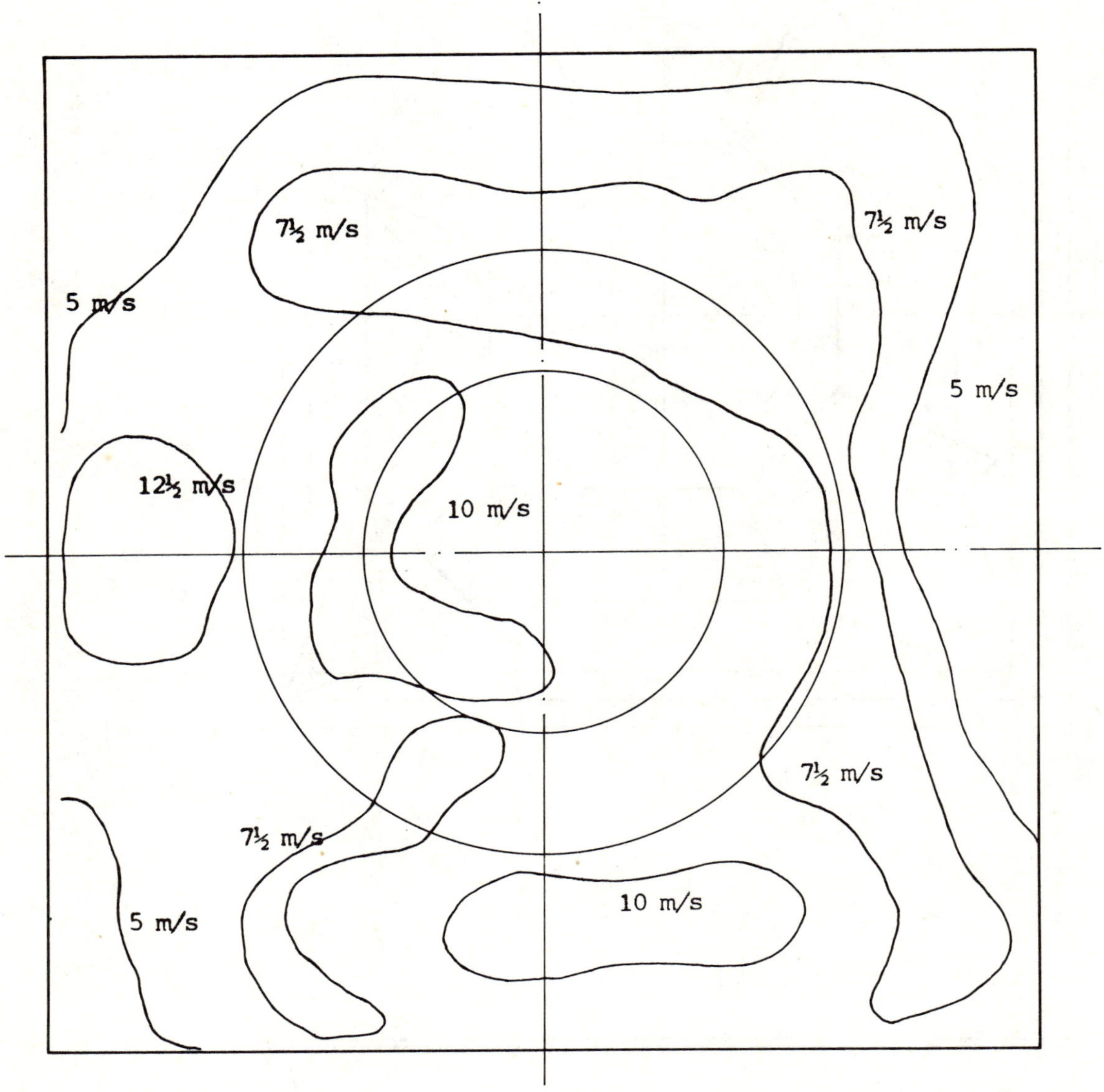

Screen in Plenum

Diffuser Deflectors Fitted

Velocities are on downstream side of screen

<u>Velocity Distribution in Plenum Chamber</u>

(Looking towards Fan)

Figure 9.

4000
3000
2000
1000
Total Pressure - Pa
Design System
Theoretical Limit
of Parallel Operation
1 Fan
2 Fans
3 Fans
4 Fans
Fan Speed 1200 rpm
Blade Angle 23½ degrees
Air Quantity - m^3/s
50
100
150
200
250
300

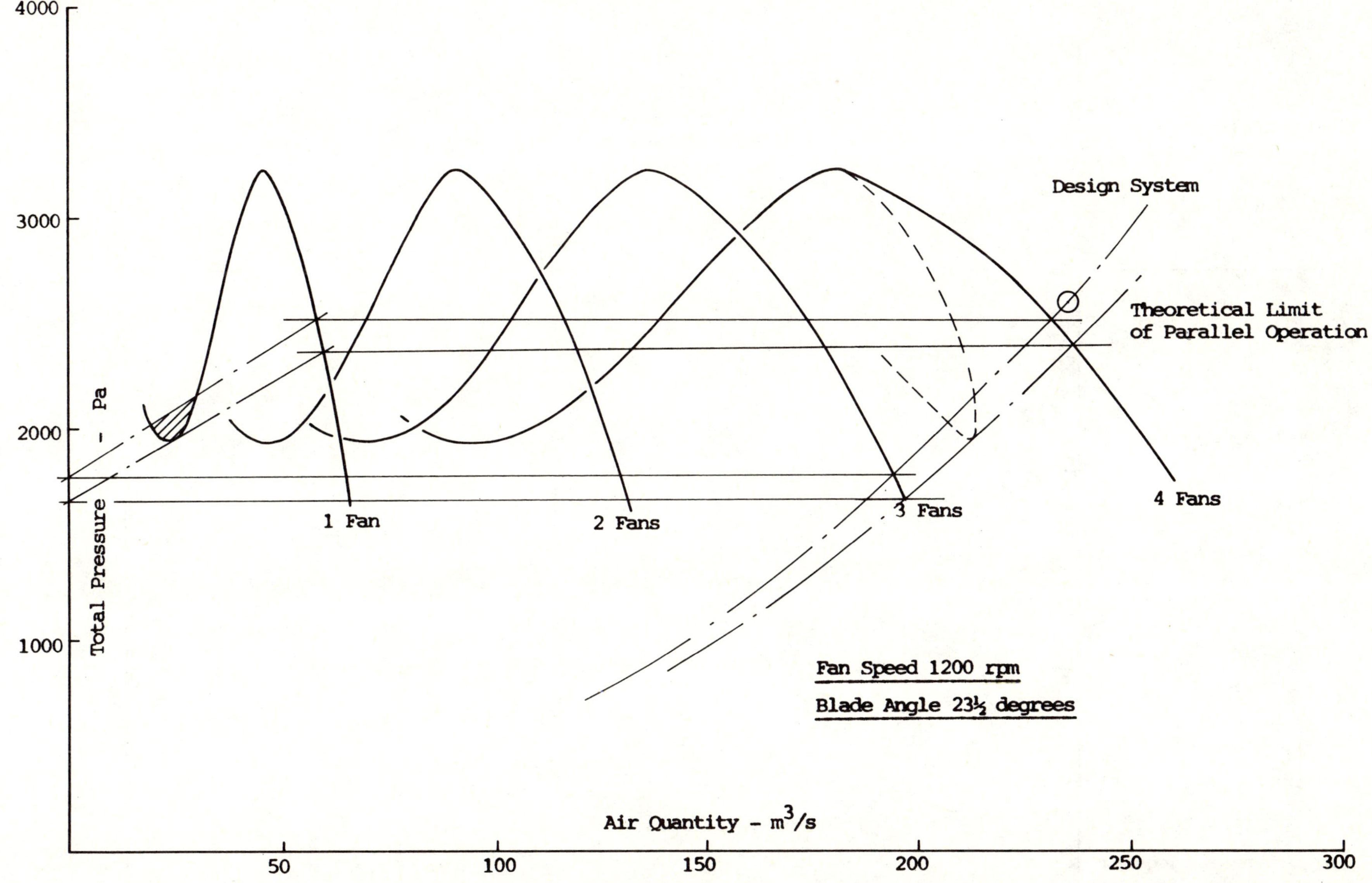

Figure 10.

250
200
150
100
Fanshaft Power - kW
10
20
30
40
50
60
70
80
4000
3500
3000
2500
2000
1500
1000
500
0
Fan Total Pressure - Pa
Denotes Test Results.
Plenum Screen and Diffuser
Deflectors fitted.
Fan Speed 1290 rpm
Blade Angle 21½ degrees
Air Quantity - m3/s

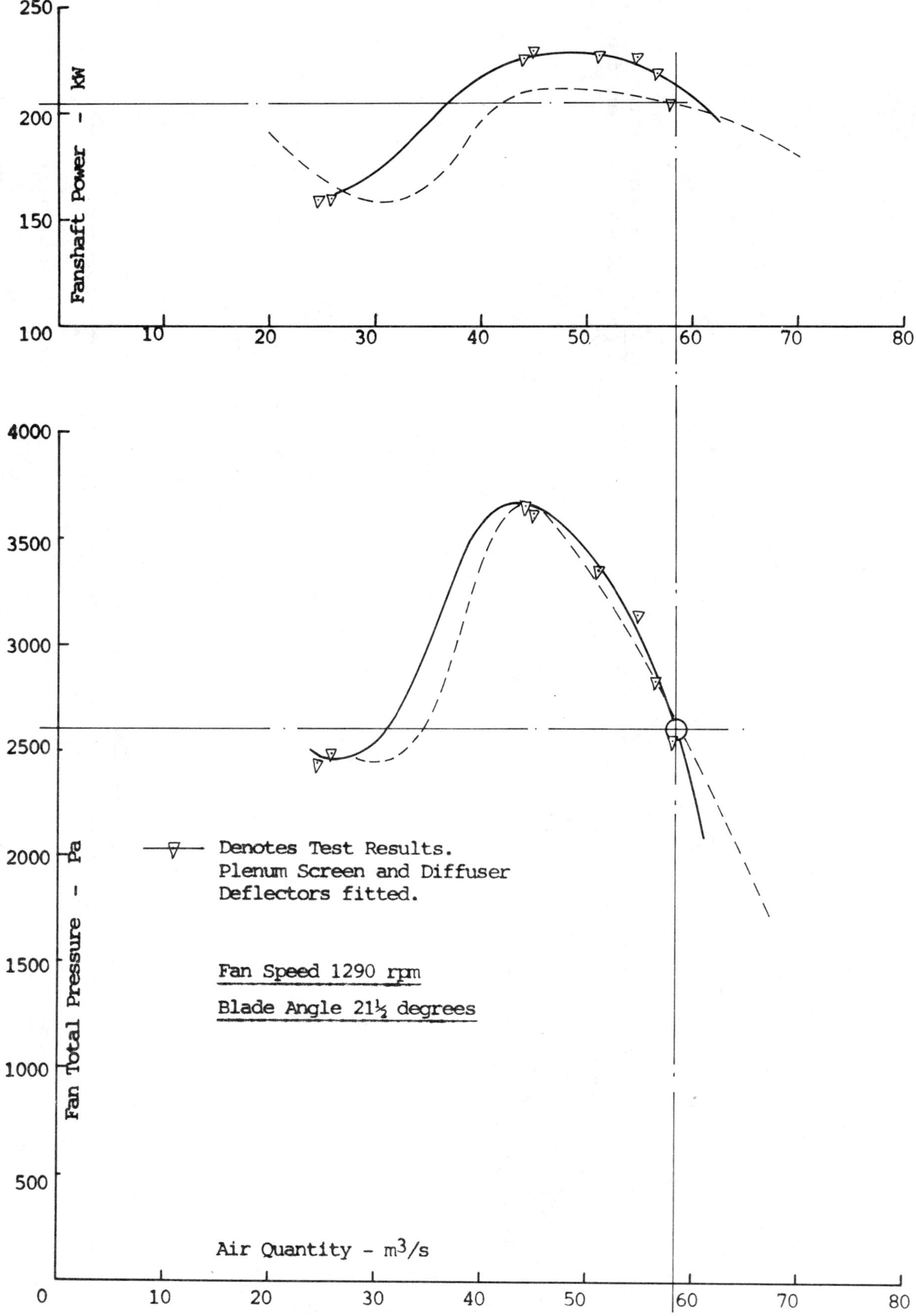

Figure 11.

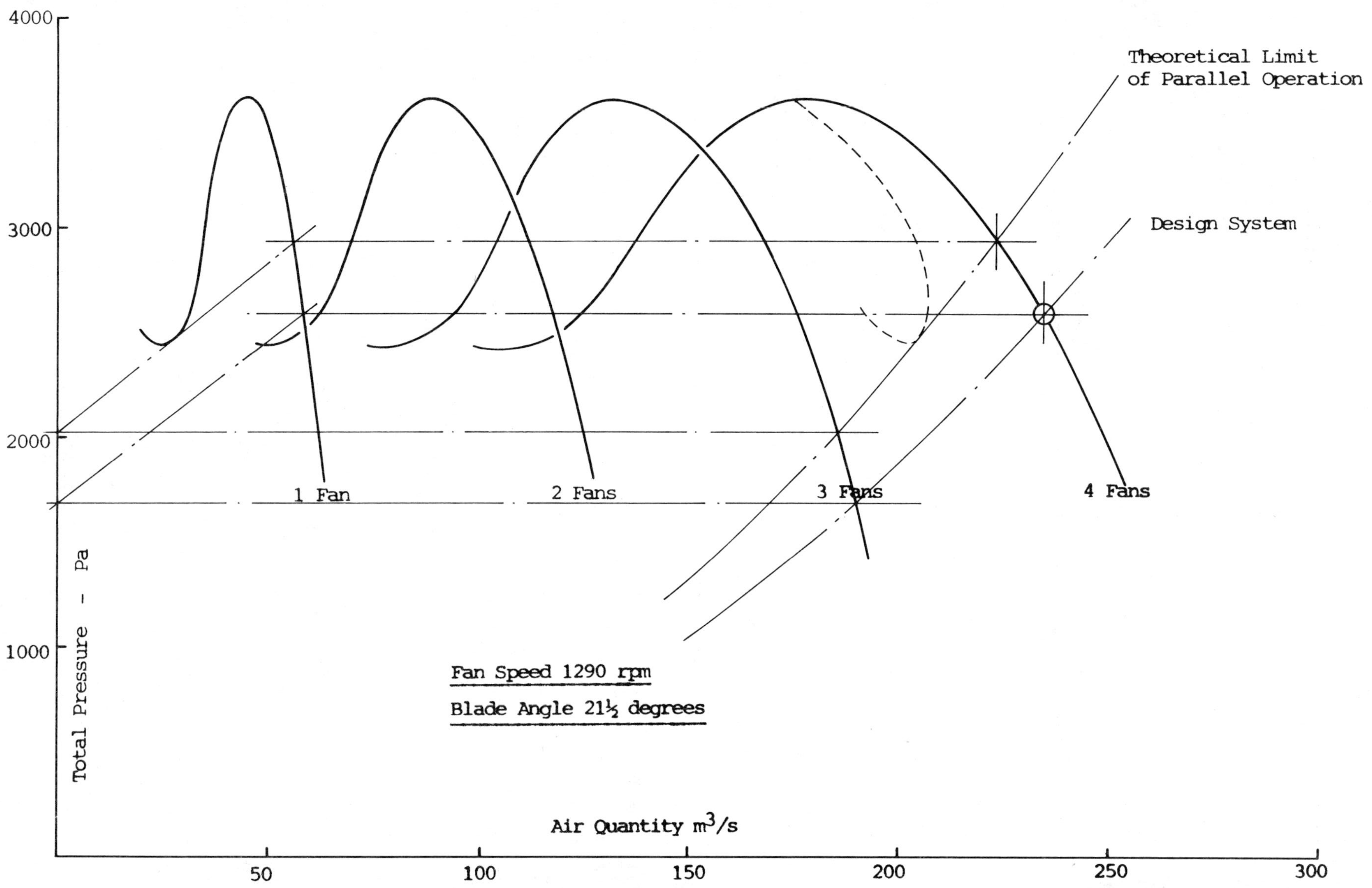

4000
3000
2000
1000
Total Pressure - Pa
Theoretical Limit
of Parallel Operation
Design System
1 Fan
2 Fans
3 Fans
4 Fans
Fan Speed 1290 rpm
Blade Angle 21½ degrees
Air Quantity m3/s
50
100
150
200
250
300

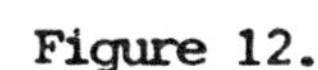

Figure 12.

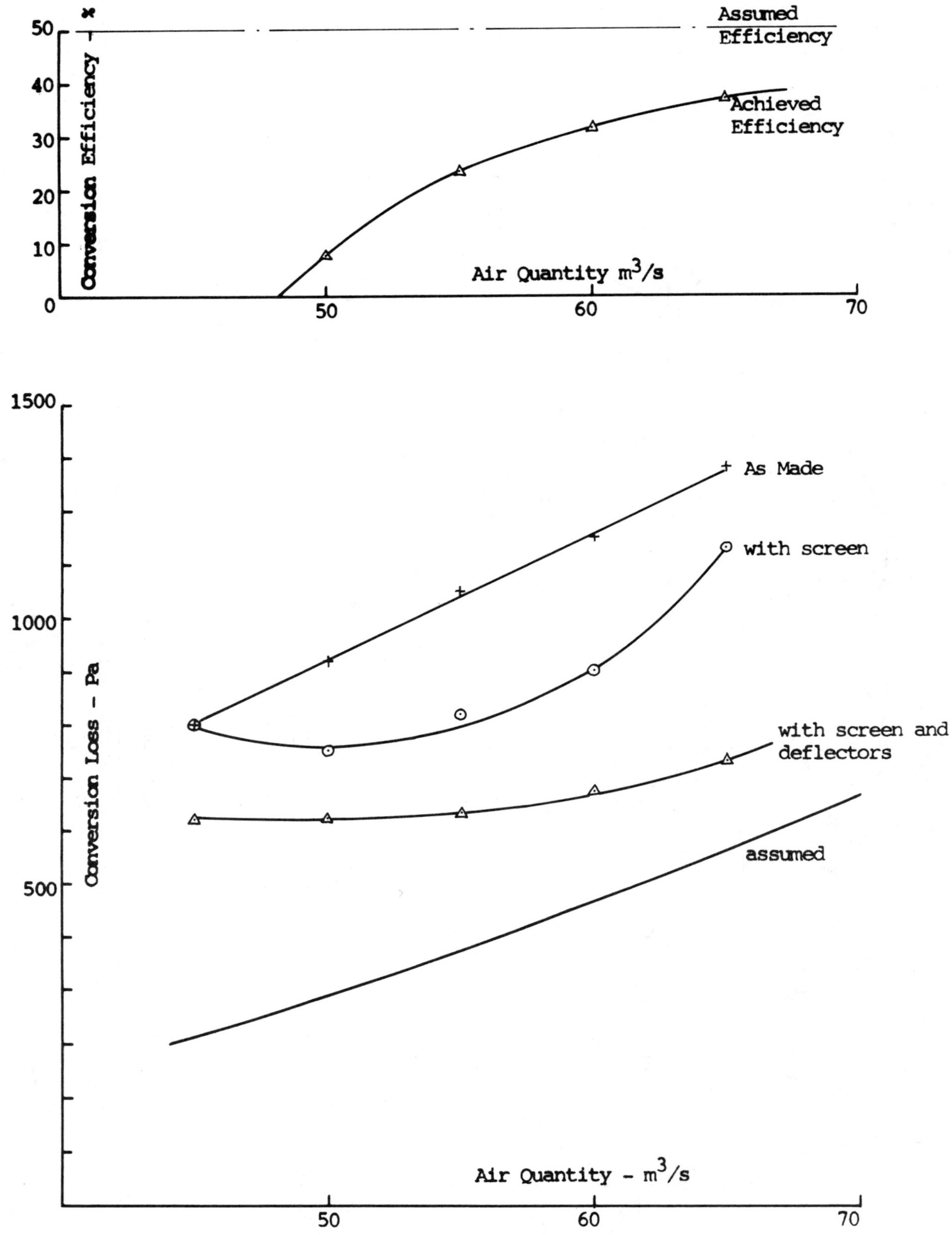

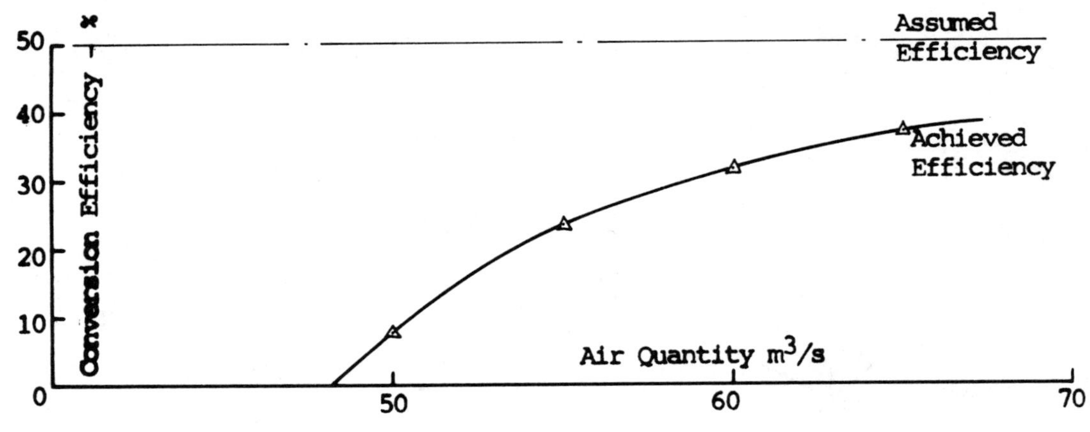

Diffuser Losses

Figure 13.

International Conference on

Fan Design & Applications

Guildford, England: September 7-9, 1982

PAPER B3

AIR INFLOW EFFECTS ON FAN PERFORMANCE IN AIR COOLED HEAT EXCHANGERS

C.M.B. Russell and J. Peachey

The Lummus Company Ltd., U.K.

Summary

Large process ACHE's (Air-Cooled Heat Exchangers) are described, as is their installation, and the poor air inlet and outlet conditions for the large propellor-type fans used. A model used to investigate air inlet effects is described, and results with different fan inlet conditions and various ground heights are presented.

Organised and sponsored by
BHRA Fluid Engineering, Cranfield, Bedford MK43 0AJ, England.

0263 - 421X/82/01 00 - 0001 $5.00
The entire volume can be purchased from
BHRA Fluid Engineering for $82.00

1. INTRODUCTION

Air-cooled heat exchangers (ACHE's) for the Petroleum and Petro-chemical industries are usually mounted in long banks; Fig. 1 is a typical example The tube bundles are arranged between headers, the tubes being 8-12 metres long. The headers are 2-3 metres wide, and usually two tube bundles are used side-by-side to form a bay, whose dimensions thus are 4-6 metres wide and 8-12 metres in the direction of the tubes. Two propellor type axial flow fans, of diameter 3½ - 4½ metres, with their axes vertical, are usually fitted to each bay. Bays are generally placed side-by-side to form rows as shown in Fig. 1. Sets of bays 20 or more wide are commonplace.

ACHE's are large, and therefore tend to require large amounts of plot area. They are commonly fitted on the top of pipe tracks, as shown in Fig. 1; grade mounting is not uncommon, however.

Fig. 1.

A bank of ACHE's using forced draft This unit is mounted on top of a pipe track.

Away from the ends of long units, the airflow into the fans will be substantially at right angles to the axes of the bays, as shown on Fig. 2. There will be virtually no airflow across Planes "A", "B" and "C" of Fig. 2.

Fig. 1. shows ACHE's with the fan in "forced" draft, driving air over the bundles. Fans in "induced" draft, above the bundles, and drawing air over them, are often used.

There is little information concerning the effect of such inlet conditions, as shown in Fig. 2, on fan performance. The "General Data" of the Airscrew Co. (1) shows fan connections as Fig. 3. They comment "Connections B and C should be avoided at all costs. Quite obviously re-siting of the fan or an alternative outlet is required............... Obviously, the way to make sure that rated performance is obtained is to install a fan as it is tested for performance". Excellent advice but impractical in this case.

Hart (2) has measured the loss of airflow in a small model ACHE, with flow from one side only; and Spiers (3) has measured the flow in a full-sized dummy ACHE, with the flow from one side only, and severely restricted. Apart from this, no serious study of the effects of the poor inlet conditions of ACHE fans seems to have been made.

1. INTRODUCTION (Cont)

The Heat Transfer Systems Company decided that the information on this subject is inadequate and a research programme was undertaken to attempt to measure the effect of actual inlet conditions on airflow.

The airflow pattern into a bank of ACHE's will be greatly affected by wind. No attempt was made in these tests to evaluate this effect.

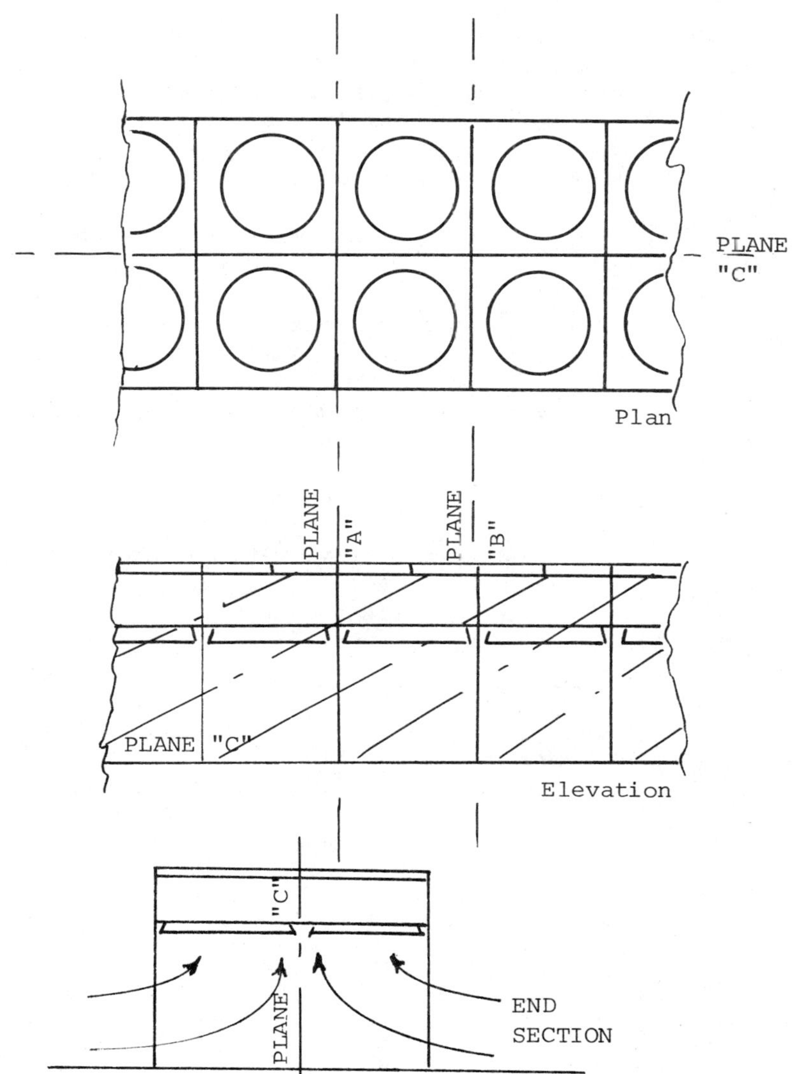

Fig. 2.

Diagram of part of a long bank of ACHE's showing the planes across which flow will not occur.

FAN CONNECTIONS

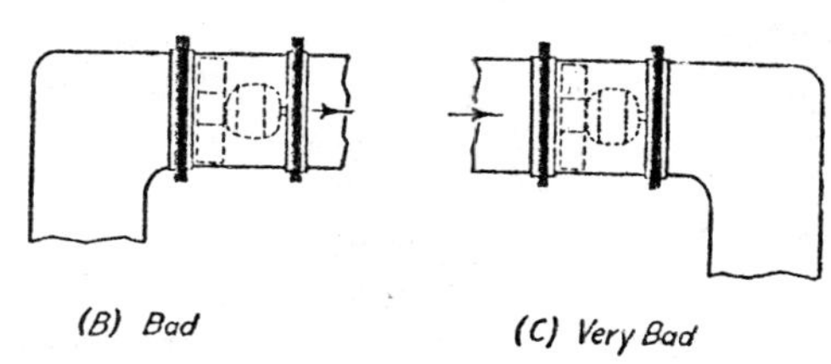

Fig. 3.

Fan connections discussed by the Airscrew Co.

1. INTRODUCTION (Cont)

1.1. Test Programme

It is difficult to undertake a programme with major alterations to the ACHE's structure at full size - as was evident from the programme of Spiers at the National Engineering Laboratories (3). A model was therefore used to permit changes to the layout to be made simply and quickly.

Satisfactory measurements of fan developed pressure in the extremely turbulent conditions of the plenum chamber of an ACHE have frequently been attempted. In the authors' opinion, no accurate measurements have yet been made, and it was decided not to attempt these.
Thus no measurement of fan efficiency was possible, fan performance being assessed in terms of total airflow only.

Five series of measurements were taken to measure the effects on airflow of:

a) Various inlet conditions, at two distances above ground, in forced draft.

b) Various heights above ground, for "grade-mounted" units, in forced draft.

c) As b) but in induced draft.

d) Tests comparative with those of Spier (3)

e) Flow visualization tests.

1.2. Model

The model scale was determined by the fan chosen, of 800 mm diameter A typical ACHE, with a tube length of 30 ft. (9.1 m), a bundle width of 9 ft. (2.7 m) giving a bay width of 18 ft. with a fan of 12 ft. (3.66 m) diameter. The scale fixed by fan diameter was 4.67:1. The leading dimensions of the model are shown in Fig. 4.

Fig. 4.

Leading Dimensions (in mm) of the model ACHE. View in direction of airflow.

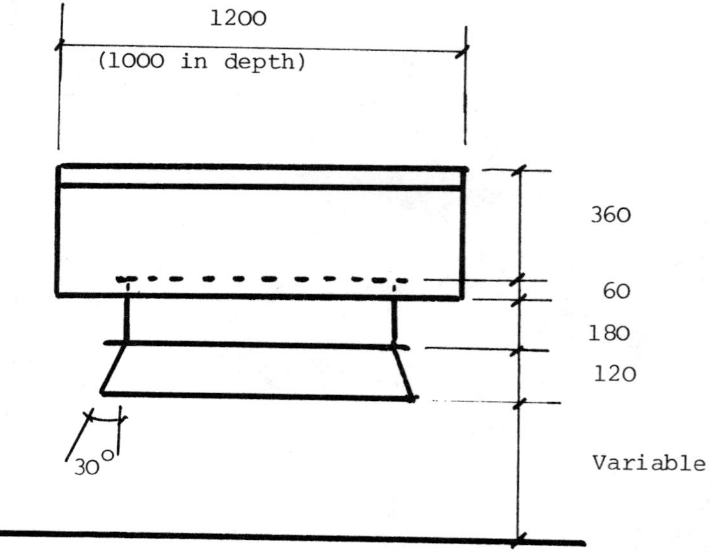

1. INTRODUCTION (Cont.)

1.2. Model (Cont.)

Mohandes (4) has shown that the effect of perforated metal on oblique flow onto a sheet is similar to the effect of a fin tube bundle, so two layers of perforated metal were used to model the fintube bundles. The sheets used have a frictional resistance of 11.3 times the approach air velocity head. Actual ACHE's have bundles whose frictional resistance is 10-20 times the approach air velocity head.

Two types of fans are typically used in ACHE's. Until recently, noise was often not a prime consideration, and fans with narrow chords perhaps 250 mm and operating at tip speeds in the 50-60 m/s range, were usually fitted. More recently, broad-chord fans, with large hubs and running at speeds of 30-40 m/s, are often used to reduce noise emission. The "Euroseries" of fans offers an excellent model of a "broad chord" fan, and a DCTLL 801-6 fan of this type was chosen. Fig. 5 is a photograph of the fan, and Fig. 6. is its characteristic curve. Also shown in Fig.6. is the orifice resistance of the bundles. The fan has 4 blades and operates at a speed of 870 rpm. The fan ring is 798 mm diameter, and the hub (incorporating the motor) 212 mm diameter. The blade chord is 213 mm at the root and 175 mm at the tip, giving an aspect ratio of 1.5 and a blade solidity of 49%. The blade angle at the tip is 27° and at the root 43°.

The most unsatisfactory feature of the fan used was the casing length which was longer than a typical ACHE fan ring. To obviate this, the fan was set into the plemum chamber, as shown in Fig. 4.

Fig. 5.

Fan used for tests.

To facilitate airflow measurement, a duct 1.2 m long was fitted to the air outlet. When in induced draft, a honeycomb was fitted to the fan outlet, to smooth the airflow in the measuring duct.

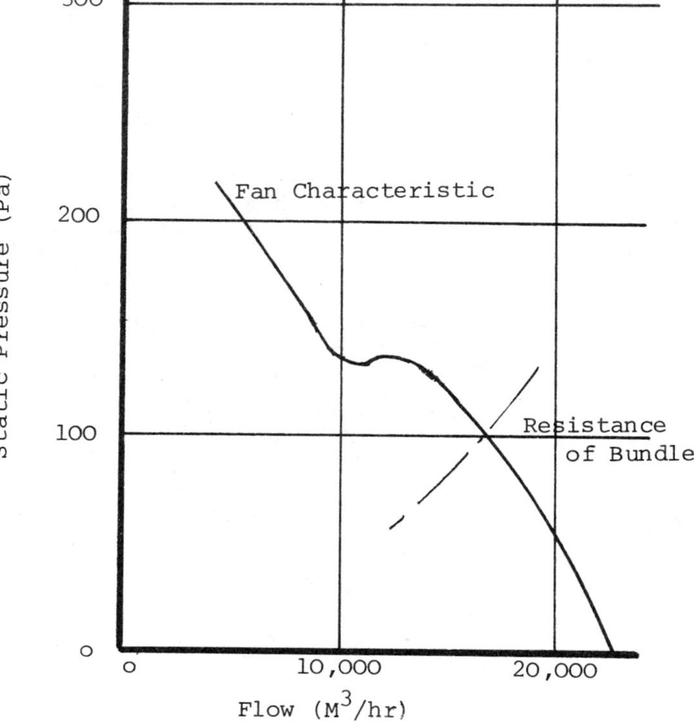

Fig. 6.

Pressure/Flow characteristic curve of fan. The "bundle" orifice resistance is shown in chain - dot.

Figs. 7 and 8 show the model. Fig. 7. shows the fan and the outlet duct, and Fig. 8. shows the complete rig with flow-directing screens.

Fig. 7 (left) and
Fig. 8. (over).

The model in forced draft with the outlet duct added (Fig. 7) and with duct and baffles (Fig. 8.)

Fig. 8.

1.2. Model (Cont)

Airflow from the model was measured using an integrating vane-type anemometer of Airflow Developments Ltd. (type AM 5000). This was systematically moved over the face of the duct for a period of 240 s. to obtain an average airflow over the whole duct. Two sets of readings were taken, and averaged.

This method of airflow measurement is not claimed to give sufficient accuracy to present absolute values of airflow; all results presented in this paper are of comparative airflow, and are presented as percentages to emphasise this point.

It can be noted that test "F" is the same condition as tested by Hart (2). Using a model with a model of a narrow-chord fan, 200 cm diameter, he reported a flow of 91.6% of maximum, comparing with the 92% of test (f).

1.3. Tests with varying Inlet Conditions

Fig. 9. summarizes the various tests made with differing inlet conditions. The first five tests A - E, were made with the "ground" at 2 fan diameters from the base of the plenum chamber. This is more than is commonly used in practice. For tests "A","B" and C the side baffles were not fitted, so air could enter the fan from three sides. The flow for test "A" was defined as standard - 100% flow, while "B" and "C" are for the fan with no cone and with a lengthened cylindrical fan ring, there being a substantial reduction in airflow.

The two sides were blanked off for tests "D" and "E", tests with and without the cone inlet being made. The introduction of the side plates did not change the flow for arrangement with the inlet cone fitted, but improved the flow from 91% to 94% for the case without the inlet cone fitted.

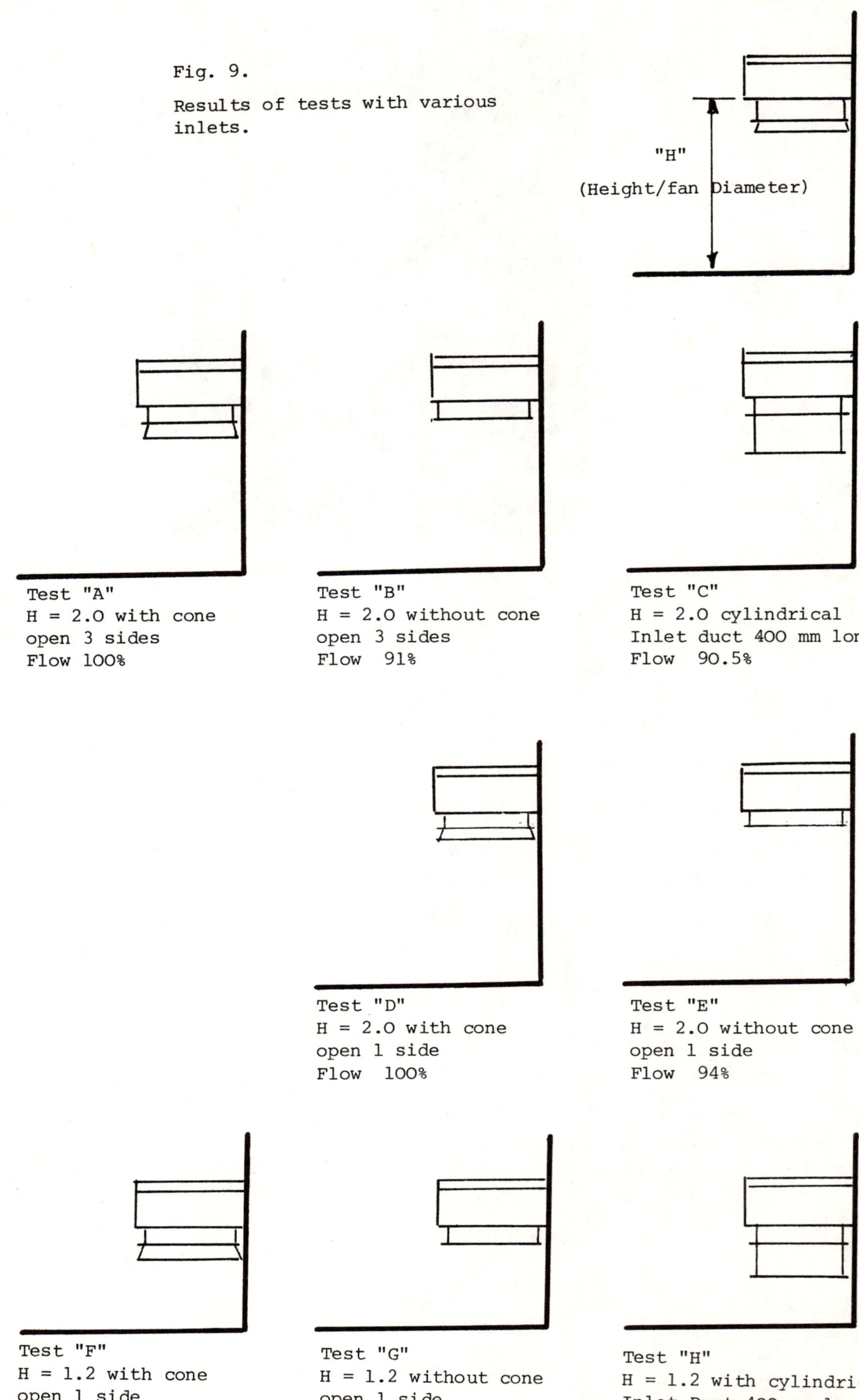

Fig. 9.

Results of tests with various inlets.

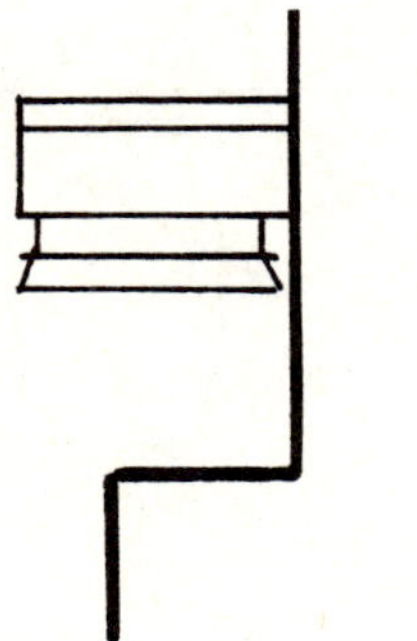

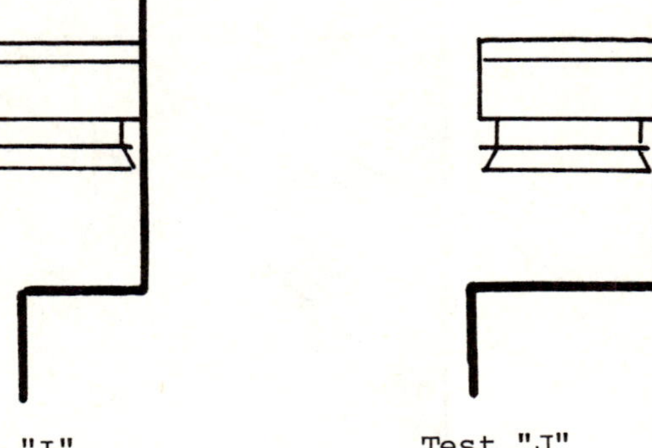

Fig. 9. Cont.

Test "I"
H = 1.2 with cone
open 1 side
simulated narrow
pipetrack
Flow - 98%

Test "J"
H = 1.2 with cone
open 1 side
simulated wide
pipetrack
Flow - 95.5%

Figs. F - J represent tests with the "ground" 1.2 fan diameters from the plenum base, which is a more commonly used height. Tests F and G correspond to tests D and E, and show reductions in flow compared to the case with better ground clearance. Test H does not compare exactly with Test C, but shows a considerable reduction in flow.

As in Fig. 1, ACHE's are frequently mounted on a pipetrack, and Figs. I and J model this case. Pipetracks are often narrower than the cooler; Test J models a track slightly wider than the cooler, and Test I a track narrower than it, of a width often used in practice. In both cases, a considerable improvement in airflow is noted, to 98% of the "standard", in the case of the narrow track.

1.4. Tests in Forced Draft

Baffles were fitted to represent the ground, and to confirm the side flow as in Fig. 8. The distance to the "ground" baffle was varied, and the flow measured. The results, expressed as a percentage of the flow with no ground are shown in Fig.10.It can be seen that the airflow falls off quite sharply as the ground clearance decreases.

1.5. Tests in Induced Draft

Similar tests to those in forced draft were made with the fan reversed, to draw air over the bundles in induced draft. The results of these are also shown in Fig. 10. There is a reduction in airflow as ground clearance is decreased, but this is not so sharp as is the case with forced draft.

1.6. Tests Comparative with those of Spiers at Full Scale

The tests made by Spiers (3) were on a dummy ACHE with a 3.66 m diameter fan, the height of the inlet cone being 3.15m above ground. Three sides were blanked off, and staging built up under the fan to give an effective clearance of 2.24 m and 1.31 m under the fan ring. These conditions were modelled and the results obtained are plotted in Fig. 11, together with the results of Spiers. It will be seen that there is good agreement between Spiers' data at full size, and the model data reported here.

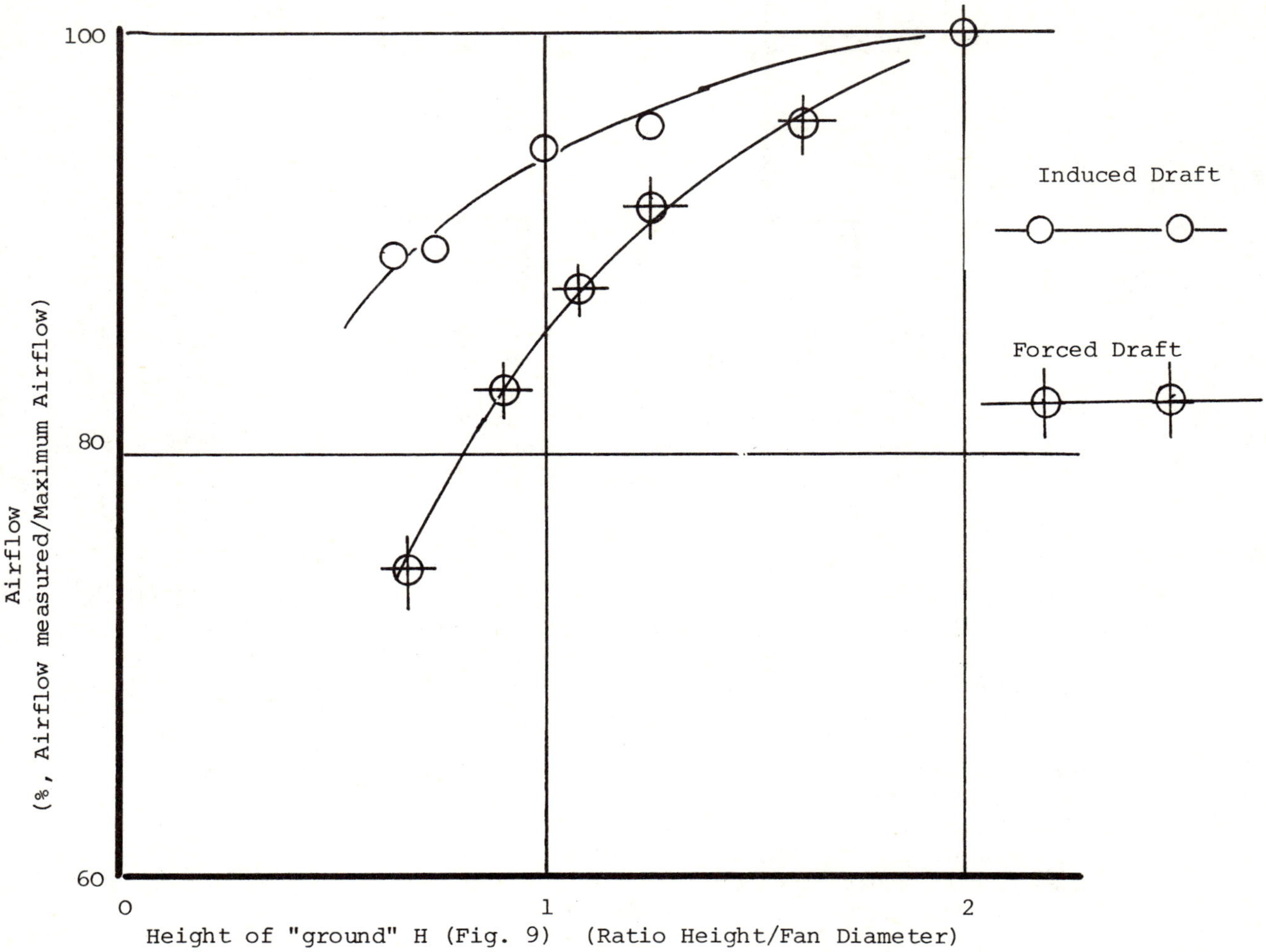

100
80
60
Airflow
(%, Airflow measured/Maximum Airflow)
Induced Draft
Forced Draft
0
1
2
Height of "ground" H (Fig. 9) (Ratio Height/Fan Diameter)

100
80
60
KEY
NEL Data (Spiers)
Model Data
Airflow
(% Airflow measured (Maximum Airflow)
0
0.2
0.4
0.6
Height above "ground"/Fan Dia.
Note - Height measured to bottom of
Fan Ring.

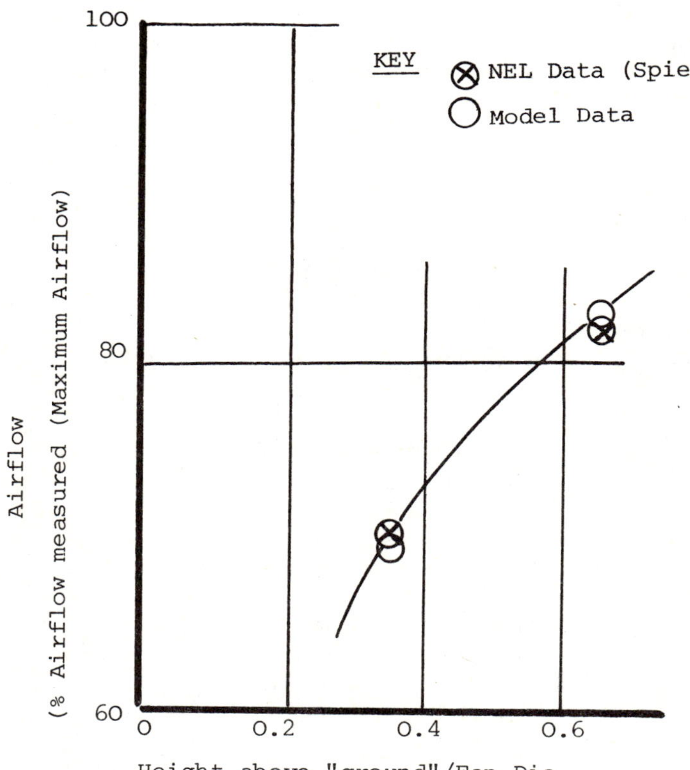

Fig. 10. (above)

Flow vs. Ground Clearance in Forced and Induced draft.

Fig. 11 (left)

Flow vs Ground clearance. Results of Spiers at the NEL on a full-size ACHE, and of the present series.

1.7. <u>Flow Visualization Tests</u>

Flow visualization tests using smoke and streamers were made. Fig. 12 is a photograph of such a test.

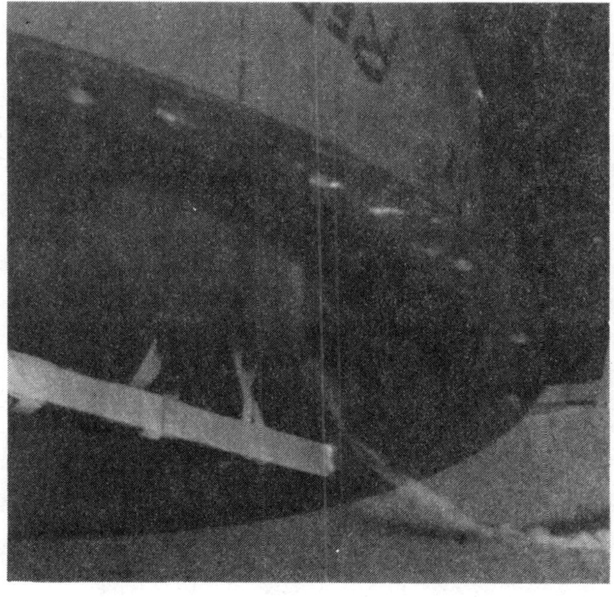

Fig. 12.

Photograph of a flow vizualization test using smoke and streamers.

Flow into the fan was seen to be very uneven, there being a flow detachment on the side of the fan near the air inlet as shown in Fig. 13. This will evidently decrease the fan efficiency; and inlet losses to the ACHE will probably be greater than expected. Flow reduction is greater in forced than induced draft (Fig.10), but the reduction in induced draft is greater than that allowed in normal ACHE design methods.

Fig. 13.

Airflow into the fan, showing flow detachment.

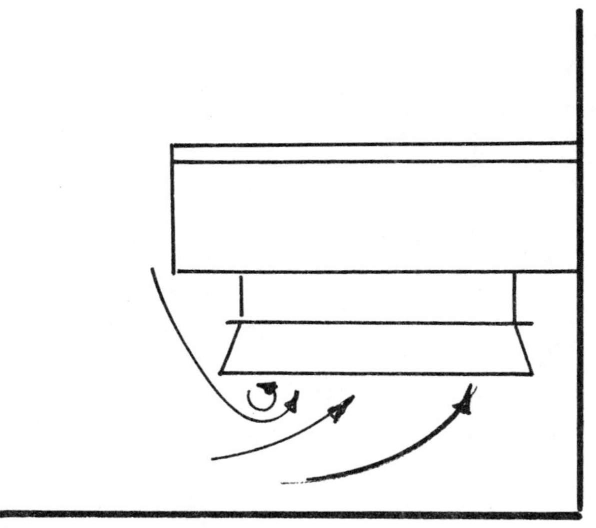

2. CONCLUSIONS

The comparison with work at full-scale is good enough to give confidence that the model is a fair representation of a full-size ACHE. However, the work reported here cannot be considered as other than a start in this field.

Until better information is available, Fig. 10 may be used to assess the loss of fan performance in grade-mounted banks of ACHE's.

3. REFERENCES

1. "General Data - Airscrew Fans", The Airscrew Co., Weybridge, U.K. p.24.
2. Hart, W-J., "Air Cooler model tests with various plenum configurations". Paper RS. 229, H.T.F.S. Harwell, U.K. June 1978.
3. Spiers, RRM, "Inlet tests on a full size air-cooled exchanger", H.T.F.S. Paper RS.361, Private Communication.
4. Mohandes, M.A. "Flow Through Heat Exchanger Banks", D. Phil. Thesis, Oxford University, Oct. 1979.

4. ACKNOWLEDGEMENTS

The work was carried out in the Osney Laboratories of the Department of Engineering Sciences of Oxford University. Our thanks are due to the staff of this laboratory for help and suggestions; particularly to Dr. T.V. Jones, Mr. J. Collins and Mr. J. Gilles.

International Conference on

Fan Design & Applications

Guildford, England: September 7-9, 1982

PAPER C1

FANS FOR THE VENTILATION OF ROAD AND RAIL TUNNELS IN THE UNITED KINGDOM

B.R. Pursall

Sheffield University, U.K.

J.F.L. Lowndes

Mott, Hay and Anderson, U.K.

Summary

In traffic tunnels over a certain length mechanical ventilation is essential to dilute and render harmless the various products produced by both petrol and diesel-driven engines and also to provide draught relief in rail tunnels.

The paper will deal chronologically with the ventilation design and application of mechanical ventilation for a number of British road tunnels, both existing and proposed, from the Blackwall and Mersey Queensway, through to the Holmesdale and other present day motorway tunnels. It will also discuss rail tunnels both rapid transit and main line.

Design philosophy will be discussed in relation to the change in transport methods, ranging from the horse to the petrol and diesel-driven car or diesel and electric train and necessitating complex ventilation system design with the use of fully and semi transverse and longitudinal systems. Particular reference will be made to the changes and that have taken place in fan design over the years and there will be an appraisal of fan performance against the orifice required and the possible savings in energy that may be obtained with more efficient fans. The philosophy of using single or multi-speed motors or continuously variable pitch fan blades for volume control, as the choice between single, large or multiple small fans will be discussed. The paper will also consider the economics of the various systems in terms of capital cost both civil and equipment and operating and maintenance charges.

Methods of control of fan systems to allow for changes in traffic flow and speed will be described and the necessity for precise monitoring of traffic characteristrics, pollution and smoke concentration to enable control to be completely effective.

Organised and sponsored by
BHRA Fluid Engineering, Cranfield, Bedford MK43 0AJ, England.

0263 - 421X/82/01 00 - 0001 $5.00
The entire volume can be purchased from
BHRA Fluid Engineering for $82.00

1.0 Introduction

With the development of the steam engine, transport of both goods and people progressed from horse-driven vehicles on roads to vehicles travelling on rails. Whereas the roads themselves were seldom straight and often in poor condition because of the crude road building techniques in those early days, rail transport could proceed along straight lines from town to town with the necessity for very few curves. Furthermore, steep gradients had to be avoided and obstructions such as hills, ravines and waterways crossed. Thus was born an increased necessity for the construction of bridges, viaducts and tunnels.

Similar considerations obtained in the development of road transport. As horse-drawn vehicles were superseded by vehicles driven by the internal combustion engine, the necessity for better roads became apparent. Obstructions still had to be negotiated but the problems of gradient and curves were not so important because of the lighter vehicles with their greater adhesion. However, tunnelling for highways was only common some time after tunnelling for railways. In mountainous areas, short road tunnels through rock formations and escarpments became numerous especially as they did not require large capital investment. Eventually there became the need for the transport of rapidly moving vehicles and heavy loads through long tunnels under rivers and waterways and for the relief of traffic congestion in cities by the construction of underpasses. There were also the long Alpine routes crossed initially by rail and subsequently by road tunnels. Here, a choice had to be made between a short tunnel at high level approached by severe gradients and sharp curves and vulnerable to winter snow, and a much longer tunnel at a lower level. The operating advantages of the latter are obvious and these tunnels are generally deep below the mountain surface. (Reference 1)

In the road tunnel, because of the poisonous nature of the constituents and the smoke issuing from the exhausts of both petrol and diesel-driven vehicles, the provision of adequate ventilation is of prime importance and this factor influences the initial design of the tunnel structure itself as well as the buildings needed to house the ventilation plant and controls. The provision of ducts within the tunnel, required to supply fresh air and/or extract vitiated air, together with the special ventilation buildings and the analysis and control equipment can amount to as much as one-tenth of the cost of the whole tunnel. The subsequent maintenance of equipment and the necessity for constant vigilance in terms of safety, which entails a large specialised staff and adequate policing of the tunnel, also involves the tunnel authority in expensive operating costs.(Reference 2)

The Thames Tunnel 1843, was the forerunner of highway tunnels in the U.K. It was 18 years in construction and was built to provide a way for pedestrians to cross the river. Originally stairs were provided at the shafts but the approaches were too expensive to build. These were finally provided when the East London Railway Company purchased the tunnel in 1866. It is still operated by London Transport today. Powered ventilation has not been used since construction.

Both the Blackwall, 1897 (1361 m) and the Rotherhithe, 1908 (1482 m) Tunnels were originally constructed without ventilation. The build up of internal combustion engines necessitated the addition of ventilation equipment (centrifugal fans) in the 1920's. These were replaced by modern axial flow fans at Blackwall in 1968 and at Rotherhithe in 1981.

The first major road tunnel built in England in the motor car age was the Mersey (Queensway) Tunnel. This was completed in 1934 (3237 m for the 4-lane tunnel plus two 2-lane branch tunnels). The potential dangers of the poisonous constituents of automobile exhaust gases were recognised and most detailed and diligent research was carried out into the subject. Considerable reference was made to the experience gained at the Holland Tunnel in New York, which at that time had been open for a few years. In view of the paucity of data available at the time the designers can only be admired for their solution. With the exception of an additional supply of air to the mid river in 1965, the system, despite the enormous increase in traffic is still in effective service today. The age of the ventilation equipment gives cause for some concern and a detailed survey is presently in progress. The main ventilation supply fans originally selected for the main tunnel were of the "paddle" centrifugal type (Walker "Indestructible"). The fans selected for the branch tunnels were also of the centrifugal type but with smaller diameters (Sturtevant)

More recently in England further subaqueous tunnels have been built. Dartford, 1963 (1429 m), Tyne, 1967 (1684 m), duplicate Blackwall 1968, (1174 m), Mersey (Kingsway), 1971 (2487 m) (and duplicated in 1973) and 2nd Dartford, 1980. These tunnels are all ventilated by the semi-transverse system, the air being provided by axial flow fans, typically of around 6 m diameter. Generally the fans are driven by 3 or 4 speed PAM motors, the highest power being the Mersey Kingsway supply fans at 300 kW. (Reference 1)

At present several motorway tunnels are in the course of design or construction. Although the longest of these is planned to be 1.1 km long, they will all be provided with a longitudinal ventilation system. In this arrangement the air passes along the length of the tunnel, the vitiated air leaving at the same portal as the traffic flow, which is unidirectional. The ventilation in this system is provided by a series of "jet" fans placed along the roof of the tunnel. These small diameter axial fans generally driven by 2-pole motors provide thrust to the air to assist the flow induced by the moving traffic. This system is particularly advantageous from the capital cost point of view as the table of indicative costs shows.

	Longitudinal Ventilation via Booster Fans £	Transverse Ventilation via Centrifugal fans £	Semi-Transverse Ventilation via Centrifugal fans £
Equipment (installed)	610,000	3,540,000	2,270,000
Civil	Nil	2,360,000	2,070,000
Ducts	Nil	3,720,000	1,860,000
TOTAL COST	610,000	9,620,000	6,200,000

HATFIELD TUNNEL
Cost Comparison of Alternative Ventilation Schemes (1979 Prices)
(Reference 3)

The ventilation system of a tunnel must ensure that the poisonous gases and smoke are diluted sufficiently to render the gases harmless in relation to the exposure times of tunnel users. The system must also maintain visibility at an acceptable level. Too much dilution will result in the construction of large, complex ventilation plant, expensive in capital, maintenance and operating costs. Too small a plant and too little dilution will result in unpleasant,(and perhaps dangerous) atmospheric conditions arising in the tunnel with reduced visibility, possibly causing accidents and certainly prompting complaints from tunnel users. (Reference 2)

With the passing of steam, the control of air quality in rail tunnels has become generally less important (diesel locomotives produce carbon monoxide and oxides of nitrogen, whilst electric trains produce ozone, together with possible changes in oxygen and carbon dioxide in the atmosphere). The main pollutant generated in rail tunnels, particularly underground systems which are used by the public, is heat which is primarily degraded traction energy. Thus ventilation for cooling is frequently required. Other major considerations which have implications regarding the ventilation system and passenger comfort are the air movements and pressure variations created by the train movements within the tunnels. These are due to the small size of tunnel in relation to the size of the train and are strongly influenced by the train speed.

Amongst the many railway tunnels which have been constructed should be mentioned those on the Liverpool and Manchester Railway (1830), this system being the forerunner of modern railway construction (the Crown Street tunnel - 265 m long, the coaches being rope-hauled to the surface, and the Wapping tunnel for goods traffic, 1930 m long and descending down a gradient of 1 in 48 to the docks; - a further 1582 m long tunnel, again rope-hauled, was subsequently constructed from Edge Hill to Lime Street). Between 1830 and 1890 over fifty railway tunnels exceeding a mile in length were constructed and amongst these were the Box, 1836-41 (2937 m), the Woodhead, 1839-45 (4848 m), the Severn, 1873-86 (7012 m), the Mersey Railway, 1881-86 and the Totley, 1888-93 (5697 m). The much poorer traction of locomotive hauled trains decrees that railways will have many more tunnels passing through hills or cliffs as opposed to roads where tunnels formerly have been largely subaqueous. During construction of these railway tunnels many working shafts are needed and in operation these become ventilation shafts. This means that generally rail tunnels are not mechanically ventilated as the piston effect is sufficient to move the air. The most notable exception is the Severn Tunnel which passes under the Severn Estuary joining Wales to England. Since its completion in 1886 its ventilation fan (originally steam driven) has been enlarged to its present size of 8.25 m diameter. It is a double entry centrifugal with a 2.75 m wide impeller capable of moving 380 m^3/s of air.

Metro systems, for transport under cities, also developed early, as for example, Glasgow (1896) and the London Underground lines from 1886, when the City and South London Railway (now part of the Northern Line) was first built, to the most recent Victoria and Jubilee lines. These are described in detail elsewhere). (Reference 4)

The most recent English system is that in Newcastle upon Tyne, the initial phases of which were opened in 1981. Here vaneless axial flow fans with special diffusers extract air from ventilation shafts between each pair of underground stations.

2.0 Design Philosophy

2.1 Road Tunnels

A natural air current flows in most tunnels and this is increased and decreased by differences in temperature between the interior and exterior, the barometric pressure differences between the two portals, the effects of wind forces at the respective portals and the speed, density and composition of the traffic moving through the tunnel. In short tunnels with one-way traffic, this air current is often sufficient to maintain satisfactory environmental conditions and much longer tunnels can also be ventilated satisfactorily when the traffic is moving rapidly, such as, for example, in motorway tunnels. However, it is usual in these longer tunnels to provide supplementary longitudinal mechanical ventilation, fresh air being induced from the entry portal, either by relatively large fans situated near there and supplying a high velocity jet of air, or by various combinations of small diameter fans installed in the tunnel itself. These fan systems normally only operate when the traffic speed is reduced or when a preponderance of heavy diesel vehicles produces smoke and impairs visibility.

In tunnels operating with two-way traffic flow, or in one-way tunnels where there is the possibility at times of two-way traffic flow, the longitudinal system of ventilation is less satisfactory because the traffic induced flow is reduced considerably by the two-way movement of vehicles. Thus, even when the natural air flows are enhanced by mechanical ventilation, the ultimate direction of the air flow will depend upon the density, type and speed of the vehicles in each lane. Under these circumstances pockets of pollutants at high concentrations could build up and visibility may become patchy throughout the tunnel. Furthermore, the longitudinal system inevitably allows the pollutant concentration to build up to a maximum at the exit portal and in long tunnels these concentrations may be quite unacceptable at discharge when dwelling houses or office buildings are situated near the portals. Here dispersion effects will determine what proportion of the pollutants reaches the buildings but the mechanics of dispersion are very complex and depend on the particular meteorological conditions, the speed and density of the vehicles and the distance of the buildings from the portals. (Reference 5)
In certain cases planning or environmental considerations dictate that a longitudinal system, otherwise satisfactory, may not be used, and the ventilation design must prevent vitiated air leaving the portal. This was the case in the North Bank Area Tunnel, 1974, near Blackfriars in the City of London. Here an exhausting semi-transverse system was installed, with 3 horizontally mounted axial fans discharging tunnel air at the roof level of a 6 storey office building above the tunnel.

Generally for two-way traffic, transverse-type ventilation systems are essential. These systems allow fresh air to leak into the traffic space from ducts usually either below or at the side, the leakage being controlled so that a regular rate of flow is maintained along the whole tunnel length. The vitiated air either passes along the traffic space (semi-transverse) to shafts situated at appropriate points where extract fans remove it, or alternatively, the vitiated air can pass upwards through slots in a ceiling duct, where the extraction is again at a regular rate and this duct is also connected to shafts where extract fans remove it (fully-transverse). The advantage of the latter is that a more positive removal of the vitiated air takes place and the longitudinal element of extract is virtually eliminated, so that in the event of fire, smoke and fumes will not pass along the traffic space but will be removed fairly quickly via the extract duct. The disadvantages are the enormous cost of constructing and maintaining a ceiling duct, together with increased extract fan capital and running costs due to the air passing in a smaller cross-sectional area than that of the traffic space. It may also be necessary, in the case of a bored tunnel, to increase the tunnel diameter in order to accommodate the ceiling duct and maintain an adequate traffic space. The practicality and cost of this may be quite unacceptable. However, in rectangular cut and cover or immersed tube tunnels the ventilation ducts can be placed at the sides or in the middle and the only restriction need be the actual width of the section, although construction costs would naturally be increased and the capital and running costs of extract fans would still be greater. In a typical dual three-lane tunnel, say 1.5 km long, the duct width required for the supply and exhaust ventilation would together be approximately equal to a single lane width of tunnel.

2.2 Rail Tunnels

2.2.1 Main Line

It is still unusual to find a main line rail tunnel with mechanical ventilation. In the case of a train traversing a tunnel at high speed the ratio of train frontal area to tunnel cross-sectional area is generally high relatively (0.2 - 0.3 for twin track and 0.5 - 0.6 for single track tunnels). Thus the induced air flow is considerable and working in conjunction with the many ventilation shafts is sufficient to maintain satisfactory conditions in the tunnel.

The emission of pollutants per unit weight of vehicle (diesel-engined) is low compared with that for a road vehicle and the specific energy consumption of rail vehicles is considerably lower than that for a typical sample of road vehicles.

The existing main line tunnel is non-critical from the point of view of thermal balance or air quality control and the aerodynamic effects of train passage are the main concern. This distinction is not true of the very long rail tunnel such as the proposed Channel Tunnel where the trains operate at high speeds. Here the vehicles experience increases in drag forces which may be far in excess of those occurring in free air. Also, when trains enter tunnels at high speeds, the tunnel air must be accelerated to a speed proportional to that of the train and these effects demand more power as the train speed increases. A prediction of the magnitude of those forces is necessary to establish the optimum ventilation design. Until recently, pressure changes in a tunnel have caused little inconvenience to passengers. However, with the advent of high speed trains, some discomfort may be experienced unless attempts are made to minimise the the generation of pressure transients. Pressure relief can be obtained by means of shafts and cross passages from one tunnel to another, if this be possible, but the designer must pay attention to the combination of vehicle velocity, blockage, vehicle length to diameter ratio, tunnel length to diameter ratio, the tunnel end conditions, the tunnel ambient pressure, the vehicle and wall roughnesses and the method of propulsion, to enable a satisfactory traction solution to be obtained which gives the minimum of problems. (Reference 6-7)

2.2.2 Rapid Transit Systems

In the early "underground" railway systems steam traction was used and in spite of the frequent ventilation openings, conditions underground were usually less than comfortable and complaints frequent.

Recent experience of subway systems has shown that today the cost of environmental control (primarily control of heat) may be as much as 8 to 10% of the total construction cost and the operating power required may be as much as 50% of that required for traction. A suitable environment must be provided for both the general public and for the operating and maintenance personnel in terms of heat, haze, odours, dust and condensate and the heat must also be extracted from the equipment and associated controls so that the life expectancy will not be diminished. A further requirement of the ventilation system is to ensure that the piston effect of the train does not cause unnecessary draughts in those parts of the underground system occupied by passengers. This refers particularly to platform cross passages and escalator tunnels. This is normally done by providing extra shafts from the tunnels to the surface of sufficient cross-sectional area to reduce the velocities to acceptable levels. Two 'emergency' cases must be considered. Firstly with a train stopped in the tunnel an adequate stream of air must be provided to remove heat deposited by the train's own ventilation, or in a hot climate, air conditioning system. Failure to do this can result in a very rapid build up of heat with possibly dangerous results. The second case is that of fire. In this case the ventilation system must provide a supply of fresh air for passenger escape and approach by firemen and must remove smoke from the system. (Reference 8, 9 & 10)

2.3 Mechanical Ventilation

Economic necessity and energy saving dictate that certain guidelines should be followed when designing the ventilation system. These are:-

1. The ventilation ducts should be sufficiently large to avoid undue high air velocities (less than about 15m/s) since power varies as the cube of the velocity.

   Also, if L = duct length
   A = duct cross-sectional area
   D = duct diameter
   O = duct perimeter (= 4A/D)
   q = leakage air quantity per unit length of duct
   Q = maximum duct air quantity (= qL)
   p = duct total pressure drop
   P = air power required (= pQ = pqL)

   then from the fluid flow equation:-

   $p = kLOQ^2/A^3 = kLOq^2L^2/A^3$ and $P = pQ = pqL$ (power), $p = kL4Aq^3L^3/3DA^3$ for uniform leakage. Thus Power is proportional to L^4/A^2 (References 11 and 12)

2. The pressure drop is proportional to the friction factor or resistance and hence all duct surfaces should be as smooth as possible, all essential bends should be streamlined or supplied with air splitters or cascades and all obstructions (even necessary ones such as screens, louvres, baffles) should be carefully designed for minimum resistance.

3. The overall efficiency of the fans, motors and transmissions should be as high as possible and the alteration of fan speed efficiently and economically carried out in accordance with the traffic flow in the tunnel. At least three fan speeds should be catered for and it may be advisable to provide for a fourth speed which would only be used for high fan outputs in the event of fire, or other emergency. Thus, the speeds may typically be high 100%, normal 80%, medium 50% and low 40%.

A much greater flexibility can be obtained by the use of continuously variable motors or fluid drives and these become more important in the fully automatically controlled tunnel. The use of variable pitch bladed axial flow fans, externally controlled and driven by one (or sometimes two) single speed motors is theoretically ideal for automatic control and would probably be the system of choice in future ventilation systems. (Reference 2)

There is little to choose in terms of overall efficiency between the modern centrifugal and axial flow fan. The backward-bladed centrifugal fan has the advantage of a non-overloading power characteristic and a smooth pressure curve which means that the fan can operate under almost any increase or decrease of pressure or quantity from the optimum (maximum efficiency) without the danger of stall, although efficiencies will be affected. This fan is also more suitable for higher pressures because of its low noise characteristics. However, pressures are usually relatively low in road tunnel systems because of the leakage method of transverse ventilation, resulting in a maximum air quantity reducing to zero at the end of the duct and this reduces the pressure drop by a factor of three compared with the constant flow case. The axial flow fan is usually more suitable for these relatively low pressure conditions and it also has the advantage of economy in installation as it can be installed in the vertical as well as the horizontal position. It also possesses a non-overloading power characteristic, but its main disadvantage is that the profile of the pressure curve is such that there is the danger of stall conditions arising if the total system resistance is under-estimated.

High speed axial flow fans discharging at say 30 m/s are used in the longitudinal system. In order to take account of possible traffic reversal or change in wind direction, these fans can readily be reversed. In rail systems, the train piston effect can temporarily reverse the air flow provided by the fan and cases exist of fans actually being destroyed. Thus, special care must be taken to protect fans from this danger. Normally non-return dampers are used, but it is possible that a specially designed fan with a non stalling characteristic could be used instead. In certain tunnels with large axial fans, automatic fan changeover systems have been installed. Each fan is erected on a carriage which is moveable. Excessive vibration, heat in bearings or transmission will automatically initiate the fan to move from its shaft position and allow a duplicate fan to take its place so that the tunnel ventilation is not impaired. The outstanding advantage of this system is that it provides 100% standby but causes very little extra expensive civil works.

2.4 Fan Operation, Tunnel monitoring and Ventilation Control

2.4.1 Fan Operation

In order to ensure reliable fan operation it is necessary for the usual functions to be monitored. These might be typically motor and fan speed, motor current, voltage and power factor, bearing temperature, vibration together with air quantity and pressure. All these data are fed to the plant monitoring system.

2.4.2 Tunnel Monitoring

2.4.2.1 Environmental

Both carbon monoxide and obscuration levels are monitored. Again the output from these instruments is fed to the processor.

2.4.2.2 Traffic

Traffic data, which are necessary to the police for normal traffic operation, are gathered by inductive traffic loops and backed up by closed circuit television.

2.4.3 Ventilation Control

To control the ventilation rate precisely under all conditions a computer based system is necessary. Signals received from carbon monoxide and visibility detectors are transmitted via a local data transmission unit (out-stations) to an electronic data processor.

The data processor controls and interrogates all out-stations, controls the switching of the tunnel ventilation in stages, presents processed outputs from the outstations to the plant monitoring system and reports system faults to the plant monitoring system.

In the event of an emergency or a high level of carbon monoxide the data processor switches on the maximum number of fans and sends an alarm signal to the local police control to enable traffic to be stopped from entering the tunnel or diverted from the approach roads.

In addition, the local police can be supplied either via the same data processor or a separate system, with information collated from speed and density monitoring devices to enable traffic flow to be accurately determined.

As each tunnel is different, an emergency plan has to be worked out as to how the various controls will operate. For example, in a longitudinal system it is advisable to stop all traffic on the upstream side of any fire but to remove traffic as quickly as possible on the downstream side. Decisions also have to be taken regarding the necessity for reversal of the air current, shutting down blowing and/or extract fans, and the best pattern of operation in a complex tunnel system.

3.0 Conclusions

The increasing consciousness of both the cost and availability of energy dictates that tunnel ventilation systems operate as efficiently as possible. Accurate and reliable measurement of tunnel pollutants has in the past been a rather elusive goal. As a result, precise control of a fan's output has been a refinement and only coarse (say 3 speed) control has been provided. This has led to overventilation. Furthermore there has been a tendency to operate tunnels at more stringent levels than allowed for in the design parameters with a corresponding increase in power consumption. It is likely that recent improvements in pollution monitoring (Ref: 13 & 14) together with precise and effective output control will result in the complete tunnel ventilation system operating at much improved efficiency.

4.0 Acknowledgements

The authors wish to thank the Directors of Mott, Hay and Anderson for permission to publish this paper.

5.0 References

1. Megaw, T.M. and Bartlett, J.V. Tunnels - Planning, Design, Construction Vol. 1. Ellis Horwood Limited, Chichester, 1981.

2. Pursall, B.R. and King A.L. The Aerodynamics and Ventilation of Vehicle Tunnels - A State-of-the-Art Review and Bibliography. (BHRA Fluid Engineering) Cranfield, 1976.

3. Lowndes, J.F.L. and Fudger, G. "A Ventilation System for Motorway Tunnels". Proc. 3rd Int. Symp. on the Aerodynamics and Ventilation of Vehicle Tunnels. (BHRA Fluid Engineering), Paper F1, March 19-21, 1979, pp 227-236.

4. Cockram, I.J. and Mahoney, T. "Fan Design and Application appertaining to Rapid Transit Railway networks". Proc. 1st Int. Symp. on Fan Design & Applications (BHRA Fluid Engineering) Sept. 1982.

5. West, A. and Pursall, B.R. "Dispersion of Pollutants from Road Tunnels". Proc. 4th Int. Symp. on the Aerodynamics and Ventilation of Vehicle Tunnels, (BHRA Fluid Engineering), Paper F3, March 23-25, 1982, pp 245-260

6. Henson, D.A. and Lowndes, J.F.L. "Design of a Ventilation System for an English Channel Tunnel". Proc. ASHRAE Annual meeting, 1977 - Hallifax Nova Scotia.

7. Henson, D.A., Gawthorpe, R.G., Pope, C.W. and Johnson, T. "Aerodynamics and Ventilation of a Proposed Channel Tunnel". Proc. 4th Int. Symp. on the Aerodynamics and Ventilation of Vehicle Tunnels, (BHRA Fluid Engineering), Paper A1, March 23-25, 1982, pp 1-14.

8. Henson, D.A. and Lowndes, J.F.L. "Economical Design in a Tropical Climate". Porc. 4th Int. Symp. on the Aerodynamics and Ventilation of Vehicle Tunnels, (BHRA Fluid Engineering), Paper G2, March 23-25, 1982, pp 295-306.

9. Parsons, Brinkerhoff, Quade & Douglas Inc. Subway Environmental Design Handbook, Subway Environment Simulation (SES) Computer Program. (prepared for the Transportation Systems Center of the U.S. Department of Transportation)

10. Danziger, N.H. "Research and analysis of subway rapid transit system aerodynamics and thermodynamics". Proc. 1st Int. Symp. on the Aerodynamics and Ventilation of Vehicle Tunnels, (BHRA Fluid Engineering), Paper G1, April 10-12, 1973, pp. 1-19.

11. Megaw, T.M. "Road tunnels in the U.K." Tunnels and Tunnelling, 4, 3. May-June, 1972, pp. 233-235.

12. Megaw, T.M. "Design of ventilation systems for road tunnels", Proc. 1st Int. Symp. on the Ventilation and Aerodynamics of Vehicle Tunnels, (BHRA Fluid Engineering), Paper A1, April 10-12, 1973, pp. 1-20.

13. Fudger, G. and Walker, E.H. "Automatic Control of a Road Tunnel Ventilation System". Proc. 4th Int. Symp. on the Ventilation and Aerodynamics of Vehicle Tunnels, (BHRA Fluid Engineering), Paper J2, pp. 383-392.

14. Wisker, S.M. and Allenby, H. "Carbon monoxide measuring in Vehicular Tunnels". Proc. 4th Int. Symp. on the Ventilation and Aerodynamics of Vehicle Tunnels, (BHRA Fluid Engineering), Paper J3, pp. 393-402.

Plate 1. The impeller of one of the Walker "Indestructible" fans in the Mersey (Queensway) Tunnel (reproduced with the permission of the County Engineer, Merseyside County Council).

International Conference on

Fan Design & Applications

Guildford, England: September 7-9, 1982

PAPER C2

FAN DESIGN AND APPLICATIONS AS APPLIED TO RAPID TRANSIT RAILWAY NETWORKS

I. J. Cockram and T. Mahoney

London Transport Executive, U. K.

Summary

1. The problems associated with locating large tunnel cooling/ventilation fans into a rapid transit rail system.
2. The solution:
   (a) Energy Conservation
   (b) Fan Drives
   (c) Quietness in Operation
   (d) Physical Size of Plant
   (e) Non-Return Dampers
   (f) Safety and Maintenance
   (g) Economics
3. Overview

Organised and sponsored by
BHRA Fluid Engineering, Cranfield, Bedford MK43 0AJ, England.

0263 - 421X/82/01 00 - 0001 $5.00
The entire volume can be purchased from
BHRA Fluid Engineering for $82.00

THE FIRST 100 YEARS OF EXPERIENCE

The Problem

To form a picture in the mind's eye of the fan requirement that the Executive needs to cool and ventilate it's 260 km of deep tubes, it takes the continual running of 120 large tunnel fans having a combined capacity of 3100 m^3s^{-1} absorbing 2000 kW to maintain the environment at it's present level.

Before proceeding, it is well to understand the theoretical considerations allied as to why there should be a need to cool or ventilate the tube system at all, when one would assume that with the frequency and piston effect of train movements, it should be self-ventilating.

Dependent upon the type of tube stock used within each tunnel, it will present a blockage ratio of 0.65 - 0.7, thus the amount of air displaced into a station by the piston effect of a single train entering is in many cases no more than the combined cubic capacity of the passageway or lift shafts and ticket hall resulting in little natural air change in the running tunnels. Estimates of air change by piston action vary between 15% and 25% (according to station design and distance between stations) of that required to remove heat to hold an average 21.1°C ambient tube air temperature.

Sketch 1

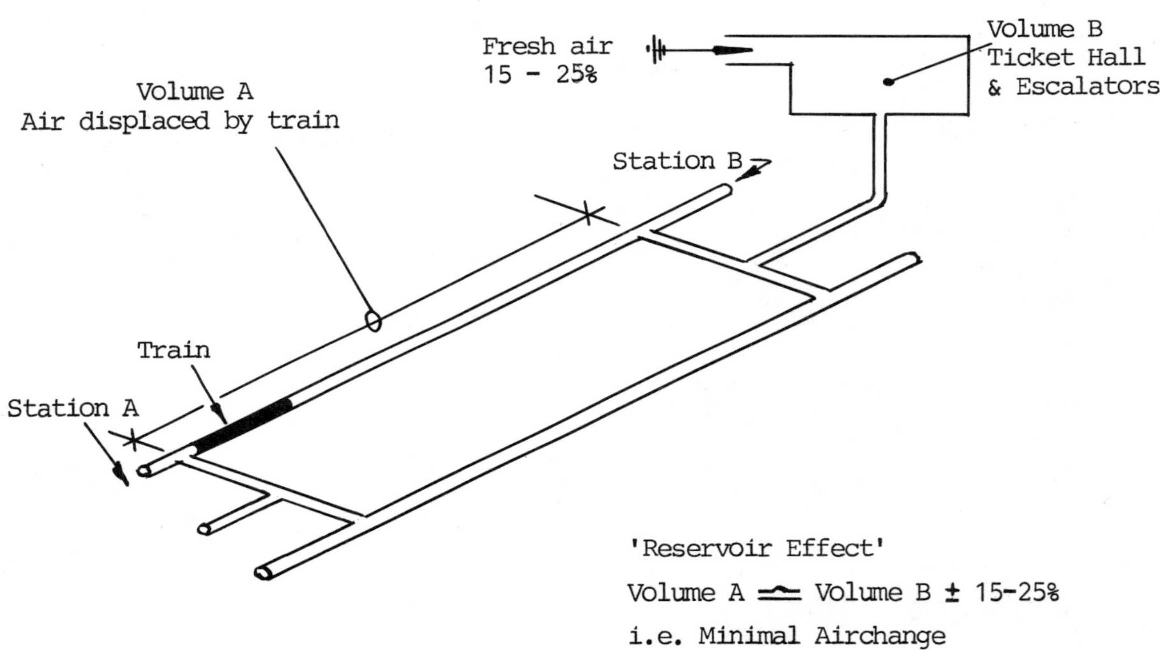

The movement of air through the tunnel by the train is taken into account when positioning tunnel cooling fans. For 'exhaust type' ventilation, either adjacent to, or within a station, the extract point is at the train entry end of the platform - this ensures that the extracted air has passed through the tunnels from the preceeding station (apart from cross movement at cross passages) and has received a maximum quota of heat. Similarly with 'pressure type' ventilation, the fresh air is supplied at the outgoing end to traverse the tunnel to the next station before escaping to atmosphere via the escalators etc. Between trains the inertia of the moving column persists and mitigates against local short circuiting.

Sketch 2

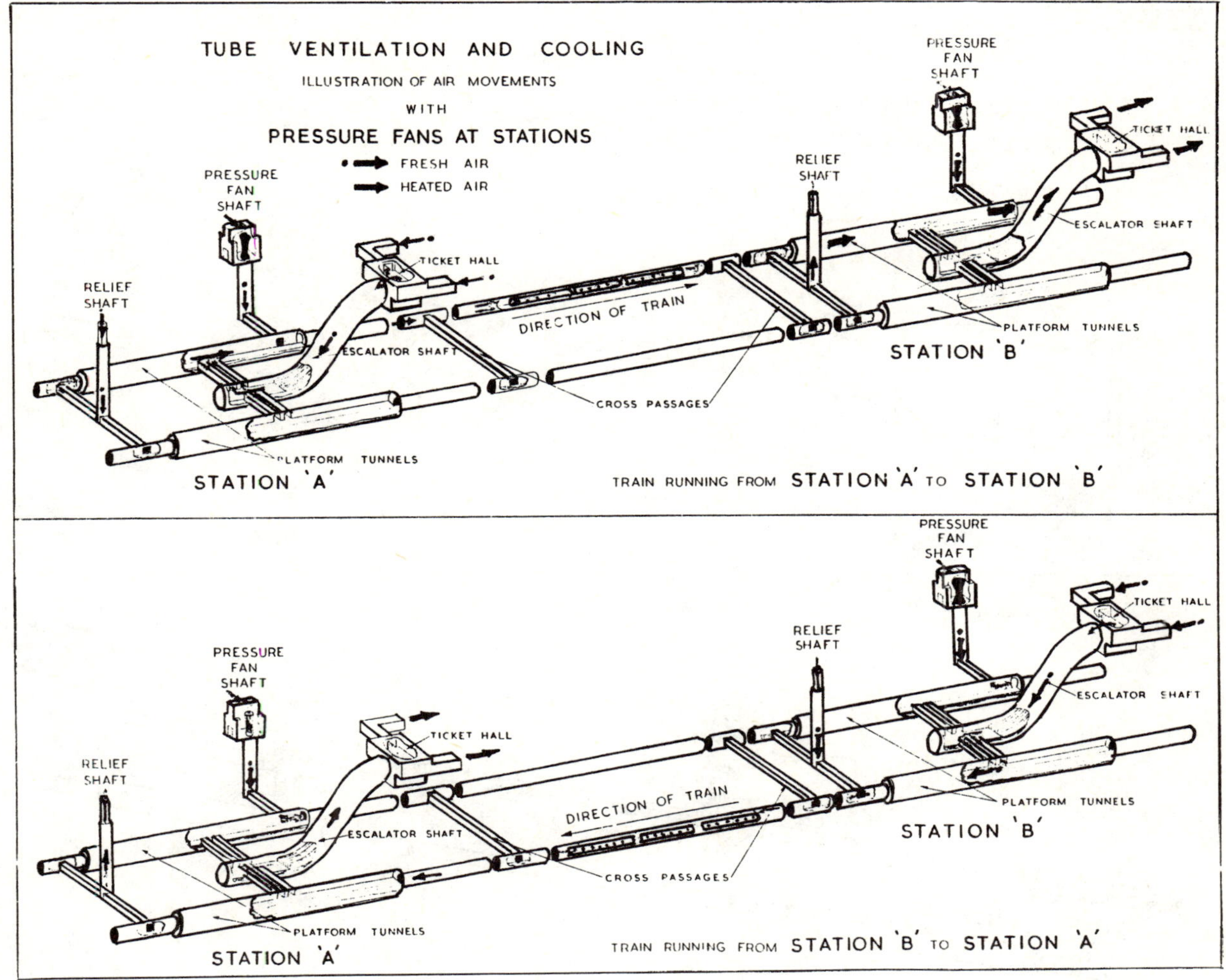

Current tunnel cooling fans found on the Victoria and Jubilee Lines, were installed mid-way between stations and run in an exhaust mode. With 73% of all tunnel fans running to exhaust, it is worthwhile to include a page as to the beneficial effect such a mode has on the system as a whole.

The earliest Tube ventilating plants were centrifugal fans of 9.44 - 16.52 m^3s^{-1} capacity running to pressure. The effect of this was to create a small air movement outwards at each station. As a result of the warm air meeting the passenger as he entered the station from the fresh outer air, it became the custom to refer to the typical tube smell.

The change of practice in recent years to run the fans to exhaust, has not in any way affected the general principle of cooling, but is considered an improvement for the passenger, who now enters the station with the fresh air drawn in by the fan. In winter he now passes smoothly from the cold outer air to the higher temperature at the platforms, without his attention being drawn sharply to temperature change of some 15°C.

Another point is the need at some stations to ventilate lavatories, shops and staff rooms - which are sometimes at basement level. With the exception of very recent stations, this type of accommodation is usually ventilated by exhausting from the rooms and drawing the make-up air in from the surrounding passages. If these passages are now filled with fresh air entering to feed the tunnel exhaust fans, then fresh air is available for the small ventilating plants referred to in this paragraph. If the tunnel fans were all run to pressure, the balance of air movement through the station would be outwards and so the basement would obviously be filled with vitiated tube air.

Sketch 3

TUBE VENTILATION AND COOLING
ILLUSTRATION OF AIR MOVEMENTS
WITH
INTERSTATION FANS
FRESH AIR
HEATED AIR
EXHAUST FAN SHAFT
RELIEF SHAFT
ESCALATOR SHAFT
TICKET HALL
PLATFORM TUNNELS
ESCALATOR SHAFT
TICKET HALL
RELIEF SHAFT
PLATFORM TUNNELS
DIRECTION OF TRAIN
STATION 'B'
CROSS PASSAGES
STATION 'A'
EXHAUST FAN

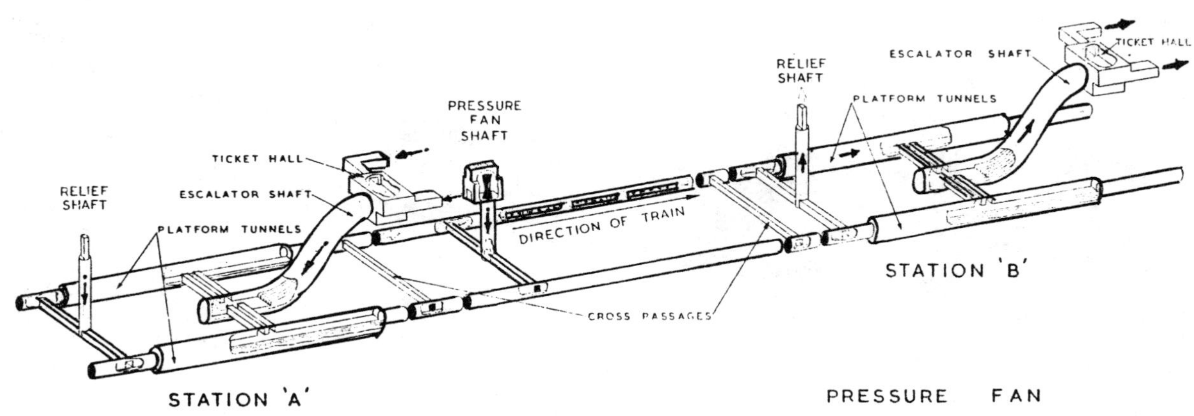

Pressure fans in winter also impart some thermal shock to the system brought about by flooding the $22^{o}C$ tunnels with fresh air sometimes as low as $-5^{o}C$ making the internal surfaces of the tunnel iron and rails 'run' with condensation. London Transport's modern tube lines are highly dependent upon automatic signalling, which is operated by the train wheels forming under dry conditions, a 'bridging resistance' between the two running rails of the section of track that the train is in.

It was found that condensation on the rails affected the system's ability to accurately interpret the resistance figure which should be some 8-10 ohms; in fact the carbon in the tube air dust coupled with the condensation formed it's own 'bridging resistance' approaching that of the design operational figure.

After 100 years of experience, the reader would think that by now we should have perfected the ideal system, but unfortunately, for every problem solved, further solutions were required to overcome the side effects.

So what are the problems?

In 1903, the first tunnel fan a centrifugal type, was installed with a capacity of $47\ m^{3}s^{-1}$ in the entrance to Wood Lane Depot to ventilate the 9.3 km from Bank to the Depot. Over the years various fans have been used from axial and centrifugal driven, both direct and indirect, with capacities up to $53\ m^{3}s^{-1}$.

Sketch 4

TUBE VENTILATION AND COOLING

PLATFORM PRESSURE

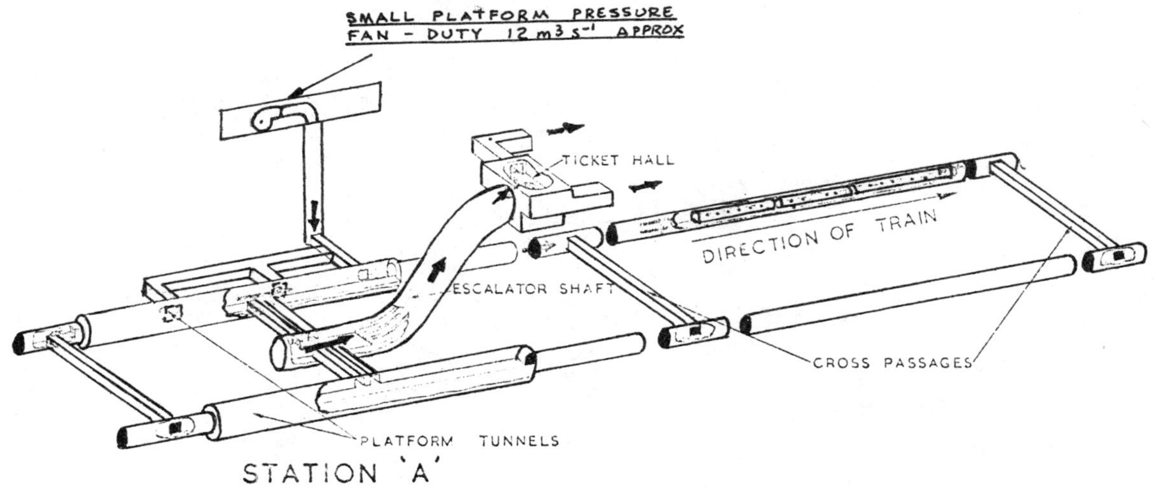

Operating resistances for such large fans are fortunately small - approximately 170 Pa, being a typical fan system resistance. For economics and convenience of installation, most of the fans are of the large axial type which can be located in line with the fan shaft airway. These fans have their drawbacks, because to avoid fan noise being transmitted to the adjacent area which could be residential, the fan and tip speed are kept quite low, resulting in a larger physical sized fan than one would expect.

When considering the construction of the fan, also it's housing and the connecting airways, it is imperative to ensure that they have the strength to resist the fluctuating surge pressures of the piston effect of the trains which could and often approach 0.41 kPa s^{-1}, when the total pressure change is greater than 0.69 kPa.

Non-return dampers are used with positive effect to stop the piston action of the trains reversing the fan. Without such precautions, the train's air volume - in the order of 90 $m^3 s^{-1}$ @ 500 Pa., would stall the fan with, as has happened, disasterous effect on the fan motors and drives.

We are now in a position to understand the WHERE - HOW and WHY tunnel fans are installed to remove the daily heat input given off by traction energy, trains lighting, passengers, station lighting and station machinery: What has to be examined now, is an effective approach which has shown through trials and experiments to be the best fan design and application for the job at hand.

Sketch 5

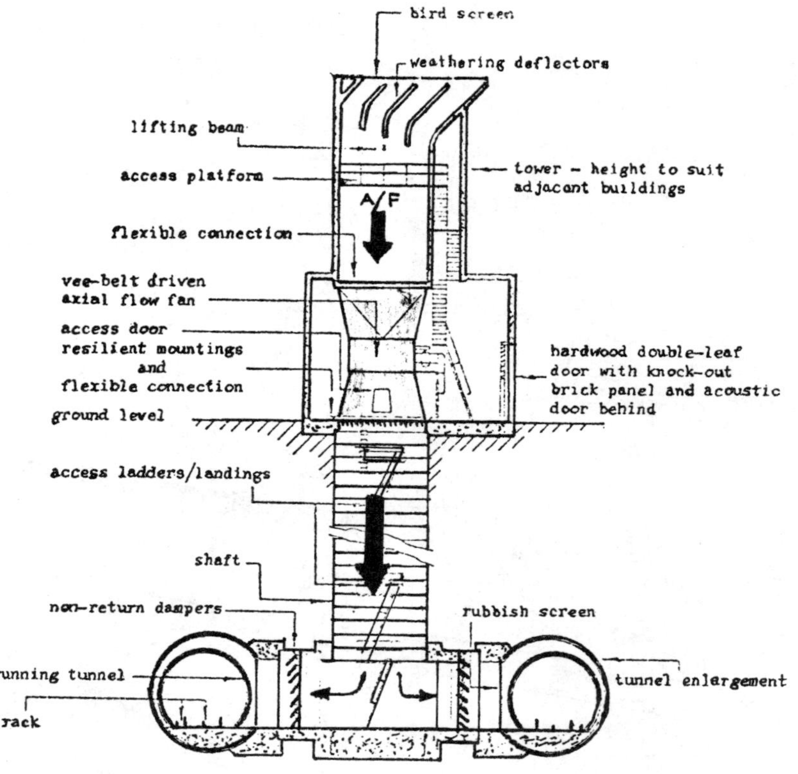

SECTION

TYPICAL TUNNEL COOLING PRESSURE FAN

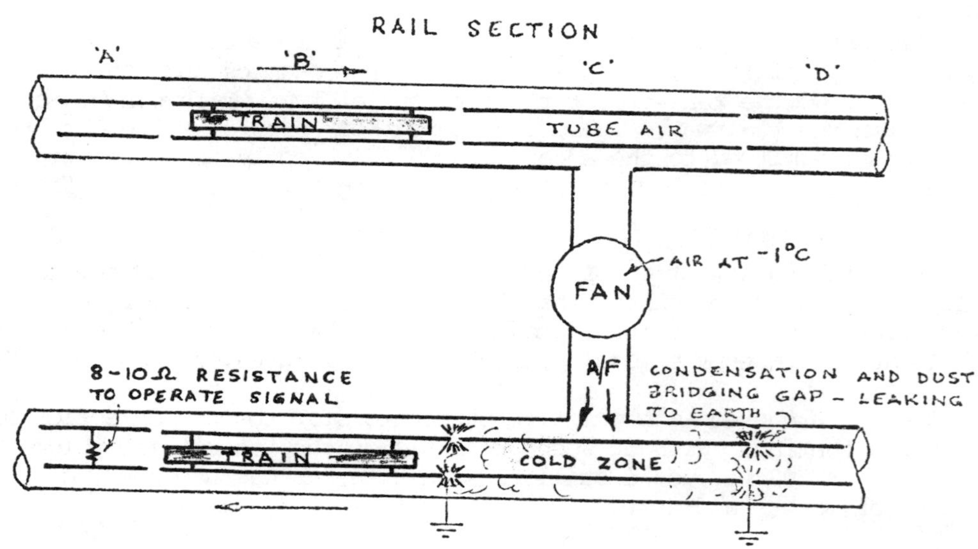

PLAN OF TUBE LINE SHOWING HOW FAN RUNNING TO PRESSURE CAUSES CONDENSATION AND SIGNALLING PROBLEMS IN WINTER

Sketch 6

NON-RETURN DAMPERS

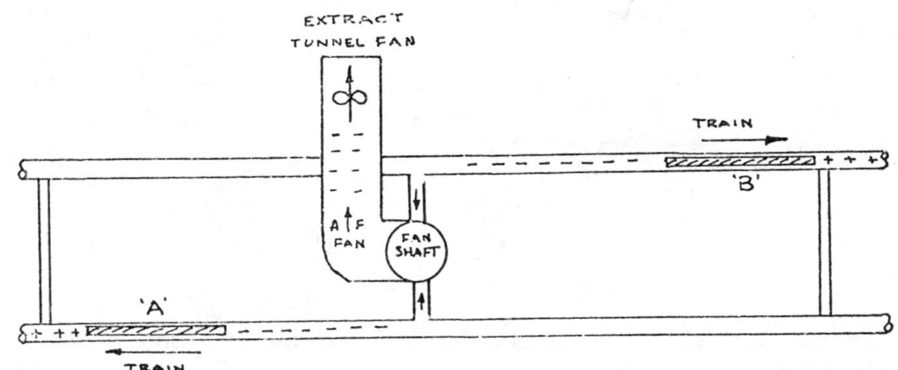

THE BASIC NEED FOR N.R. DAMPERS

TRAINS 'A' & 'B' INDUCE UNDER CERTAIN RUNNING CONDITIONS A HIGH NEGATIVE BACK PRESSURE ON THE FAN WHICH IF LEFT UNCHECKED WOULD STALL THE FAN AND DAMAGE THE MOTOR.

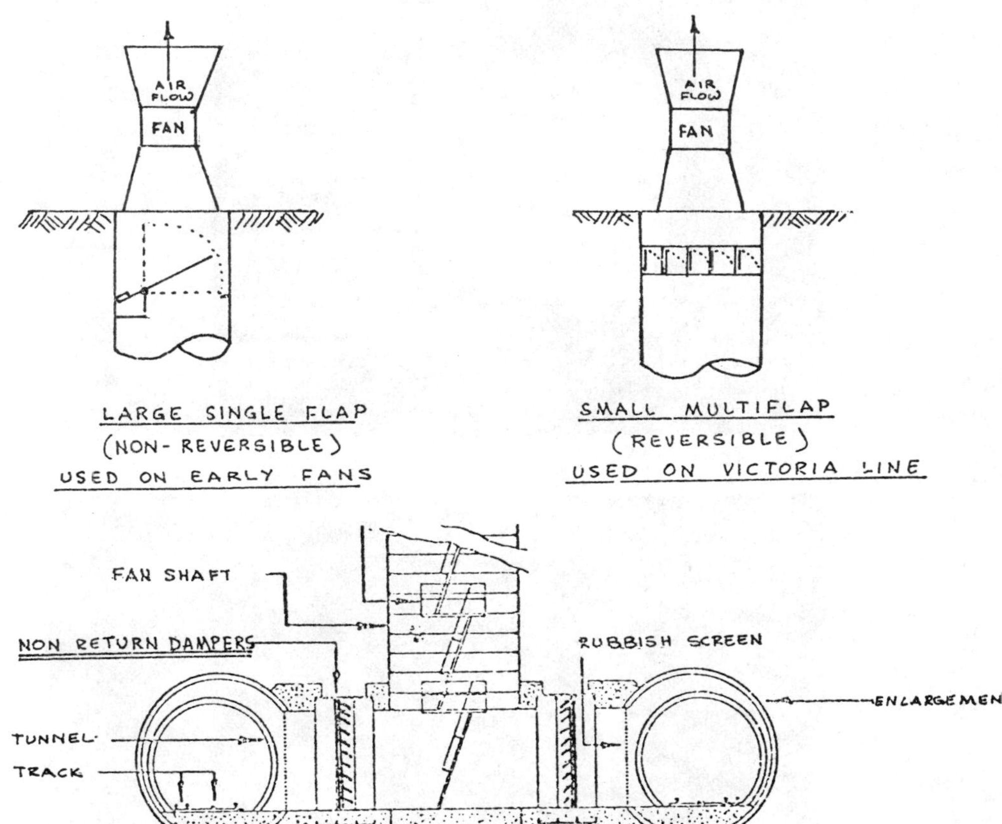

LARGE SINGLE FLAP
(NON-REVERSIBLE)
USED ON EARLY FANS

SMALL MULTIFLAP
(REVERSIBLE)
USED ON VICTORIA LINE

MULTIFLAP NEOPRENE DAMPERS MOUNTED ON A FLOWFORGE GRATIN COMPLETE WITH ACCESS DOORS FOR TRACK PERSONNEL CURRENT ARRANGEMENT ON NEW FANS

Sketch 7

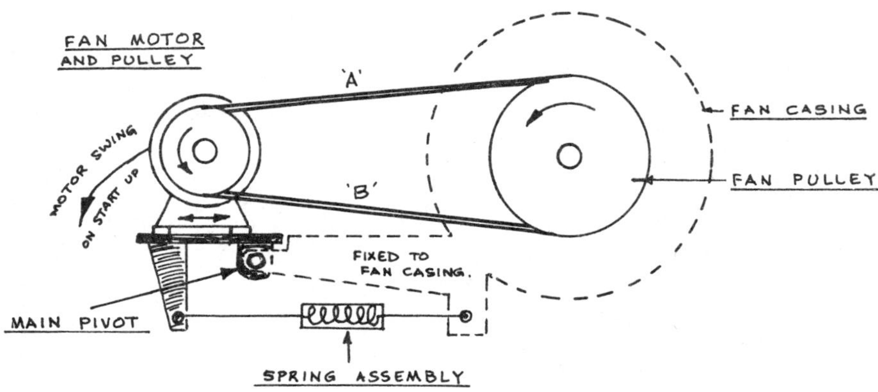

Motor Damping Assembly - To reduce 'over-tension' in A and 'under-tension' in B on start-up, the motor swings on main pivot and is damped by spring assembly, thus increasing distance between pulleys thereby taking up excessive belt slack.

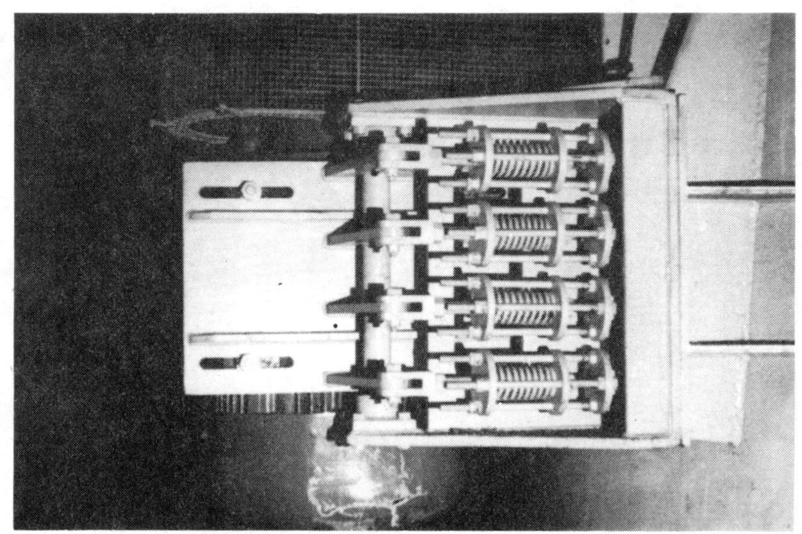

The Solution

(a) Fan Drives

To transmit such a heavy torque put out by an instantaneous start motor of 24 kW plus and be able to withstand the 'normal running - surge - feathering - normal running' cycle constantly being imposed, a special drive arrangement as shown ....., was designed to accept and equalise the large differential belt tensions throughout the cycle, thus cutting down the periodic maintenance required.

From the sketch, it can be clearly seen that the springs set in the 'pivoted assembly', move and adjust to equalise the belt tension. The drives in standard form are rated to accept 100% overload in the event of a belt failure (in a multi-belt installation) and the pulleys are 'deep grooved' for added insurances against belt stretch and groove jumping.

Motors are usually 2 speed, being either dual wound or pole amplitude modulating (P A M) so that they can be run at reduced speed in winter to conserve energy and run-up through low speed into high speed in summer, thereby reducing the frictional wear on the belts.

When considering the resultant loss in efficiency of a belt drive related to the running and maintenance costs, other methods of direct transmission are always being examined. With the advent of highly sophisticated electronic technology available, various options have been investigated, but unfortunately rejected.

Thyristor Control:- This form of drive produced an infinitely variable motor speed and was suitable for both size and application on an initial examination. Unfortunately as the thyristor control circuitry was tailor made for the project, all the development charges would have been passed on by the manufacturer. Additionally, it would not have comprised 'off the shelf components', which are so necessary for future maintenance.

Inverter Control:- This initially looked a better proposition, however, this too proved unrealistic, in that a 24 pole motor was needed in order to comply with the fan duty required.. Consequently, this led to the frame size being too large to be accommodated within the hub diameter.

Hydraulic Drive:- This was considered and tried (even though it incurred an additional electrical loading of some 5%) in the anticipation that it would (a) reduce the high construction/building costs required for a conventional fan tower i.e. it could be nearer the 'source' and integral with the airway tunnel. The very small pump chamber being located near the surface where it was considered easy to maintain and silence. However, it was eventually found that the sound pressure level could not be reduced to the severe criteria laid down, which very simply summarised was NC45 within 3 metres of the building housing both the fan and pump unit. Added to this air noise was the 'resonance' from the oil in circulation. Despite the use of an acoustic chamber for the pumping set and pipeline attenuators, the nuisance value to adjoining properties could not be significantly reduced.

(b) Energy Conservation

Everybody without exception, is looking towards energy conservation and experiments are being conducted at present to thermostatically control tunnel fans to ensure that they do not operate when the tunnel air drops to 14 ^{o}C. Obviously, this can only work on the fringes of the network where train intervals are less frequent and 'train assisted' ventilation maintains a low stable figure throughout the year.

As with most design changes which are brought about, 'side issues' must be looked at - the most immediate one being that of safety. It is a widely accepted fact that due to the 'piston effect', trains move a considerable amount of air, the average being 88 $m^3 s^{-1}$ for the London Transport Tube Network which as stated, has a blockage ratio of 0.65; The standard duty of an inter-station fan being in the order of 47/50 $m^3 s^{-1}$.

However, once a train/s is/are stopped in either one or both running tunnels for any length of time, then the air movement is obviously reduced, if not arrested completely. Passengers will naturally begin to suffer 'heat stress' to some degree and it is here that problems arise with thermostatically controlled fans. The 'stack effect' of the warm column of air will not move quick enough to energise the thermostat, consequently, the temperature differential per unit time between the running tunnels and the fan inlet will increase. Passengers may therefore require to be de-trained sooner . . . . . the present guideline period being 30 minutes . . . . . anything less could result in chaos if the service was restored within that period.

(c) Quietness in Operation

It has already been stated that the tunnel fans are to be run at a slow speed - some 3-4 revs. per sec. to achieve a slow tip speed with resultant quietness and that the fan when in operation does not emit any pure tones. Careful thought in each case must be given to this aspect of the problem, in order to avoid complaints by neighbours, whose windows overlook a fan-discharge point.

A sufficient reduction of the noise level can usually be accomplished by taking care to mount all fans and motors on suitable vibration-absorbing pads, insulating all holding-down bolts and particularly by fitting fire-retardant canvas connections at all places where the fan joins the masonry of the discharge shaft, or any ducting or shaft lining that leads up to the point of discharge. It is also desirable to arrange for up-cast discharge wherever possible, rather than turning the air at right angles at the top and discharging sideways. This involves some consideration of rain entry, but the design can be arranged to deal with this more easily, than to discount any noise that may be emitted horizontally in the direction for instance, of a block of flats.

The modern axial-flow plants have belt-driven motors. Thus any inherent noise from this source is damped out by the belts, before reaching the air stream.

(d) Physical Size of Plant

The size of a large axial flow tunnel fan would be approximately 6 m high and 3 m diameter including fairings on inlet and exhaust. Whether or not such a large item of plant has to be hidden or made a feature of, is largely dependent upon it's location and pressure placed on the Executive by local planning and other Statutory Authorities.

(e) Non-Return Dampers

Brief mention should be made of these all important dampers. To withstand the continued train surge pressures of $\pm$ 500 Pa various damper configurations and materials were tried over a period of years, in order to reach the high standard of performance that is achieved with the current design.

(f) Safety & Maintenance

With such large fans situated in the airway, special care is taken when designing the fan and it's associated housing in order that maintenance personnel are not injured by the spinning rotor which 'free-wheels' in the airstream due to the piston action of the train.

When maintenance is to be carried out on the moving parts of the fan i.e. rotor, bearings, pulleys and belts etc., the fan rotor is 'lashed' to avoid movement and if need be, the airway sealed to avoid large volumes of air effecting the work and health of the personnel.

A friction brake was investigated to lock the rotor for maintenance, but it was felt that it's efficiency over a period of time would diminish and on the other hand when the rotor is 'lashed' by the operating staff, the feeling for their own personal safety is self satisfied.

Maintenance work could obviously be carried out during non-traffic hours, which would add extra costs to the on-going budget . . . . . incurring overtime and unsocial hour payments, limited work being therefore undertaken.

The following photographs give an indication as to the extent and skill that has been taken in placing such fan houses in various environments:-

(g) Economics

The feasibility at every stage must be examined, considering whether or not a new tunnel fan is to be installed into the existing system, or as has recently been the case, taken out completely and the airway opened up as a Draught Relief Shaft.

In the early 1920's a large 47 m^3s^{-1} fan would cost in the region of £2,500.00 many of these are still in service today. To install such a fan at today's price would cost between £70/80,000.00 and consideration is now being given to either re-engineering or replacing these long standing 'Goliaths' with 4 short casing 1200 diameter axial flow fans mounted in parallel for £8,000.00 and look to replacing them after 5 years service.

It would appear that there is merit in the adoption of smaller fans from a maintenance point of view when considering the 'down-time' of a major fan plant, due to say a minor component failure, when only 25% of the fan capacity is lost if one small axial fails.

Economically it must make sense, due to the fact that the fan housings can be reduced in size and in some cases located where a larger fan house would be impossible to site. Additionally, with judicious application of airway connections 'train moved air' can be utilised as a cooling medium, at the expense of mechanical forced ventilation.

Noise produced from such small high revolution fans is an important factor - silencers it would appear to be the answer, but once again, there is a drawback caused by the 'contaminating effect' of the tunnel dust, which would after a short period of time, drastically effect the life and performance of the silencer.

Overview

In the light of the previous pages, how can one summarise the evolution of the tunnel fan which has ranged from giving consideration in 1879 to a steam driven fan; putting the largest size fan into the largest size hole available; leading to the current position where every aspect is examined . . . . . starting from the financial premise that 'we cannot afford it'through to the dire consequence of the detrimental effect on the passenger environment if we do not. Compromise to the benefit of the passenger is the only solution.

Various options for fan designs have been tried and considered in one form or another:-

Centrifugal	v	Axial
Belt	v	Direct Drive
Upstream	v	Downstream guide vanes
A C	v	D C Voltages
Variable	v	Fixed blade pitch angle
Pressure	v	Exhaust modes
Electric	v	Hydraulic drives
Single direction	v	Reversible flow (electric)

Early tunnel fans were thought to have an asset life of 25 years, yet some 60/70 year old fans are still doing an admirable job and with regular maintenance carried out on them (or as aforementioned, being re-engineered), who knows how long they will last.

As to the future, when looking at long term maintenance and running costs, 'train assisted' ventilation may be more viable by increasing the number of inlets/outlets from the running tunnels to atmosphere, so that the trains have a better chance of 'scavenging' the heat from the tunnels.

With prices of property and land within the Central London area being at a premium, the cost of acquiring such a large number of sites, together with the associated civil engineering/tunnelling work involved, would require careful consideration.

Sketch 8

TRAIN ASSISTED VENTILATION

TRAIN ASSISTED VENTN. SCHEME
STANDARD TUNNEL FAN SCHEME
RUNNING TUNNELS
S
F

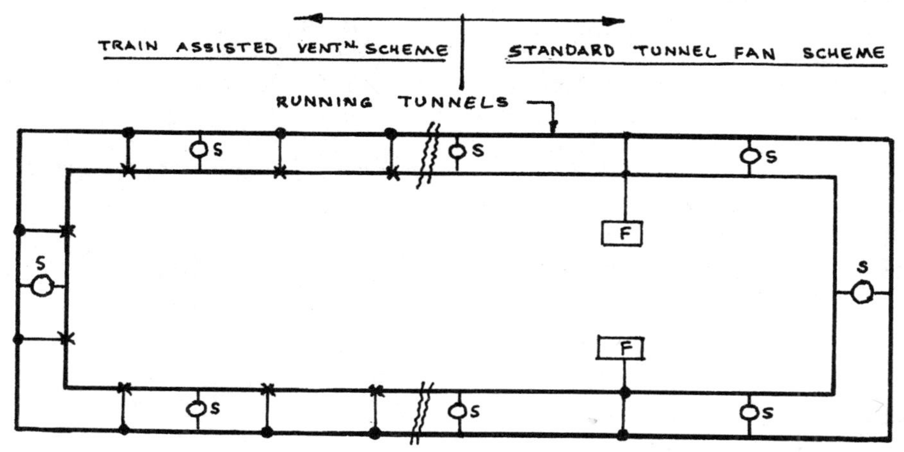

BASIC TUBE SYSTEM

F TUNNEL COOLING FAN —O— STATION —X— TRAIN ASSISTED POINTS

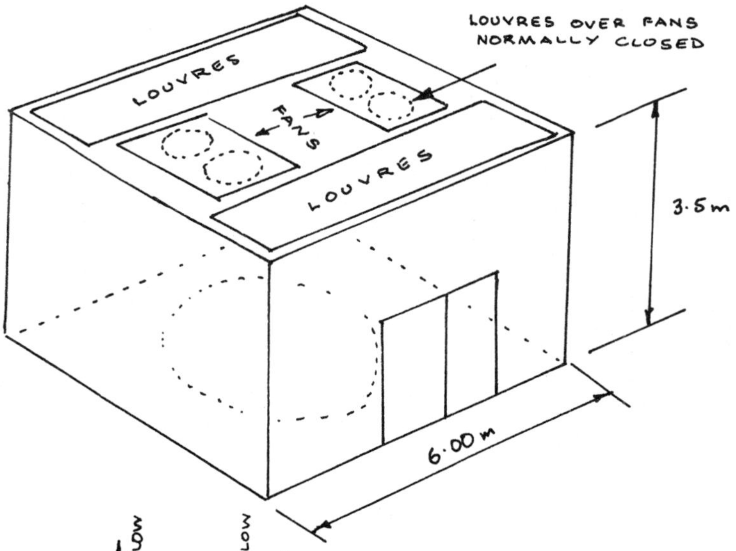

LOUVRES OVER FANS NORMALLY CLOSED
LOUVRES
FANS
LOUVRES
3.5m
6.00m

ISOMETRIC VIEW OF TERMINAL

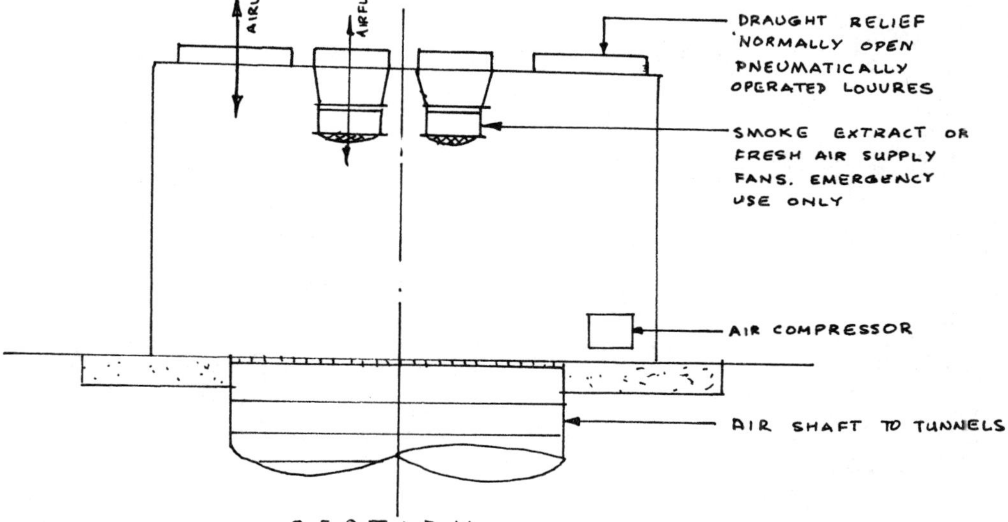

AIRFLOW
AIRFLOW
DRAUGHT RELIEF NORMALLY OPEN PNEUMATICALLY OPERATED LOUVRES
SMOKE EXTRACT OR FRESH AIR SUPPLY FANS. EMERGENCY USE ONLY
AIR COMPRESSOR
AIR SHAFT TO TUNNELS

SECTION

London Transport has at present 26 disused stations, many of which are being opened up as ventilation/draught relief escape exits - using the disused lift shafts as airways.

Such a large organisation as London Transport, with it's 278 Tube Stations, 64 Bus Garages, 21 Railway Depots, 3 major multi-acre Works and their associated offices and Computer suites, has many conventional fan assisted ventilation systems, all of which, apart from the problems of removal of diesel fumes from unheated bus parking areas, require no further elaboration.

With the many sophisticated fan designs having integral electronic control on the market at present,DESIGN and MAINTENANCE should always be inseparable when considering any job: It is a pointless exercise to give one's client a fan design, correct in application, but impossible to maintain, due to lack of access or skilled maintenance personnel.

Acknowledgement

The authors wish to thank the London Transport Executive for permission to publish this paper.

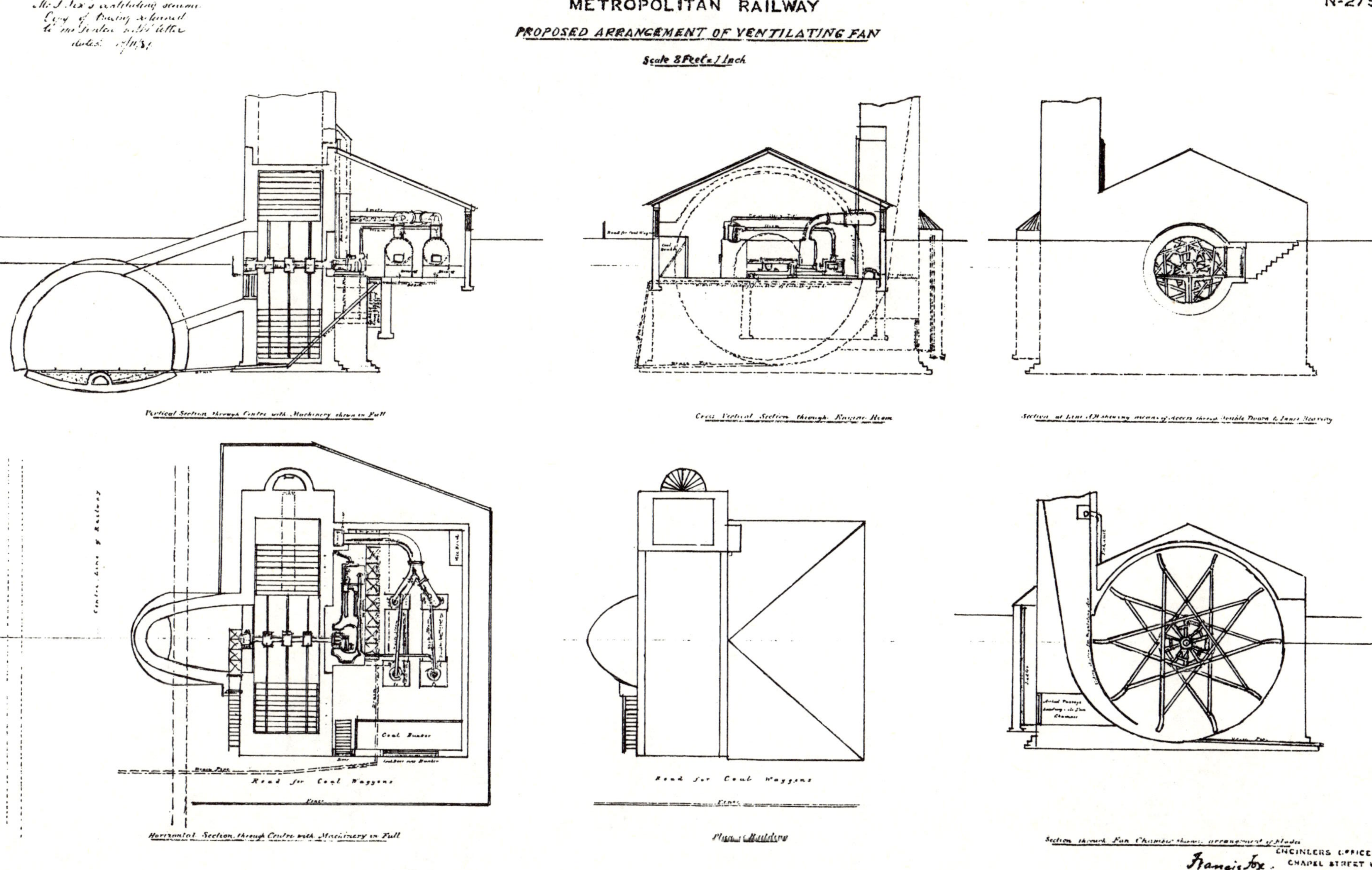

No 2754
METROPOLITAN RAILWAY
PROPOSED ARRANGEMENT OF VENTILATING FAN
Scale 8 Feet = 1 Inch
Vertical Section through Centre with Machinery shown in Full
Cross Vertical Section through Engine Room
Horizontal Section through Centre with Machinery in Full
Plan of Building
Coal Bunker
Road for Coal Waggons
Francis Fox.
October 28th 1879
ENGINEERS OFFICE
CHAPEL STREET W
EDGWARE RD

International Conference on

Fan Design & Applications

Guildford, England: September 7-9, 1982

EXPERIMENTAL STUDY ON A PHYSICAL MODEL OF THE FAN PERFORMANCE FOR LONGITUDINAL VENTILATION SYSTEM OF ROAD TUNNELS

D. M. Fontana and A. De Lieto Vollaro

University of Rome, Italy

Summary

The procedures to calculate the thrust or the head of fans in a longitudinal ventilation system of road tunnels, are based on several expressions often not in complete accordance.

In addition the experimental measurements indicate notable differences from the best theoretical predictions (from 30 to 40%).

Main purpose of this paper is the determination with measurements on a physical model, of the fans effective performance and the reasons for the differences from the values obtained by using the theoretical expressions.

Organised and sponsored by
BHRA Fluid Engineering, Cranfield, Bedford MK43 0AJ, England.

0263 - 421X/82/01 00 - 0001 $5.00
The entire volume can be purchased from
BHRA Fluid Engineering for $82.00

NOMENCLATURE

Ag = tunnel area

Av = inductor tube area

φ = area ratio $\frac{Av}{Ag}$

Uv = fan air velocity

Ug = tunnel air velocity

ψ = velocity ratio $\frac{Ug}{Uv}$

η = efficiency

ρ = air density

Fv = thrust

P = pressure

n = revolutions per minute

Δ = difference

1. INTRODUCTION

The increasing use of inductive ventilating systems in vehicle tunnels has increased the interest in furthering our knowledge of the inductive mechanism, especially in order to obtain more reliable data to predict performance, especially that of the pressure head and the ventilation efficiency.

Various one dimensional models have been proposed to provide an acceptable solution to this problem.

E. Rohne [1] propers a relationship founded on the momentum equation which puts in evedence the concept of "thrust" of the fan, so that:

$$Fv = \Delta Pg \cdot Ag \qquad (1)$$

The meaning of these symbols can be found in the nomenclature with ΔPg being the pressure generated by the inductor fan in the tunnel. The thrust, Fv, in turn is found by means of the relationship:

$$Fv = \rho AvUv(Uv - Ug) \qquad (2)$$

deduced from the simplified theory of the ducted propeller so obtaining in the end

$$\Delta Pg = \rho\varphi U^2 v(1-\psi) = \rho\varphi U^2 g \frac{(1-\psi)}{\psi^2} \qquad (2a)$$

U. Meidinger [2], K. Wieghardt [3], F. Reale [4], proposed, on the other hand, relations deduced from the equations of energy, momentum and continuity in various hypothesis. In particular, referring to fig. 1, U. Meidinger assumes a "two-current" model with a section uniformity increasing and an internal one converging. Further hypothesizing the possibility of motion whenever reversible between sections 1 and 2, he obtains:

$$\Delta Pg = \rho \frac{U^2 v}{2}(2\varphi - 2\varphi\psi) + \frac{\varphi^2 - 2\varphi^3 + 2\varphi^3\psi - \psi^2\varphi^2}{(1 - \varphi)^2} \qquad (3)$$

F. Reale in an entirely analogous model hypothesizes the irreversability of the external downflow between sections 1 and 3 while ignoring the increase of pressure head thereby making $P_3 = P_1$, and so obtaining:

$$\Delta Pg = \rho U^2 v \frac{\varphi}{1-\varphi}(1-\psi)^2 = \rho U^2 g \frac{\varphi}{1-\varphi}\left(\frac{1-\psi}{\psi}\right)^2 \qquad (4)$$

One notes that for $\varphi/4 \rightarrow 1$ the formulae 2a, 3, 4, become identical.

In conditions differing from these, the 2a, 3, 4, furnish very different values for ΔPg.

The experimental data obtained both on scale models and on full-scale measurement models, result widely scattered, as such they do not allow a full validation of any of the above cited relationships.

Moreover, the experimental data often give values of ΔPg that are inferior to those predicted by all the above mentioned equations and this is unacceptable for the design of a ventilation system.

The "Comité des Tunnels Routières" [5] recomends the use of the formula (2a) with the introduction of a corrective factor η

$$\Delta Pg = \eta\rho\varphi U^2 v(1-\psi) \qquad (5)$$

The experimental variation of factor η is between 0.5 and 0.9, and it was further noted that it was influenced, among other things, by:

- the position of the inductor fan inside the tunnel;
- configuration on the inductor fan and especially the presence of deflector grills of the outflow jet;

- the roughness of the walls;
- the position of the inductor along the length and section of the tunnel

Experimental data is always scattered and is sometimes in fundamental disagreement with the above factors. This is partly caused by the intrinsic difficulty of the phenomenon and lack of understanding of certain fluid-dynamic aspects of the phenomenon.

The object of this research, of which some first results are here presented, is to contribute to our knowledge and limit the degree of indeterminability in evaluating inductor fans performance.

2. EXPERIMENTAL APPARATUS

The experimental apparatus (see fig. 2) was constructed with the intent of obtaining information on the performance of an inductor fan inserted in a tunnel.

The latter is simulated of a tube of square cross-section whose sides are 0.63 m long, made of perspex, so allowing the visualization of the fluid-dynamic fields.

The tube is constructed in 1.5 m long blocks mounted on castors. The fan is situated about 2m from the opening section and 6m from the outlet. The tunnel tube is made of a circular tube of heavy PVC, with an internal diameter of 15,5 cm.

Inside there is placed an electric motor that can drive a propeller or an axial fan. The axis of the inductor is situated in the vertical plane of the conduit and its position in this plane can be varied throughout.

The section under examination is arranged to allow the measurement of static pressure and of velocity on the internal points of the transvers sections of the outer conduct for various position along its axis as well as its ingress and egress sections. For measurements of the static pressure and velocity of the air superior to 3-4m/s a miniature Pitot tube was used; for the velocities inferior to 3m/s an Alnor hotwire anemometer was used. The pressure was measured using a differential micromanometer equipped with a micrometric adjustment capable of determining the pressure with an accuracy of ± 0.05 Kg/m^2.

It was considered opportune to obtain direct determination, of the thrust, Fv, effectively furnished by the inductor. To such an end the inductor was mounted on a dinamometric balance (fig. 3) by means of which Fv could be measured with the precision of ± 10 gr.

A stroboscope, thanks to the transparency of the walls, permitted the determination of the rotational speed of the propeller.

In the present study a three blade propeller was used, being 10 cm think and 15.5 cm in diameter. A mercury thermometer placed at the opening of the conduct as well as nearby barometer permitted the measurements of temperature and pressure necessary for the determination of the air density.

3. MEASUREMENTS

The purpose of the first series of measurements was to study the possible variation of the phenomenon from the hypothesis of "unidimensional fluid motion" and the effect of velocity fluctuations on the value of thrust Fv and of pressure ΔPg; in pursuit of this, measurements were taken at sections 3, 4 (see fig. 1) to determination the profiles of the average values in time of velocity and static pressure by the use of an hot wire anemometer connected to an oscilloscope, the fluctuations of velocity in various points of the above cited sections were also recorded.

The thrust, Fv, of the fan and the rotational velocity of the propeller were also measured.

The measurements were taken under various conditions of flow obtained by varying the revolutions of the propeller in the field between 6000 and 14.000 r.p.m.

4. EXPERIMENTAL RESULTS

In fig. 4-5 are reported an example at velocity profiles measured in the sections 3,4 respectively in horizontal and vertical plans for revolutions at the propeller near the mean tested values.

In fig. 6 are shown the curves of static pressure against the same modalities.

In fig. 7 reported the values of static pressure obtained along the axis of the tunnel at the transverse sections for several values of n.

On the table are shown all experimental values of velocity and static pressure in sections 3 and 4.

r.p.m.	U_g m/sec	U_v m/sec	P_{1g} Kg/m^2	P_{2v} Kg/m^2	P_{2g} Kg/m^2	F_v gr
6000	1.47	5.58	0.13	0.624	0.25	64
7200	1.66	6.67	0.17	0.87	0.27	86
8100	1.85	7.73	0.21	1.07	0.30	112
9000	1.92	9.27	0.23	1.25	0.32	133
9700	1.96	10.43	0.24	1.60	0.35	169
10200	1.99	11.20	0.25	1.75	0.37	178
10500	2.01	11.54	0.25	1.80	0.38	188
11300	2.09	12.36	0.27	1.93	0.40	225
12500	2.47	13.51	0.38	1.99	0.42	275
13200	2.76	14.70	0.48	2.02	0.51	317
14000	2.82	15.20	0.50	2.05	0.53	345

Table

The values are averaged on all the transverse area of the duct (Ag), in the sections 1-4 and they are averaged respectively on the transverse area section of the inductor tube (Av) and in the remaining section of the duct (Ag - Av) in the sections 2-3. Also the table report the values of thrust, Fv, measured.

In fig. 8 are shown an typical oscillogram of the temporal fluctuation of internal velocity expressed as average values, for n = 10500 r.p.m. in the section 4.

5. INTERPRETATION OF RESULTS

In fig. 9 the thrust (Fv) is plotted against speed (U) and compared with Ug and Uv, the mean values of $\bar{U}g$ and $\bar{U}v$.

As one can see the experimental values for thrust are always considerably lower than those furnished by the theoretical expression, with (4) the best approximation.

From the formula (4) another new expression was derived where in the evaluation of kinetic energy flux and of momentum equations, the variation of velocity in space and time are taken into account. That is to say at a point P at time t on the section S, we have:

$$U(P,t) = \bar{U}(P) + U'(P,t) \qquad (6)$$

where $\bar{U}(P)$ indicates the mean value in time: thus obtaining

$$\bar{U}(P) = \bar{\bar{U}}_S + \bar{U}^*(P)$$

where $\bar{U}_s$ is mean value in time and in the section.

$$\int_t U'(P,t)dt = 0; \qquad \int_A \bar{U}^*(P)dA = 0 \tag{7}$$

The equations of momentum and of continuity, keeping (7), allow to calculate Fv with the expression:

$$Fv = \rho\Big[\bar{\bar{U}}^2_{sv}A_v + \bar{\bar{U}}'^2_s(Ag-Av) - \bar{\bar{U}}^2_{sg}Ag + \int_A \bar{U}^{*2}_v(P)dA + \int_A \bar{U}^{*'2}(P)dA - \int_A \bar{U}^{*2}_g(P)dA +$$

$$+ \frac{1}{t}\int_A\int_t U''^2(P,t)dAdt - \frac{1}{t}\int_A\int_t U'^2_g(P,t)dAdt + \frac{1}{t}\int_A\int_t U'^2_v(P,t)dAdt\Big] \tag{8}$$

then giving

$$\alpha_v = \int_A \bar{U}^{*2}(P)dA \; ; \quad \alpha_g = \int_A \bar{U}^{*2}_g(P)dA \; ; \quad \alpha' = \int_A \bar{U}^{*'}(P)dA \; ;$$

$$\beta_v = \frac{1}{t}\int_A\int_t U'^2_v(P,t)dAdt \; ; \quad \beta_g = \frac{1}{t}\int_A\int_t U_g(P,t)dAdt$$

$$\beta' = \frac{1}{t}\int_A\int_t U''^2(P,t)dAdt \; ;$$

the expression become:

$$F_v = \rho\Big[\bar{\bar{U}}^2_{sv}A_v + \bar{\bar{U}}'^2_s(Ag - Av) - \bar{\bar{U}}^2_{sg}A_g + \alpha_v + \alpha' - \alpha_g + \beta_v + \beta' - \beta_g\Big] \tag{9}$$

The value of the coefficient α was derived from the profile of measured velocities; for the coefficient β it can be deduced only to an order of magnitude, from the assumption of sinusoil fluctuations whose amplitude is equal to that measured. In fig. 9 the variation in thrust with speed predicted by expression (9) is given. One observes better correlation with the experimental data.

The pressure trend shown in the fig. 7 appears in disagreement, not only with the hypothesis of Meidinger ($P_3 > P_1$) but also with that of Reale ($P_3 = P_1$). It's clear, from fig. 6, how the hypothesis of isobaric pressure in section 3, for all the theoretical approaches cited, is far from being valid.

In effect the pressure is stationary in the (Ag-Av) section duct, while in the inductor tube section (Av), this pressure reduces from the rim of the tube towards the propeller axis.

This trend is generated by centrifugal effect produced by the tangential velocity.

In this hypothesis the pressure reduction (ΔPv) is connected with the tangential velocity component Ut by the expression

$$Pv = -\int_r^R \omega^2 r dr$$

where $\omega r = Ut$.

The experimental values of revolutions (n), of axial velocity (U_V) and the propeller pitch, (P) allow one to calculate a theoretical limit (infinite blade propeller) for the tangential velocity component and also the pressure differences for several values of the radial distance. For example at 11.000 r.p.m. the tangential velocity has been calculated as:

$$Ut = 10 \text{ m/s}$$

and keeping (10) with the experimental corrections in regard to the number of the blades proposed by Anastasi [8], one has been obtained:

$$\Delta Pv(r = 5\text{ cm}) = -\ 2.2 \text{ Kg/m}^2 \ ;$$

$$\Delta Pv(r = 8\text{ cm}) = -\ 1.97 \text{ Kg/m}^2 \ ;$$

in accordance with the experimental values measured:

$$\Delta Pv(r = 5\text{ cm}) = -\ 1.96 \text{ Kg/m}^2 \ ;$$

$$\Delta Pv(r = 8\text{ cm}) = -\ 1.82 \text{ Kg/m}^2 \ .$$

In the remaining transverse section 3 (Ag-Av) the pressure is constant, but the values are not in accordance with the Meidinger's (reversibility) and Real's ($P_1 = P_3$) hypothesis.

In effect the pressure values increase between the sections 2 and 3, but in the section 2 the mean value is smaller than in the section 1.

Using the corrections proposed, in the expression

$$F_v = \Delta P_{3,1} A_g + \rho \left[\bar{\bar{U}}^2_{sv} A_v + \bar{\bar{U}}'^2_s (Ag-Av) - \bar{\bar{U}}^2_{sg} A_g + \alpha_v + \alpha' - \alpha_g + \beta_v + \beta' - \beta_g \right] \ ; \qquad (10)$$

gives results which agree reasonably with the experimental data. (fig. 9) The other corrections as to the hypothesis ($P_1 = P_3$) generated by the real air flow into and out of the inductor tube inlet, gives agreement of the same order.

6. CONCLUSIONS

In the experimental situations examined the differences between theory and measurement are caused by inexactness of the hypotheses for section 3.

Experimental evidence would suggest the following:

- in the section at the end of the inductor tube, tangential velocity effects and therefore the fan characteristics are important. Deflector grills of the outflow jet are consequently of considerable influence.

- in the other sections the phenomenon appears essentially hydrodynamic and connected to the friction losses between the sections 1 and 3 in the "external air flow". In this case the geometric characteristics are important. Particularly the section configurations, the section ratio (Ag/Av), the distribution of the inductors along the length and section, the fluid dynamic parameters, are important.

The drag and the mixing etc. do not appear important for the phenomenon characteristics. After all this phenomenon advantageously can be studied in regard to the hydrodynamics aspects, systematically and separately.

The "Istituto di Fisica Tecnica dell'Università di Roma", is operating towards this purpose and we hope for new results within some time.

At the end of this work, the authors thank heatily Mr. Gianfranco Mattiacci for the cooperation in specially in regard to the physical model realization.

REFERENCES

[1] Rohne, E. : "Nongitudinal ventilation of vehicle tunnels and axial fans". Schweizerische Bauzeitung. N. 48 (November 26 th, 1964).

[2] Meidinger, U.: "Longitudinal ventilation of vehicle tunnels with axial fans". Schweizerische Bauzeitung. N. 28 (July 9th, 1964).

[3] Wieghardt, K.: "Belüftungsprobleme in U-Bahn und Autotunnel". Sschiffstechik, 9, (1962).

[4] Reale, F.: "Ventilazione ad induzione di gallerie stradali mediante ventilatori assiali". Ingegneri n. 51 (November-December, 1968).

[5] Reale, F.: "Indagini sperimentali su modelli di impianti di ventilazione ad induzione per gallerie autostradali". La Termotecnica, n. 9 (September, 1970).

[6] Reale, F.: "Fundamentals and applications of induction ventilation systems of vehicle tunnels". BHRA-Aerodynamics and Ventilation of Vehicle Tunnels. University of Kent, England, April 73.

[7] Parolini, G, Fontana, D.M.: "Model study of a new system of longitudinal ventilation for long motorway tunnels". BHRA-Aerodynamics and Ventilation of Vehicle Tunnels. University of Kent, England, April 1973.

[8] Anastasi, A.:"Macchine a fluido". Ed. Cremonese, 1956.

[9] Abstract from the symposium of the "Comité des Tunnels Routiers". Vienna 1979.

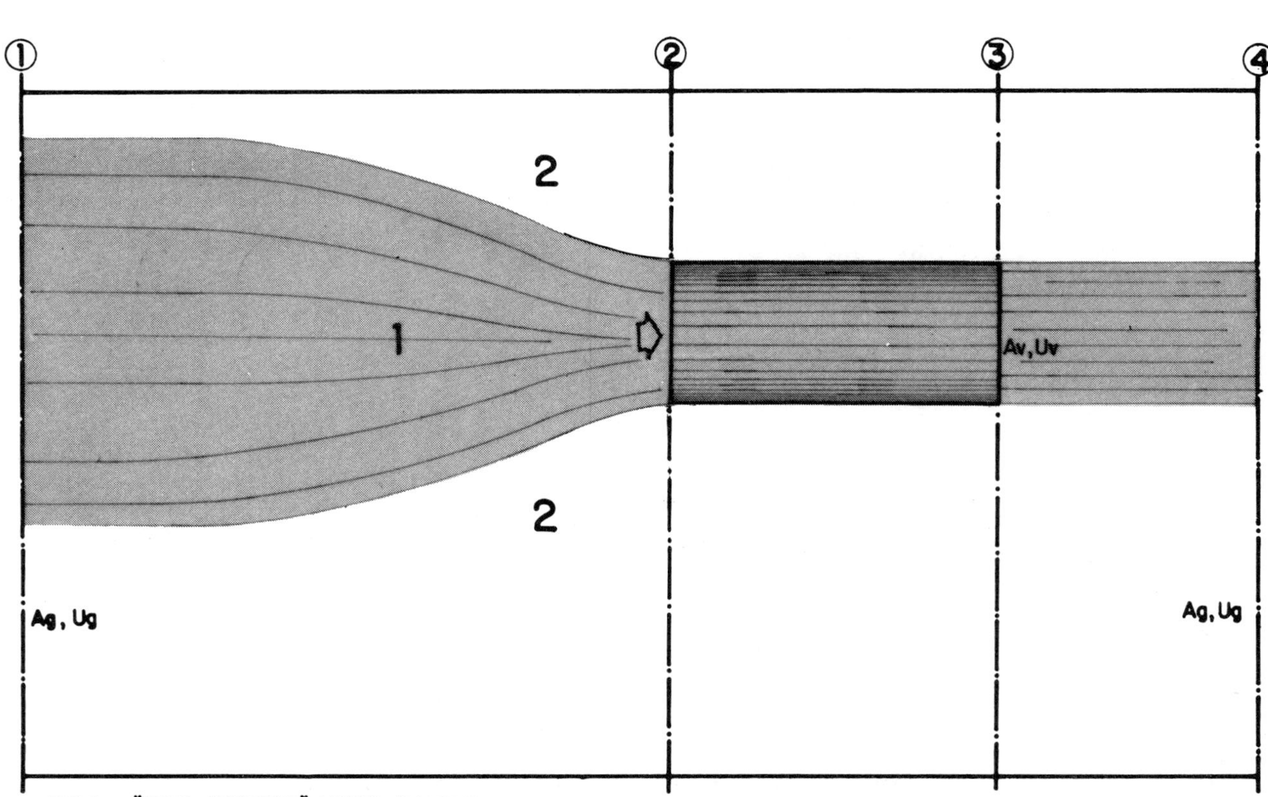

FIG.1 "TWO-CURRENT" MODEL LAYOUT

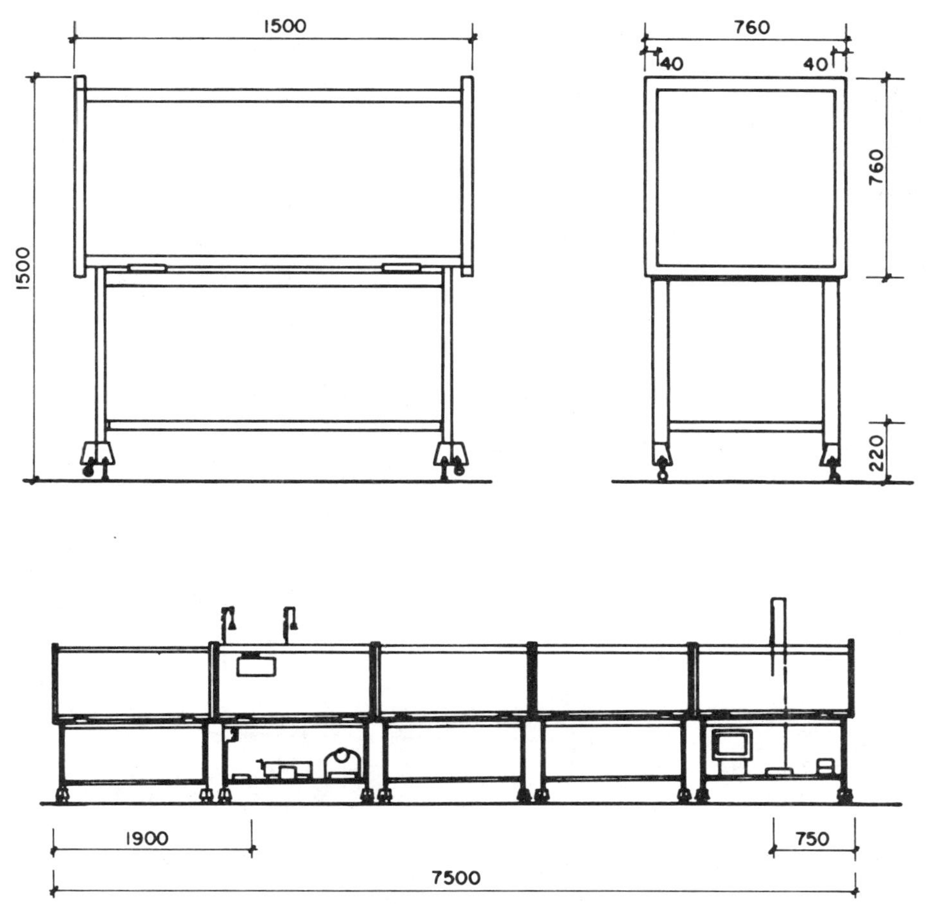

FIG. 2 PHYSICAL MODEL : LATERAL AND FRONT VIEW

FIG. 3 DYNAMOMETRIC BALANCE

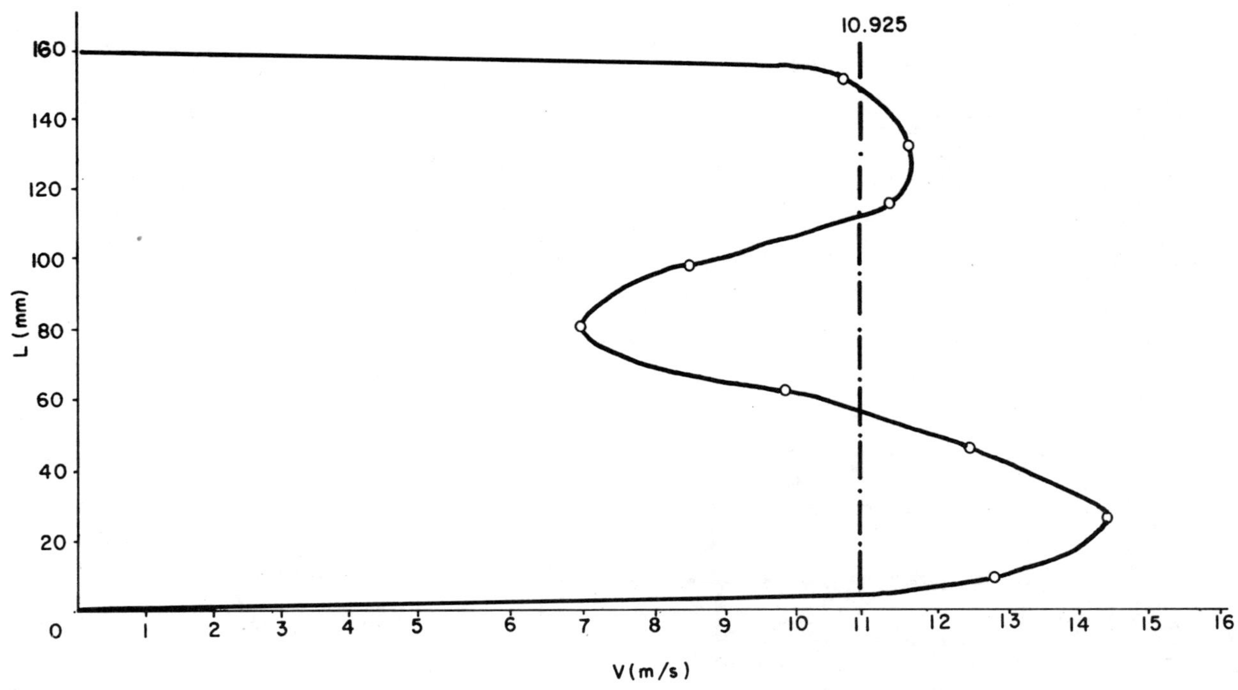

FIG. 4A TYPICAL AXIAL VELOCITY IN THE CROSS SECTION AT THE END OF THE PROPELLER DUCTED (10'500 r.p.m.)

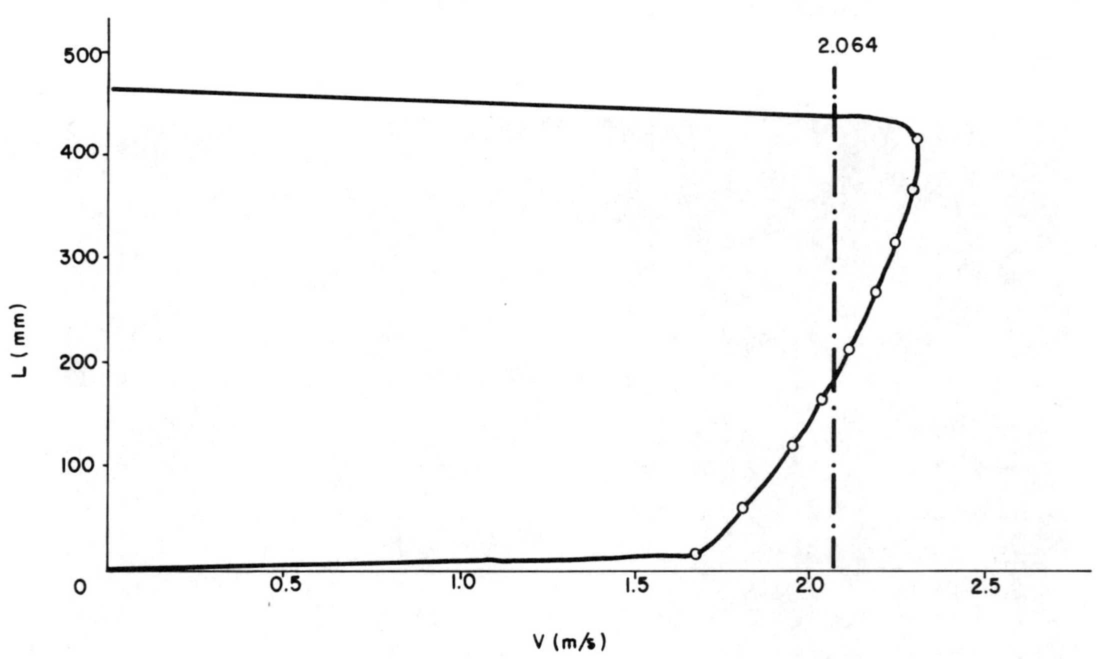

FIG. 4B TYPICAL AXIAL VELOCITY IN THE CROSS SECTION AT THE END OF THE PROPELLER DUCTED (10'500 r.p.m.)

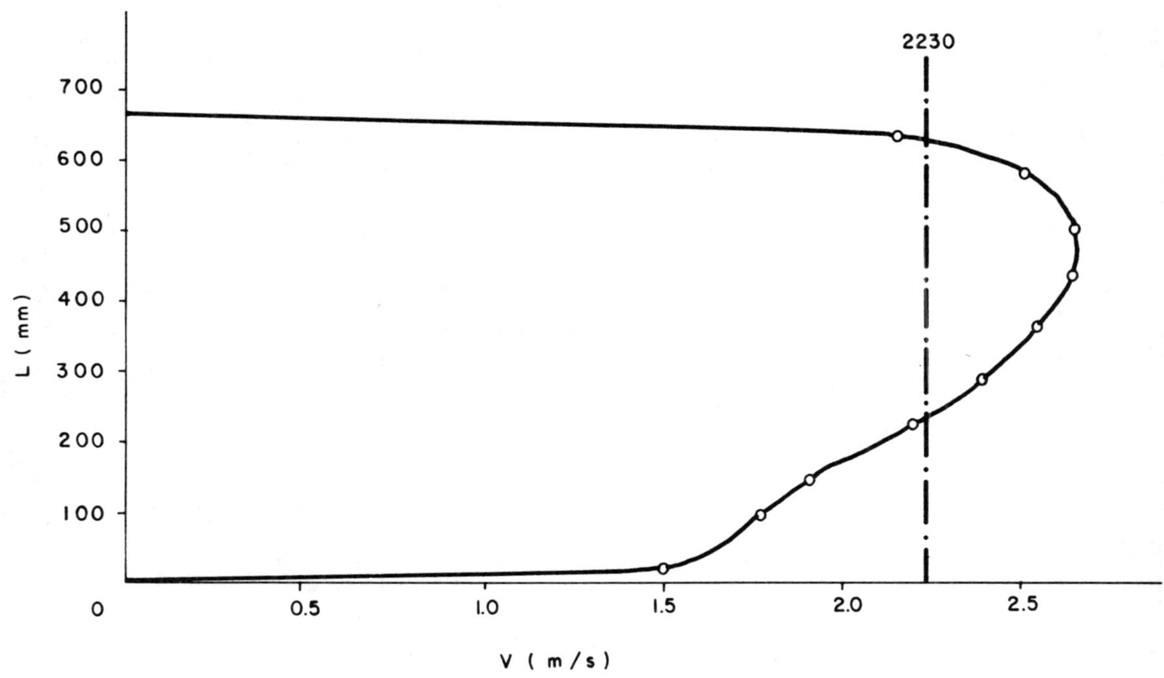

FIG.5 TYPICAL AXIAL VELOCITY IN THE CROSS SECTION AT THE END OF THE DUCT (10'500 r.p.m.)

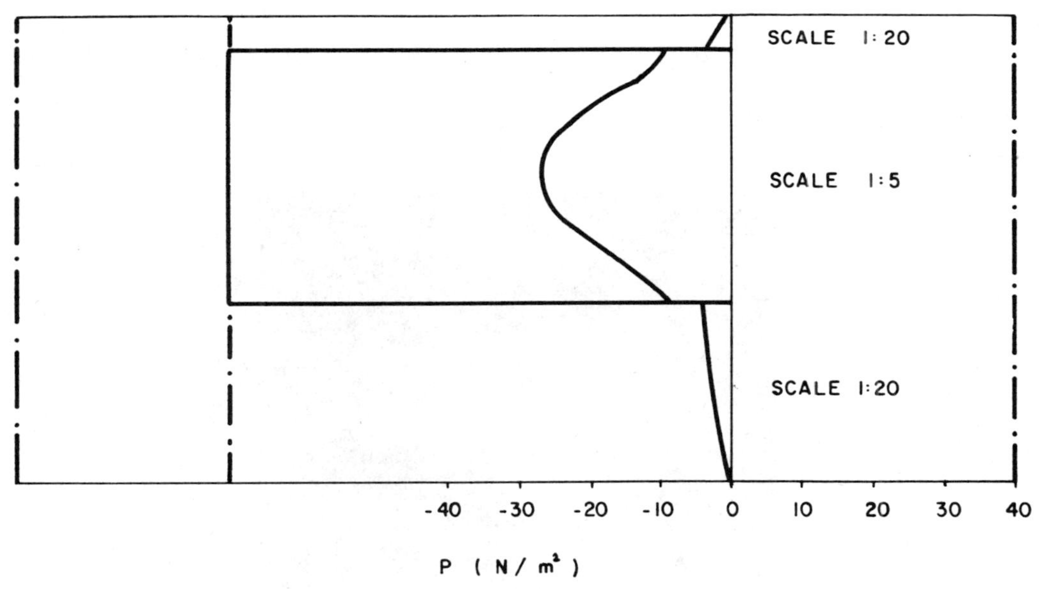

FIG. 6 TYPICAL STATIC PRESSURE IN THE CROSS SECTION AT THE END OF THE PROPELLER DUCTED (10'500r. p.m.)

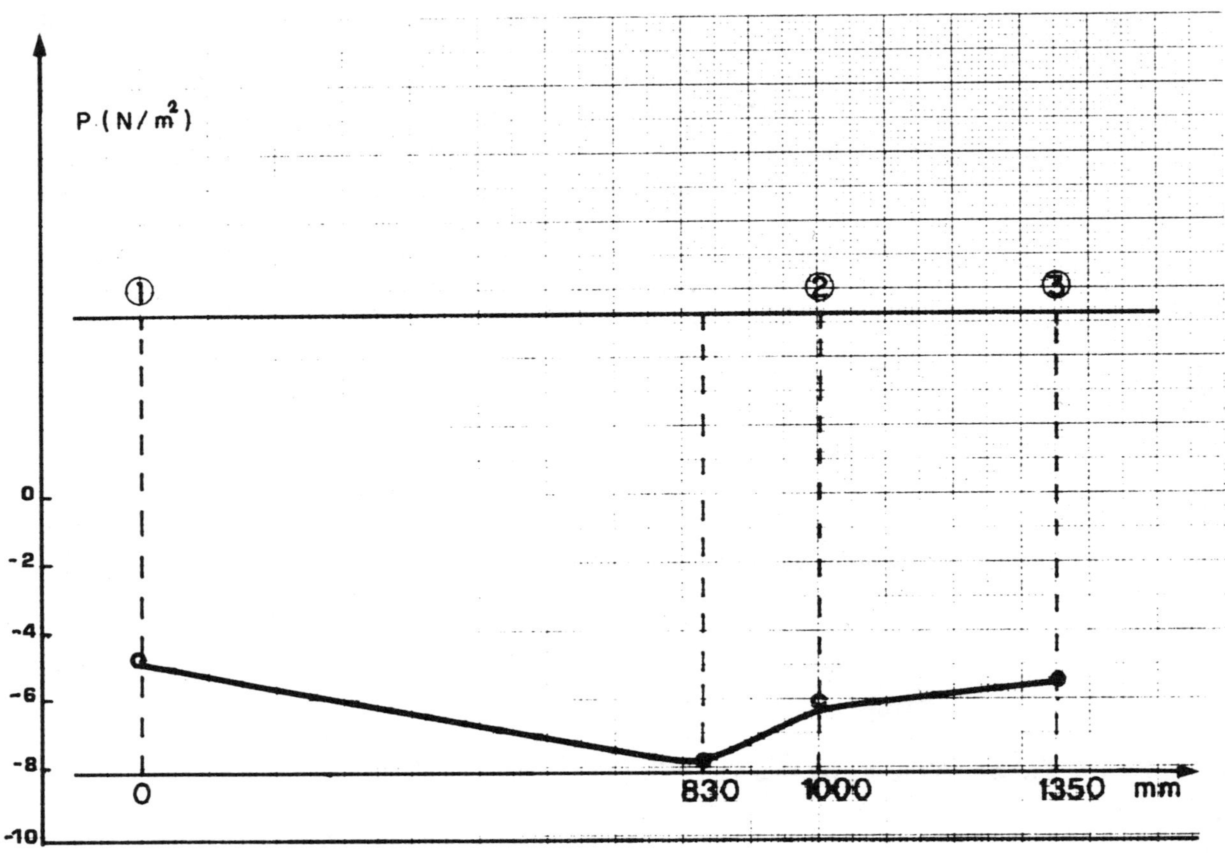

FIG. 7 STATIC PRESSURE ALONG THE AXIS OF THE TUNNEL (14'000 r.p.m.)

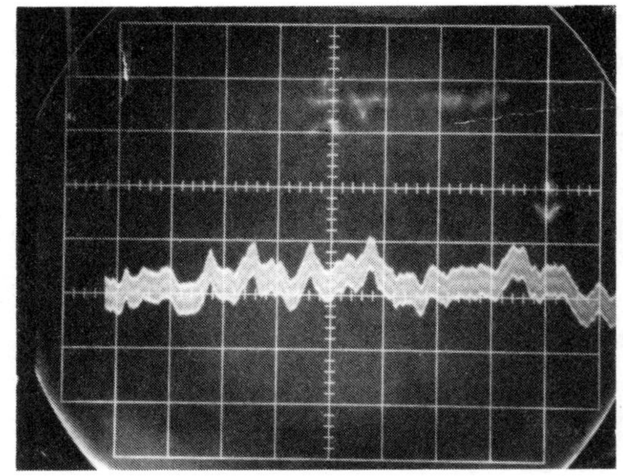

FIG.8 TYPICAL OSCILLOGRAM
THE X AXIS IS DEVIDED IN 0.2 SEC
" Y " " " " 0.2 mV

Fv 5.0
4.8
4.6
4.4
4.2
4.0
3.8
3.6
3.4
3.2
3.0
2.8
2.6
2.4
2.2
2.0
1.8
1.6
1.4
1.2
1.0
0.8
0.6
0.4
0.2
NEWTON
0
1
2
3
4
5
6
7
8
9
10
11
12
13
14
15
REVOLUTIONS x 1000 / MIN.
C.I.T.R. formula
Experimental data
Corrected formula

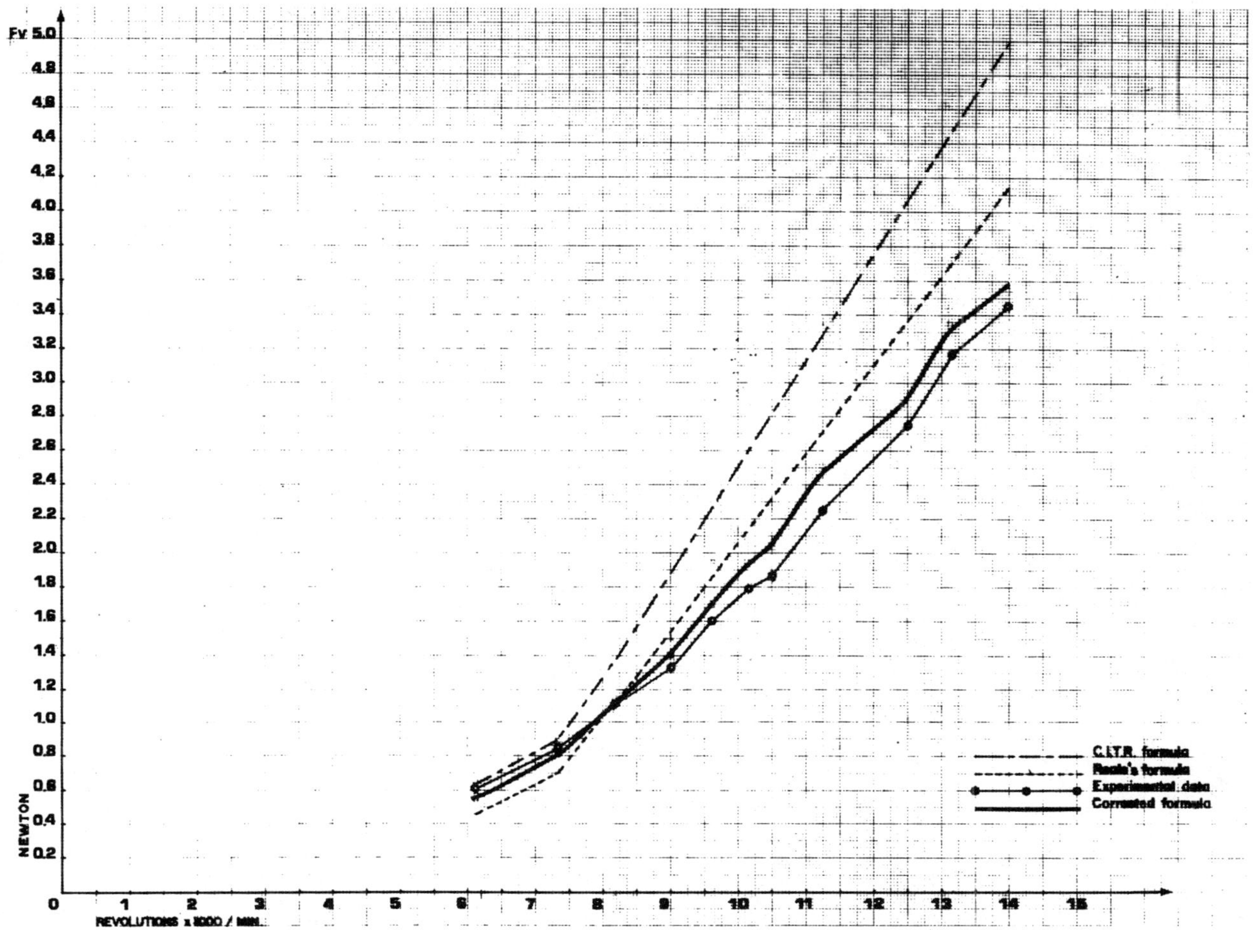

FIG. 9

International Conference on

Fan Design & Applications

Guildford, England: September 7-9, 1982

PAPER D1

FANS FOR DOMESTIC GAS APPLIANCES IN THE 1980's

A. Wharf

British Gas Corporation, U.K.

A.H. Middleton

Southampton University, U.K.

Summary

This paper is about the selection and development of fans for gas appliances; fans to supply air for combustion and fans for the circulation of warm air in the home.

It traces some of the fan assessments sponsored by the British Gas Corporation during the 1970's and describes the development of a quiet toroidal fan for use in domestic appliances. The development of a grease lubricated bearing of unconventional design is also described.

The paper gives some of the leading considerations in the design of a laminar flow fan where the preservation of laminar flow whilst work is being done on the air provides a basically quiet aerodynamic compression process.

Organised and sponsored by
BHRA Fluid Engineering, Cranfield, Bedford MK43 0AJ, England.

0263 - 421X/82/01 00 - 0001 $5.00
The entire volume can be purchased from
BHRA Fluid Engineering for $82.00

1. INTRODUCTION

It was in 1973 that the first oil crisis shocked the Western world and it is from this time that the conservation of energy has become a matter of growing importance for industry and also for the domestic consumer. In the domestic scene, new house building methods leading to the more economical use of fuel have been gathering momentum, and the 1980's and 1990's will see further rapid changes and even more highly insulated houses. These changes in building methods and also the need for a greater number of smaller dwellings in the UK, such as studio flats, starter homes, are having a significant effect upon the design of gas appliances.

Appliances will be of even higher thermal efficiency (Ref.1), some fully condensing the exhaust water vapour (efficiency 90-95%) and with their installation and flueing arrangements (Ref.2) matched to the needs of the new house construction methods. Higher efficiency appliances will also be needed to replace those boilers in central heating systems that are reaching the end of their lives. The use of fans in these new breeds of appliances will enable problems in appliance design and installation and flueing to be solved more readily.

2. FAN DESIGNS TO MATCH WITH MODERN APPLIANCES

It was well before the oil crisis of 1973 that this fan story starts and when a comprehensive survey of air moving devices for gas appliances was first undertaken. It was believed that there would be a growing need for powered flueing systems as the years progressed and that this would be combined with more positive ventilation systems. Additionally, it was thought that warm air systems in such universal use in the USA, with their ability to cater for complete air conditioning, would gradually take over more of the market from the wet central heating system.

Specifications for air moving devices for supplying combustion air and also for the circulation of warm air were developed.

Fans for use in the home must be very quiet. They must have long life, preferably equal to that of the appliance, and be of high reliability and require minimum maintenance. They must be small and match easily to the appliance, whether a gas fire, boiler or heater. They would need to be capable of "high" pressure rise and certainly much greater than the 25 - 75 Pa of then current equipment for supplying combustion air if they were to provide the appliance designer with the opportunity to optimise his heat exchanger designs, and also enable long small diameter flues to be used. Such flues would allow the appliances to be sited anywhere in the home and well away from outside walls. Another desirable property would be a steep pressure rise/flow characteristic so that the effects of wind pressure on the flue terminal in those cases where an unbalanced flue is used would be minimal and combustion performance therefore unaffected. The relatively flat characteristic of the normal bladed centrifugal fan puts it at a disadvantage in this respect.

A fan specification meeting the needs outlined above for supplying combustion air, sited at appliance inlet and with an appliance of a heat output of about 7kW is given in Table 1.

TABLE I

Fan Specification
Appliance Output 7kW
Fan to supply Air for Combustion

Air Pressure Rise
At least 500 Pa

Life
Matched to a 10 year appliance life or 50,000 running hours with at least 7,000 stop/starts (modulated control) or 500,000 stops/starts (on/off control).

Size
As small as possible but maximum dimension not to exceed 170mm.

Drop Test
To withstand 300mm drop on to concrete floor.

Air Flow Rate
14 m^3h^{-1} at 15°C

Shaft Attitude
To be capable of running satisfactorily with the shaft horizontal, vertical or in any other attitude.

Electrical Supply
240 volts (+12% - 15%): 50 Hz Acceptable Air flow variation ± 10%.

Reliability
The bearings, electrical and mechanical components to have reliabilities consistent with a fan reliability above 99% per year.

Air Inlet Temperature
-10°C to + 30°C

Cost
To be based on 50,000 units/yr and be competitively priced in relation to appliance and installation cost.

Resistance to Outdoor Dust
Design to be based on ASHRAE dust at .4 - .6mg/m^3. Any filters used must hold 3 microns and larger particles

Noise
Total noise from an appliance in a living room using fans not to exceed the following limits. The motor and fan must be free from audible tones.

Octave Band Centre Frequency Hz	125	250	500	1k	2k	4k	8k
Sound Power Level dB re 10^{-12}W	54.5	47	42	38	35.5	34	32.5

For kitchen application an additional 8dB is allowed.

With 5dB appliance casing attenuation the fan specification in a living room is:-

Octave Band Centre Frequency Hz	125	250	500	1k	2k	4k	8k
Sound Power Level dB re 10^{-12}W	59.5	52	47	43	40.5	39	37.5

Many different types of air moving device, including some of unconventional design, were considered against the requirements for the combustion fan. Perhaps the most unusual device to be evaluated in practical tests was a positive displacement machine based on the well understood diaphragm technology of the domestic gas meter. It was planned to drive it with a linear motor but this project foundered because of the large space needed for the pulsation damper. The fan types considered are listed in Table II, with comments upon their expected advantages and disadvantages. A reliability evaluation was also undertaken.

TABLE II

Advantages and Disadvantages

Air Mover Type	Advantages	Disadvantages
Axial Fan (3,000 rpm)	Low noise	Multi-stage required. Too long and costly. Possible tonal noise difficulties.
Centrifugal Fan (3,000 rpm)	Well understood technology.	Multi-stage required; too large if single stage.
Centrifugal Fan (6,000 rpm)	Few simple parts only. Small dimensions, Easy motor cooling.	Noisy. Expense and reliability of frequency doubling circuit. Less tolerant to out of balance.
Cross Flow Fan (3,000 rpm)	Simple construction. Few parts. Effectively single stage. Low air and casing noise.	Possible aerodynamic instability. Near limit of pressure rise. Side entry possibly makes motor cooling more difficult. Integration with motor and bearings not easy in space available.
Toroidal Fan (3,000 rpm)	Small. Single stage. Integrates well with cooled motor and bearings. Can easily provide pressure rise.	Noisy. Requires acoustic treatment.
Roots Blower (1,500 rpm)	Low casing noise. Good reliability. Vertical flow characteristic, no matching problems.	Too large. Many parts, including gears and extra bearings. Requires safety valve. Low frequency pulsations.
Roots Blower (3,000 rpm)	Motor matches well with blower.	As above, except reduced size. Greater air and gear noise.
Rotary Vane (1,500 rpm)	Low casing noise. No matching problems with vertical flow characteristic. Fewer bearings than Roots Blower.	High air noise. Potentially poorer life than other types. Motor does not integrate easily. Requires safety valve.
Rotary Vane (3,000 rpm)	As above.	As above.
Diaphragm Blower	Small. Integrates well with linear motor. Diaphragm technology well understood for gas metering.	Potential unreliability with diaphragm deterioration and wear. Needs pulsation damper. Requires safety valve. Assembly probably too large.

2.1 Toroidal and Cross Flow Fan Designs

The choice for the combustion fan was narrowed down to the Toroidal design(Fig.1) and the Cross Flow design (Fig.2), each capable of providing the pressure rise required at nominal 2 pole speed of 3,000 rpm, but each with their own individual problems. The toroidal fan is basically a siren and would need extensive acoustic treatment and perhaps aerodynamic development to fit it for use in the home, whereas a cross flow fan, whilst perhaps requiring some acoustic work, presented problems of instability connected with the characteristic vortex and needed aerodynamic development. The acoustic problem was eventually judged to be the more tractable and the toroidal fan (Ref.3) was chosen for further development.

2.2 Laminar Flow Fan Design

Fans for the circulation of warm air used in existing appliances are of the bladed centrifugal and cross flow designs and are usually run at slow speed to ensure adequate quietness. They must move larger quantities of air than the combustion fan. They must deal satisfactorily with house dust. The slower speed leading to physically large dimensions is no disadvantage and the fans match well with the dimensions of the sheet steel heat exchangers normally used.

However, a proposal for a laminar fan (Fig.3) was made by the Cranfield Institute of Technology (Ref.4) and some of the development work carried out in association with Southampton University and British Gas is described in this paper. The laminar fan is attractive because it is inherently quiet and free from tonal noise, provided that the laminar flow can be preserved during the compression process.

2.3 Bearing and Electric Motor Considerations

It was recognised that as the aerodynamic compression process is quietened then noise from the bearings and the electric motor becomes more obtrusive.

Because of anticipated deficiencies in existing types of bearings a new type of grease lubricated bearing was proposed, where the grease is pumped by self-generating action to give full film lubrication. This process is virtually noiseless. An outline of the development of this bearing is given in the paper.

Normally, when low noise is required, electric motor drives are isolated from the housings of fans by being flexibly supported. In the integrated fan and motor design, flexible mounting was not considered possible without unreasonable cost and complication, so particular effort was made to minimise the generation of magnetic noise at source.

The designs of the higher speed motors available in the early 1970's were not considered to be reliable or quiet enough to meet the specification. Since that time, a number of different designs have appeared and they will provide the fan designer with greater opportunities to match his product more readily to the different applications.

3. TOROIDAL FAN DESIGN AND DEVELOPMENT

Early in the 1970's, apart from the possibility of testing a few production examples of toroidal fans, all of which were noisy because of rotor-stator interaction, there was little information available to enable the characteristics of different sized fans to be predicted. A few papers had been published on side channel pumps which gave some initial assistance.

3.1 Flow and Pressure Rise Considerations

The flow of air in a toroidal fan is indicated in Fig.4. Air is drawn in through the inlet port in the stator which is an unbladed half toroid. On entering the bladed half toroidal rotor centrifugal force throws the air outward and by virtue of the toroidal shape induces a flow pattern in which the air spirals around the toroid moving from rotor to stator several times within one rotation of the rotor. The effect is of a multi-stage fan with only one physical stage, and is ideally suited to the relatively high pressure and low flow requirement of a fan to supply combustion air. Fig.5 shows the performance of the T1200, T500 and T120 fans.

The inlet and outlet ports are separated by a blockage piece or stripper in the stator. It is the interaction of the flow from the rotor on the blockage piece which produces the siren-like noise characteristics.

Development tests showed that the size, shape and angle of the ports was important. Angling of the inlet passage and the outlet passage in such a way that there was minimum change of flow direction as the air entered the spiral flow pattern from the ports and left it at the exit was found to be important (Ref.5). Generous port area is desirable to reduce flow velocities and well rounded edges on the ports also help to reduce flow generated noise. Port size is limited by the need to fit each into a quarter of the cross-sectional circumference of the toroid. In the direction around the toroid, excessively long ports reduce the active length of the toroid and cause loss of performance.

In the interests of minimising flow noise generated at the ports, a maximum gross flow velocity at exit of 20 m/s has been used. Local flow velocities around the port area can be greater than this and must be controlled by detail design.

Many other features affect the flow performance of the fan. A powerful influence is exerted by the blade tip to blockage piece clearance (analagous to the cutwater clearance in a centrifugal fan). Increasing this clearance not only reduces pressure rise at a given flow by increasing leakage between inlet and outlet ports across the face of the blockage piece, but it also has a great influence on noise generation. The final setting has to be a compromise. There are also other important leakage paths which affect performance. Significant leakage can occur at the inner and outer circumference of the toroid. In our designs, this has been controlled by means of relatively small clearance leakage paths, typically 0.25mm gaps with a path length of 7 - 10mm.

The effect on performance of blade number was investigated. This is intimately tied up with blade thickness. Our investigations showed that if thick blades (3 - 5mm) were used, performance was reduced because too much of the volume of the rotor became filled with metal. Too few blades, whether thick or thin, were also disadvantageous. It was found that about 24 blades about 1mm thick were suitable, but the blade number is not particularly critical around this number.

3.2 Fan Noise Control

Most toroidal and side channel pumps sold for commercial use are provided with integral inlet and outlet silencers, usually consisting of sections of duct lined with acoustic absorption material. For the gas appliance applications these were too bulky. Some means of silencing at source was needed. Most effort was put into reducing the blade passing frequency tone and its harmonics, as these would be primary contributors to the annoyance which might have been caused by the fan noise, even with a closed air system.

Taking the fan in isolation, the control of casing radiated noise was all-important, because the British Gas specification applied to that. However, consideration of gas appliances as complete systems showed that airborne noise in inlet and outlet pipes was also very important because that noise could be conveyed to other parts of the appliance which may have low effective sound reduction indices, such as pressed steel combustion chambers and heat exchangers, usually made of thin sheet steel.

A silencing "vane", as shown in Fig.6, was designed. It is an extreme example of a "gradual entry and exit device", giving three dimensional control of the rate of approach of the moving vane to the stationary blockage piece. It is thought that there are two main sources of noise in a conventional toroidal fan, without silencing vane. These are believed to be the sudden cut off of flow from a rotor pocket at the exit port position as its mouth is closed by passage over the blockage piece, followed by the sudden discharge of high pressure air contained in a rotor pocket as the blade leaves the blockage piece and enters the low pressure part of the toroid. The vane is designed to deal with both of these sources. In the exit port area the vane tapers in plan view and because of its "dihedral" angle the blade tip to stator gap, in elevation, closes relatively slowly. The closing of the rotor pocket mouth is thus made relatively gradual. The close clearance of the rotor blade to the blockage piece at the centre of the vane is maintained (at about 0.5 - 0.8mm) to maintain the seal between inlet and outlet ports, but as the pocket of air trapped between the blades and the vane is conveyed over the blockage piece, a controlled leakage around the tapered edges of the vane is allowed to lower the pressure in the blade pocket relatively gradually as the pocket becomes exposed to inlet pressure.

The above is a simplified version of the action of the vane. Small details in its construction and position can alter the noise output by 10 dB at blade passing frequency, so the critical details must be controlled in production.

For a given rotor blade tip clearance, the silencing vane produced a noise reduction of 20 - 25 dB at blade passing frequency, and allowed the design specification to be achieved, as shown in Fig.7.

4. ELECTRIC MOTOR DESIGN

It was recognised early in the design study that motor noise was a possible problem, particularly electromagnetic noise radiated as a result of vibration of the stator. The first toroidal fans built had performance targets of 33 m^3h^{-1} at 1000 Pa 14 m^3h^{-1} at 500 Pa both requiring motors with outputs exceeding 25W. Capacitor start and run motors were considered desirable. Special motors were designed, but making use of existing laminations where possible. A conventional technique for reducing motor noise is to reduce flux density, but this would have increased motor size. Noise control was therefore concentrated on using a stiff stator lamination pack with adequate back iron and carefully chosen stator slot and rotor bar numbers. The motor rotor used "buried" bars, set 0.1mm below the surface of the rotor and skewed by one pitch to minimise slot ripple frequency components.

A further influence on motor generated noise comes from the bearings. To realise the advantages of a low noise motor design, it is essential to ensure that the rotor rotates concentrically within the stator, on a bearing which is free from radial slackness. The special bearings developed for the fan have this feature.

5. SELF PUMPING GREASE BEARINGS

The bearing consists of a stationary ball pin retained in the fan housing (Fig.8.) Grooves are formed on the surface of the ball pin which cause the grease to be pumped inwards from the outer diameter or "equator" of the ball to the pole. The ball engages an ungrooved cup which is retained in the shaft. The grease is retained in the rotating reservoir surrounding the cup. When the shaft is rotated, centrifugal force tends to throw the grease outward in the reservoir, thus leaving a void in the centre. For that reason, provided the grease does not "run" when left standing, it is not necessary to have a rubbing seal between grease cap and ball pin stem. A radial clearance of 0.2mm is used. This can accommodate small misalignments.

The ball is not grooved at its centre, thus providing a restriction to flow. As a result, the pressure generated causes the cup to lift off the ball and grease flow occurs. There is a small hole in the centre of the cup through which used grease is discharged, to be returned to the main body of grease by centrifugal action. The shaft runs on a pair of bearings with a light axial pre-load (1.5 - 2kg) applied to prevent slackness at rest and to allow for differential thermal expansion of shaft and housing.

The spherical seating of the cup is made about 0.02mm larger in radius than the ball, which allows space for the grease film when the ball and cup are concentric. It is desirable for the cup and ball materials to form a good bearing pair to cater for the initial start condition, but a measurable thickness of pumped grease film is generated from 100 rpm upwards, and does not decay completely until after the bearing has stopped.

The success of the bearing has depended as much on the development of special greases which will withstand the high shear rate as on the other details of development.

Self pumping grease bearings inevitably consume power because of their pumping action. Fig.9 shows the power consumption and the lift generated at a speed of 2850 rpm, and various loads for a 12.5mm dia. bearing. Fig.10 gives the bearing stiffness.

So far, long running endurance tests on fans have been running for 37,000 hours and 60,000 on/off cycles, and are still going.

Advantages of the bearings are:-

(a) they allow stable operation in any shaft axis direction;

(b) by preventing any slackness in the bearing system, such as would be present in conventional sleeve bearings, they allow the motor to run as quietly as possible;

(c) because there are no rubbing seals, they cannot wear and allow grease to leak out;

(d) the bearings themselves produce no measurable noise and no discreet vibration components such as are possible with rolling element bearings.

A computer program has been written to predict the performance of this type of bearing.

6. SOME CONSTRUCTIONAL CONSIDERATIONS FOR TOROIDAL FANS

To make a toroidal fan at an acceptably low price, great efforts must be made to produce a design which can be manufactured easily, and combine this virtue with the other features sought by the specification.

The design of the main housings has a considerable influence on noise because they are the main load carrying paths.

The functional requirements of the structural housings(Fig.1)are to:-

- retain the bearings to give motor rotor-stator concentricity;
- prevent axial creep which would destroy the axial preload on the bearings or the alignment of the shaft;
- provide transverse stiffness to resist the cantilevered loads on the bearing ball pins resulting from unbalanced magnetic pull from the motor;

- provide stiffness to minimise radiation of motor noise caused by stator vibration;

- provide stiffness to contain the internal sound field in the fan;

- allow manufacture to sufficient accuracy without post forming machining to give the desired rotor clearances, interference or adhesive bonded fits;

- make the assembly air tight;

- withstand the temperatures experienced in a gas appliance;

- be cheap.

Whilst plastic moulding offers promise from the point of view of accuracy of manufacture, it is doubtful whether it can deal with the potential creep problem, and it may be difficult to provide enough stiffness to resist the internal sound field.

The early T1200 size fan was based on aluminium diecastings and proved very successful. However, it met immediate resistance from fan manufacturers because they were not used to working with diecastings of the size required.

The T500 fan was designed from the start to be made from pressings and deep drawings. The optimum solution may be a mixture of diecasting and deep drawing, and designs have been produced of this type. Later designs of a smaller version have used diecastings throughout, with almost no machining required.

It would be ideal if the fan could be assembled without any need to adjust clearances. It appears to be impractical to achieve this with realistic manufacturing tolerances. Some setting, using jigs, is required during assembly. The axial clearance of the fan is set by adjusting the position of the motor end ball pin, then the bearing preload is adjusted by setting the position of the fan end ball pin. This should ensure a very small scatter on aerodynamic performance and noise, provided that all the component parts are made to correct dimensions.

7. LAMINAR FAN DESIGN AND DEVELOPMENT

The concept of the laminar fan (Fig.3) is deceptively simple. It is capable of providing a quiet compression process and it was originally proposed for use in association with a high output heater and annular combustion chamber where flame stabilisation was achieved by the transition from laminar to turbulent flow. Air is propelled through the rotor by a combination of friction drag and centrifugal action. Laminar flow is maintained between the discs to minimise noise. The performance of the fan is independent of surface roughness of the discs.

The laminar fan rotor consists of a stack of thin discs, which could be made of any sufficiently stable sheet material. Practical fans have been made with plastic and metal discs. The discs are spaced about 1mm apart and are rotated within a conventional volute.

Two sizes of fan suitable for hot air circulation have been built, with duties of 500 m^3h^{-1} at 870 Pa and 335 m^3h^{-1} at 60 Pa. Rotor sizes were 280mm dia. x 80mm width and 210mm dia. x 65mm width. The aerodynamic performance of the 280mm dia. fan is shown on Fig.11, together with that of a smaller experimental fan. Fig.12 shows the in duct sound power radiated by these two fans. A comparison between the total acoustic power radiated from the 280mm dia. laminar fan and a commercial centrifugal fan of similar performance is given on Fig.13.

To realise the full potential of the quiet operation of the laminar fan, it is necessary to be very careful about selection of motor drive, because this can easily become the loudest noise source.

8. DISCUSSION

Field trials on 40 prototype T500 toroidal fans built using the grease bearings were carried out in heating appliances during the heating seasons 1976 - 1980, and the results fully supported the claims for the bearing, motor and fan. The fan was limited to an air inlet temperature of about 35°C because of grease limitations. A number of fans were subjected to temperatures nearing this value and ran quite satisfactorily in trials at Watson House during the summer periods 1977 - 1979. Since then, a grease has been developed which provides immunity against air inlet temperatures of about 45°C.

A smaller range of toroidal fans supplying air for combustion has been developed recently to match with appliances in the range 1 - 5 kW and the problems of siting fans alternatively at appliance outlet aspirating flue products have also been addressed. When operating at higher efficiency and in the fully condensing mode, appliances will have reduced outlet temperatures but probably not below 45°C - 50°C. Additionally, even though there would probably be drainage from the main heat exchanger, the fan must cope with at least some of the water droplets which will be of a mildly acidic nature. The higher temperature together with the wetness and slightly acidic condition combine to present a more onerous environment than for the fan sited at the appliance inlet. However, with suitable corrosion prevention treatment and drainage arrangements, these problems can be satisfactorily overcome. It is also apparent that the high pressure available from the toroidal fan can be used beneficially in purging and clearing the appliance/flue system of water droplets.

The laminar fan is not at present being used in the Gas Industry, although it is thought to have a place for the circulation of warm air in both large and small systems. However, a small number of fans with impellers 13cm long and incorporating 76 discs have been built recently for use in the aircraft industry, where they replaced propeller type fans. Several smaller fans have been made with performance suitable for cooling and air handling operations in, for instance, office machinery.

9 CONCLUSIONS

Fans providing air for combustion need to be more powerful than those used previously if they are to match satisfactorily with the smaller domestic appliances and long flue systems which will be used in the new highly insulated houses of the 1980's.

Toroidal fans have adequate pressure rise for the provision of combustion air; they are small and the special silencing arrangements needed so that they can be used in the home have been described. The work has also covered aerodynamic/acoustic development, together with attention to the bearings and the electric motor.

It has been shown that the laminar fan is inherently quiet in operation. It is well suited for the circulation of warm air in the home. However, the development of production routes and automatic methods of assembly is required if such fans are to be manufactured economically in large numbers.

10. ACKNOWLEDGEMENTS

The authors wish to thank British Gas and the Wolfson Unit for Noise and Vibration Control at Southampton University for permission to publish this paper. They are indebted to Professor S.P. Hutton, Head of Mechanical Engineering, Mr.R.W. Woolley, Gas Bearing Advisory Service Manager, Dr.K. Binns, Reader in Electrical Engineering and Mr.A.K. Molyneaux, Research Student at Southampton University, for their contribution to the work described.

They are indebted to Professor I.E. Smith at Cranfield Institute of Technology for his help and encouragement during the development of his laminar fan invention.

11. REFERENCES

1. Patrick, E. A. K.: "Gas in the home - a glance into the future". Journal of the Institution of Gas Engineers, Vol 21. No. 7/8, July/August 1981. pp. 299 - 309,

2. Bennett, W. J.: "Flues - past, present and future". Journal of the Institution of Gas Engineers, Vol 19, October 1979. pp. 515 - 533.

3. Hutton, S.P., Binns, K.J., Middleton, A.H., Woolley, R.W.: "The development of an integrated design of motor driven fan with very low noise level". IEE Conference "Small Electrical Machines" (London,30th March 1976).

4. Smith, I. E.: "Improvements in or relating to fans". UK Patent 1461776. 1977.

5. Middleton, A. H.: "Improvements in peripheral blowers". UK Patent 1422194. 1976.

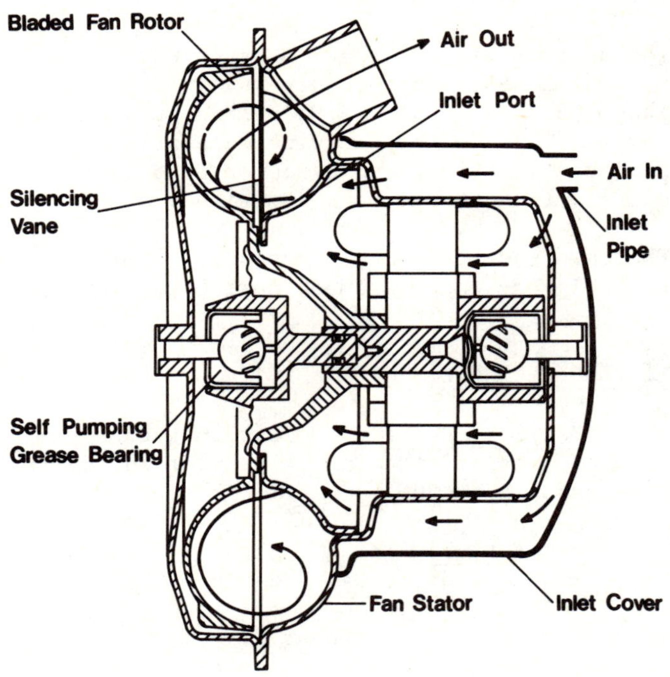

FIG.1 TOROIDAL FAN

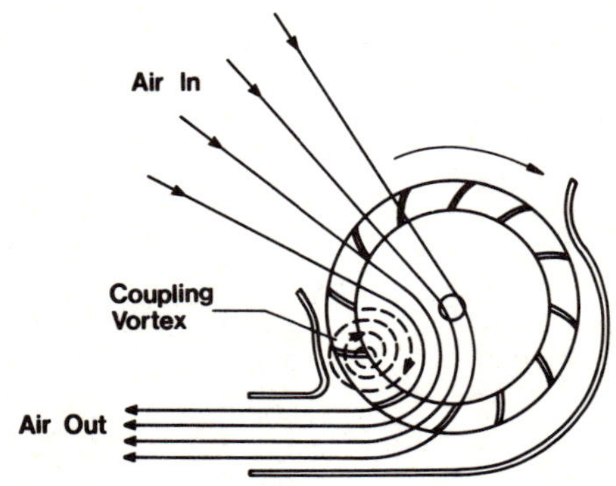

FIG.2 CROSSFLOW FAN

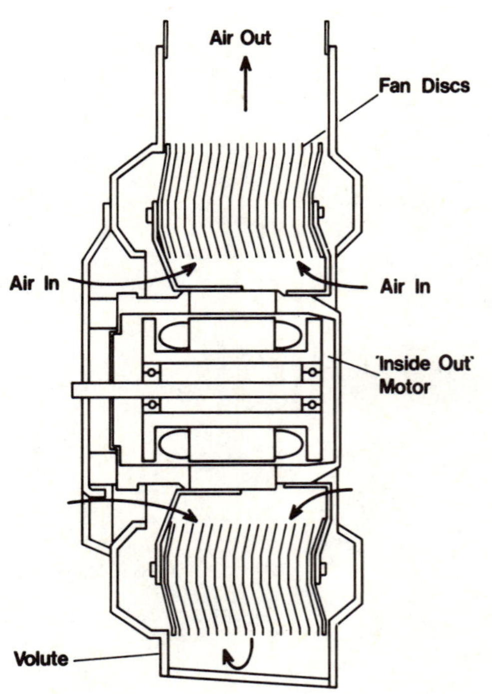

FIG.3 LAMINAR FLOW FAN

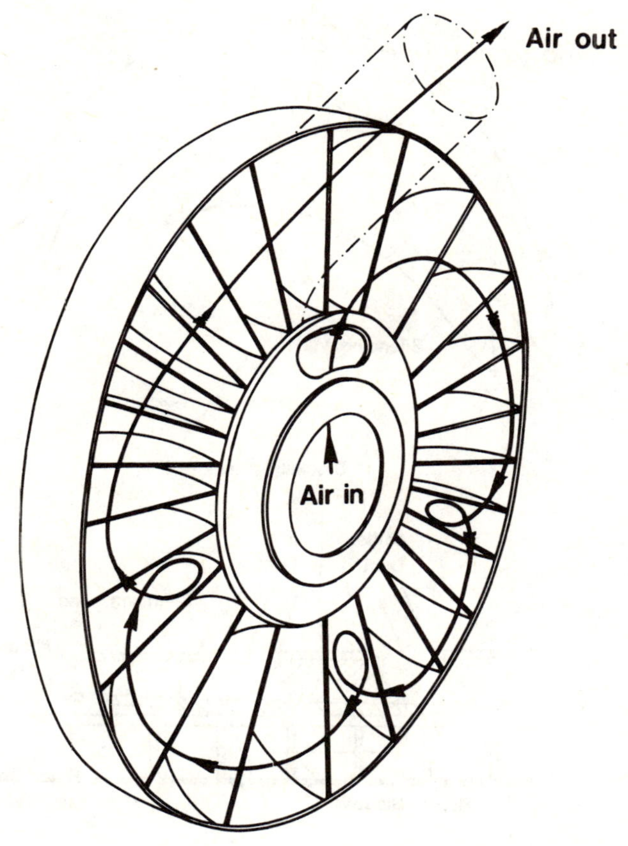

FIG. 4 TOROIDAL FAN – AIR FLOW PATTERN

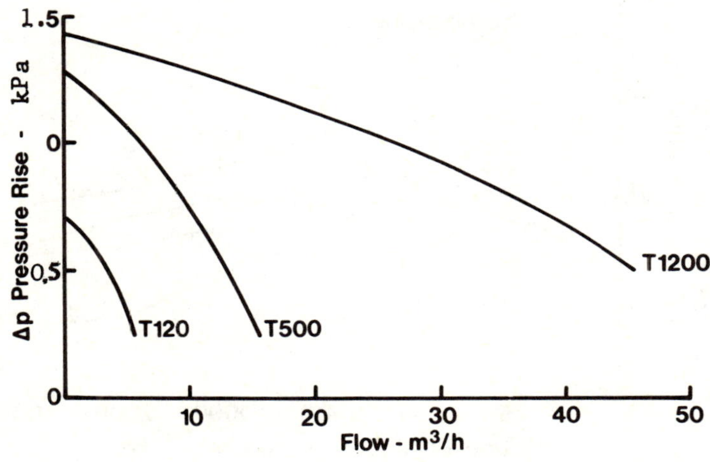

FIG.5 PERFORMANCE CHARACTERISTICS OF T120, T500 & T1200 FANS

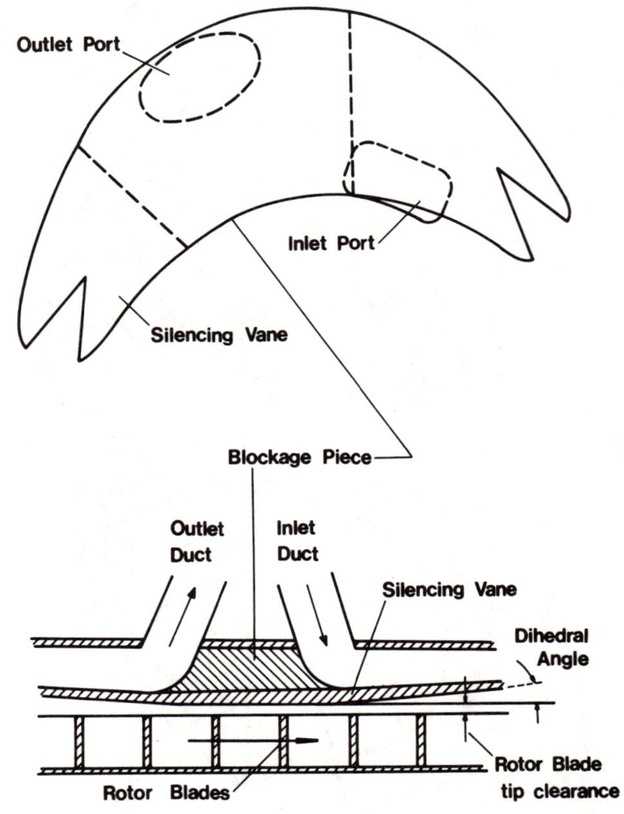

FIG. 6 SILENCING VANE ASSEMBLY

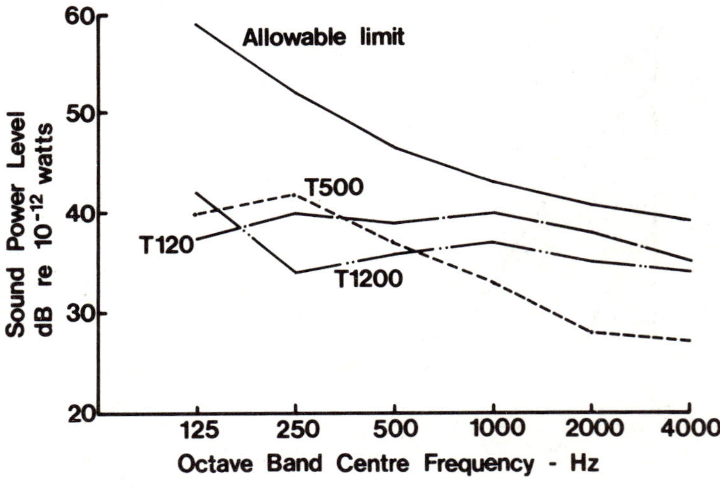

FIG.7 NOISE CHARACTERISTICS OF TOROIDAL FANS WITH SILENCING VANE

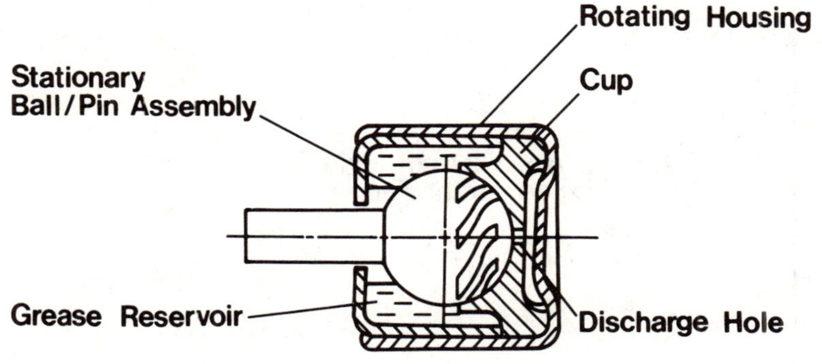

FIG.8 CROSS-SECTION OF
SELF-PUMPING GREASE BEARING

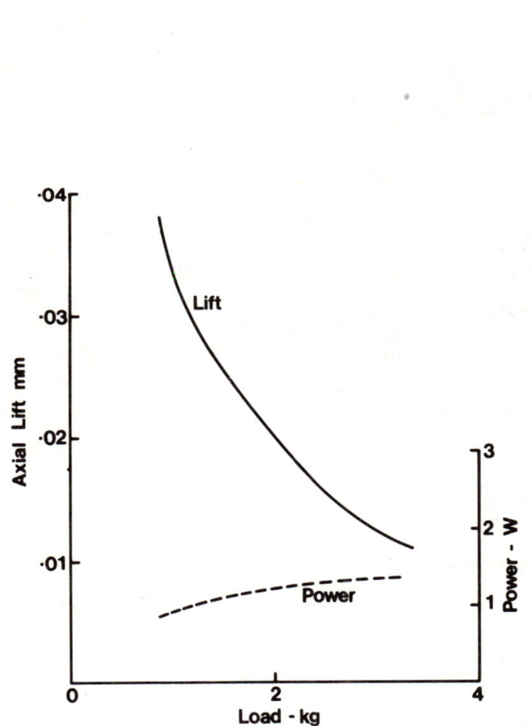

FIG.9 POWER/LIFT LOAD CHARACTERISTICS
(12·5mm Bearing)

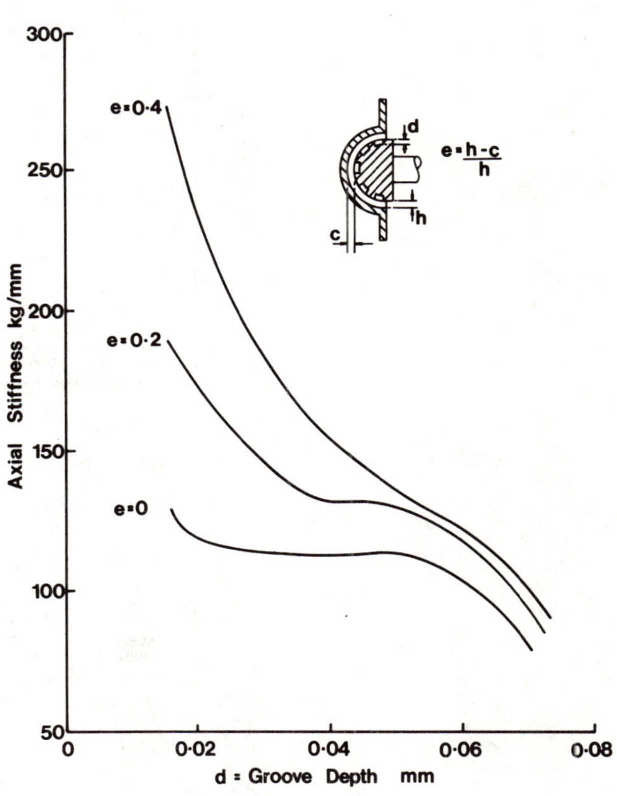

FIG.10 BEARING STIFFNESS CHARACTERISTICS
(12·5mm bearing)

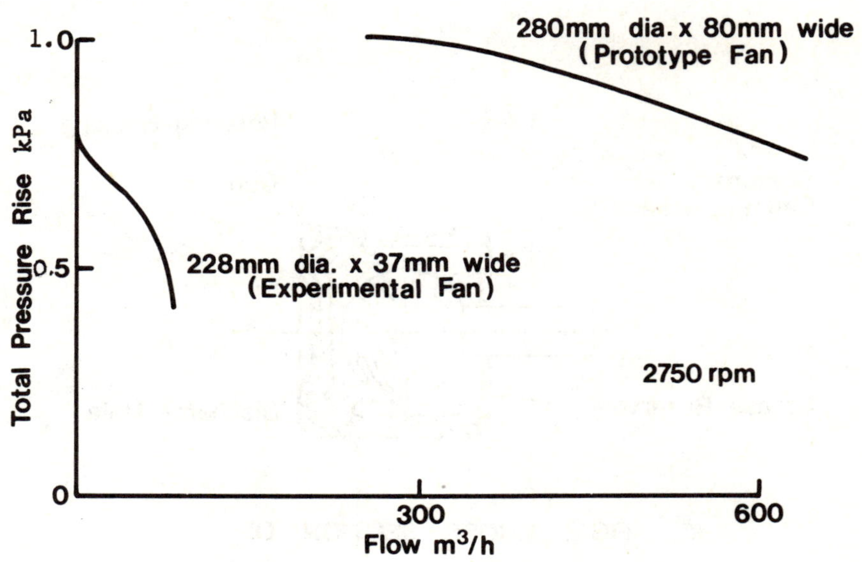

FIG.11 PERFORMANCE OF EXPERIMENTAL & PROTOTYPE LAMINAR FANS

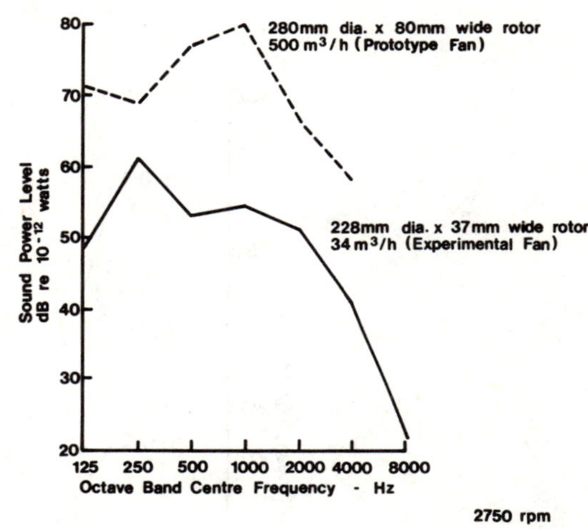

FIG.12 IN DUCT LAMINAR FAN SOUND POWER LEVELS

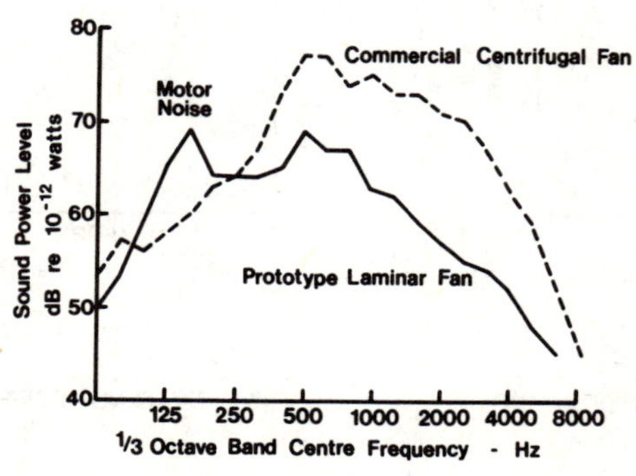

FIG.13 COMPARISON OF OVERALL SOUND POWER LEVEL OF COMMERCIAL CENTRIFUGAL FAN & PROTOTYPE LAMINAR FAN

International Conference on

Fan Design & Applications

Guildford, England: September 7-9, 1982

PAPER D2

PUBLIC TRANSPORT GARAGE (PARKING AREA) VENTILATION

I. J. Cockram and J. F. Bearman

London Transport Executive, U. K.

Summary

After numerous tests and experiments on the best way to overcome the problems associated with Bus Garage Ventilation, a simple solution was arrived at - being the most effective and economic.

Organised and sponsored by
BHRA Fluid Engineering, Cranfield, Bedford MK43 0AJ, England.

0263 - 421X/82/01 00 - 0001 $5.00
The entire volume can be purchased from
BHRA Fluid Engineering for $82.00

Introduction

London Transport has, at this time, 64 operational bus garages with bus allocations ranging from 30 to 170. Some are hole-in-the-wall establishments, while others are barn-like structures. The only thing most have in common is that they are cramped or awkward shapes, being developed from extensions of Horse Bus Stables, Re-constructions of Tram Sheds - either directly or via a Trolley-Bus Conversion, or built on land left over from something else. The parking diagram ..... Figure 1 gives an idea of what is expected, although admittedly it is theory rather than practice; anyway, it is a far cry from some other European transport installations which have a place-for-every-bus and every-bus-has-its-place.

Most of the ex.Tram Sheds had continuous louvres along the roof ridges, but surprisingly the Old Bus Garages had nothing more than an occasional louvred vent and no other ventilation. Limited natural vents were incorporated into the garages built in the late 1920s and early 30s, such as Upton Park, whilst the last new garage built before the war, Gillingham Street Victoria, had mechanical ventilation. This was restricted to the basement parking area and comprised openings in the wall 1m off the floor, with special emphasis on the 'climbing ramp'. The premises involved in the 1936-40 Tram to Trolley Bus conversion retained the louvred roof vents.

The post-war Tram Conversion commenced in 1949 and extended to 1952. The programme included the reconstruction of 7 Tram Depots, 6 Bus Garages, all on the existing sites and 5 Bus Garages on new sites; the last being probably the most spacious and best arranged of all the premises housing the bus fleet. The design of the Parking Area at all these garages was purposely designed to give the longest unsupported spans possible on the site and some approach 61m. By this time Garage Ventilation was considered right from the inception of the design of each garage. Where possible and to keep the Building Services (Mechanical) installations as simple and uncomplicated as practicable - still a fundamental rule; the Garage Configuration was designed so that the entrance arrangements, as well as complying with operational requirements, gave a generous natural airflow through the Parking Area. Where this was not possible,and this applies to nearly all the re-constructions because of site restrictions, ventilating shafts were incorporated. These comprised a structural shaft, with a diaphragm supporting axial flow fan(s), in a position for easy maintenance, picking up just above floor level and discharging above the roof and clear of surrounding premises. The fans were sized to give, in combination, an air movement of $0.28\ ms^{-1}$ through the garage i.e. vertical cross-sectional area x air movement = volume, location being in the back walls at the opposite end to the entrances - which functioned as the main air inlets, occasionally supplemented by wall or roof vents. The object as stated, was to promote a positive air movement in the single direction from the front to the back of the garage - so that as buses nearest the doors left the garage, the atmosphere was cleared progressively towards the rear and the crews of each new bus rank were able to work in 'nearly agreeable' conditions. There was one major exception to this basic formula which will be discussed later. It may be said that after nearly 30 years these arrangements, both natural and fan-assisted versions, worked effectively and there is no record of any dis-enchantment during the whole of this period.

The Trolley Bus Depots converted to diesel buses in the 1959-62 Programme were all generously provided with un-powered roof vents before the conversion and needed no mechanical provision. Bus Garages modernised during the 1960s were also laid out in such a way that it was not considered necessary to provide additional ventilation in the general parking area, although in 3 cases, special arrangements had to be made to ventilate the areas where buses were 'stacked', to wait their turn to be re-fuelled at the 'run-in'.

Environmental Conditions

The early 1970s brought an increasing awareness of the health & safety (small initials at that time) aspects of the work. Following a staff complaint to a Factory Inspector at Sutton Bus Garage, some correspondence ensued, referring tc Section 63 of the Factory Act 1961, which culminated in a joint visit (December 1971), during the early morning 'run-out' by the London Transport Executive's Medical Officer, Scientific Adviser, Rolling Stock Department, Building Services and the Factory Inspectorate, who also brought along a H M Chemical Inspector and a mobile laboratory. The result of this was that we were informed that the 'concentrations of the common toxic exhaust constituents, identified and determined were individually below the threshold limit values, when averaged over a 40 hour working week and the total exposure time for the permanent garage staff amounted to only 10-12 hours per week. Measurements were not made of Acrolein or Aldehydes, which were identified as being present in the fumes'.

The Factory Inspector's report stated that it was generally agreed that the 'fumes generated' were offensive, whilst hinting that although they could not strictly be proved to be actionable, it might be a gesture to do something and offered a suggestion as to what that 'something' might be ..... Figure 2 refers.

A similar complaint to the Factory Inspector arose shortly after, at Tottenham Bus Garage and again it was demonstrated that the fumes so generated could not be regarded as having a concentration of toxic constituents, beyond the threshold value.

A new Garage Modernisation Programme was being considered soon after this and in view of the increasing attention being given to Health & Safety (now with capitals), it was decided to install Mechanical Ventilation as a standard provision of Parking Areas, when premises were dealt with.

Defining The Problem

In order that the policy should be implemented in the most effective and economic way, one of the team was asked to approach the problem, which he defined as 'the Build-Up of Toxic Exhaust from Buses, particularly during run-up periods (early morning and afternoon)' from first principles.

It is not proposed to detail his work here, but only to 'summarise' it in Question & Answer form:-

Two questions were asked of the Engineer responsible for the design of power units (i) During the early morning visits, it had been noticed that Buses were stationary for periods of 15 minutes or more (before run-out time) and with their engines running. Was this necessary? or could it be shortened by more disciplined operation? The reply stated that it was necessary, both for engine warming and for replenishing the vehicle air reservoir. (ii) What were the engine speeds? and what was the volume of exhaust gases discharged? We were advised that the engine speeds were 2 minutes at 33.3 revs. s^{-1} and 13 minutes at 6.6 revs. s^{-1}, with an engine swept volume of 12 litres. The mean volume of exhaust gas for each engine was calculated to be 0.061 $m^3 s^{-1}$ for the 15 minute period.

$$\text{i.e.} \left[2 \left[12 \times \frac{33.3}{2} \right] + 13 \left[12 \times \frac{6.6}{2} \right] \right] \div 15 \times 10^{3}$$

These figures would obviously differ for other capacity engines.

The Scientific Adviser was asked to give the constituents of exhaust gas and the Senior Medical Officer (Environmental) to give the threshold values of these constituents. The combined answers being as set out in the following Table:-

Constituent	ppm. in Undiluted Exhaust	ppm. Dilution required	Dilution Volume m^3s^{-1}
CO	1000	100	0.613
CO_2	90000	5000	1.1
Aldehydes	20	5	0.245
Formaldehydes	11	5	0.134
Oxides of N_2	400	5	4.9
SO_2	200	10	1.23
Water Vapour	remainder	-	-

Oxides of Nitrogen and the Aldehydes are the Constituents which cause smoke and irritation.

Acrolein - forms part of the Aldehydes (it is stated to be the simplest of the unsaturated Aldehydes). It is a colourless liquid having a boiling point of 52.4°C, with a disagreeable 'tear exciting' odour. Tail pipe gas temperature 32-38°C.

From the above it can be seen that a replacement volume of 4.9 m^3s^{-1} i.e. $(0.061 \times \frac{400}{5})$ per engine, will prevent any of the toxic elements in the exhaust, reaching a threshold value. This figure was then applied to the 'run-out' requirements in each case.

To determine the dilution volume for a particular garage, the maximum hourly scheduled 'run-out' is obtained from the Bus Operating Department; this figure is then divided by 4 so as to give the maximum number of vehicles, which will be standing with their engines running at any given moment. The result multiplied by 4.9 will then give the optimum dilution rate and therefore the required fan capacity in m^3s^{-1}.

It will be noted that neither the physical size nor indeed the bus capacity of the garage is of any consequence, since there is just as likely to be a more intense 'run-out' from a small garage as from a large one - this is where the calculation differs from the usually accepted air change procedure.

Having now arrived at a dilution rate, how is it achieved and what are the ways and means?

In association with the calculations, some experiments were conducted to see what happens to 'exhaust' after it leaves the bus tail pipe ..... Figures 3 & 4 refer. Observations were made at a number of garages - it was considered essential that to get the 'feel' of the operation, the Design Engineer should attend before the start of the 'run-out' and observe the whole procedure.

From the experiments it was deduced that the chief force affecting the gases was 'buoyancy', which takes effect as the tail pipe velocity decays. By installing a fan in an inspection pit and forming an 'adjustable-slot' at garage floor level, attempts were made to determine if there was a 'practicable pick-up velocity', to capture the fumes in underfloor ducts - thus avoiding them passing through the 'breathing zone'. Inlet velocities of up to 10.16 ms^{-1} were tried, but in each case the bulk of the discharge escaped collection.

To assess the suitability of natural roof ventilators, two louvres were installed at Hounslow Bus Garage in northlight glazing. As in earlier experiments, 'smoke' was introduced into the exhaust gas in order to observe and photographically record the movements. Vehicles were positioned in locations which allowed for the 'throw' of the exhaust into the ventilated bays. Despite a number of tests with varying combinations of one and two engines running, engines idling or accelerated and with increasing volumes of 'smoke', at no time was 'smoke' issuing from the vents in sufficient quantity to enable photographs to be taken; visually it was just possible to discern a faint 'whisp' from the upper part of the ventilators. A limited amount of extraction took place, but unfortunately the lower part acted as a fresh air inlet and effectively cooled the rising exhaust gases, thereby depressing same.

The tests were carried out in late June 1973, in warm weather - 21/24°C, with a fresh breeze blowing across the face of the ventilators. It is unlikely that there would have been an improvement in cooler weather and with adverse wind conditions no extraction at all could have been expected. It was therefore concluded that natural ventilators alone as such, were unsuitable for diesel fume extraction - but used in conjunction with mechanical units and judiciously located, could be employed as fresh air inlets.

From the results of this work, it was decided that the basic design of Parking Area Ventilation should be by 'roof extraction' to boost the 'buoyancy' effect and the volume related to the dilution rate required by the bus 'run-out' intensity. There should be air inlets - carefully placed to avoid short circuiting and to get an air movement into 'dead' corners or areas.

Installation

The first installation to be carried out using these principles was at Peckham Bus Garage .....Figure 5 refers and this has been the prototype for subsequent installations.

Peckham was a new garage built on an existing site during the 1950/51 period, having an allocation of 140 vehicles; major inspections and overhauls being undertaken additionally. It has a single span roof of concrete barrel vault construction with 47 circular ventilators incorporated. The bus 'run-out' was a maximum of 48 per hour, or 12 with engines running for the 15 minute warm up. The fan extraction to give dilution was 61.5 $m^3 s^{-1}$. Twelve 0.914 m diameter propeller fans, each of 5.19 $m^3 s^{-1}$ in glass fibre reinforced polyester housings were selected and these ran at 9.5 revs.s^{-1} with a sound level of 66 dBA.

The fans are controlled from a panel located in the Night Foreman's Office and are started in 'groups' to avoid overloading the electrical supply and at predetermined times - being switched off automatically. Whilst there is provision for manual selection and for operation out of the 'set' time periods, this is only for emergency use by the operation of a keyed switch. The automatic starting and running sequence is we find important, as it frees the supervisory staff from an additional responsibility, it also covers the change in 'shift' which occurs at the morning 'run-out' period.

The installation went into service in August 1973 and the series of photographs ..... Figures 6 & 7 refer, show the effectiveness, during a morning 'run-out' - in early 1974. The results were generally satisfactory.

As the full effect of the Health & Safety At Work legislation has resulted in a demand for better working conditions all round, a 3-year programme to provide Parking Area Ventilation was authorised on a similar basis for all garages. The 63rd installation (totalling 922 fans) has recently been completed.

The number of fans, their positions, rating and sound levels is specified in each case. Where applicable, consideration is given to the location and effect of the smoke partitioning - which is required to give a maximum of 3700 m^3 of unsegregated roof space. Equally, if the fans can be incorporated in the smoke ventilation schemes now being demanded, then this is done if there is no prejudice to the fume exhaust function, otherwise smoke actuated vents are installed.

The number of stages in the fan starting sequence is selected so that the maximum current does not exceed 100 amps., thus saving installation costs and cable sizes.

This is usually achieved by pre-set pneumatic switches, but one series of panels has the motor driven 'cam' type. Operationally there is little difference, although for a larger number of stages, the motor driven units can be accommodated in a smaller space - with the corresponding reduction in cubicle size.

A number of fan manufacturers were considered having regard to performance, weight, sound levsls, cost and their inclination to undertake the installation work. The fans installed are of two main types produced by three manufacturers - 55% being of the aluminium vertical discharge type and 45% of the glass fibre cowl type. The most desirable size being 710 mm running at 14.3 $revs.s^{-1}$ with an extract rate of 3.8 m^3s^{-1} at Free Air and 63 dBA at 3 m. All are fitted with safety cages beneath and designed for maintenance from inside (due in the main to roof conditions); the vertical discharge models have weathering flaps. Noise limits have generally been restricted to a maximum of 66 dBA at 3 m. However, where residential buildings are close to the site, this limit is reduced to 61 dBA so as to avoid 'arousing' the neighbours too early in the morning!

The three years of experience we have had with this type of installation, has highlighted the difficulties. These are mainly related to those fans fitted with non-return flaps, in that the flaps jam with packed snow and the flap bearings ice-up during winter. Additionally, the flaps generally jam with dirt. Fans of all types have suffered from the safety cages clogging with dirt - this can obviously only be overcome by regular cleaning.

A New Look At Underfloor Extraction

Sometime after the tests described earlier, proposals for Bus Stations were being considered, especially a new garage in the basement of a development, which has a bus station and a bus terminal working within it as a combined unit. Consideration was again given to the use of underground extract points, or even the use of low 'bollards' opposite bus stands.

The one exception to the 1949/52 Bus Garage Programme mentioned at the beginning, was Loughton - this is at the extreme north-eastern corner of the Executive's area of operations, on the edge of Epping Forest. It was built on a new sloping site, so that the Parking Area floor is a 'raft' raised on stilts, allowing the economic installation of a large underfloor horizontal airway, with grilles in the floor of the garage along the rear wall. A single centrifugal fan rated at 26.5 m^3s^{-1} extracts from this airway. Because development in the area has not been as vigorous as was originally envisaged, the garage capacity was found to be excessive. Part of the covered Parking Area has therefore been let - with the result that half the underfloor duct is sealed-off and 5 grilles serve the operational area. Tests showed that the total volume extracted through the grilles is 16.08 m^3s^{-1} at a face velocity of 5.08 ms^{-1} ..... Figures 8 & 9 refer, (which is approximately one eighth of the exhaust pipe terminal velocity).

The diagram ..... Figure 10 refers, shows the results of tests, carried out with a bus at varying distances from a grille. It will be noted that the overall 'extraction' is reasonably good, due to the grille being adjacent to the end wall and any fumes missing 'direct pick-up' at the grille are ultimately extracted through a general air movement - the photographs illustrate this quite well. It is considered that if the grilles were in the centre of the Parking Area, the fumes would pass through the 'breathing zone' twice - this effect is obviously worse than the high level extraction position, where only one 'pass' is made.

The installation is only effective very close to the grilles because of the very high volumes passing through them, giving a correspondingly high pick-up velocity. Volumes on this scale could not be provided at the required 'pitch' in a garage without introducing prohibitively high quantities of fresh air and the use of excessive fan power, as well as the adoption of fixed parking positions for buses.

It was considered that the results of an installation of this arrangement would not justify the high capital and running costs involved. We have however, still not given up and are seriously considering the use of low profile 'powered bollards' in Bus Stations, where the bus standing position is fixed by the queue pens.

Summary

Generally the recommendations are:-

1. Study the way the garage is operated at first hand, or if a new site, study similar premises already in operation.

2. Decide what effect can be achieved by locating the entrances & openings, in relation to the prevailing winds etc.

3. Determine the chemical, medical and engineering criteria.

4. Design for simplicity.

The fume exhaust extraction systems described, apply to the Executive's standard unheated parking areas. A heated parking area would present a different problem bearing in mind the need for energy conservation.

Having said all this, let us take another look at the Factory Inspector's proposal for Sutton ..... Figure 2 refers. Should buses discharge their exhaust at ground level? Perhaps this is another legacy from the horse buses! Although there are problems associated with the extension of the exhaust pipe to discharge at roof level, insulation, access, height clearance and the possibility of bodywork/window 'streaking', the difficulties are not insuperable. Our Bus Engineers have not therefore entirely rejected the suggestion for the next generation vehicle.

Acknowledgements

The authors wish to thank the London Transport Executive for permission to publish this paper.

OPEN YARD
COPPER SMITH
BATTS
ELEC
R/ SHIFT
STORE
TYRE FITTER
TYRE STORE
COMM
NOTICES
DRYING ROOM
HOUSE (BOARDS PROPERTY)
LOCKERS
P/CYCLES
RECREATION ROOM
IMPREST STOCK STORE
OIL STORE
STORES OFFICE
ADVERTS
STORE
LAV
MESS ROOM
M. LAVS
HEATING ROOM
CLERK
G.O.M.
FIRST AID
G.E.M
TRAFFIC OFFICE
AFC
LOBBY
CONDUCTORS
LOCKER ROOM
OVERALLS
W & B
BLINDS
FEM LAV
BOILER FUEL
FUEL STORE
DOCKING AREA
STORE
8
FITTERS SHOP
FORE-MAN
COACH-MAKER
PAINTER
FUELPUMP TEST
MACHINE AREA
UNIT CLEAN
KITCHEN
STORE
STORE
E Ex
CANTEEN
REST ROOM
COLD STORE
SCALE
DRAWN
GARAGE CODE
DATE
APPROVED
GARAGE
95 VEHICLES

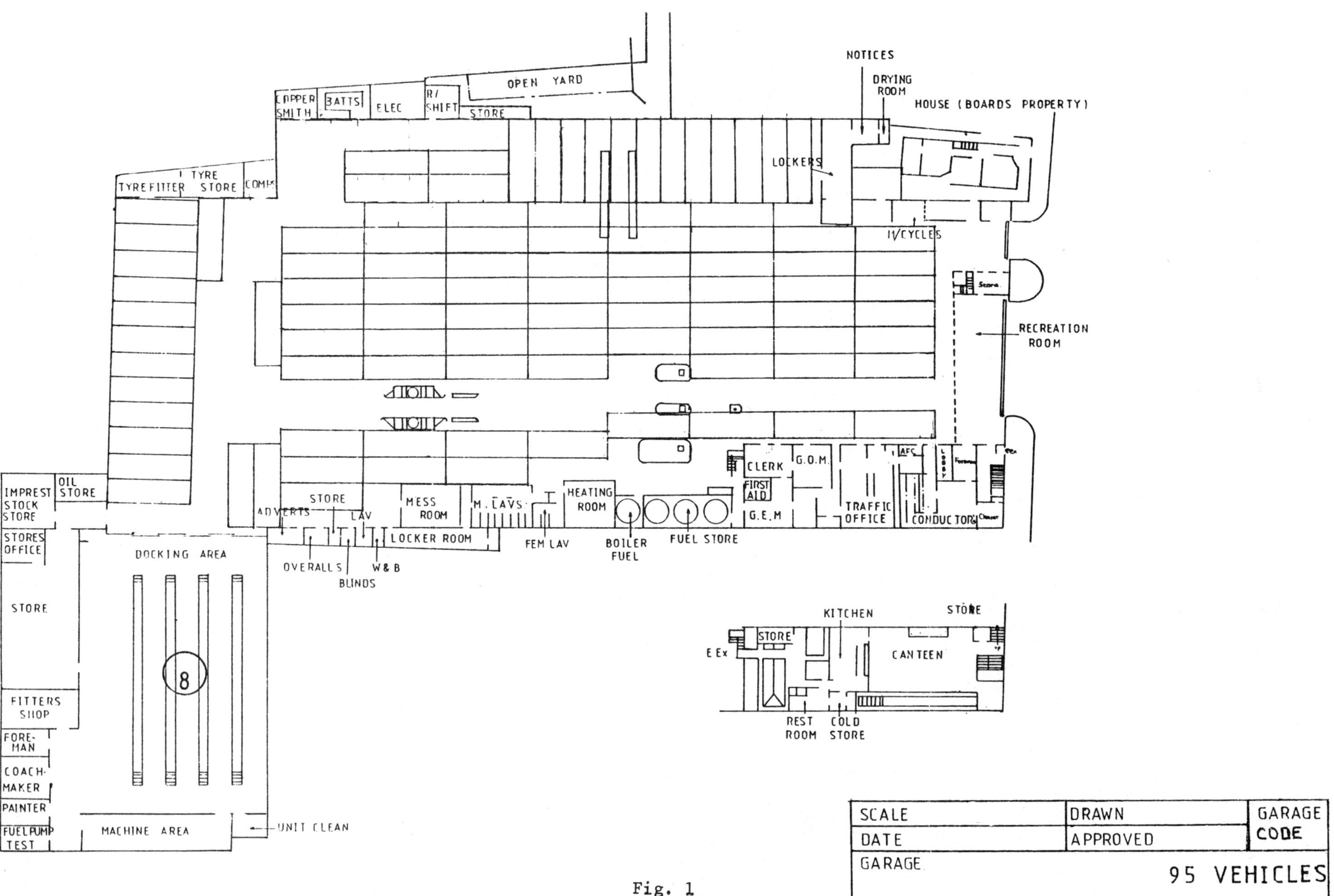

Fig. 1

1. Detachable exhaust extensions attached on entry to garage with suction pads or magnets.
2. Roof bay partitioned to act as collection hoods.
3. Roof fans.

Fig. 2

Fig. 3 Hounslow - chimney effect

Fig. 4 Hounslow

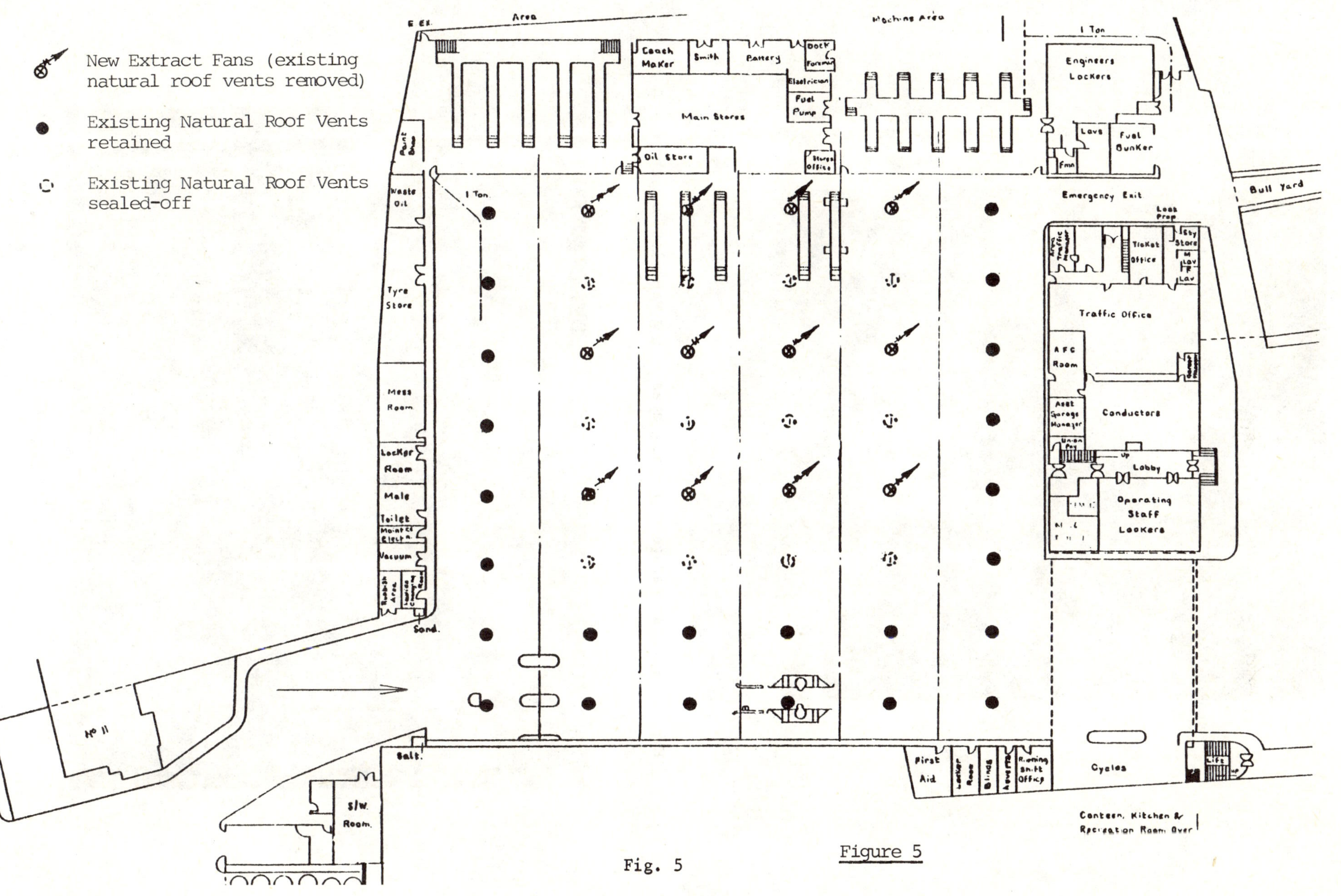

Fig. 5

Figure 5

Fig. 6 Peckham garage - parking area - new ventilation scheme

Fig. 7

Fig. 8 Loughton
All grilles open
Bus at 20ft (6m)
Running: tickover
Grille vel. 1200ft/min

Fig. 9 Loughton
1 grille open only
Bus at 20ft (6m)
Running: tickover
Grille velocity: 2500ft/min
Grille size (open) 4ft x 2ft

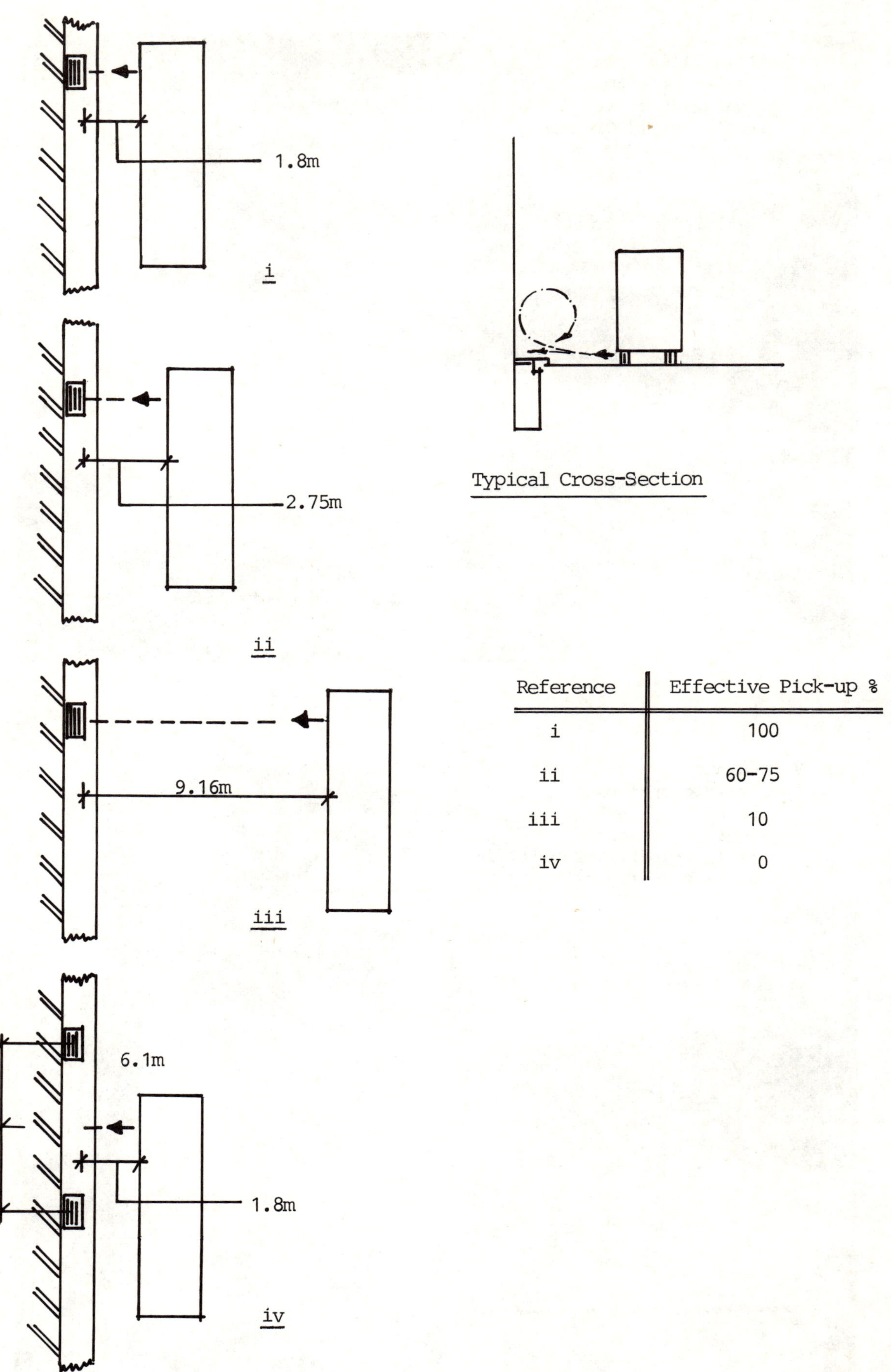

Reference	Effective Pick-up %
i	100
ii	60-75
iii	10
iv	0

Fig. 10 Underfloor parking area ventilation - test positions of buses

International Conference on

Fan Design & Applications

Guildford, England: September 7-9, 1982

PAPER E1

HYBRID - RECIRCULATION VENTILATION FOR INTENSIVE LIVESTOCK PRODUCTION

J.E. Owen

University of Reading, U.K.

Summary

Ventilation systems incorporating single phase propeller fans are widely used for intensive livestock housing. However such systems, particularly those incorporating fan speed control, have led to ill health and poor productivity among animals housed. These problems can be largely associated with poor equipment and poor design, which in turn can be associated with the economic and site constraints particular to agriculture. This paper describes a ventilation system that has been developed at Reading University which overcomes many of the problems normally found and demonstrates the fundamental considerations involved in the systems design. Despite its low capital cost the system has proven to be accurate, economic and is applicable to a wide range of buildings.

Organised and sponsored by
BHRA Fluid Engineering, Cranfield, Bedford MK43 0AJ, England.

0263 - 421X/82/01 00 - 0001 $5.00
The entire volume can be purchased from
BHRA Fluid Engineering for $82.00

1. INTRODUCTION

Livestock farmers have found through bitter experience that many of the currently available ventilation systems are quite inadequate, problems of poor air distribution and lack of control of ventilation rate being common. The result is that buildings used for the production of livestock are often draughty, temperature control is poor and the inside conditions are rank. As a result animal performance is detrimentally affected and production costs of feed and fuel (when heating is used) are excessive.

The conventional systems used usually incorporate single phase propeller fans which extract air from the ridge or sidewall of buildings, air inlets are usually long slots in the ridge or sidewall with internal flaps and external cowls for wind protection. In recent years the reasons for the poor performance of such systems have been clearly recognised, however the non-uniformity of the buildings used by producers has made it very difficult for manufacturers to develop systems which are satisfactory for the wide range of different building sizes, shapes, stocking conditions, types of animals and internal layouts that are used. In addition manufacturers often seem quite unaware of the basic principles involved, pay little or no attention to the basic requirements of the animals, use only the most crude methods of determining ventilation and heating requirements and rarely design a system for the particular application.

Research undertaken at Reading University in conjunction with farmers therefore concentrated on the development of a system of ventilation which could not only overcome the normal deficiencies of conventional systems but which would also be widely applicable. This research led to the development of the Hybrid Recirculation ventilation system which is described in this paper.

2. DESIGN BASIS

2.1 Mode of ventilation rate control.

Those familiar with the ventilation of livestock buildings will be well aware of the very large change of ventilation required between hot and cold conditions. The maximum ventilation rate required will frequently be 50 times greater than the minimum required and in some housing situations 200 times greater. However for most of the time ventilation rates will be below 50% of the maximum installed capacity and it is the control of ventilation at low levels of ventilation that is of paramount importance.

Fig.1 demonstrates the effect of achieving different levels of ventilation control on the temperature lift, i.e. the amount the heat from animals alone will raise the house temperature above the outside temperature, in a pig building. Also shown is the supplementary heating required. The effects are obvious and massive. However there are many systems currently being marketed which cannot achieve the good control that is required in cold conditions, in particular conventional systems using speed control of propeller fans are often inaccurate and particularly wind prone.

Because of the above considerations the Hybrid Recirc system uses recirculation for precise control at low ventilation rates and fan speed control, step control or even natural ventilation at high ventilation rates, when the control of ventilation is relatively unimportant. The basic recirculation unit design is shown in Fig.2, propeller fans with total pressure developments of 150-200 Pa are used and their pressure capability basically decides the unit sizing. The unit is used to pressurise the building and distributes air via an inflatable polyethylene ducts (see 2.2), such ducts requiring an internal static pressure of about 25 Pa for inflation. Air spin from the fan, which leads to unsymmetrical air distribution in the building, is rectified by the incorporation of an air straightener down stream from the fan. Fig.3 shows the results of tests on a 630 mm unit to determine the relative amounts of fresh and recycled air that pass through the unit when the recirculation flaps are at different angles, in a laboratory situation (without external wind effects) fresh air entry can be reduced to virtually zero, the reduction on fan performance i.e. the reduction in the total amount of air that can be moved by the fan with free intake and discharge, is about 9%.

Auxiliary units consisting simply of a fan and straightener are used in conjunction with recirculation units to make up the total ventilation system capacity.

2.2 Air distribution.

The basic theory of the fluid dynamics of air movement is sufficiently well developed for the airflow patterns within buildings, of the type used by livestock farmers, to be designed rather than guessed. Unfortunately few manufacturers appear to have the ability to use such theory for this purpose and the airflow in buildings often produces draughts, poor air distribution and poor air mixing (which results in large temperature gradients). In buildings with conventional long slot inlets and extract fans, inlet velocity will need to be maintained at all times if these problems are to be avoided, this generally means that air inlet size must be altered to suit the particular ventilation rate at any particular instant in time. Manual control of inlets is not sufficiently accurate and automatic systems have been developed. However if they have to be designed and fabricated specially for each different building they are expensive. In addition when large variations in ventilation rate have to be accommodated the slot sizes required at low ventilation rates are so small that very good draughtproofing has to be provided if the cracks around doors, etc., are not to interfere with the airflow.

To overcome some of these difficulties and to give the system the flexibility to be used in many buildings, the Hybrid Recirc system incorporates inflatable polyethylene ducts for air distribution. These ducts can be accurately designed, are lightweight, virtually frictionless up to about 100 m in length, very flexible in application and are low cost. Using these ducts in association with recirculation means that airflow patterns can be established and maintained at all times, air velocities at stock level can be controlled, and thorough air distribution and mixing achieved. The results of the investigation of the design of such ducts by Carpenter (Ref.1) and others has proved invaluable for design purposes, however their results have to be used in conjunction with the theory of inlet air jets, Farquharson (Ref.2), Nevins (Ref.3), Walker (Ref.4) and others.

2.3 Control system.

The inadequacy of control systems in the past has been amply demonstrated by Barrett (Ref.5) who in a series of simple investigations showed the inaccuracy of most commercially available speed controllers. In order to overcome the deficiency of the generally available control systems, development work on a controller was started in 1979. The control concept was based on an analysis of the mode of operation required of the system, modelling and simulation studies of the response of internal conditions within buildings to changes in outside conditions.

Based on these studies an electronic control system was developed which was based on a control action related to both the rate of temperature change occurring within a building and the deviation from the set temperature (error). The combination of the error and rate of change is used to make the appropriate control adjustments, varying the control response e.g. the time the recirculation flap motor runs, accordingly. The use of a digital system makes it straightforward to provide logic to interlock auxiliary fans and heating with the recirculation flaps. A special logic design allows the controller to be built without recourse to the use of microprocessors, thus reducing cost. The entire controller is assembled on a single circuit board and all external connections are made by screw terminals, to give improved reliability. If the controller fails it is automatically isolated from the system, the flaps are driven open and all fans forced to full speed.

Multiple sensors can be used in any number required and both sensors and controller can be factory calibrated and require no field calibration, controller gain however sometimes needs to be adjusted in the field (particularly to accommodate different heating systems), this is done by replacing a wire wrap connection rather than adjusting a potentiometer which again increases reliability.

Reliability has been mentioned a number of times and is of major importance. While the environment within which equipment has to operate is particularly onerous, the failure of any component which shuts off the ventilation can rapidly lead to the death of livestock. Therefore having developed and tested a prototype control system, subsequent work concentrated on designing a production system which would give highly reliable performance in the field. This was largely a matter of ensuring the enclosure and sealing of the system and components against the ingress of dust and moisture.

Fig.4 shows the basic control system components which can either be built up in a modular fashion or integrated for specific dedicated control needs.

3. SYSTEM PERFORMANCE MONITORING

In 1975 in conjunction with farmers a number of prototype systems were designed and installed in various buildings, these were designed to be operated on a day to day basis by the farmers without any interference from ourselves. The Ministry of Agriculture, Fisheries and Food (MAFF) conducted a series of short term monitoring runs on some of the systems (work done by ourselves in 1978 concluded that short term monitoring is just as valuable as long term). This was done under normal working conditions and the results provided a great deal of the information required to ratify the systems' capabilities and to identify problems.

3.1 Temperature performance monitoring.

The long term monitoring of ventilation rate is a problem still waiting to be fully resolved and is the subject of much investigation currently. However a very good indication of the success of a system at controlling ventilation rate can be obtained from temperature monitoring of the inside conditions within a building when cold conditions prevail outside. Their results of such monitoring were sufficient to indicate:

(a) the potential temperature lift that could be achieved in such buildings with the recirculation method of ventilation rate control,

and (b) the inadequacy of then commercially available control systems (which were fitted to the systems that had been installed) at holding the temperatures close to the set temperature for the buildings.

Monitoring of internal house temperatures also gave a good guide to the amount of air mixing that was being created by the airflow pattern from the duct. Fig.5 shows the results of a series of spot temperature readings taken in a building at a low ventilation rate with outside air at a temperature of 8°C. In this case the effect of introducing the ventilation air by recirculation from a duct at high level can be seen, there is a temperature gradient from floor to ceiling, the floor level temperatures being higher than those at the roof. This is most important since in many buildings the temperatures are inverted the other way, in these circumstances animals are often too cold while the roof is losing more heat than is necessary because of the elevated temperatures at ceiling level. The beneficial effect of reducing the ceiling temperatures on energy consumption has been demonstrated in industrial premises where savings in heating costs of up to 25% have been reported (Ref.6). The beneficial effects of recirculation on temperature gradients in poultry buildings has also been demonstrated by Moulsley & Fryer (Ref.7). The Hybrid Recirc system can be expected to achieve such benefits. Fig.6 shows the effect of the Hybrid Recirc system on the temperature gradient occurring in a laboratory and a factory fitted with the system is currently being monitored to see if the same effect can be achieved.

Since 1980 the Hybrid Recirc system has bee produced and sold commercially and again the system performance has been monitored by MAFF, however the commercial version of the system incorporates the controller that has been developed at Reading rather than other commercially available controllers. Fig.7 shows an example of the result from a pig building with heating, the level of control being achieved can be clearly seen and it can be compared with the results which were being simultaneously recorded in an adjacent identical room fitted with a new commercial high speed jet ventilation system.

3.2 Airflow pattern in different buildings.

Fig.8 shows some of the different buildings in which the Hybrid Recirc system has been installed, in each room the airflow pattern shown in the diagrams has been confirmed by simple smoke tests. This figure primarily demonstrates the wide applicability of the system to a range of building shapes and layouts. The main constraint on the systems use is when it has to be retro-fitted to buildings with insufficient headroom for duct installation.

The ventilation control mode adopted, using speed control of auxiliary fans in warm conditions, also allows a further advantage to be gained i.e. the use of increased air velocity over livestock to provide additional cooling of the animal. As fan speed is increased to increase ventilation rate in hot conditions, the air velocity over livestock can also be increased. The use of fan speed in this way is however not detrimental to the systems ability to achieve low ventilation rates in cold conditions since auxiliary fans are shut down at such times and the uninflated polyethylene ducts form an effective non return valve.

In this system with the recirculation duct dominating the airflow pattern, air outlets can be located in the most convenient position for contaminent removal and where they are least prone to outside wind effects, where possible low level below slat air outlets (i.e. in the side wall of tanks used to collect faeces and urine) are used with simple wind baffles. The need for large areas of wind baffles on inlets is unnecessary, since it is only the discrete openings for the recirculation and auxiliary fan units that need to be protected.

3.3 Energy consumption.

Energy consumed in intensive livestock buildings can be considered either in terms of the feed energy the animals consume or in terms of fuel energy for running equipment and heating if it is required. Determining improvements in feed energy utilization requires long term experiments with animals if statistically viable results are to be produced and such experiments have not been undertaken because of their expense. Monitoring in detail of energy consumption of different items of equipment within a building, particularly in working buildings on farms is expensive and difficult. Also since sites may differ quite considerably, for example climatic conditions, fuel type used, building stocking and management may all vary, it is difficult to draw general conclusions without undertaking the monitoring of a large number of installations. Again the resources required to do this are not available.

Reliance has therefore been placed upon feedback of information from farmers, who have confirmed that the Hybrid Recirc system can reduce the heating capacity required in existing buildings and that fuel costs can be reduced. One farmer who was unable to maintain temperatures in his small pig rearing rooms, even with 9 kw of heating per room, can now do so with only 4 kw per room. Comparison between two virtually identical winter half years showed a saving of 2770 litres of propane in three such rooms when the conventional fan speed control systems were replaced by Hybrid Recirc.

Another farmer using the Hybrid Recirc system in an end room of his pig fattening house in the winter of 1982 was able to maintain an inside air temperature in that room of 16°C at outside temperatures down to −15°C using pig heat alone. The four other rooms in his building were at −1°C inside when it was −15°C outside. The benefit was that his pigs grew faster, reaching bacon weight 7 days sooner, in cash terms this meant the farmer saved £625 in food for the 418 pigs reared over the 20 week period.

4. CONCLUSIONS

The proper ventilation of livestock buildings is of paramount importance to the health and wellbeing of the housed animals. The economic constraints of the livestock production enterprise however do not allow expenditure on expensive conventional ducted ventilation systems. Despite the fact that the systems need to be low cost they must however be reliable, accurate, and able to withstand the harsh environment in which they operate. Because of the cost constraint most manufacturers have either ignored the livestock building market or have produced cheap, usually inadequate, systems.

This paper has attempted to demonstrate a ventilation system which is designed on a sound engineering basis, with an understanding of the constraints imposed by farm conditions and the needs of the livestock. In developing this system there has been a constant awareness of cost and the need to allocate expenditure to the most important features of the system i.e. those affecting its performance, rather than on frills e.g. the major portion of the production cost is devoted to the control system, the part that ensures the system performs properly. The result is that a modular system has been developed which can be factory built to reduce cost, but still retains the flexibility to be used in many different situations.

While it has not been possible to engage in detailed recording of the system's performance in relation to energy consumption, or to conduct experiments to determine the performance of livestock housed in buildings fitted with the system, monitoring is confirming the system's capability to control temperature closely and to produce a stable and predictable airflow pattern in many different buildings. Currently the system is undergoing the real test, that is use in real farm situations. In the last year 30 installations have been made and no problems have been reported which have not been simply rectified. Farmers who have been using this system are confirming its capabilities in the most appropriate way, they are installing it again.

5. REFERENCES

1. Carpenter, G.A.: "The design of permeable ducts and their application to the ventilation of livestock buildings". J.Agric.Engng.Res.17, pp. 219-230, 1972.

2. Farquharson, I.M.C.: "The ventilating air jet". J.I.H.V.E. 19, 1, pp. 149-469, 1952.

3. Nevins, R.G.: "Air diffusion dynamics". Business News Publishing Company, Birmingham, Michigan, 1974.

4. Walker, J.N.: "Review of the theoretical relationships of isothermal ventilating air jets". Trans. A.S.A.E., pp.517-522, 1977.

5. Barrett, M.: Personal communication, 1974.

6. Anon: "Warehouse heating costs halved". Energy Management, January 1978.
"Air recirculation saves 23% in space heating costs". Energy Management, June 1978.
"Record company has large energy saving fans". Energy Management, April 1981.

7. Moulsley, L.J. & Fryer, J.T.: "Air recirculation in a broiler house using large propeller fans". N.I.A.E. Dept.Note DN/En/928/10003,1977, Silsoe, Bedford, England.

FIGURE 1. EFFECT OF MINIMUM VENTILATION RATE ON TEMPERATURE LIFT AND HEATING REQUIRED IN A FLAT DECK FOR 160 WEANER PIGS.

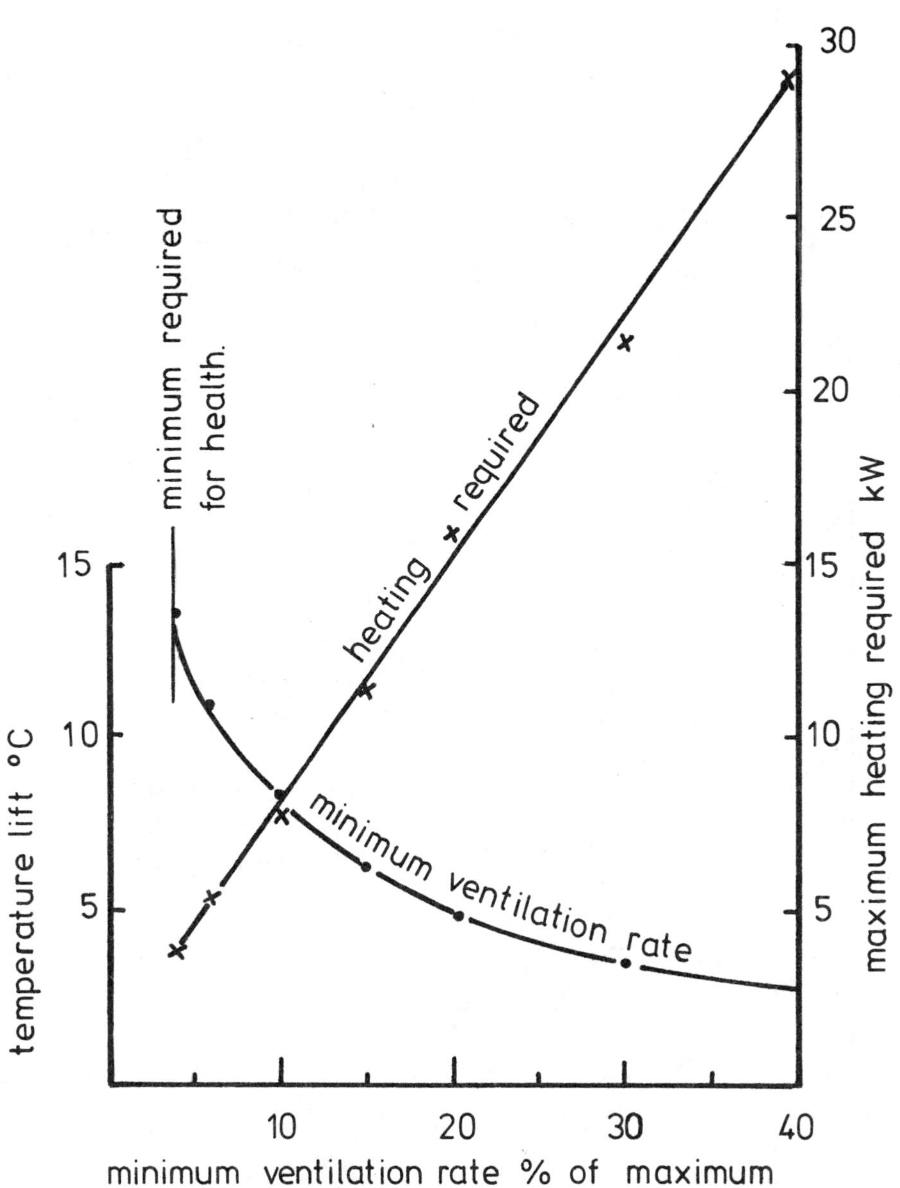

FIGURE 2. RECIRCULATION UNIT.

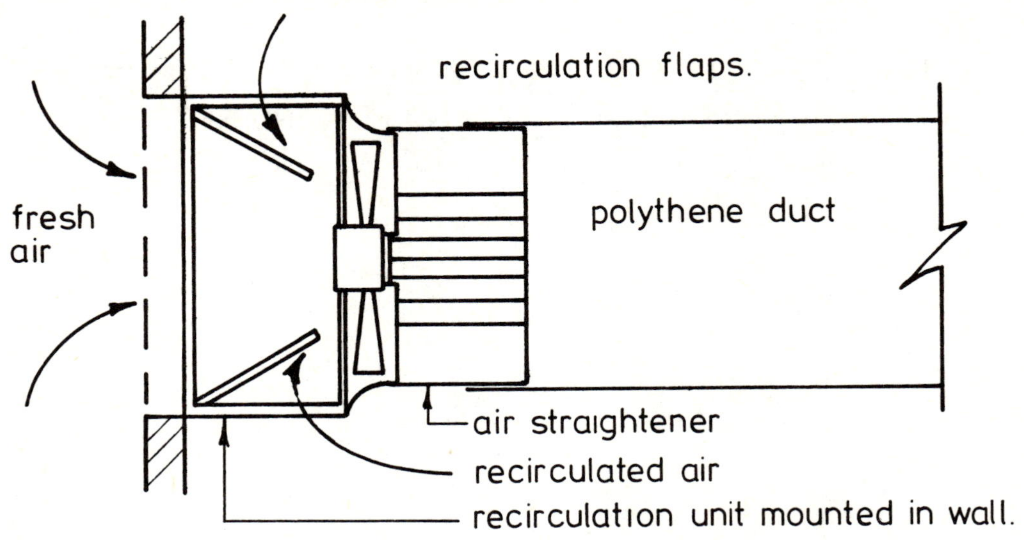

FIGURE. 3. FRESH AIR FLOW THROUGH A RECIRCULATION UNIT AT VARIOUS FAN SPEEDS.

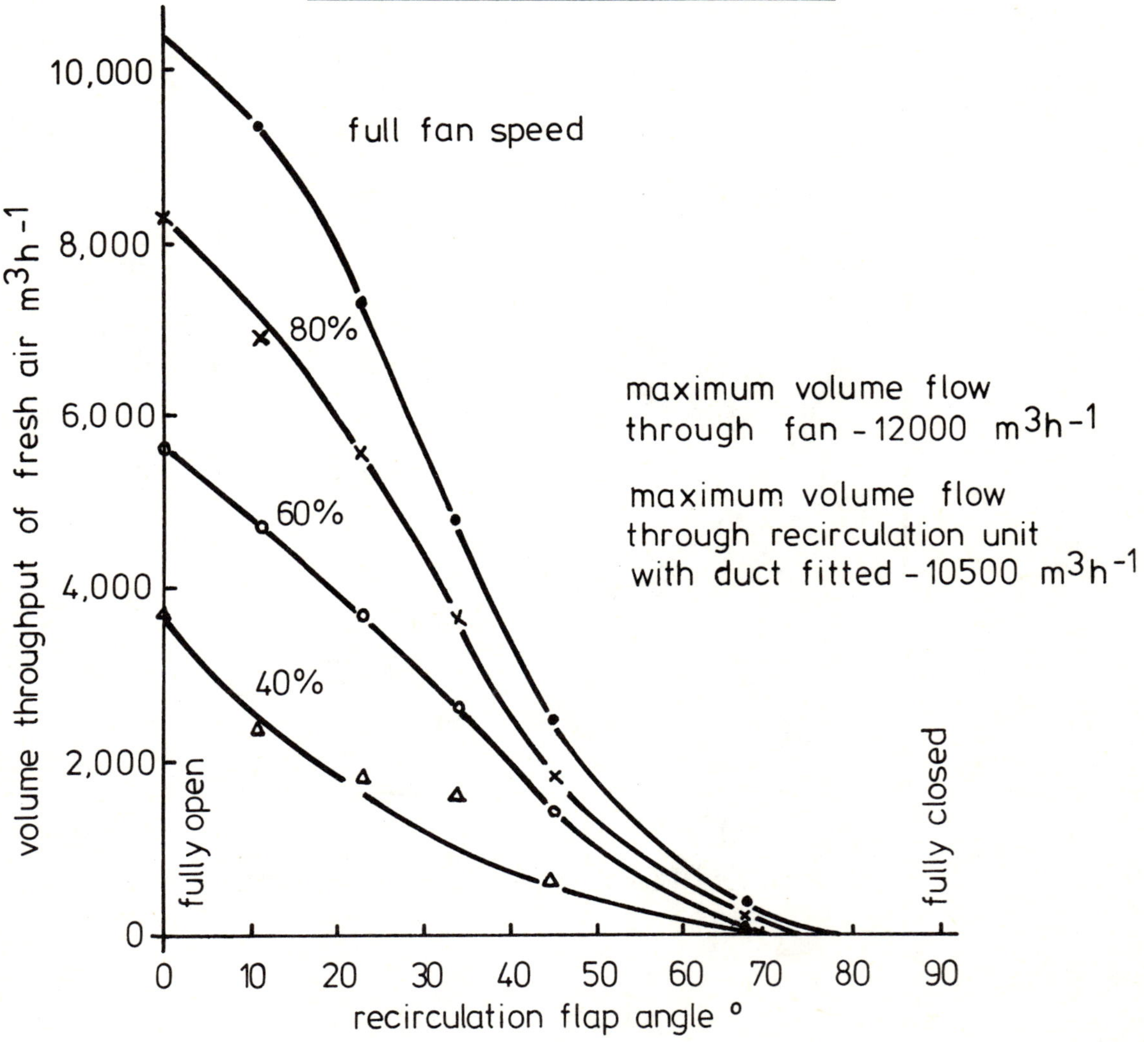

full fan speed

80%

60%

40%

maximum volume flow through fan - 12000 m^3h^{-1}

maximum volume flow through recirculation unit with duct fitted - 10500 m^3h^{-1}

fully open

fully closed

volume throughput of fresh air m^3h^{-1}

0, 2,000, 4,000, 6,000, 8,000, 10,000

0 10 20 30 40 50 60 70 80 90

recirculation flap angle °

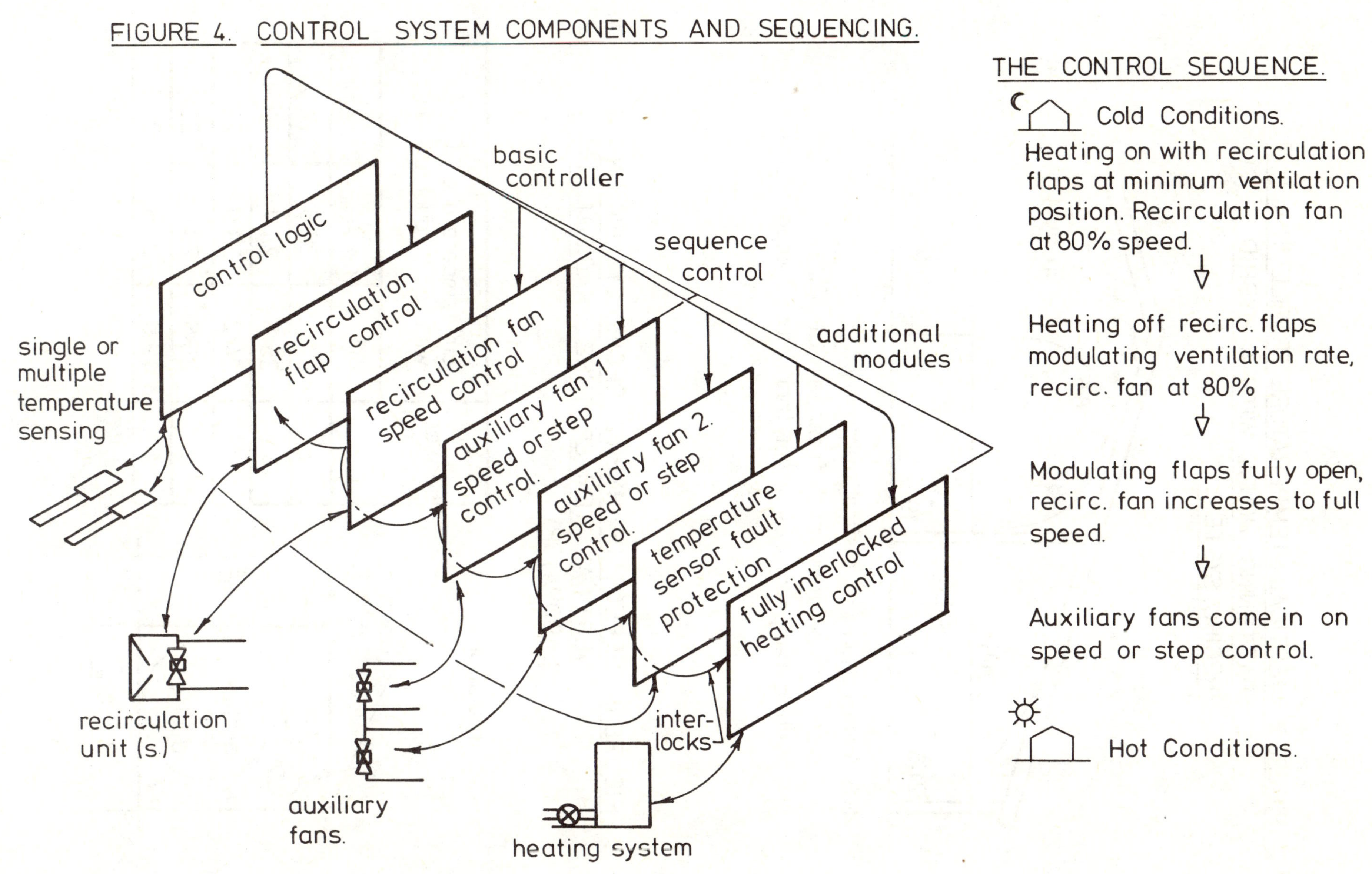

FIGURE 4. CONTROL SYSTEM COMPONENTS AND SEQUENCING.

FIGURE 5. VERTICAL TEMPERATURE PROFILE IN A PIG FATTENING HOUSE WITH A HYBRID RECIRC. SYSTEM.

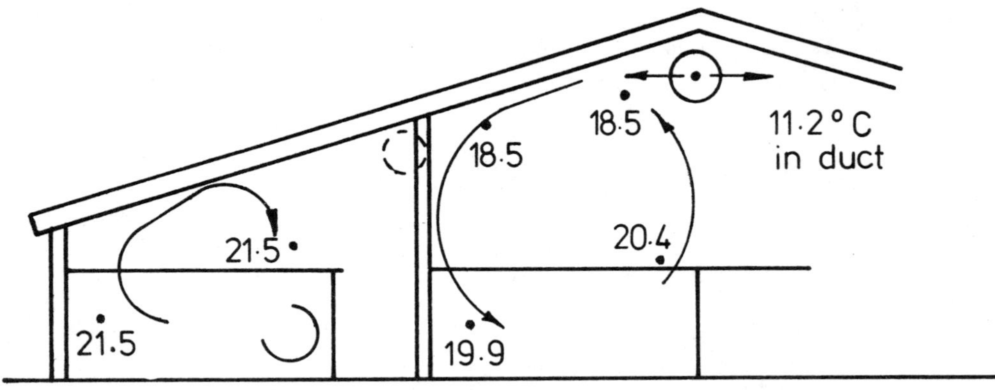

N.B. Ventilation via central duct only.
part recirculation, part fresh air at 8°C.
ceiling temperatures lower than floor temperatures.

FIGURE 6. EFFECT OF HYBRID RECIRC. SYSTEM ON VERTICAL TEMPERATURE GRADIENT IN LABORATORY.

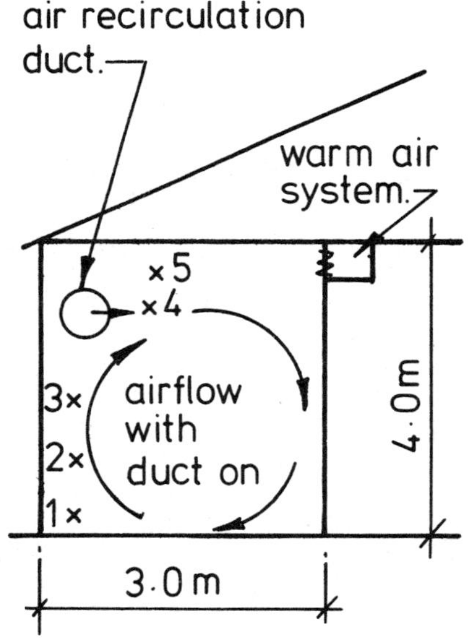

sensor location	air temperature °C with recirc. duct system.		
	off	on [for 7 mins]	on [for 15 mins]
5	37	25.6	24.2
4	35.5	24.7	23.9
3	24.2	23.9	23.7
2	20.5	23.0	23.1
1	19.0	21.5	22.3

FIGURE 7. SAMPLE OF TEMPERATURE RECORDS FOR TWO 2nd STAGE FLAT DECK ROOMS FOR 360 PIGS.

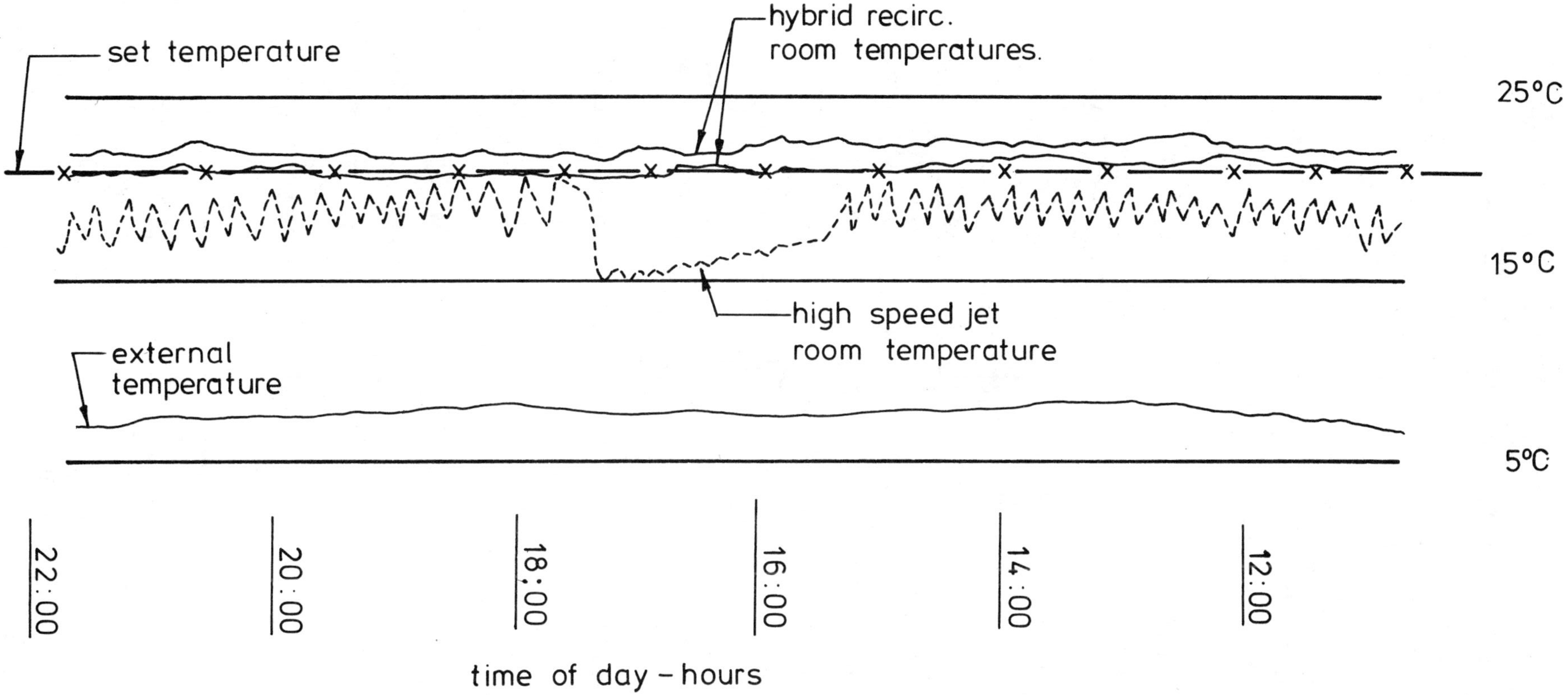

FIGURE 8. SOME EXAMPLES OF THE APPLICATION OF THE HYBRID RECIRC. SYSTEM.

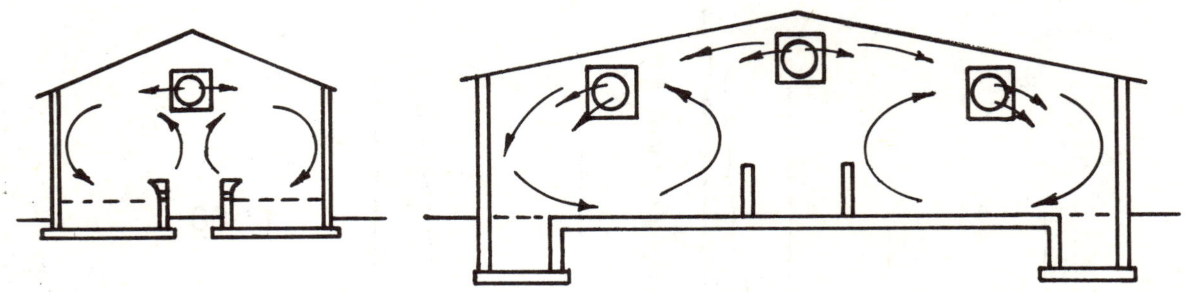

a) Flat deck for young pigs.

b) 2nd stage flat deck or fattening house for pigs.

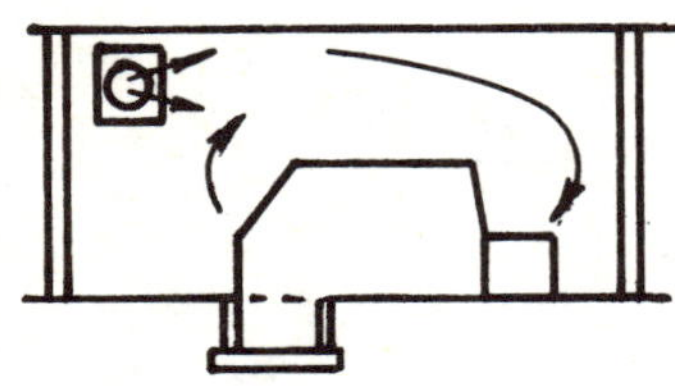

c) Farrowing room.

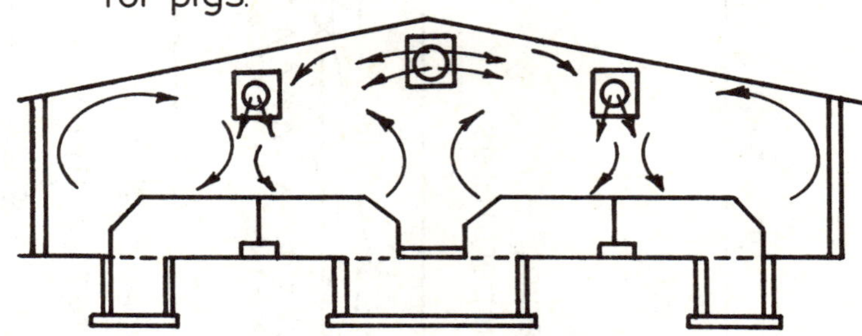

d) Dry sow house.

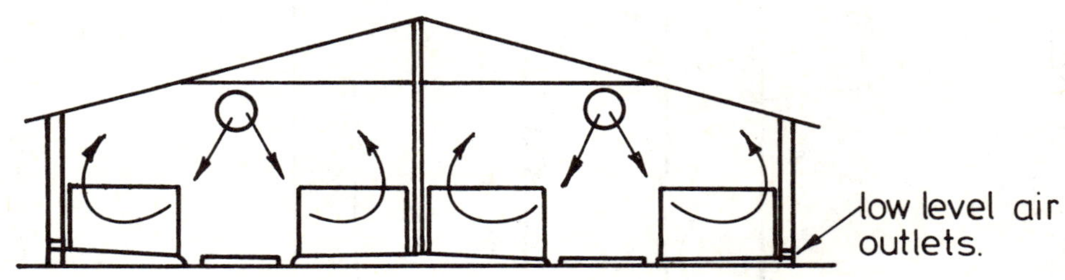

e) Calves up to 50 kg.

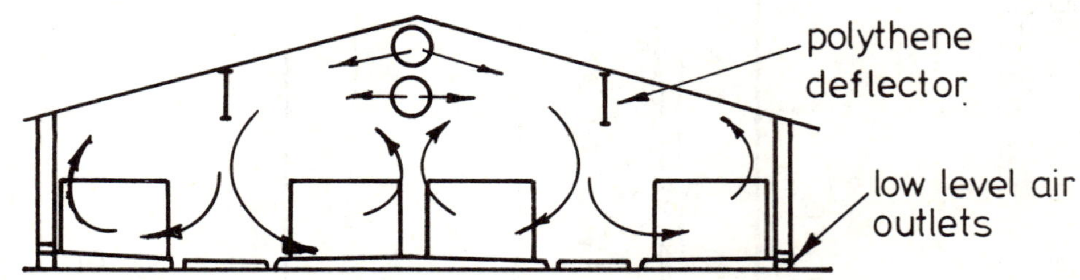

f) Calves up to 150 kg.

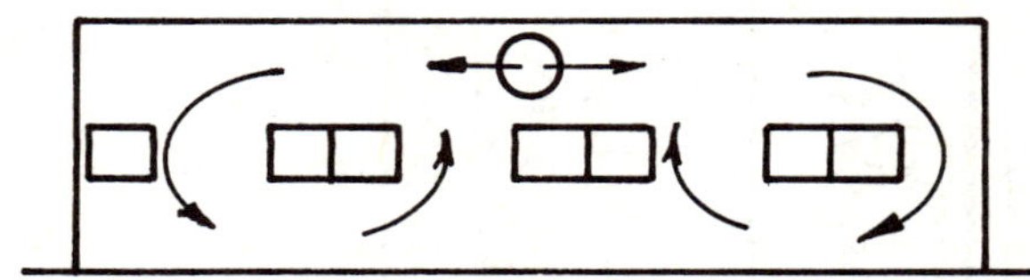

g) Experimental rabbit room.

International Conference on

Fan Design & Applications

Guildford, England: September 7-9, 1982

PAPER E2

SPECIFICATION & SELECTION OF FANS FOR THE PETROCHEMICAL INDUSTRY

M. J. Rex and G. Thomson

Snamprogetti Ltd., U.K.

Summary

The paper will discuss the need for, and the use of a Purchaser's Specification, and illustrate typical specifications and data sheets used in the Petrochemical Industry. The requirements of a good specification will be reviewed and will cover such aspects as duty requirements, technical features for fans, drivers and auxiliary equipment, testing and inspection and the Purchaser's need for documentation. Technical evaluation of bids will be discussed and the importance of operating range, shape of Head/Capacity curve and integration into the process system. Typical applications will be described and the approach of a Petrochemical Contractor to obtaining a satisfactory installation.

Organised and sponsored by
BHRA Fluid Engineering, Cranfield, Bedford MK43 0AJ, England.

0263 - 421X/82/01 00 - 0001 $5.00
The entire volume can be purchased from
BHRA Fluid Engineering for $82.00

1.0 INTRODUCTION

It is desirable before attempting to purchase equipment, that the buyer knows and understands the application and has an idea of the type of goods required. If the vendor is aware of the buyer's application and goods envisaged, there is the possibility that goods can be supplied which will fulfil the duty and be acceptable to purchaser, vendor, owner and others involved. A poor application description coupled with an unclear description of the goods required will lead to unsatisfatory performance, initially when bidding the equipment, when negotiating the supply and finally in operation. The specification can in its broadest sense be taken to include all the essential requirements (excluding commercial terms) between purchaser and vendor. A review of the required contents shows a clear division into general aspects which can be applied to a variety of purchased items and a 'Technical' aspect which is unique for the purchased item. A purchase order for a complex fan of high value will refer to a number of specifications viz:-

- The technical specification - could include technical specifications for instrumentation and electrical equipment supplied as part of the fan contract.
- A documentation specification.
- Painting specification.
- Packing and shipping specification.
- Purchaser's inspection requirements.
- Instructions for quotation of spare parts.

The specifications, once written, can be revised to reflect the equipment as ordered, and then further revised if necessary to reflect details as delivered. Thus the documents become an easy way of conveying key information throughout the delivery and commissioning period, and subsequently as a record. They can be used between purchaser and vendor and as informative documents to:-

- The Owner - if different from the Purchaser.
- Purchaser's staff associated with the fan, eg. electrical engineer, civil engineer, inspector/expeditor, shipping department, planning engineer.
- Sub-contractors - particularly when the fan is part of a large complex and site erection is carried out by others.
- Vendor's staff associated with the contract, sales department to project department to design department.

The Technical specification is usually initiated by the Purchaser; however, the Vendor may also have his own specification which should be used if the purchaser's specification is inadequate or absent.

2.0 TECHNICAL SPECIFICATION

In many instances the "Specification" can be reduced to a single sheet - the data sheet (Fig. 1). For more complex installations or where the vendor's scope is to include a number of items the specification will contain Data Sheets and a Narrative (Fig. 2 shows typical cover sheet). The application of the fan should be clearly described, often a couple of words are adequate, as, "Combustion Air Fan" conveys much more than "Fan" or "Air Fan". In other installations it will be necessary to provide more detail, particularly when cyclic and variable duties are concerned.

The narrative which may accompany the data sheet should be concise and composed in a logical fashion conveying:-

2.1 Application - describe the fan as clearly as possible and its position in the process. Identify if the plant is dusty or if corrosive or toxic chemicals abound. Consider whether the duty is to be cyclic, starting against closed or open dampeners, or if there are special flow rates and fluids to be handled during plant start-up and shutdown.

2.2 Duty - state the main head and flow requirements for normal operation and the limits of operation that are likely, maximum temperature, flow, head, and minimum flow. If there are a large number of required operating points the fan will be fitted with a capacity control device; consider suction or discharge dampeners, inlet guide vanes or variable speed.
Consider carefully the measurement points specified so that no ambiguity exists over inlet filter/silencer, guide vane or dampener losses, and losses due to seal or cooling flows. Ambient air temperatures have a dramatic effect on fan powers and pressures and in cold climates consideration should be given to installing heaters at the inlets to air fans, not only to prevent ice formation but to reduce installed motor power.

2.3 Purchaser's special requirements - these must be kept to a minimum and should state only what features are necessary from the purchaser's/owner's experience. Too frequently owner's specifications are added to purchasers specifications giving a mass of paper which is largely repetative and often irrelevant. Before issuing the specification the purchaser must carefully vet the contents of outside documents, collate the information and then either produce a single new document combining features of value, or select the most complete specification and add an addendum sheet.

If the equipment is to be on automatic start the purchaser needs to clearly specify the maximum torque that can be developed by the driver. Under exceptional circumstances motors can develop greater than 18 x Full Load Torque if restarted when still rotating. This very high torque is a transient but its existence needs to be acknowledged.

Any special requirements with respect to the driver installation and co-ordination should be identified. Attention must be given to requirements for the coupling to accommodate axial thrust, axial movement of shafts or to provide "limited end float".

Inlet and discharge orientations and configurations should also be described and the seller informed if it is intended to place a sharp bend on the fan discharge or to have asymmetrical inlets on double flow fans.

2.4 Scope of supply - consider first the hardware, fans, couplings, gearbox, driver, lube oil system, auxiliary drivers, bedplate, foundation bolts, dampeners, guide vanes and actuator, filters, silencers, sealing system, cooling water system, seal packing, insulation lugs, insulation, access doors, and of course, instrumentation. An integral part of the supply are the services required which include drawings, manuals, material documentation and test certificates, perhaps lateral and torsional vibration analysis, assembly of free issue components, standard and special tests, supervision on site and assistance at start-up. Decisions are required with respect to who is to produce a fan/driver general arrangement drawing, and who should co-ordinate the supply of motor(s).

2.5 Testing and Inspection requirements - define what standards are to be used for any fan performance test and clearly state the permissible tolerances. If special materials are required identify the certification requirements before the order. Clarify the position with respect to test and inspection of sub-contracted items eg. electric motors, auxiliary steam turbines, lube oil pumps.

2.6 Performance Guarantees - clearly state acceptable tolerances and any absolute limits. Tolerances quoted in various test codes can be misleading, particularly when relating a power guarantee with a fan which is at its lowest acceptable tolerance on flow and head. The commercial part of the purchase order will define responsibilities when machine guaranteed performance is not met.

2.7 Documentation requirements - this can often be a separate specification which can be standardised to cover various items of equipment. The number of prints, reproducibles and manuals required should be stated. To the purchaser, the data contained on drawings is important to allow plant design to proceed and a programme for submission of drawings should be agreed; the importance of receipt of documentation to a purchasing contractor is often emphasised by tieing progress payments to receipt of drawings. A well prepared drawing schedule will assist the Vendor as it can show dates when documentation for free issue items (drivers, motors etc.) will be available.

3.0 OTHER SPECIFICATIONS

The extent and necessity for specifications other than the technical specification will depend on the complexity of the fan installation and the end user. It must be borne in mind that the fan is often a very small component in a large plant, but if it fails the loss of production is out of all proportion to the value of the fan; for this reasons owners and purchaser's specification's have to be rigorously applied, particularly with respect to instrumentation and protection devices.

4.0 TECHNICAL EVALUATION

All technical information relating to the bid requires to be entered on the data sheet; these areas are usually clearly indicated on the sheets. Failure to complete areas marked most certainly delay evaluation of the bid and possibly render it null and void if other vendors have complied with the bid requirements.

For technical assessment the bid is split into the following sections, namely:-

- Installation List if a Special Duty
- Completed Data Sheets
- Design Features
- Extent of Supply
- General Arrangement and Sectional Drawing
- Shop Tests
- Performance Curves
- Performance Guarantees
- Power Economy

The above sections are expanded under the following sub-headings:-

4.1 Installation List

In support of an offer the vendor should include a relevant installation list, and should identify similar units in operation. A vendor's past experience is particularly important when considering fans with high heads, large flows, extremes of temperatures, or when high rotational speeds are required. The ability to demonstrate experience with the "odd-ball" duty will establish confidence with the evaluating engineer. However, ensure that the reference lists quoted are correct and up-to-date.

4.2 Completed Data Sheets, which include the bulk of the technical information, and form a summary of relevant supporting data.

4.3 Design Features

4.3.1 Materials of Construction

These require to be clearly stated for all major components of the fan and specified in accordance with one of the Internationally accepted standards - ASTM, DIN, BS or equivalent. (Ref. 1)

The acceptability of the materials directly in contact with the gases, although thoroughly checked by the purchaser's Metallurgy Department, relies on the vendor's metallurgical expertise gained on similar duties borne out by the Installation List of equipment in operation.

The corrosion allowance and thickness of inlet box and main casing require to be stated.

4.3.2 Fan Casing Design
Reference requires to be made to the casing drawing which should show inlet box (if required), the position of the suction and discharge, the area of fan inlet, the number of branches, and the drilling of the flanges. The size of the inlet and outlet ducts and the drillings should match exactly with the contractor's ducting.

In addition to the above openings, the fan casing will require to be fitted with access doors conveniently situated for maintenance purposes.

The Fan Casing should be of a suitable design to accommodate withdrawal of the impeller through the sideplate on the casing, and in the case of large double flow fans, the casing should be axially split to enable withdrawal of the complete shaft assembly without disconnecting ducting. Positive sealing is required on all joints to contain the gas being compressed.

Permissible nozzle forces and moments on the fan casing should be stated. In the event of the Purchaser having to spend extra computer design time and install extra fittings and equipment to achieve the loading values stated, the extra cost estimated by the Purchaser would be taken into account when the bid was commercially reviewed.

Although the vendor's proven design for the casing is acceptable, when reviewing the Casing Drawing the Purchaser will pay special attention to the stiffness of the casing and the thickness which should include at least a 3mm corrosion allowance.

4.3.3 Gas Seal Type
The seal type to be selected for the duty must be of proven design and detail furnished on an accompanying drawing for the more sophisticated application. The Drawing requires to clearly show all component parts, method of sealing, methods of attaching the seal unit to casing and shaft and a heat shield in the event of high gas temperatures. Any utility requirements should be clearly stated showing, fluid, pressure, and flow rate.

The reference number of the seal assembly requires to be specified on the Data Sheet, in order to avoid confusion over seal design if the bid should become an order.

4.3.4 Bearings and Lubrication Systems
For a valid assessment of competitive bids the following information requires to be stated:-

. Type of Bearing - whether roller/ball or self aligning axial split journals. Type of thrust bearing and loading.

. Method of bearing sealing against the environment and against lubricant leakage.

. Type of lubrication whether force fed, oil, or grease, and stating manufacturer.

4.3.5 Type of Cooling.

. Axial and Radial Thrust loads imposed on the bearing assembly by gas loading and coupling lock-up.

. Bearing Mounting.

. Cooling Discs.

. Instrumentation.

. Radiation Shields.

In addition to the above, a detailed drawing of the bearing assembly showing terminal connections, instrumentation, and ancillary equipment is required. The sectional drawing of the fan unit shall be consulted for ease of disassembly of the bearings without removing fan from the ducting.

When using antifriction bearings a statement on the life expectancy is required. With rolling element bearings an L10 (Ref. 2) life of 25000 hours is usually expected; any deviation should be clearly stated.

The vendors proven standard is acceptable provided the bearing design has been used on a similar duty in the past.

All ancillary equipment incorporated in the vendor's lubrication system, is best shown on an extent of supply diagram as requested on the data sheet.

Any instrumentation supplied should be to the hazardous classification prevailing in the area of the unit and quoted by the purchaser.

4.3.6 Rotating Assembly

The following information requires to be stated:-

. Diameter of Impeller.
. Tip Speed.
. Critical Speed.

The Fan Sectional Drawing requires to be consulted to check the impeller, blade type, fixing of the impeller to the shaft, inlet configuration and whether the impeller is between bearings or overhung. The blade design proposed should be in accordance with the vendors proven design based on experience as stated on the Installation List. The method of construction of the impeller requires to be clearly shown on the sectional drawing together with the relevant position of the inlet cone.

In the case of high temperature fan applications, the tip speed requires to be down rated by a factor to account for the lower design stresses.

4.3.7 Flow Regulation

The flow regulation range desired should be stated by the purchaser on the data sheet, while the method of achieving the control range is usually left to the vendor. The influence on power requirements, running costs and capital will be judged accordingly by the contractor.

The proposed method of flow control, and if automatic, the actuator type must be described and the design of the control system should be shown on a drawing accompanying the bid to clarify extent of supply and detailed design.

The philosophy governing whether the vanes should fail open or closed in the event of a power failure would be with the purchaser and decided on order placement.

As previously, the drawing number of the assembly must be stated at the appropriate position on the data sheet.

4.3.8 Gear Design

When using gearing to obtain the rated speed to give the required aerodynamic performance, the appropriate design codes and service factors require to be entered in the data sheet. The gearbox should, of course, be sized to transmit driver full load power and torque. In addition, a separate gear data sheet, in which detail design features are requested, requires to be completed. Reference to this document is made on the fan data sheet. In order to check dimensions and gear internals, outline and sectional drawings are requested.

Gear units incorporating rolling element bearings are fitted with splash lubrication and temperature indicators, while at the other extreme, pressure fed, gear units incorporating journal bearings, require to be fitted with pressure switches, an auxiliary lubrication pump and perhaps additional equipment to start up an auxiliary lube pump in the event of a main pump failure. For large power ratings the oil system will be fitted with oil coolers, and the design codes of manufacture need to be stated such as ASME VIII, Tema R (Ref. 1).

As stressed before, all instrument and electrical equipment must comply with the electrical hazardous area classification stipulated by the purchaser and relevant instrument specifications.

In the event of an order, a cooler data sheet would require to be completed in full.

4.3.9 Couplings

The type of coupling utilised varies with the driver eg. a steam turbine could be fitted with a flexible Metastream type coupling while an electric motor driven unit would be fitted with a "limited end float" gear coupling. With the latter, an oil lubrication system would be required if grease were to be unacceptable.

Special applications, where the fan is driven from each end by different prime movers sometimes require Clutch Couplings to be used to permit the automatic changeover from one prime mover to the other.

In the Petrochemical Industry it is essential that the couplings are fitted with non-sparking coupling guards.
Drawings require to be provided with the bid showing a section of the coupling, mounting and lubrication type. The coupling type, design code, service factor and lubrication all require to be entered in the data sheet.

4.3.10 Drivers

Normal procedure is for all drivers to be "free issue" to the Vendor. However, in order that the correct driver is obtained by the Purchaser, the fan vendor must produce torque versus speed curves for the rotating assembly with an open fan inlet at rated speed, the moment of inertia of the rotating assembly, maximum power absorbed values and the shaft critical speed. The last should be at least 20% above the maximum possible speed which can be achieved by the driver.

When using an electric motor as a driver, the maximum absorbed power at rated speed would be multiplied by a factor to obtain the nameplate rating of the motor; the rotating system needs to be sized to transmit this rating.

If steam turbine drive is to be used, the Purchaser would provide the turbine unit and gearing. The gearing is required in order that the turbine would run at a rotational speed giving maximum blade efficiency and subsequently saving steam consumption. It would be the responsibility of the fan vendor to produce the complete general arrangement drawing showing driver/gear/coupling/fan. All drawings relating to the drive equipment would be furnished by the Purchaser. If all the equipment is required to be installed on a single baseplate, it would be the fan vendor's responsibility to fit drivers and associated equipment.

In addition to the fan system, if the vendor is requested to supply the drivers, the responsibility will be with the vendor to ensure that all speed and power conditions are met. The turbine would be sized to drive the fan at rated speed delivering the maximum absorbed power at the minimum inlet and maximum exhaust steam conditions and the fan vendor needs to establish these conditions from the purchaser. All gearing and driver data sheets will require to be completed in full.

If a 'V' belt system is used to connect the driver to the driven equipment, the belts shall be of the Anti-Static, oil resistant type conforming to BS 1440 (Ref. 3) or equivalent. The fan vendor shall furnish the complete 'V' belt system including driver and fan pulleys and belt guard.

4.4 Extent of Supply

The evaluating engineer will tabulate the extent of supply of equipment, material tests, non-destructive tests, mechanical and performance tests, drawings and operating manual in the correct language, and the co-ordination responsibilities. It must be stressed that no ambiguity should exist in this area, as commercial and possibly legal problems could result after order.

When bidding equipment with ancillary equipment, and in particular instrumentation, it is good sense to produce a schematic drawing as shown (Fig. 3) in the symbology relevant to the project. This diagram would then be added to the data and would constitute part of the order document.

4.5 Drawings

In order to back up the technical descriptions provided in the bid, drawings of the fan together with all the major components associated with the unit require to be provided. It is essential to provide sectional and outline drawings in order to display all the relevant technical features of the unit together with the overall dimensions as emphasised in preceeding paragraphs.

In summary provide:-

. Completed data sheets
. Performance curve
. Outline drawings of complete unit
. Sectional assembly drawings
. Bearing assembly
. Seal drawings
. Schematic of flow control system
. Gearbox outline and sectional drawings
. Lube oil schematic
. Coupling documents

The section of the data sheet entitled Vendor's Drawing Commitment requires to be completed at the bid stage. This document focuses on the delivery dates of each drawing after order and the quantity of drawings to be supplied; these are agreed by the vendor at time of bidding. In the event of the bid becoming an order, the vendor would be held to these dates and failure to do so could possibly effect stage payments.

In order to help the fan vendor incorporate information relevant to the purchaser for integration of the unit into the plant design, a documentation specification provided by the purchaser should highlight all the information which is required to be seen on a particular drawing.

4.6 Shop Tests

Mechanical and Non Destructive Tests on Fan Components. These are stipulated by the Purchaser on the data sheet at the bid stage.

In the event of an order, the Purchaser will require to be supplied with all the necessary material certificates, weld certificates, tensile and Charpy (if required) test results; radiographic plates and the like, necessary to verify the Seller's compliance with the data sheet.

Tests on Fan and Auxiliaries Fans used on Petrochemical Applications shall be subjected to the following unless otherwise stated on the data sheet, namely:-

. Static and Dynamic Balancing - the minimum balancing speed shall not be less than 750 rev/min or at rated speed if lower than 750 rev/min.

. Overspeed Test- when a turbine is used as driver, the overspeed test shall be conducted at turbine trip speed.

. Operational Check - on Inlet Guide Vanes and Inlet Dampeners when supplied and on any lubrication system.

. Mechanical Run - this shall be at full load and preferably full speed for 2 hours during this test; vibration, sound and bearing temperature levels shall be recorded.

Performance Test. When specified on the fan data sheet, the Vendor would carry out performance at his test facility in accordance with BS 848 Part 1 (Ref. 4) or agreed equivalent Code.

The test results shall be furnished by the Vendor in the form of a constant speed characteristic curve of head versus capacity also indicating absorbed power and efficiency.

In the case of variable speed, inlet vane, or inlet dampener control fans, a series of characteristics covering the operating range shall be required together with the torque/speed curve.

It should be stressed that the tolerances on head, efficiency and absorbed power stipulated by the Purchaser can be more stringent than that defined in the particular code.

4.7 Performance Curve

At the bid stage, the Vendor requires to provide the following curves:-

. flowrate versus head (0% to 125%)
. flowrate versus efficiency
. flowrate versus power
. flowrate versus power considering the effect of suction dampers from open to closed position (as shown on Fig. 4).

Assuming the system resistance curve is proportional to the square of flow, it can be seen from Fig. 4 how much turndown can be achieved by the use of suction dampers without encountering pressure pulsations at appreciably reduced flows.

Special attention must be paid to head/flow curves with a more pronounced "hump" or unstable region than shown in the example, particularly when operating two or more fans in parallel. Operating points to the left of the maximum point on the head curve, may rise to instability causing a reduction in the total flow and possibly combined with reverse flow through one of the fans connected in parallel.

On any fan system, fan instability can be avoided if a careful selection of the fan together with the fan system curve can be made, so that the fan unit operates to the right of maximum head when using a radial fan and with an axial fan to the right of the surge line.

4.8 Guarantees

The Vendor shall require to guarantee the capacity, static discharge pressure and power absorbed as specifed on the data sheet. In addition to achieving the aerodynamic performance within the envelope of the tolerances stated in the data sheet the sound pollution produced by the fan unit must be kept to a sound pressure level acceptable to the purchaser.

The performance test specified on the individual data sheet, shall be in accordance with BS 848 Part 1, (Ref. 4) or equivalent codes. The sound pressure level spectrum between 63Hz - 8000Hz resulting at 1m from the fan discharge or suction must be guaranteed by the vendor.

4.9 Power Economy

In addition to the in depth evaluation of the fan unit as described in previous sections, an economic penalty is imposed on the less efficient equipment with the most efficient unit being the basis of the evaluation.

The extra power costs accrued over usually a 2 year period, can be added to the total cost of technical acceptable units and commercial judgement taken on the total cost.

5.0 INTEGRATION OF VENDOR'S EQUIPMENT INTO PURCHASER'S PROCESS SYSTEM

The object of procuring a piece of equipment is to use it to fulfil a particular role within the framework of the Purchaser's Process System.

In order to achieve this function, the unit requires to be provided with the necessary utilities and control by the Purchaser.

Information furnished by the vendor, such as static and dynamic loading imposed by the unit, enables the Purchasers Civil Department to design foundations while allowable nozzle loads on terminable flanges enables the Piping Stress Department to design pipe runs to the unit.

The Piping drawing office gathers information from all engineering departments and compiles a schematic diagram showing piping connections, extent of supply limits, pneumatic and electrical connections, and control necessary to ensure that the fan unit will fulfil the process requirement and the safe running of the fan, driver and ancillaries.

It is imperative that all equipment furnished is in accordance with the design codes governing during a particular contract, especially piping where mismatching of pipework could occur.

A typical schematic diagram showing the necessary utilities, and control logic for turbine/gear/fan/motor train is as shown in Fig. 5.

6.0 TYPICAL FAN APPLICATION (Refer to Fig. 5)

In the particular application shown (Fig. 5), the induced draught fan is used to control the pressure in the exhaust ducting for Flue Gases from a wall fired reformer furnace.

In view of the pressure drops resulting from the heat exchangers situated within the exhaust ducting the natural draught action would not function. Induced movement of the gas is therefore required to be provided by a fan.

The pressure controller located on the top of the exhaust ducting transmits a pneumatic signal to the suction dampener actuator producing a constant depressed pressure in the exhaust duct at any flow within the operating range of the fan.

The controlled pressure of the flue gas, is maintained constant to prevent blowout of the furnace in one extreme, while at the other ensuring that the vacuum drawn by the fan is not excessive thus pulling air through the ducting and furnace joints, consequently lowering the thermal efficiency of the plant, and introducing the possibility of refractory collapse.

In order to remove the optimum amount of heat from high temperature flue gases, systems as described above, including heaters and induced flow fans, require to be utilised in order to increase plant thermal efficiency to todays high standard.

Ref:-

1. American Society of Mechanical Engineers Boiler and Pressure Vessel Code Section VIII, Part 1.
   Tubular Exchanger Manufacturer's Association Classification R.

2. Anti-Friction Bearing Manufacturer's Association rating life based on testing identical ball bearings for a specified number of revolutions and load until 10% show evidence of fatigue damage.

3. British Standards Institution BS 1440
   Endless V-belt drives.

4. British Standards Institution BS 848
   Fans for general purposes. Part 1 Methods of testing performance. Part 2 Fan noise testing.

ISSUE	DATE	CH.			BLOWER DATA SHEET	JOB Nº	Sheet of

REVISIONS

PURCHASE ORDER Nº.	MANUFACTURER:	SIZE
CUSTOMER	ITEM Nº.	Nº. REQ'D
ADDRESS	PLANT LOCATION	
SERVICE	DRIVER TYPE:	

OPERATING CONDITIONS		1	2	3
Fluid analysis	gas Nº.			
	% Vol.			
	% Vol.			
	% Vol.			
	% Vol.			
	% Vol.			
	% Vol.			
Total				
Molecular weight				
Viscosity at suction T	cPs			
Corrosive compounds				
Erosion caused by				
Relative humidity	%			
Suction conditions	gas Nº.			
Temperature	°C			
Pressure	ata			
Flow	Nm3/h			
Discharge conditions				
Pressure	ata			
Tot. diff. head	mm H_2O			
Tot. diff. head	gas m			
Efficiency	%			
Absorbed power	KW			

PERFORMANCE	
Proposal curve Nº.	*
Flow design	*
Head design	*
Blower RPM	*
Guar. abs. power KW	*
Sound level - db	*
Min. flow - m³/h	*
Rotation facing coupling end	*
Water cooling	*
Bearing	*
PEDESTAL	*
Total water req'd m³/h	*
Water temp. °C	*
Water press. ata	*
WR2	*
Moment/torque (on flanges) kgm	*
Axial force kg	*
Radial force kg	*
Specific speed	
Driver power CV	*
Driver RPM	*

SHOP TESTS	Req'd	Witnessed
Mechanical run		
Performance air		
Performance gas		
Hydrostatic		

Operating: single ☐ parallel ☐

Flow regulation type:

Installation: indoor ☐ outdoor ☐ Roof ☐

Leakage: allowed ☐ not allowed ☐

Altitude above sea

Min/Max ambient temperature °C

SITE TESTS	
Performance	
Hydrostatic	Ata
Design press.	Ata
Design temp.	Ata
Max all. water press.	Ata

CONSTRUCTION AND MATERIALS

* Blower type

* Casing mounting Centerline ☐ Foot ☐ Vertical ☐

* Split: Axial ☐ Vertical ☐

* Seals type

NOZZLES	SIZE	RATING	NUMBER	POSITION
* SUCTION				
* DISCHARGE				

Suction vanes type

Discharge vanes type

* Casing thickness: Min/Max

* Corrosion allowance:

* Lubrication type:

Weights: Blower	*
Gear	*
Driver	*
Gear codes: AGMA	*
Service factor	*
Mechanical eff. %	*
Lubrication type	*
Coupling type	*
1st critical speed RPM	*
Test curve Nº.	*

MATERIAL CODE

* S-steel	Casing	S	C	A	X
* C-11+13% Cr	Blades	S	C	A	X
* A-alloy	Shaft	S	C	A	X
* X	Shaft sleeves	S	C	A	X

		GENERAL ARRANGEMENT DWG Nº *
		FAN SECTIONAL DWG Nº *
		SEAL ARRANGEMENT DWG Nº *
		CLUTCH DWG Nº *
		COUPLING DWG *
* TO BE COMPLETED BY	GEAR DWG Nº	* IMPELLER DWG *
VENDOR AT BID STAGE	GEAR DATA SHEET Nº	* BEARING ARRANGEMENT DWG Nº *
	DRIVER DWG Nº	* LUBRICATION SYSTEM DWG Nº *
	DRIVER DATA SHEET	* FLOW REGULATION EQUIPMENT DWG Nº *
	FAN CASING DWG Nº	* INLET SILENCER DWG Nº *
		OUTLET SILENCER DWG Nº *

DA.

DR.

CH.

APPROVED BY

SP 00006

Figure. 1

	CUSTOMER	JOB	UNIT
	PLANT LOCATION	SPC. N°	
	UNIT	Fg. - Sh./ /	Rev. 0

TECHNICAL SPECIFICATION

SERVICE ITEM No.

QUANTITY

ENQUIRY No. PURCHASE ORDER No.

ATTACHMENTS

CODES AND GENERAL SPECIFICATIONS

JOB SPECIFICATION

THIS SPECIFICATION COMPRISES OF THE FOLLOWING:-

1. GENERAL
2. DATA REQUIRED
3. SUPPLY LIMITS
4. MATERIALS OF CONSTRUCTION
5. SEAL DESIGN
6. BEARINGS AND LUBRICATION SYSTEMS
7. COUPLINGS
8. DRIVERS
9. METHODS OF FLOW REGULATION
10. JOINT RESPONSIBILITY
11. DRAWINGS AND DOCUMENTS
12. SHOP TESTS
13. POWER CONSUMPTION GUARANTEE
14. AVAILABLE UTILITIES DATA
15. FAN DATA SHEET

FIG. No. 2

0	Emissione - Issue				
Rev.	Descrizione Description	Prep'd	Chk'd	Appr'd	Date

Figure. 2

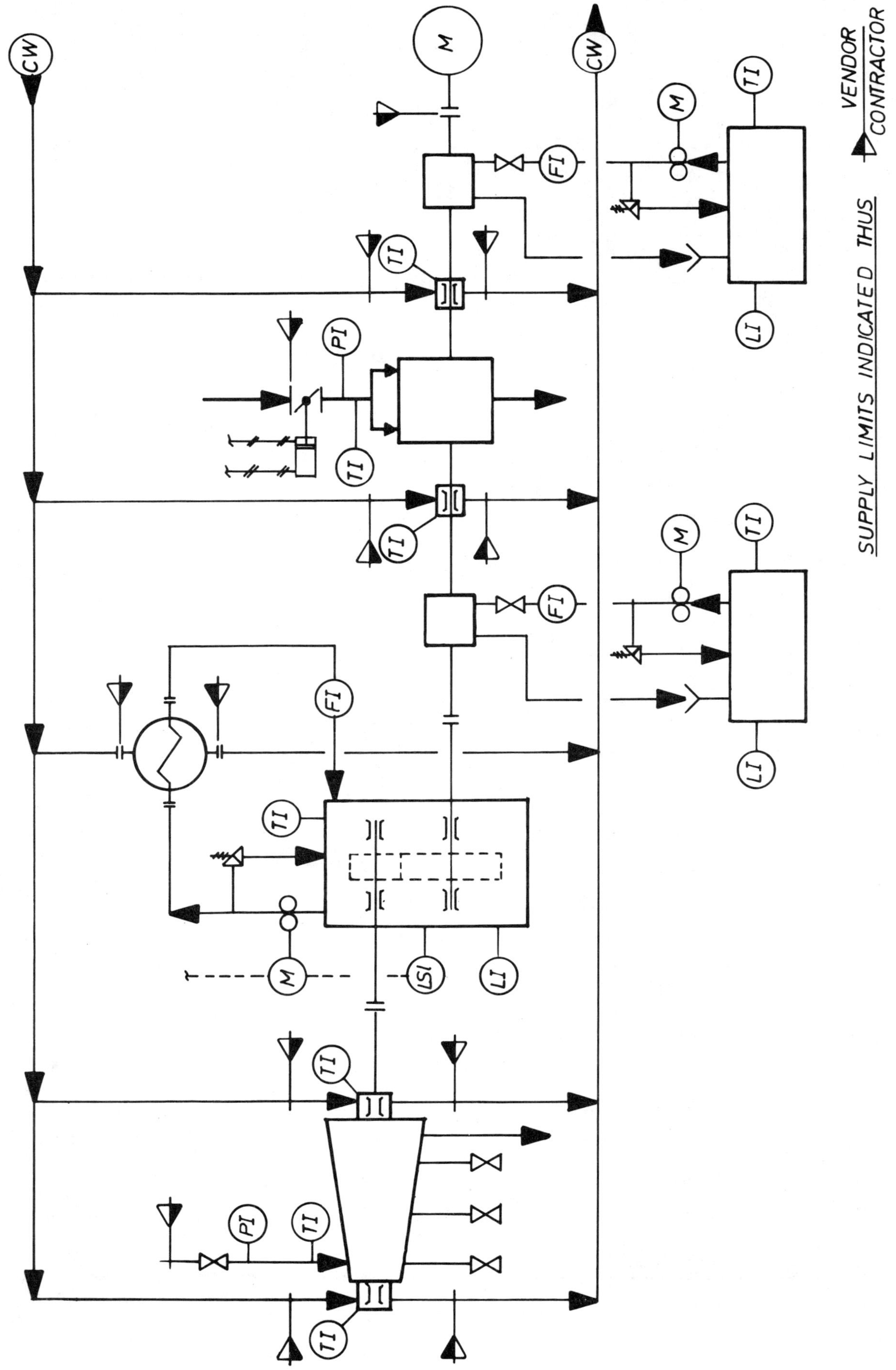

CW
PI
TI
TI
TI
M
TI
FI
LSI
LI
TI
PI
TI
TI
M
FI
FI
CW
M
M
LI
TI
LI
TI
SUPPLY LIMITS INDICATED THUS
VENDOR
CONTRACTOR

Figure. 3

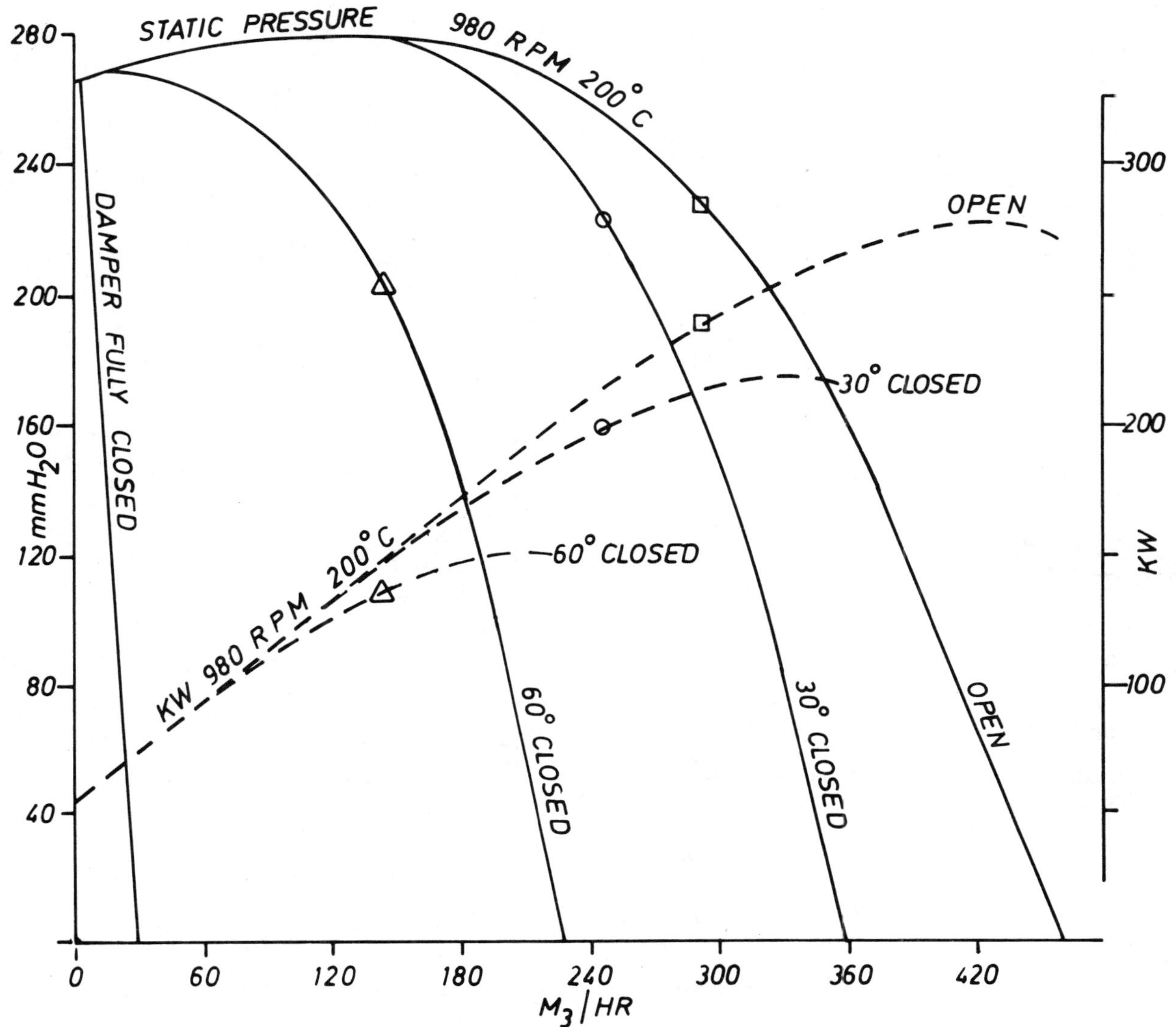

Fig.4.

Main characteristic showing the effects of head/power and flow with different suction damper settings.

△ - Dampers 60° Closed
○ - Dampers 30° Closed
□ - Dampers Fully Open

CW
LS
PI
HTI
Th
PIC
PI
FLUE
GAS
ST
ST
M
SE
ST
QA
I
PSI
M
ST
ST
ST
TI
LI
QA
HTS
I
CW
MC
LC
SV
I
HC
HS

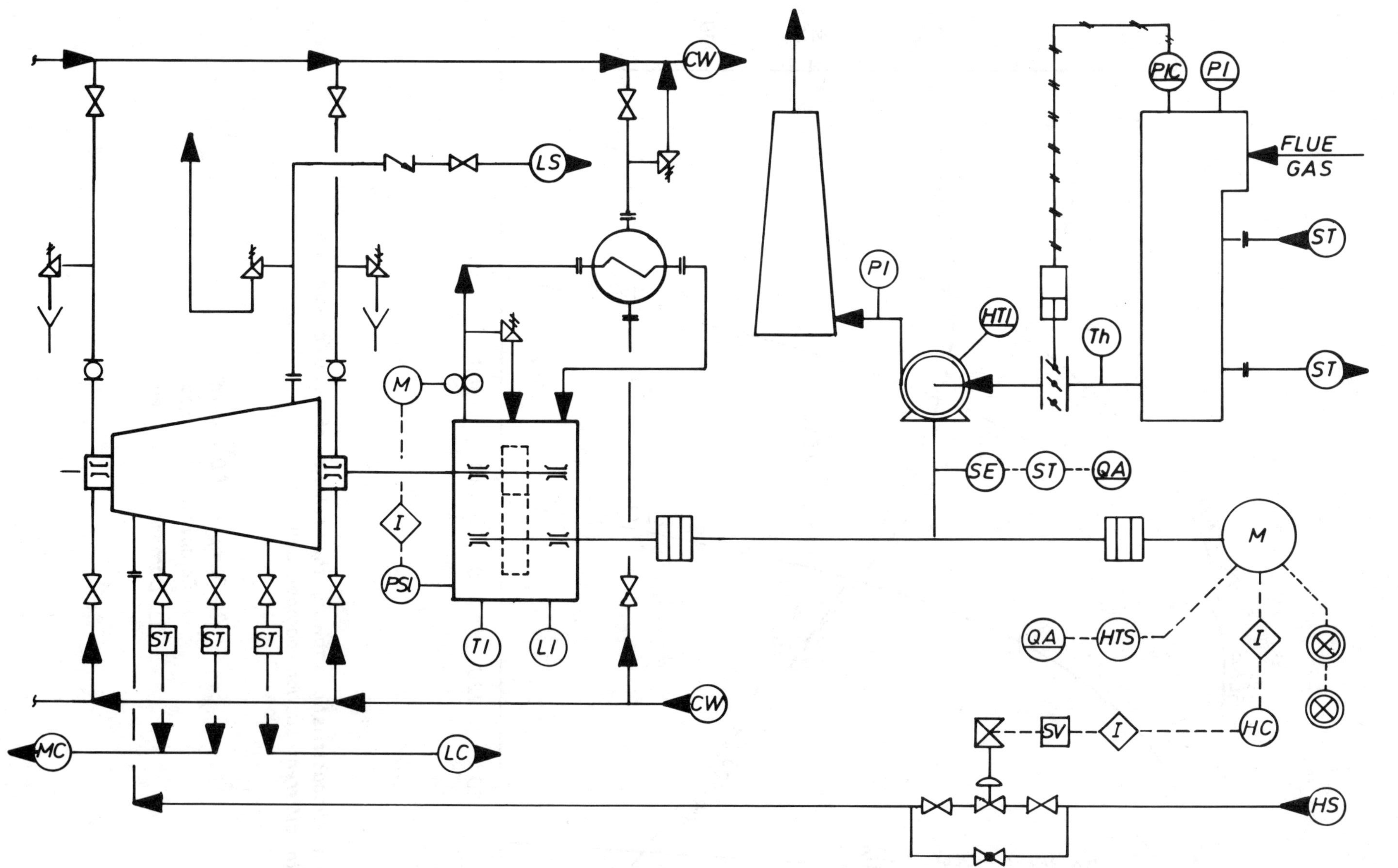

Figure. 5

International Conference on

Fan Design & Applications

Guildford, England: September 7-9, 1982

PAPER E3

DEVELOPMENT OF INERT GAS BLOWERS

J.G. Campbell and W.A. Brown

Davidson & Co. Ltd., Belfast, U.K.

Summary

This paper generally describes the function of an inert gas system and the development of Davidson and Company Limited inert gas fans to meet the modern stringent requirements for performance and reliability. An outline of the operational experiences which led to the selection of reliable materials to combat the arduous corrosive duties involved is given.

Organised and sponsored by
BHRA Fluid Engineering, Cranfield, Bedford MK43 0AJ, England.

0263 - 421X/82/01 00 - 0001 $5.00
The entire volume can be purchased from
BHRA Fluid Engineering for $82.00

NOMENCLATURE

C	....	Gyroscopic Couple
W	....	Impeller Dead Load
R1	....	Inboard Bearing Reaction
R2	....	Outboard Bearing Reaction
f	....	Frequency
fr	....	Running Frequency
fo	....	'Critical' Frequency

INTRODUCTION.

PURPOSE OF INERT GAS SYSTEMS ON OIL TANKERS.

The ever present danger of fire and explosion on board oil tankers has in the past ten years focused attention on operational safety of such ships in service. The consequences of such a fire or explosion represent grave danger to both life and property not to mention the disastrous economic results and pollution of the environment.

A flammable mixture of air and hydrocarbons in cargo spaces is the common cause of such a disaster, however the presence of this mixture can be prevented by blanketing the surface of the cargo with an inert gas.

In general the inert gas system protects against a tank explosion by introducing inert gas into the tanks to keep the oxygen content low and reduce to safe proportions the hydrocarbon gas concentration of the tank atmosphere.

The inert gas supply may be treated flue gas from the main or auxiliary boiler(s), from one or more separate gas generators or other sources or any combination thereof.

This paper concentrates on the former supply i.e. the inert flue gas system.

GENERAL DESIGN PARAMETERS FOR THE INERT GAS SYSTEM.

A typical arrangement for an inert flue gas system is illustrated in Fig. No.1. It consists of flue gas isolating valves located at the boiler uptake points through which pass hot, dirty gases to the scrubber and demister. Here the gas is cooled and cleaned before being piped to blowers which deliver the gas through the deck water seal, the non return valve and the deck isolating valve to the cargo tanks. A gas pressure regulating valve is fitted downstream of the blowers to regulate the flow of gases to the cargo tank. A liquid filled pressure vacuum breaker is fitted to prevent excessive pressure or vacuum from causing structural damage to cargo tanks. A vent is fitted between the deck isolating/non-return valve and the gas pressure regulating valve to vent any leakage when the plant is shut down.

To deliver inert gas to the cargo tanks during cargo discharge, de-ballasting tank cleaning and for topping up the pressure of gas in the tanks during other phases of the voyage, an inert gas deck main runs forward from the deck isolating valve for the length of the cargo deck. From this inert gas main, inert gas branch lines lead to the top of each cargo tank.

FUNCTION OF INERT GAS BLOWERS.

Inert gas blowers are primarily used to draw the cleaned flue gas from the scrubber and deliver it to the cargo tanks. Regulation 62 (chapter 11-2 of the International Convention of the Safety of Life at Sea (S.O.L.A.S.) 1974 Resolution A.418 (X1) requires that at least two blowers shall be provided which together shall be capable of delivering inert gas to the cargo tanks at a rate of at least 125% (per cent) of the maximum rate of discharge capacity of the ship expressed as a volume.

In practice, installations vary from those which have (a) one large blower and one small blower whose combined total capacity complies with the above regulation, to (b) those in which each blower can meet the total capacity.

The advantage claimed by the former is that it is convenient to use a small capacity blower when topping up the gas pressure in the cargo tanks at sea.

The advantage claimed by the latter is that if either blower is in-operative, the remaining one is capable of maintaining a positive gas pressure in the cargo tanks without extending the duration of the cargo discharge.

DEFINITION OF CUSTOMER DUTY.

The parameters specified by inert gas system manufacturers generally fall within the following range.

(1) a pressure head between 1600 to 2400mm.water gauge.

(2) a suction depression at fan inlet between -500 to -1000mm water gauge and 100mm water gauge over-pressure at tanks.

(3) a volume capacity up to 30,000 m^3/Hr.

Other factors are also considered when calculating the fan design duty e.g. inlet gas temperature, gas composition and density, sea water content, acid formation and the ship's electrical supply frequency.

GAS COMPOSITION

As mentioned earlier the inert flue gas system uses boiler exhaust gases as its inert medium. The oxygen content of the exhaust gas is maintained below 5% usually by a two level boiler combustion control system. Typically the main components of the gas by volume are as follows:-

O_2	....	4%
CO_2	....	13%
SO_2	....	0.03%
NO_2 & NO	....	0.01%
N_2	....	Remainder

The Gas is saturated with water vapour from sea water which is used to clean the gas in the scrubber unit. Actual water entrainment is of the order of 0.125% by weight.

It should also be noted that the gas entering the fan also contains solid particles of carbon and ash which are by products of the boiler. Concentration of these flue gas solids are of the order of 8 mg/mm^3 by weight.

CUSTOMER CONSIDERATIONS.

Davidson & Co. have supplied marine boiler and ventilation fans for the past seventy years and inert gas blowers for over twenty years. It is on the basis of the above accumulated experience, continuous research and development and close consultations with the inert gas system manufacturers and oil tanker operators, that it is now proposed to examine all aspects of inert gas fan equipment.

FAN TYPE

841mm Dia. PSC.08/5 Inert Gas Fan.

GENERAL DESIGN

As can be seen from the typical performance range already mentioned, a relatively narrow width high pressure type impeller is required. This type of impeller produces a performance characteristic as shown in Fig. (2a).

To maintain efficiencies over the required volume ranges it is necessary to vary the impeller width from 100% to 50% range giving a performance envelope as shown in Fig.(2b).

This facility is readily available with the PSC.08-5 design of inert gas fan as the basic performance envelope can be covered with the same full width casing.

The Davidson inert gas fan has a "between bearing" configuration, i.e. the fan impeller is mounted between two bearings on a stepped shaft and connected to the prime mover via a flexible coupling.

This Arrangement is illustrated on Fig. (3).

The main components of this fan are as follows:-

(1) Combined one piece casing/inlet box coupled with an integral bearing support structure.

(2) Backward Laminar bladed impeller.

(3) Stepped shaft.

(4) Inlet cone.

(5) Shaft seals.

(6) Bearings.

(7) Flexible coupling.

(8) Fabricated fan/motor or Fan/turbine underframe.

(9) Glass Flake or Rubber internal lining on casing/inlet box assembly.

Fan performance can only be reliably predicted on the basis of information established from extensive, rigorous research and testing programmes.

The minimum space requirements essential on board ship has set diameter and speed parameters to achieve a simple, compact and cost effective direct driven unit.

IMPELLER DESIGN.

The present impeller design in inert gas fans is the result of a long development period, both in the Design Office and particularly in service, (See Fig.4.)

Having decided on the basic design parameters as outlined previously, it then remains to select suitable materials to manufacture the impeller.

In the early sixties to accommodate the centrifugal stresses generated by the high speed impeller, it was found that a high yield strength material was necessary. The most favoured materials at that time were the high strength low alloy steels. These steels were quite attractive at the time because they were easily formed and welded and provided an impeller of high strength and integrity with very high safety factors on centrifugal and dynamic stress loadings.

There was however an obvious disadvantage. The unprotected steel would be immediately susceptible to the extremely arduous working environment caused by the gas handled by the inert gas system. To overcome this problem various forms of resistant coatings were tested and a stoved epoxy resin coating was selected.

This combination of high strength steel with epoxy coating went into service and operated quite successfully for a number of years.

However during this time a certain inherent problem became apparent. This was due to the inadequacy of the epoxy coating to withstand impact and abrasion. The most common fault was damage to the coating occurring during handling. Small defects in the coating could cause quite severe corrosive damage to the impeller and in certain cases complete failure occurred.

Investigations were carried out to find suitable alternative materials. Stainless steel designs were checked and rejected basically due to inadequate strength properties and corrosion resistance to sea water and chlorine products found in inert gas system atmospheres.

At this time aluminium bronze (To BS.2033:1963 10 per cent Aluminium Bronze (Copper-Aluminium-Nickel-Iron) Rods, Sections, Forging Stock and Forgings) offered a solution in that it combined good corrosion resistance with sufficiently high strength. Initially the material proved to be more difficult to fabricate and plate manufacturers advised hot pressing of the impeller conical sidesheet and stress relieving of the finish welded impeller. Although this seemed to cure the manufacturing difficulties, results in service were quite unsatisfactory and a number of impellers failed. These failures were found to be due to poor quality plate which was quite common at that time, but more significantly to a severe loss in ductility directly attributable to the heat treatment processes during manufacture of the impeller. This loss in ductility led to cracking in the hot worked plate and in the welded areas of the impeller.

Concurrent with these problems on the inert gas fans, significant developments were being made in fans designed specifically for chemical processes in wet sulphur dioxide gas handling applications. It was found that a new stainless steel derivative called 'Ferralium' was giving marked improvement in impeller life even over Incoloy 825. Ferralium is a duplex stainless steel i.e. Austenitic-ferritic and was primarily developed for marine applications. Typically in service it gave 50% extra life over Incoloy 825 and ten times the life of 316 type stainless steel.

Ferralium inert gas fan impellers proved very successful in service except for a few curious cases. A small number of fans were reported to be suffering from heavy vibration presumably due to out-of-balance. General inspection showed the impellers to be quite clean. However a closer inspection of the bottom half of the fan indicated that a large quantity of contaminated sea water carryover had collected due to a blockage in the drain system. This sea water had in turn eroded that section of the impeller which had been submerged, thus causing the resultant out-of-balance and vibration.

Another material which proved totally successful in hydrochloric acid plants was Titanium. This is not surprising as this material is practically inert, however, the initial cost of the fan was prohibitive and subsequently only one trial unit was sold for service.

It was at this stage, that an alternative supplier of aluminium bronze was found who could give consistently good quality plate. An extensive design exercise was then carried out to develop an impeller which would eliminate the manufacturing processes which gave rise to the failures on the earlier aluminium bronze impellers. It was also imperative that the new impeller design matched the efficiency of the previous fans. It should be said that although this aluminium bronze was metallurgically superior an extensive quality control standard had to be developed with the plate supplier to produce plate to the required flatness and thickness tolerances necessary for the satisfactory manufacture of high speed blowers of this type.

This resulted in the PSC Design which consists of straight backward inclined laminar blades welded between a parallel backsheet and sidesheet, thus eliminating the need for hot forming work. The simple blade profile allowes for machine cut weld preparation giving full penetration welds.

Since 1978 several hundred of these impellers have been manufactured and are continuing to give satisfactory service.

SHAFT AND BEARING DESIGN.

For inert gas handling applications shaft and bearing design is influenced by quite a few parameters.

Conventionally for fan design work, having established a suitable arrangement, the shaft must be selected to provide an adequate safety cover on resonant or natural frequency. The arrangement of the shaft and impeller can be predetermined to give optimum loading conditions and therefore maximise service life of the overall fan unit.

For the particularly arduous duties involved with these fans we can only consider two possible impeller/shaft arrangements, namely, impeller mounted between bearings on a two bearing shaft (Fig.5a) or impeller overhung on a two bearing shaft (Fig.5b). This decision is in fact quite simple to make by considering the loading effects on the bearings.

There are three basic loading factors common to both arrangements. These are, dead loads due to the weight of the rotating assembly and fan axial thrust, dynamic loads due to impeller unbalance and gyroscopic loads due to precession of the rotating assembly caused by normal ship motion. As can be seen from the two arrangements, the gyroscopic couple load is the same in each case, as the distance between the impeller and the nearest bearing is similar due to the fact that the span in arrangement 'a' and the overhang in arrangement 'b' must be kept to a minimum distance which is predetermined by the impeller, casing and bearing pedestal layout.

On the other hand, by considering dead loads and dynamic loads it can be seen from a simple resolution of forces that the bearing nearest the impeller in the overhung arrangement carries 50% more load than that in the between bearing arrangement. This means that for the between bearing arrangement a bearing which is more reliable can be selected with a longer life and obvious advantages in service and maintenance. It should also be said that the overhung shaft arrangement will always require a larger bearing size to give an equivalent critical frequency cover and in most cases bending stresses at the inboard bearing to the impeller are higher.

Having established that a between bearing arrangement provides a better long term proposition, the shaft must then be assessed to give safe cover between shaft natural frequency or 'critical' frequency and running frequency. The shaft proportions are suitably designed to take account of the bearing span, impeller weight and operating speed. At this stage shaft materials must be selected with suitable properties for stiffness and corrosion resistance.

In the development of the inert gas fan, shaft material selection followed a similar pattern to impeller material selection. Satisfactory materials were only found after operational experience. Originally mild steel was used with the coated alloy steel impellers. This material suffered corrosion problems especially in the region of the shaft seals. Attempts made to sleeve the shafts with stainless steel, improved the general corrosion situation, but fatigue failures occurred due to stress concentrations at the shrunk on sleeve. Stainless steel shafts were then tried with reasonable success but became suspect after vibration problems occurred when shafts distorted due to instability of the material after machining.

Then with the arrival of satisfactory aluminium bronze impeller material an electrolytically compatible material was required and the most suitable choice was in fact aluminium bronze to (BS 2874:1969: Copper and Copper Alloys Rods and Sections). This combined both high strength and stiffness with excellent corrosion resistance. Also, being a compatible material it obviated the need for sleeves and complicated hub mounting arrangements.

Having determined the shaft material and properties, calculations were carried out to determine the shaft natural frequency. Typically most fans operate at speeds in the subcritical range, and the system exhibits a first order frequency response characteristic. In fans handling clean air it is possible to reduce the margin between running frequency and natural frequency to 15%. However, in applications handling wet dirty gas where high dynamic loads due to unbalance are experienced, it is necessary to increase this margin to a minimum of 40%. (See Fig. (6)

Davidson inert gas fans are sold with a range of impellers with the heavier impeller/shaft combination having a typical ratio of critical frequency to running frequency of 1.5.

BEARINGS

Two Cooper type split roller, grease lubricated bearings are fitted, this design permitting easy inspection and rapid replacement of complete bearing or bearing races without the need to dismantle other components.

COUPLING

The coupling supplied on the standard inert gas fan using electric motor drive is from the "Holset" type heavy duty series.

A key feature of this flexible coupling is its multi-rubber block construction which uses the maximum volume of rubber working under compression to give high shock absorption and damping properties. Some of the proven advantages are as follows:-

(1) Control of shock and torque fluctuations extends the life of driving and driven machines.

(2) Radial, axial and angular misalignment capability.

(3) Flexible elements protected from external damage.

(4) Silent operation.

(5) Electrical insulation between driving and driven machines.

(6) No lubrication or maintenance servicing under normal operating conditions.

When steam turbines are used as the prime mover a "SPACER TYPE - METASTREAM" "M" series coupling is used.

This type of coupling was originally designed for oil refinery service but has now spread to all areas of power transmission and particular marine applications.

The "M" series is designed to accommodate angular and lateral misalignment and axial displacement created by the thermal expansion of the turbine. Another design feature is a non-sparking construction by using monel membranes together with non-ferrous parts.

CASING/INLET BOX

In the PSC inert gas fan design the casing and inlet box provide a combination of very important functions. The casing itself is proportioned to give a complementary volute to develop the full performance of the impeller with the maximum aerodynamic efficiency. The flush mounted inlet box is sized to produce the optimum inlet velocity to the impeller via the aluminium bronze inlet cone. The overall fan is sized to include the nominal loss factor of the inlet box. The casing and box are manufactured throughout in 10mm mild steel. The generous plate thickness combined with the integral stiffening provides maximum rigidity to withstand vibration due to aerodynamic pulsation throughout the performance range of the fan. This also results in a very high degree of sound attentuation.

CORROSION/EROSION PREVENTION.

As in all aspects of inert gas system design suitable precautions must be taken to prevent damage caused by the contaminated gases handled by the units. It has been found in the past that severe damage can occur to the casing with inadequate protection. The damage occurs due to the corrosive and abrasive content of the wet gas leaving the impeller at very high velocities and impingeing on the casing scroll.

It has been found more economically practical to use coatings on steel for the casing and box fabrication rather than the exotic materials used in the impeller construction. There is one component which is an exception to this and that is the inlet cone. This item is subject to heavy localised erosive attack and as such it is manufactured as a one piece casting in aluminium bronze.

INTERNAL AND EXTERNAL COATINGS.

There are a number of suitable coatings available for protecting the steel structure.

The present standard internal coating is a glass flake epoxy type which is applied on a shot blasted surface in the form of a primer, base and top coats to a thickness of between 35 and 40 microns. Customer requirements occasionally request an internal rubber lining which can be supplied in 3mm to 6mm thick rubber and includes spark and vacuum testing.

The present standard external coating is a Red Oxide primer paint. Additional or special paint finishes can also be supplied on request.

BEARING SUPPORTS.

Integral with the casing and box are deep section outrigger bearing supports. The primary function of this feature is to provide bearing housing and casing seal housing landings which can be machined in one set up. This ensures shaft alignment to these components and eliminates the various wear and maintenance problems associated with misalignment.

These outriggers also allow the casing and box to be rigidly supported on the underbed.

SEALING.

An absolutely vital component of the casing/box design is the shaft seal. There are two seals, one on the inlet box and one at the casing side sheet. Because of the fan configuration the inlet box seal is under suction and must prevent the ingress of atmospheric air which would in fact defeat the purpose of the inert gas system. The casing seal is, on the other hand, under pressure and it must prevent the escape of contaminated gases into the non-hazardous area where the fans are situated.

The Davidson inert gas fan seals are extremely effective combination type seals. The primary sealing member is segmented carbon ring held in constant contact to the shaft a by a garter spring. The secondary member is a 'Headland' lip type seal mounted on the shaft with the lip contacting a static keep plate.

Where the water washing facility is required the drive side casing seal has an additional cast iron drain chamber to facilitate the drainage of water which unavoidably escapes during the water washing cycle. This sealing arrangement is illustrated on (Fig.7).

The testing of this sealing arrangement was witnessed by one of the major inspecting authorities.

UNDERBED.

The underbed is designed to support the casing, the fan rotating assembly and the electric drive motor. It consists of a fabricated motor support base and bearing support base

fully welded to a deep section channel underframe. The support bases have integral machined pads which ensure the ease of alignment via the flexible coupling drive connection.
The heavy rigid construction enables the complete unit to be handled in one piece and it can be deck mounted without being affected by flexing which occurs on large tanker structures. When necessary the unit can be mounted on anti-vibration mounts without the need for further stiffening normally required with this type of structure.

WATER WASHING FACILITY.

The water washing facility consists of stainless steel pipework fitted internally to the casing/inlet box and directed at the fan impeller. To achieve maximum cleaning coverage the circular pipe has a 'fish tail' formed nozzle. An external mounting plate and pipe connection is positioned on the side of the inlet box.

ANTI-VIBRATION MOUNTINGS & FLEXIBLE CONNECTIONS.

The anti-vibration mountings and flexible connections perform a dual function in isolating the fan vibrations from the ship and vice versa.

CONCLUSION

Although the company has been involved in marine applications and also inert gas applications from around 1935 the previous discussion covers the more significant design changes in Davidson inert gas fans from the early sixties. Over this period the company has manufactured more than one thousand inert gas handling fans.

The manufacturing and operational experience gained over this period has produced what many fleet operators believe to be the optimum design to accommodate the arduous environment inherent in the inert gas system.

In order to overcome these conditions the design offers two important advantages. The first is that the impeller and shaft are manufactured throughout in aluminium bronze. Excellent results have been obtained with this material with minimal evidence of corrosion or erosion. Also, despite gradual deterioration in the acidic environment it can be regularly inspected to anticipate replacement, thus preventing sudden catastophic failure.

The other advantage in the design lies in the between bearing shaft design as mentioned earlier. Rotational equipment mounted on board ship suffers gyroscopic loads in additon to the normal rotating and unbalance loads experienced on a solid foundation. As these loads cannot be eliminated then the design must reduce their effects on the bearings as much as possible. The accepted and proven method in obtaining this improvent in bearing life has been to mount the impeller between the bearings.

In summation it must be said that success in this fan design has been obtained by operational experience and close consultation and co-operation with the customer. Although new technologies have been investigated in attempts to find reliable materials the present design uses only conventional fan technology to achieve its total reliability.

ACKNOWLEDGEMENTS

The Authors wish to thank the Directors of Davidson and Company Limited for their permission to publish this paper and their colleagues in the Company for their assistance in the preparation of the paper.

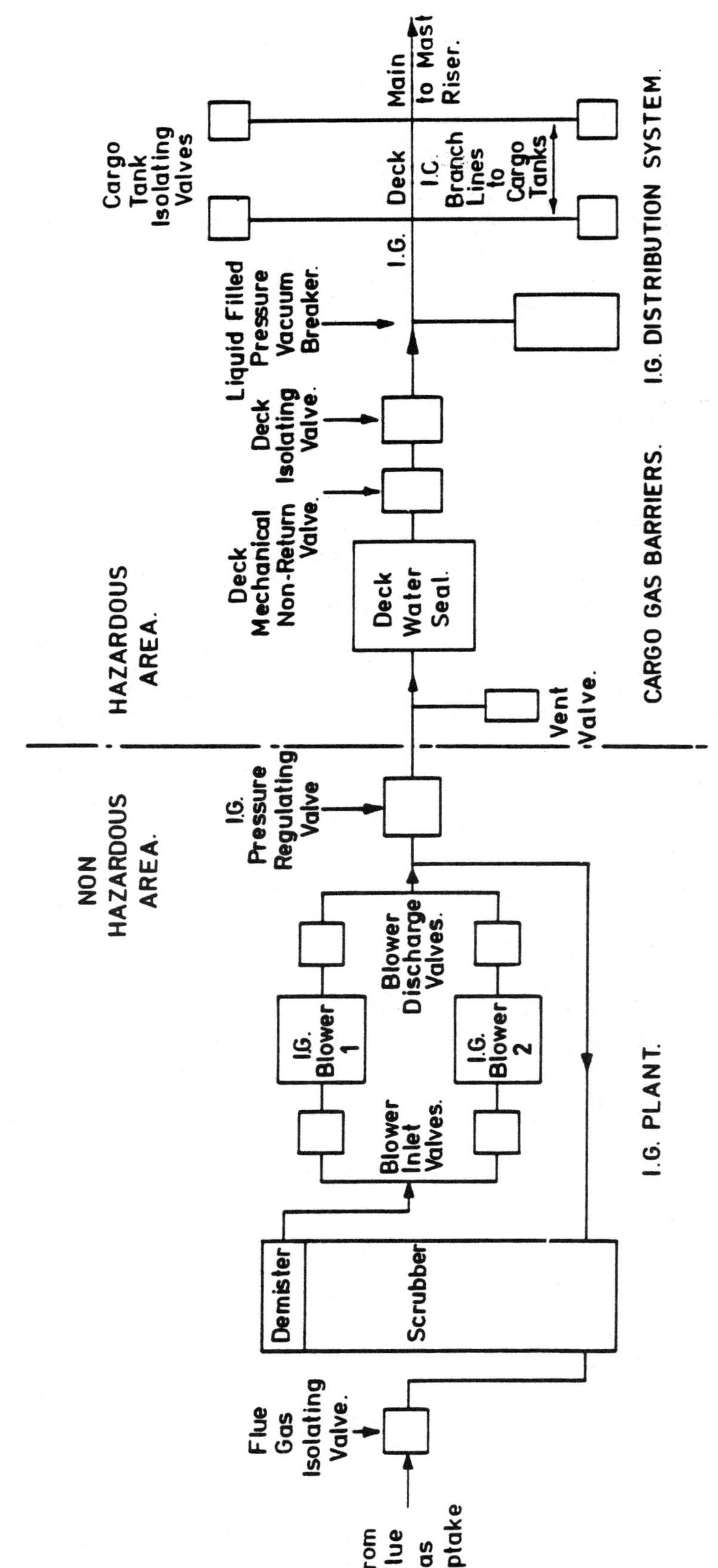

TYPICAL ARRANGEMENT OF INERT GAS SYSTEM FIG. 1.

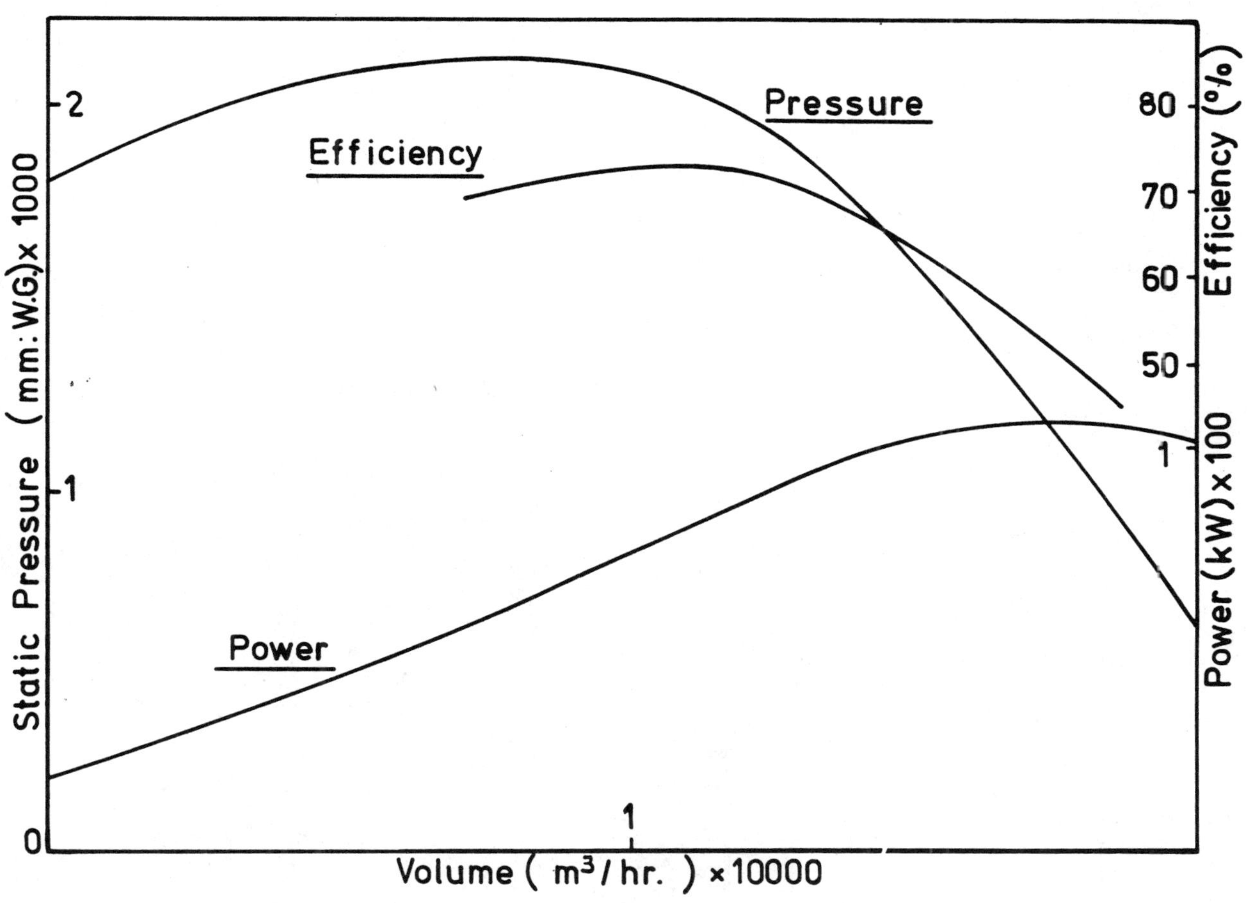

PSC 08/5 Performance Characteristic Fig. 2(a)

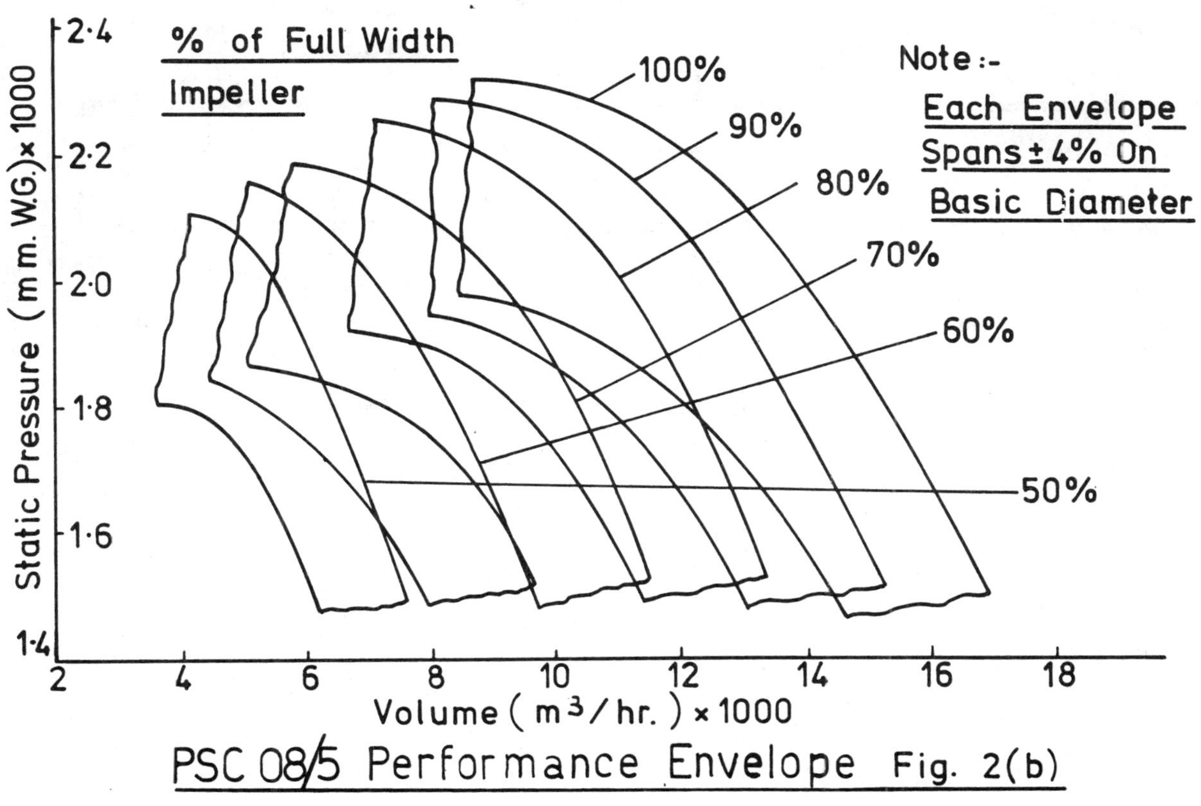

PSC 08/5 Performance Envelope Fig. 2(b)

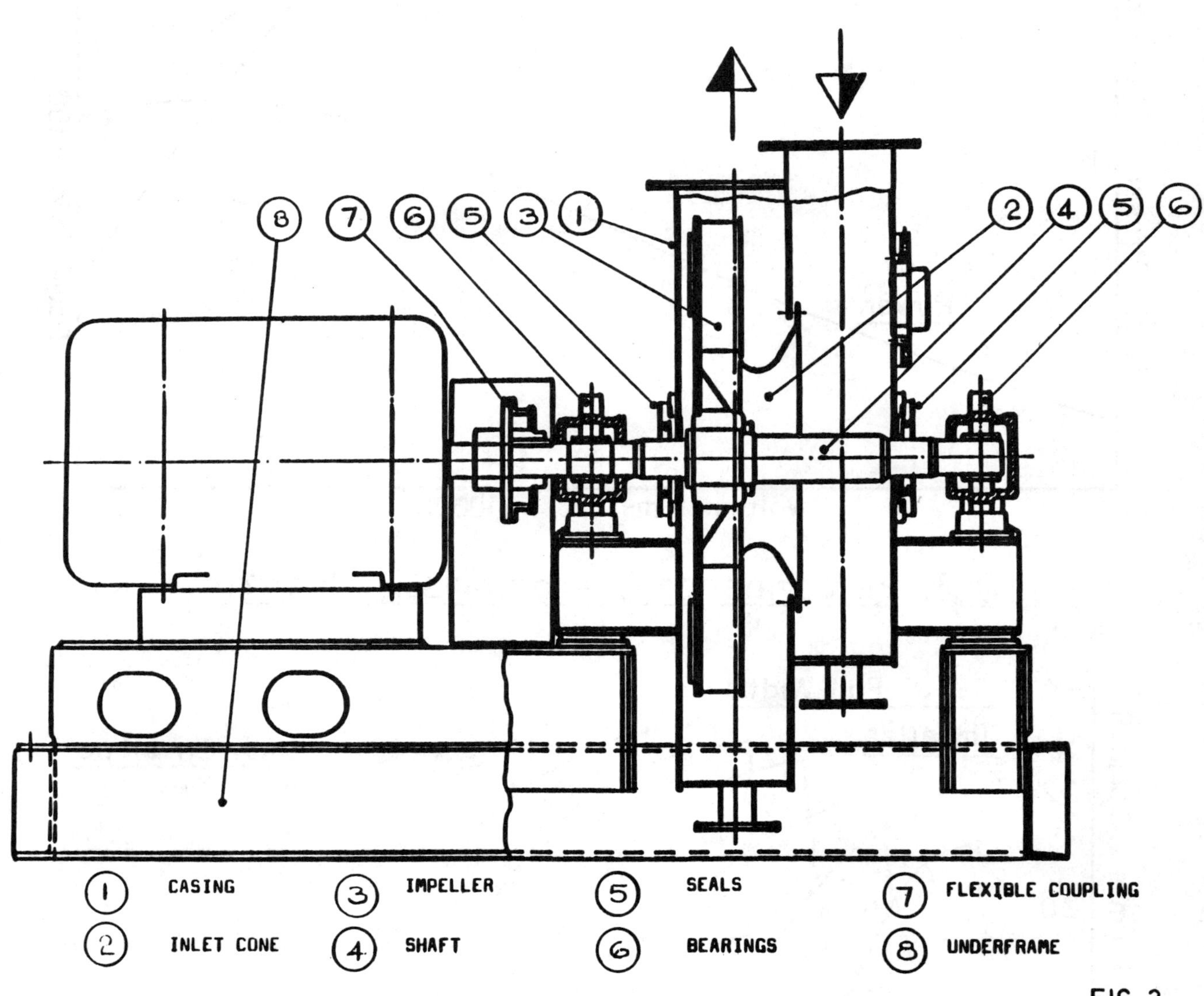

FIG. 3.

GENERAL ARRANGEMENT OF INERT GAS FAN.

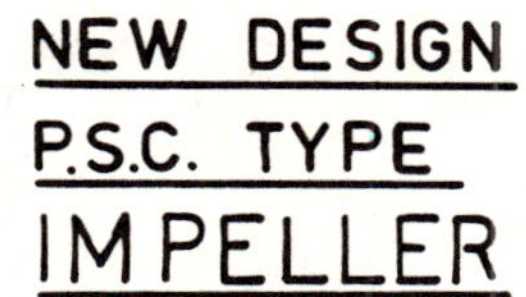

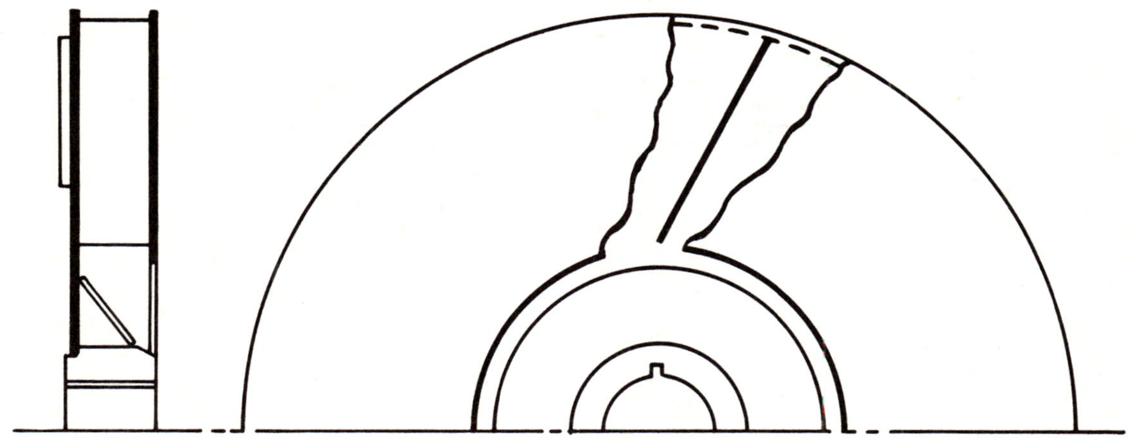

FIG 4(a)

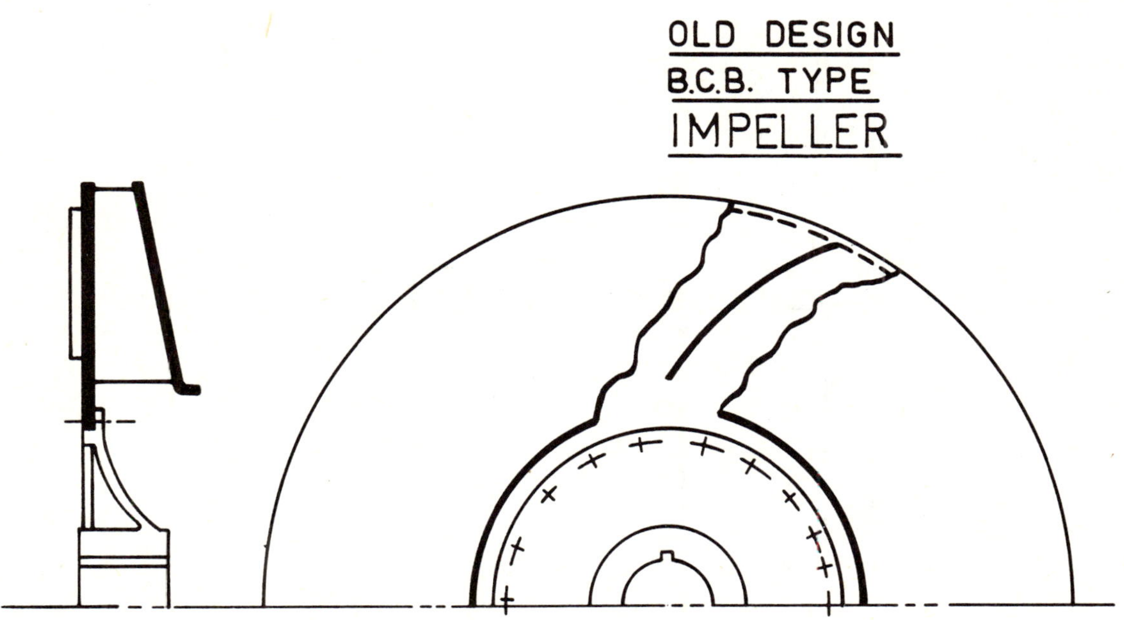

FIG. 4.(b)

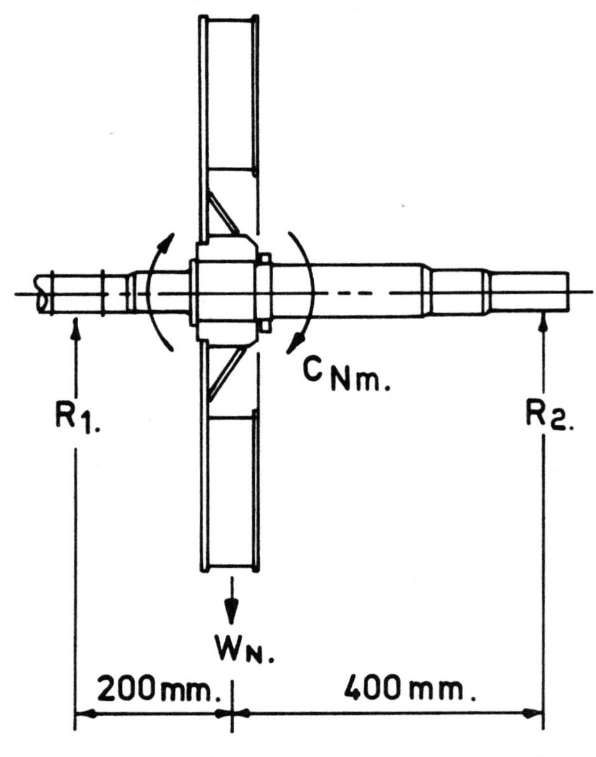

Between Bearing Arrangement FIG. 5 (a)

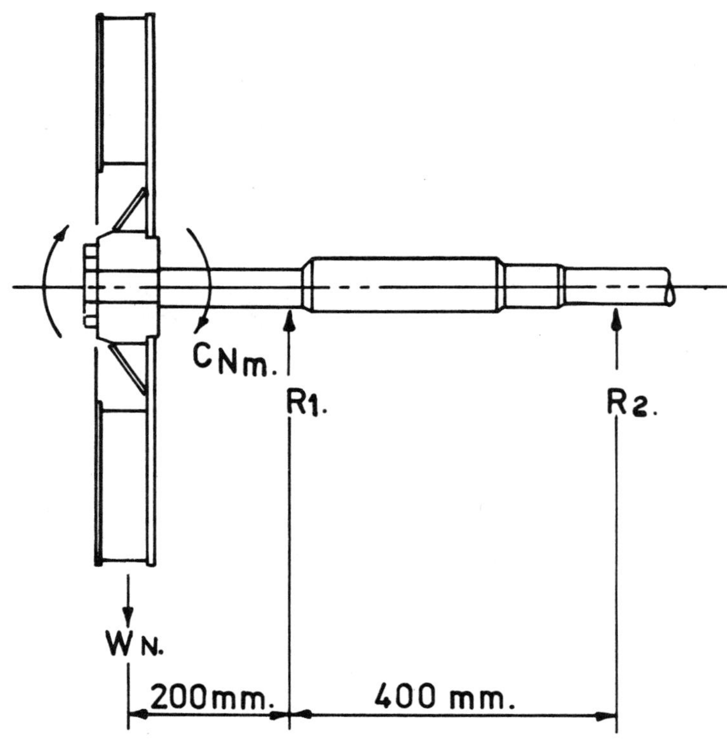

Overhung Arrangement FIG. 5(b)

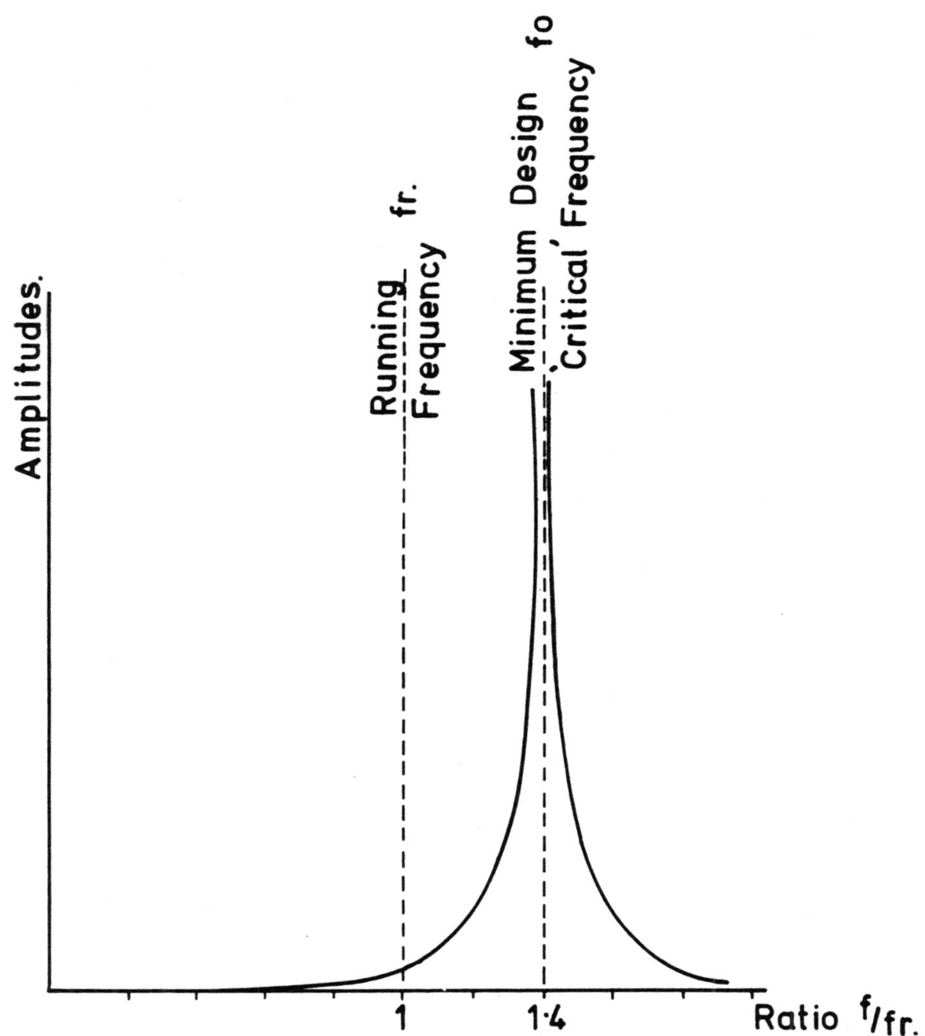

CRITICAL FREQUENCY DESIGN PARAMETERS FOR DAVIDSON & Co. I.G. FAN SHAFTS.

FIG. 6.

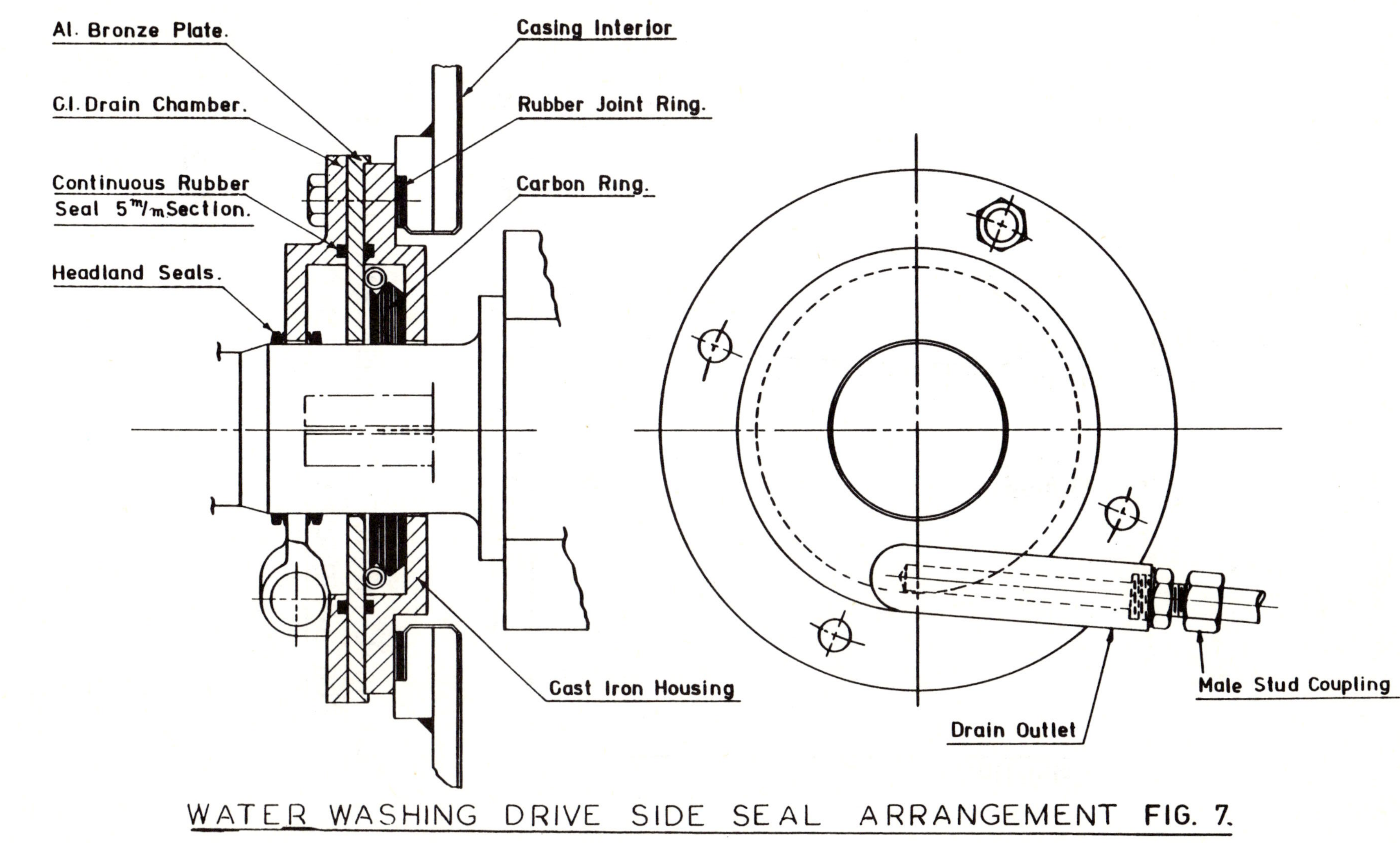

WATER WASHING DRIVE SIDE SEAL ARRANGEMENT FIG. 7.

International Conference on

Fan Design & Applications

Guildford, England: September 7-9, 1982

PAPER F1

LARGE WINDTUNNEL FANS

S. E. Richardson

Permali Gloucester Ltd., U. K.

Summary

Permali Gloucester Limited has been involved with the design and manufacture of windtunnel fans with wood blades for over 40 years.

In the 1950's and 1960's most of these fans were for relatively small windtunnels with fan diameters in the range 1.2 m to 3.6 m.

During the mid 1950's one windtunnel fan unit was made with a diameter of 6.4 m. A description of this unit with details of the manufacturing technique is presented.

After 1970 more larger windtunnels were being considered and Permali were involved with some of these. Fans were made having diameters of 6.1 m, 13.1 m and 12.2 m.

A description of the manufacturing processes for the 13.1 m diameter fan is given. This fan was built for a full size motor vehicle testing windtunnel constructed by General Motors at their Technical Center in Detroit. Reference is also made to the installation procedure.

Six fans, each of 12.2 diameter, comprise the power unit in the largest windtunnel in the world at the NASA/Ames Research Center in California. Permali were responsible for the structural and manufacturing design of the blades for these fans and then carried out the manufacture. Descriptions are given of the testing involved at the design stage and the manufacturing processes.

Some reference is also made to the testing requirements of the fans after installation in the windtunnel.

Organised and sponsored by
BHRA Fluid Engineering, Cranfield, Bedford MK43 0AJ, England.

0263 - 421X/82/01 00 - 0001 $5.00
The entire volume can be purchased from
BHRA Fluid Engineering for $82.00

1. INTRODUCTION

Permali Gloucester Limited has been involved with the design and manufacture of rotor blades, mainly of wood construction, for over 40 years. In the early days and during World War II these blades were mainly for aircraft propellers, but after the war other applications were developed.

These applications ranged from helicopter rotor blades (as a logical development for aircraft), to commercial fans, primarily for water cooling fans; even some experimental wind turbine generator rotors were made.

Due to involvement in the aircraft industry it was natural for specialisation to be made with customer designed fans for wind tunnels.

This paper traces the development from the early days of fixed pitch aircraft propellers to the world's largest axial fan unit with six fans each of 12.2m diameter.

2. HISTORICAL BACKGROUND

The design and manufacture of fixed pitch propeller blades started, of course, well before World War I and the basic technique has not changed even today although improvements have been made in adhesives and coating materials for surface protection.

During World War II variable pitch propellers were essential and this led to the development of detachable blades with their individual root attachments capable of supporting the large root loadings.

The favoured type for propeller blades used compressed wood made from thin veneers of a suitable species of timber (usually Canadian birch) bonded with a phenolic resin adhesive under heat and pressure to give a material comparable in strength to mild steel but only some 16% of its weight. This material could then be turned into a conical root shape with a suitable thread cut on to which a steel adaptor could be screwed.

After the war, during the original development of helicopter rotors an alternative method was devised. This consisted of a rectangular root section on to which was bonded flat metal plates. It has been found that metal could be bonded by using a special type of adhesive requiring heat and pressure and this process was adapted for metal to wood bonding by using thin veneers which would stand the high temperature (150°C) required.

3. SMALL WINDTUNNEL FANS

During the 1950's there was an extension of the aeronautical development that had taken place during the war and this led to requirements for trained personnel. Many technical colleges and universities had aeronautical departments and these soon found the necessity for adequate windtunnel facilities.

Although not all super-sonic windtunnels have fans, often being what is known as "blow down" tunnels, in practically all slow speed tunnels the fan is a vital component.

The Company supplied many of those that were required, the majority being purpose designed and manufactured.

The small models were usually based on the configuration of the original fixed pitch propeller and this type was normally suitable for fans up to about 2m in diameter. As a very approximate guide the working section area of a small slow speed tunnel is half that of the fan area so that a windtunnel using a 2m diameter fan has a working section area that is of suitable size for the models to demonstrate aerodynamic principles at the university level.

However, larger and more sophisticated tunnels were built requiring fans somewhat larger than 2m and it then became necessary to have multi-bladed fans (more than 4 blades) so detachable blades were required. The design of variable pitch propellers with compressed wood blades and steel root adaptors was adopted for some of these fan blades.

Depending on the application, (provided that the rotational speed is not too high) natural wood blades were more economical so models were built of this type with the root retention based on helicopter blade experience. These fan blades had flat metal plates, usually high strength Aluminium alloy, bonded to a substantially rectangular root section. Such blades were then clamped, with two or more bolts, between two metal discs to form the complete fan impeller.

4. AIRCRAFT RESEARCH ASSOCIATION FAN

During the middle 1950's the aircraft industry found the need for an independent windtunnel so that it did not have to rely on the tunnels at the Government research establishments.

The design eventually developed was for a trans-sonic windtunnel (with a maximum working section speed of 450m/s) requiring a fan to absorb 18,600 KW. The design of the fan was a co-operative effort between ARA and the Company. The aerodynamics of the tunnel circuit led to the detailed aerodynamic requirements of the fan which gave a diameter of 6.5m with two impellers each having 20 blades. These blades had an aerodynamic length of 1.06m with a tip chord of 0.6m. Fig.1 shows a general view of this fan (1).

A natural wood blade was considered most appropriate to keep the centrifugal loading down and mahogany was the final choice, being the standard material for fixed pitch propellers at that time.

It was found that the most appropriate type of root for this size of blade and hub design should be based on the bonded plates scheme but in this case with forged steel plates.

The basic method of manufacture for all these wood blades is to start with suitable material grade and select it and, if necessary, make up boards to the required width by edge gluing. Such boards are then planed to a common thickness, (usually 20mm) and the lamination shape, which is established during the design phase, is cut out.

Laminations are glued to form a block of the approximate shape of the blade profile. It is common practice to use a resin glue, resorcinol based, that has gap filling properties and is not subject to the effects of moisture.

After the main block is glued the root assembly is fitted. This particular fan for ARA had steel root plates of L-section machined from forgings. The bonding adhesive requires heat (at 150°C) and pressure so it is only practical to bond a thin veneer to the plate surface before the assembly is cold glued to the block. Additional retention bolts were used to clamp the plates to the main block.

The blade profile shape is achieved by machine shaping using a three dimensional copying machine, the finished section being achieved by hand work. The blade surface is then protected by the addition of a glass cloth cover bonded with epoxy resin.

Each blade is checked for balance measured against a master value and extra weights are added at the root to give a balanced set.

The original manufacture of these blades took place during 1955 and the windtunnel was opened in early 1956. A spare set of blades was manufactured soon afterwards and these two sets were used until 1981. A further set was made in 1981 for current use and it is intended that a selection or the original blades should be refurbished to be held as spares.

Frequently the question arises as to how the operational life of a fan can be forecast. There is no better method than to be able to show that a similar model has operated for over 25 years.

5. GENERAL MOTORS FAN

In the early 1970's General Motors found that they required more windtunnel capacity than they could conveniently hire by using aircraft tunnels. It was decided to build at their Technical Center, Warren near Detroit a slow speed windtunnel especially designed for testing road vehicles. The design chosen was a closed circuit tunnel some 120m long x 45m wide having a working section 21.3m long x 10.5m wide x 5.5m high with a fan 13.1m diameter driven by a 3000KW motor.

The working section of this tunnel is large enough to take full size models or even production cars and most of the testing is carried out at a wind speed of 24.5m/s (55 mph) the legal maximum road speed in the U.S.A. The tunnel also has the capability of testing models of large commercial vehicles, usually at one third scale, when a maximum air speed of 67m/s (150 mph), or three times road speed, is required. An outline sketch of the tunnel is shown in Fig.2 (2).

The design of the fan is similar to that of the fan shown in Fig 3. Six blades each with an aerodynamic length of 3.8m have an integral root block 1.8m long x 1.2m wide x 0.76m thick. The six root blocks are retained between two steel hub plates, 5.5m in diameter, by 12 bolts. The hub plates have concentric grooves cut into the inner faces to increase the friction grip on the blades when clamped tight.

The fan can run at a maximum rotational speed of 300 rev/min. The finished assembly is shown in Fig 4.

Various production problems were encountered due to the size of the blades. The root thickness, 0.76m, required some 40 laminations to be glued together. It was not possible to accomplish this in one operation as the glue must be spread, the block built up and pressure applied within about 30 minutes. It was found necessary to make 4 sub-blocks each of 11 laminations thick which could be glued within the required time span. These sub-blocks then had to be re-surfaced before being glued together.

The standard technique of shaping was employed. Straight forward machining was used for the flat surfaces and the plan form profile was also machined. The rough edges of the glued block were then axed away and finally the section shape was formed by the use of power and simple hand tools such as spokeshaves and planes to give the correct blade angles and contour shapes. These section shapes were all checked with reference to profile templates previously cut to the required dimensions, and the areas between the sections blended to smooth contours.

Considerable effort was made to achieve the appropriate tolerances for the blade retention holes and the mate holes in the hub plates. It can be appreciated that boring a cluster of 12 holes, 82mm diameter, through a block of timber 0.76m thick is a somewhat hazardous operation to ensure that they break through in the correct position to mate with the corresponding holes in the steel plates.

It was arranged that this fan should be fully assembled in this country before being broken down, crated and shipped to the U.S.A. The various hub components were made in different localities subsequently to be brought together at one workshop for assembly, leading to some transport problems.

Check measurements were made by both the Company and the customer (in fact by engineers from Vauxhall Motors, General Motors' U.K. Company) at this assembly stage. Very close limits were achieved for such a large fan with wood blades, the tips being correct within about 2mm for sweep and 3mm for track.

Towards the end of the construction of the tunnel the blades had to be fitted to the hub which had been assembled to the motor previously positioned during the building of the main shell. A small aperture had been left at the top of the circular duct section through which the blades could be lowered.

During this phase the blades had their tips trimmed to the appropriate length to fit the duct radius as made. A problem does occur with such large ducts to ensure that they are reasonably circular so that a minimum tip clearance can be achieved.

After the blades had finally been assembled, the last balance check was made. It was found that a mass of only 9kg was required to be added at the outer radius of the hub plates to compensate for the out-of-balance recorded. This balance mass is very small compared with the total impeller mass of some 25,000kg.

6. AMES RESEARCH CENTER FANS

Ames Research Center, a department of NASA, situated at Moffett Field, California has the world's largest windtunnel. (It is even recorded as such in the Guinness Book of Records).

This windtunnel, known as the 40 x 80, was originally built in 1944 with a closed circuit, the working section being 40ft high x 80ft wide (12.2m x 24.4m). The fan section consists of six axial flow fans each 12.2m diameter in two rows of 3 fans with a total area of 24.4m x 36.6m. The original fans each had six wood blades and were similar in design to the General Motor fan previously described.

Some years ago it was decided to rebuild the tunnel by adding another open section leg to give a dual facility using the same fan section. The new section then gives a through tunnel with a working section of 24.4m x 36.6m, but the original closed circuit feature was retained. This modification required the fan section to be rebuilt and it now consists of six new fans, each of the same nominal diameter as previously (12.2m), but these fans each have 15 variable pitch blades and are powered by 16,400KW motors. When the open section tunnel is in operation the fans deliver a maximum volume of 50,000m^3/s. Fig 5 shows an outline diagram. (3)

The re-design of the tunnel was undertaken by Ames themselves and they laid down the basis for the design of the fans, including the detailed aerodynamic design of the blades.

The variable pitch hub design is relatively unsophisticated (at least in comparison with aircraft propellers) and pitch change is obtained by an electric motor which driving through a gear box rotates retention shafts on to which the blades are fixed.

The limitation of the electric circuitory requires that the fans are run up to full rotational speed, 180 rev/min, while in flat pitch and then the pitch change mechanism comes into play to increase the air flow through the fans requiring greater power.

Permali Group were awarded the contract for the manufacture of the blades including the structural and manufacturing design and certain on site technical services.

The design of the blades submitted to and approved by Ames was based on the type, previously referred to, having a compressed wood root fitted to a steel adaptor capable of being bolted to the Ames designed blade retention shafts. Due to the size of each blade the main aerodynamic section was designed for manufacture in Canadian Sitka Spruce, the transition between the two portions being achieved by scarf jointing. Fig 6 shows some completed blades before despatch and Fig 7 an assembled fan.

Considerable development testing was carried out before the design was finalised and testing was even continued throughout the manufacturing period.

Full scale root pull tests were undertaken on double ended specimens to check the adaquacy of the root and adaptor design. Small scale fatigue tests were carried out on a coupon samples of the type of compressed wood to be used before full size root specimens were built and subjected to fatigue loading to demonstrate the expected life of the blades.

Production blades were subjected to static bending loads to verify that excess deflection would not occur in service. These tests also enabled strain gauges, that were fitted to four blades, to be calibrated. At this stage tests were also carried out to determine the natural frequency of a selection of blades.

The manufacturing processes are very similar to those already described except that in the initial stages the specially made compressed wood has to be scarfed to the natural wood.

After the main block was glued together the root cone was turned and the thread cut to allow the steel adaptor to be screwed on to the assembly. The retention holes, positioned to mate with the corresponding holes on the retention shafts, were drilled after the adaptor was installed.

Due to the difficulties in making these large ducts truly circular, whether in steel as for these fans or concrete such as for the GM fan, it is usual to manufacture the blades slightly longer than required and trim them on site before installation.

A survey of the different modes of Natural Frequency of every blade has been made after the blades were installed on to the hub.

As previously stated, four blades had strain gauges applied at the manufacturing stage and these were calibrated against applied bending loads. This will enable a full survey of the operating loads to be recorded during the development running. At the date this paper was written there are no results available as the tests have not been carried out. Possibly the results may be available at the date of the Conference.

The whole programme of re-designing and re-building this very large windtunnel has taken many years and is now in its final phases. The fan section with its six fans is very impressive and it is extremely unlikely that any other windtunnel larger that this will ever be constructed.

7. CONCLUSION

Wood was perhaps the first major material to be used for rotor blades. Many years ago, about the time of World War II, new materials were being developed which seemed to have superior properties and lent themselves to mass production techniques. This led to the view that wood fans would soon become obsolete. However, over the last 30 years there is evidence to show that in the realm of large or very large fans, particularly for windtunnels, wood is still one of the most appropriate materials to be considered for the blades.

It is ironic to consider that the designs for future jet aircraft and even space vehicles, some of technology's latest developments, are being based on data obtained by the help of simple fan blades made of wood in a similar manner to the way aircraft propeller blades were produced before World War I.

REFERENCES:

1. AIRCRAFT RESEARCH ASSOCIATION LTD. - TRANSONIC WIND TUNNEL - Brochure for Opening Ceremony - May 1956.

2. GENERAL MOTORS CORPORATION - Press Release - September 1980.

3. JOURNAL OF AIRCRAFT - AMERICAN INSTITUTE OF AERONAUTICS AND ASTRONAUTICS - Improving Large-Scale Testing Capability by Modifying the 40- by 80-ft Winch Tunnel - K.W. Mort, P.T. Soderman, W.T. Eckert - Vol.16 No. 8 August 1979.

Fig 1 ARA FAN UNIT

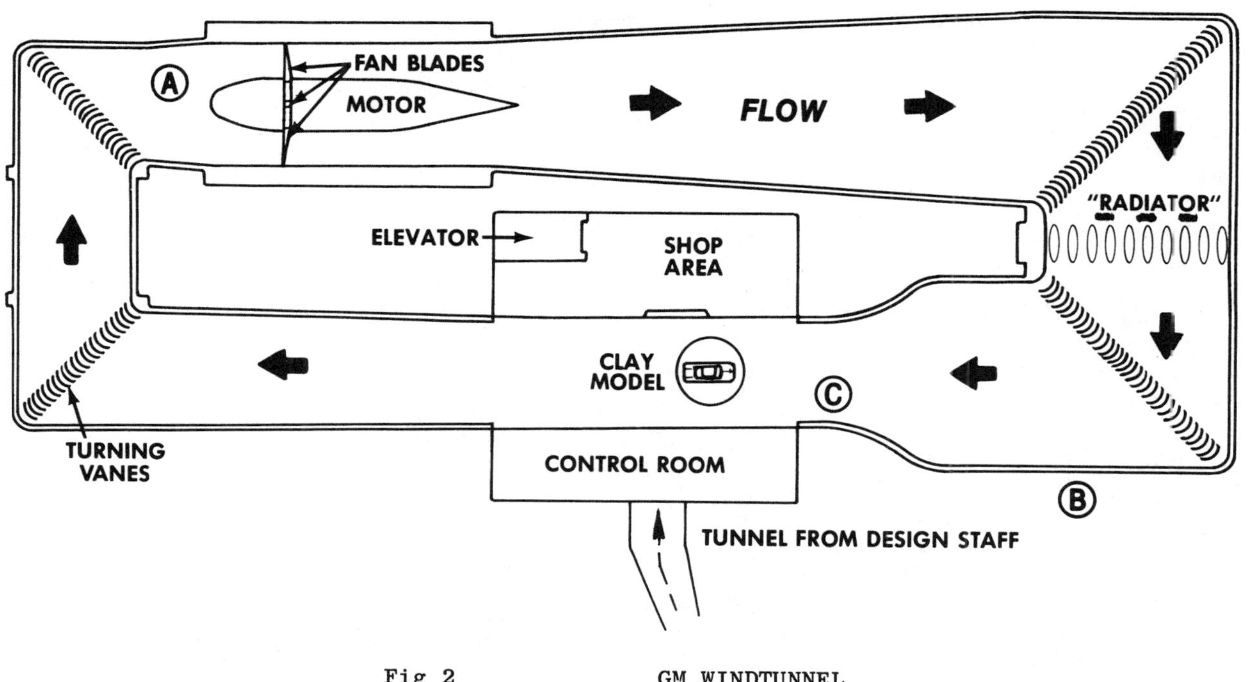

Fig 2 GM WINDTUNNEL

Fig 3 FAN ON BALANCE RIG

Fig 4 GM FAN

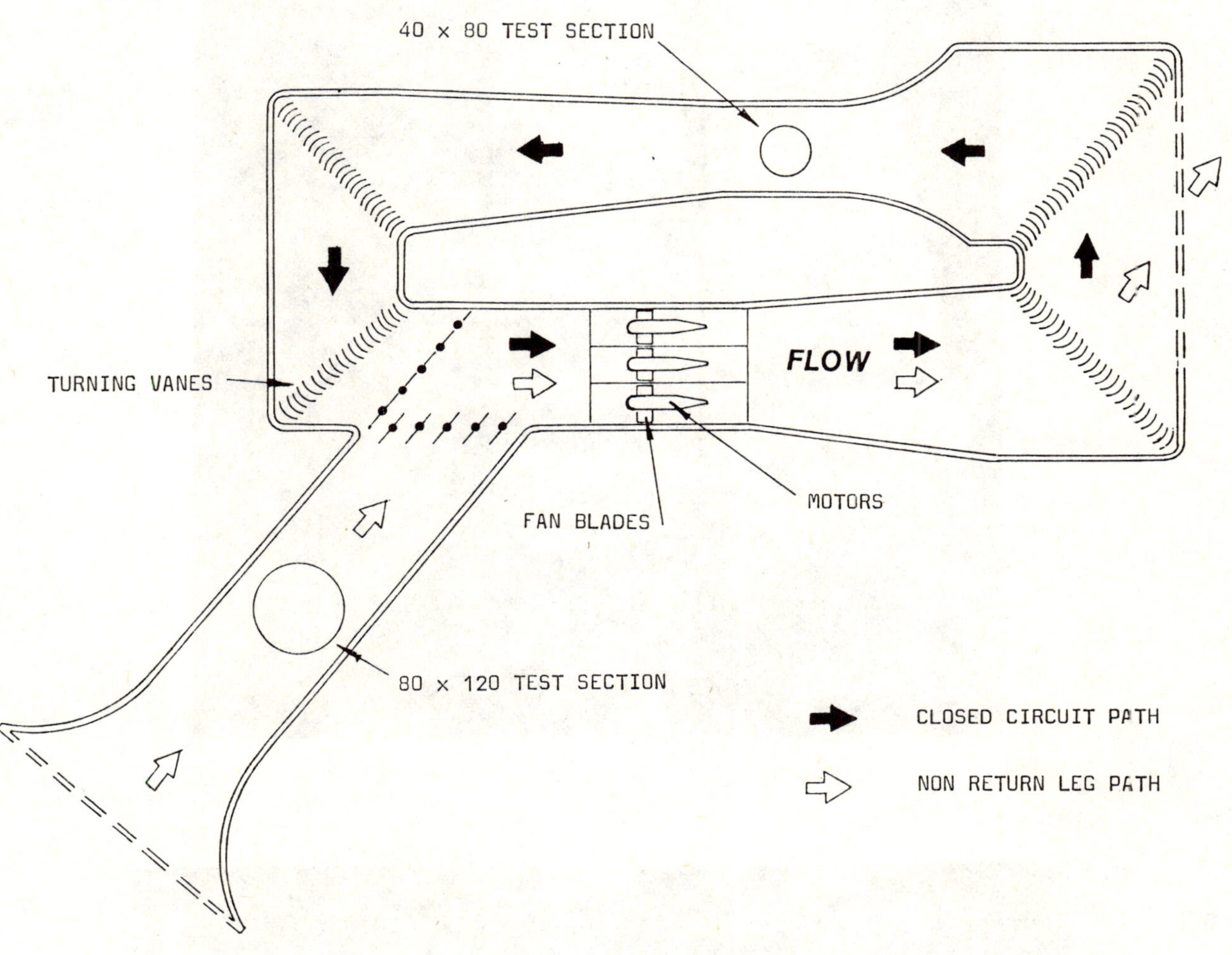

Fig 5 **AMES WINDTUNNEL**

Fig 6 AMES FAN BLADES

Fig 7 AMES FAN UNIT

International Conference on

Fan Design & Applications

Guildford, England: September 7-9, 1982

PAPER F2

FACTORS AFFECTING THE SELECTION OF FANS FOR AUTOMOTIVE ENGINE COOLING APPLICATIONS

W.R. Stapleford

Motor Industry Research Association, U.K.

Summary

The important need to reduce energy consumption and costs of manufacture has heightened the importance of efficient design of cooling systems in vehicles. A most significant feature is the selection of an appropriate fan, involving a prediction of the installed fan performance.

The achievement of the thermal performance target, for a cooling system with the minimum of power consumption and noise output, depends on (1) the amount of heat to be dissipated, (2) the maximum allowable engine coolant temperature and vehicle ambient operating conditions, and (3) the available space in which to install the system. Theoretically, the required system airflow can be calculated from a consideration of these points, from which the correct balance of fan and ram-induced airflow can be determined. Such calculations rely on the accurate knowledge of engine heat rejection as a function of load and speed, radiator airflow and pressure drop, fan airflow and pressure rise, and finally the installation resistance i.e. taking into account pressure losses due to the air inlet (including ornamental grille) and outlet, underbonnet layout, and the location of the radiator.

This paper outlines a rational approach to the selection of a suitable fan, leading to a cooling system which will meet its design performance with the minimum of further development.

Organised and sponsored by
BHRA Fluid Engineering, Cranfield, Bedford MK43 0AJ, England.

0263 - 421X/82/01 00 - 0001 $5.00
The entire volume can be purchased from
BHRA Fluid Engineering for $82.00

1. INTRODUCTION

In the face of space constraints imposed by current body design trends related to energy savings, together with noise legislation and, in some cases, increasing engine outputs, the selection of a fan most appropriate to a given duty, and the prediction of its installed performance, are becoming increasingly important.

In order to select a fan for an automobile engine cooling system, two approaches can be made, the distinction being drawn between primary and secondary measures. Primary measures must be planned and carried out during the design and development stage of a vehicle, and for this a precise knowledge of the other components of the cooling system and its installation is required. Secondary measures are concerned with improving the performance of an existing system.

The intention of this paper is to outline a rational approach of considering all the factors governing the selection of a suitable fan, leading to a cooling system which will meet its design performance with the minimum of further development.

2. DESIGN PROCEDURE

The starting point for any engine cooling system design is to obtain the relationships of the heat rejected to the coolant, and the coolant flow rate*, with engine speed (at full and part-load). This information together with the limiting ambient air temperature and the maximum allowable engine coolant temperature, is then used to calculate the required amount of cooling system heat dissipation capacity. From a knowledge of the maximum space available, the radiator frontal area can be fixed and by using a core construction having the best possible combination of a high overall heat transfer coefficient and a low air pressure drop, a radiator with the required heat dissipation capacity, combined with the minimum contribution to the overall system resistance, can be installed.

The distribution of air static pressure about the surface of a vehicle, due to the forward motion of the vehicle, should also be investigated as the differential pressure head thus created can assist or impair the flow of air through the cooling system. For example, in the case of a system mounted at the front of a vehicle (ie, with a forward-facing inlet) air inlets should be sited in the region of the highest pressure, and air outlets should be located in a low pressure region in order to obtain the maximum benefit from the dynamic pressure. Due to the fact that the performance of a cooling system is defined over a range of engine speeds, loads, and vehicle road speeds, the dynamic pressure usually cannot be solely depended upon for engine cooling system airflow, but, dynamic pressure should always be utilised.

After the component arrangement and airflow circuit layout have been determined, the airflow/pressure characteristics of the installed system must be evaluated in order that the fan may be selected. The total system resistance is often more than double that of the radiator core alone, and the increase in resistance which is caused, for example, by close proximity of the fan to the radiator core or engine, obstructions in the incident airstream, etc, is difficult to predict. Fan selection is made by use of known test data, with extrapolation if necessary using the 'fan laws'. These data can be obtained by carrying out tests on representative fans, or from manufacturers' specifications, or from other sources. In general, this type of performance data, measured under ideal operating conditions, of no obstructions, minimum tip clearance, and optimum position in the fan cowl, must be de-rated to take into account the effects of a particular installation. With due consideration to these installation effects, the fan type, size and speed must be chosen consistent with the lowest power consumption, noise level and overall cost.

---

* This paper concerns itself only with the more common designs of water-cooled systems. However, the general principles of fan selection advocated can be equally applicable to air-cooled arrangements.

Once a prototype vehicle has been built, the cooling system can be fitted in order to evaluate such aspects as thermal performance, noise and efficiency. From the point of view of noise, the fan drive and its vibration isolation, fan material, blade resonances, resonances in neighbouring panels, and vibration transmission to other parts of the vehicle, must be analysed. During the development phase, where original cooling system components are fabricated and changed to improve the design, noise measurements can be made to determine whether reductions in noise have been achieved. With regard to the aerodynamic design and cooling system performance, the success of the design, and any subsequent modifications, can be conveniently verified on a chassis dynamometer in a full-size wind tunnel, ideally with climatic control.

3. COOLING SYSTEM DESIGN REQUIREMENTS

Early in the design stages of a new vehicle, or, improvements to an existing vehicle, requirements are defined by chassis and body design engineers for cooling system configuration and size layout. In order to meet such requirements, and minimise the subsequent time-consuming and costly development, a rational approach to cooling system design is necessary, particularly as regards the selection of a suitable fan to provide the correct balance of fan and ram induced airflow. Although it is important to achieve the highest possible figure for fan efficiency, this cannot be considered in isolation, as the major design parameters of a cooling system are:
(1) the airflow rate, determined by the amount of heat dissipation required, and,
(2) the static pressure, determined by the total installed system resistance that must be overcome to maintain the airflow. Because the radiator-face area may be constrained by vehicle styling and installation considerations, its core depth, ie, the dimension that affects the air pressure drop, will relate to the heat dissipation capacity required to meet the cooling system performance targets. Therefore, before it is possible to design or select a radiator with the minimum air-pressure drop, some knowledge must be gained of the amount of heat rejected under the operating conditions of the engine being considered.

3.1 Engine heat rejection.

Fig 1 shows a diagrammatic representation of an engine and its associated cooling system, from which the heat rejected to the coolant (H) can be defined as:-

$$H = m \times c_p \times (Tc_1 - Tc_2) \qquad (1)$$

where m = coolant mass flow rate, kg/s

c_p = specific heat of coolant. (For water = 4.19 kJ/kg. K)

Tc_1 = outlet temperature of coolant, K

Tc_2 = inlet temperature of coolant, K

There are of course a considerable number of variables relating to the design of engines, such as cylinder size and jacket design, gas flow conditions, mean-effective pressure, compression ratio, etc, all of which affect heat rejection [1], but a discussion of these effects is outside the scope of this paper. However, it is unlikely that any of these variables could be materially altered at the design stage of a cooling system.

Apart from the influence of different designs, for a given engine, heat rejection can be conveniently expressed as a function of brake-power output. This affords a basis of comparison for all engines, irrespective of actual sizes. For preliminary design calculations, and in the absence of actual measured values, the heat removed by cooling for an engine under full-load conditions can be derived from:-

$$H = k \times P$$

where $k = \dfrac{\text{ratio of heat rejected to coolant}}{\text{heat to useful work}}$

P = engine brake-power output

In the engine speed range between 40 and 100 per cent of maximum speed, typical values of k at full-load are 0.6 - 0.8 for spark-ignition engines [2], and 0.55 - 0.90 for diesel engines [3], giving an average value of 0.7 for both types of engine. For part-load operation, k will increase, and typically for spark-ignition engines could approach unity [2].

In addition to the heat rejected directly from the engine, via the coolant, in some cooling-system designs account must be taken of the extra heat load imposed on the system arising from the cooling of the engine oil, automatic transmission oil, etc, via an oil-to-water heat exchanger located in the coolant circuit. This could amount to an additional 25 per cent [4].

Fig.2 shows the typical engine-heat rejection/speed envelope for various types of road vehicle.

3.2 Factors affecting system heat transfer

In order to specify a suitable radiator it is necessary to determine, for the critical cooling condition, the required radiator heat dissipation rate (kW/ K), given by:-

$$D = \frac{H}{(Tc_1 - Ta_i)} \quad (3)$$

where Tc_1 = coolant temperature at radiator inlet, K

Ta_i = air temperature at radiator inlet, K

In the absence of air recirculation through the radiator the air entering the radiator does so at ambient temperature and, $Tc_1 - Ta_i$ is usually referred to as the 'extreme temperature difference' (ETD).

However, it is necessary to relate equation (3) above, to specified cooling system performance targets such that 'D' is expressed in terms of the maximum allowable air temperature and also the boiling point of the coolant. The boiling point (Tc_b) is a function of the system operating pressure, coolant composition and altitude; thus once a value has been accepted as the practical limit to high temperature operation then the ambient air temperature at which the coolant would just boil under the opposing actions of absorbing heat from the engine and dissipating heat through the cooling system, becomes important. This ambient temperature, which is an index of cooling system performance [5], is usually expressed in terms of the quantity, 'air-temperature-to-boil (Ta_b)', and defined as:-

$$Ta_b = Tc_b - (Tc_1 - Ta_i) \quad (4)$$

The value of Ta_b can only be obtained during vehicle tests (ideally in a full-scale wind tunnel) with various simulated vehicle speeds and loads, when the whole system has achieved stability for a particular operating condition. Combining equations (3) and (4) to give ($Tc_1 - Ta_i$) in terms of H, and D, gives:-

$$Ta_b = Tc_b - \frac{H}{D} \quad (5)$$

from which $$D = \frac{H}{Tc_b - Ta_b} \quad (6)$$

It can be seen from equation (6) that radiator dissipation is the quantity which must be determined to satisfy a target Ta_b value for a given engine heat rejection and coolant boiling point. The choice of values for Tc_b and Ta_b is usually based on experience, and in the case of Tc_b is usually fixed by the engine manufacturer*. With regard to Ta_b a value can be chosen by the vehicle manufacturer, such that the vehicle is capable of operation at this temperature throughout its entire speed and load range. Experience has shown that this latter requirement is excessive, particularly for cars and light commercial vehicles, from the point of view that some vehicles present critical cooling conditions when idling, and others when under maximum power, or when operated at maximum torque in indirect gears. The critical operating condition is a function of vehicle and cooling system interaction, as different system designs can produce quite different critical cooling points in the same vehicle.

Table 1 gives examples of vehicle manufacturers' typical air-to-boil targets for a small-to-medium saloon car and a heavy truck, where it can be seen that the targets vary with gear and engine conditions, and also with the maximum likely ambient temperature the vehicle is likely to be subjected to.

4. DETERMINATION OF SYSTEM AIRFLOW AND INSTALLATION CONSIDERATIONS

4.1 Airflow and pressure drop

Heat exchanger manufacturers' published data on the overall heat transfer coefficients of their radiators, enable the system airflow requirement to be calculated to give the required heat dissipation as established from equation (6). Performance curves available from manufacturers generally take the form shown in Fig. 3, to give, at constant waterflow (Fig 3(a)), core heat-dissipation and pressure drop as functions of airflow in terms of unit core-area and temperature differential. The temperature differential commonly used is the mean coolant temperature minus the air inlet temperature, which is referred to as the mean temperature diffence (MTD).

The relationship between MTD and ETD (ie the terms in which equation (3) is stated) is given by:-

$$MTD = ETD - \frac{\Delta T}{2} \qquad (7)$$

where ΔT = radiator coolant temperature difference between inlet and outlet at design condition.

It must be remembered also that the coolant velocity will influence dissipation but this effect will be relatively small providing the water flow rate through the radiator water-tubes is above a minimum value. As Fig. 3(b) shows, this means that relatively moderate mass-water flows will permit near to the maximum heat transfer on the water-side and additional water flow will not add significantly to the thermal capacity of the radiator.

Therefore, after taking into account the effect of variations in water flow, the required radiator specific heat dissipation, D_s (kW/ K MTD/m^2), will be:

$$D_s = \frac{D}{A_r} \qquad (8)$$

where D = required system heat dissipation (as defined in equation (6)), kW/ K MTD

A_r = face area of radiator core, m^2

---

* For design purposes, it is general automotive practice to use the boiling point of the coolant at system pressure. However, for heavy duty diesel engines a cooling system that is dependent on a pressure cap is not recommended [6] , and usually the maximum coolant temperature is limited by the sump oil temperature, which could be 20 per cent higher than the coolant outlet temperature.

Vehicle styling and space considerations are the major influence on radiator frontal area, but, where possible, it has been suggested [7] that to minimise fan power and noise output the highest practicable radiator-face area must be used. A square core equal to or slightly larger than the fan diameter is ideal in order to provide optimum fan coverage. If other considerations determine a mis-match between radiator and fan, then it is likely that this will result in airflow maldistribution. A consequence of this non-uniformity of airflow is that the radiator-face area will be under-utilised. However, where a limit is placed on frontal area it will be necessary to increase the radiator depth. Hence, consideration must be given to optimising the radiator bulk volume, ie, the product of the face area of the core, and the depth of core (the latter being related to the number of water-tube rows, and affecting the air pressure drop) should be optimised. Having optimised the radiator bulk volume, it follows from equation (8) that, as soon as A_r is fixed, D_s can be determined, and the radiator-face air velocity V_r, and radiator-core pressure drop, ΔP_r can be deduced from radiator manufacturers' data similar to that shown in Fig. 3. Obviously it is important to strike a balance between system heat dissipation capability and air movement capability, bearing in mind that D_s and ΔP_r vary as $(V_r)^n$, where typically [8] the exponent (n) for radiator heat dissipation is 0.6, and for core pressure drop is 1.7. This implies that a radiator with a high ratio of specific heat dissipation to pressure drop coefficient must be chosen.

4.2 Installation principles

The installed system airflow will depend on the extent to which the 'air-head' produced either by the 'ram' effect (due to the forward motion of the vehicle), or by a fan, or by a combination of both, can overcome the total system resistance. Fig. 4 shows the relationships between radiator specific heat dissipation, radiator and total system resistance, and radiator-face airflow, from which the design air-head duty point can be identified. It follows that the accurate prediction of the air-head duty point will depend on the precise determination of air-head and system resistance. However, because of wide variations in vehicle dimensions and styling, it is not easy to provide a definitive mathematical solution of the air-side aspects, although various attempts using both theoretical and experimental methods to achieve this have been made [2 9-15].

In order to appreciate the problems involved it is necessary to analyse the aerodynamic aspects of a typical water-cooled system installation, as shown diagrammatically in Fig. 5, for the case of an idealised combined fan and ram dependent system. Considering first Fig. 5(a), it can be seen that a 'stream-tube' of air, which will subsequently enter the inlet to the radiator, can be identified at some distance ahead where the ambient air is not disturbed by the approaching vehicle. At this stage the air has a velocity relative to the vehicle which is equal (and opposite) to the velocity of the vehicle, V_o, and the stream-tube has a cross-sectional area denoted by A_o, the 'capture' area. The resistance of the cooling system causes the air to decelerate, so that the stream-tube expands progressively to the area of the inlet, A_i, while the velocity decreases to V_i. Assuming a continuity of airflow through the inlet (with ornamental grille) diffuser, radiator, fan and cowl, and engine bay outlet it follows that with no leakage*,

$$V_iA_i = V_rA_r = V_fA_f = V_eA_e \quad (9)$$

---

* ie, in a situation where the air is constrained to flow through the radiator only

where V_i = air velocity at inlet, m/s

A_i = area at air inlet, m^2

V_r = radiator-face air velocity, m/s

A_r = radiator-face area, m^2

V_e = air velocity through outlet, m/s

A_e = total outlet area, m^2

V_f = air velocity through fan, m/s

A_f = $\pi/4\ (D_f^{\ 2} - D_h^{\ 2})$ = fan area, m^2

D_f = fan cowl diameter, m

D_h = fan hub diameter, m

Unless a sealed duct/diffuser is provided between air inlet and radiator, not all the air passing through the inlet will also pass through the radiator, thus the respective areas of leakage, A_i, and radiator-face, A_r may encompass differing proportions of the originally entering stream tube. Also, the capture area, A_o, will be dependent on the proportions of ram and fan involved; ie, if A_o is expressed as a ratio of A_r, then it has been shown [13] that, for the same air inlet conditions, A_o/A_r, could vary between 0.25 for a high vehicle speed condition, to ∞ for a stationary vehicle condition. Under ram-dependent conditions, ie, low values of A_o/A_r, this was shown to result in an increase in air inlet losses attributed to high local velocities and angle of incidence changes at the grille.

It follows also, from Fig. 5(b), that for steady-state airflow conditions, a pressure balance exists, ie, the cooling air-head equals the system resistance as follows:-

$$P_d + \Delta P_f = \Delta P_i + \Delta P_r + \Delta P_e \qquad (10)$$

where P_d = ram dynamic pressure due to vehicle motion $(\Delta C_p . q_o)$, kPA

ΔP_f = fan static pressure rise, kPa

ΔP_i = inlet static pressure drop, kPa

ΔP_r = radiator static pressure drop, kPa

ΔP_e = air exit static pressure drop, kPa

ΔC_p = static air pressure coefficient difference (= $Cp_i - Cp_e$), non dimensional

Cp_i = static air pressure coefficient at inlet = P_i/q_o

Cp_e = static air pressure coefficient at air exit = P_e/q_o

q_o = dynamic head = $\frac{1}{2}\rho\ V_o^{\ 2}$

ρ = air density, kg/m

Where the area of the fan cowl is smaller than the face area of the complete radiator or part of the radiator, then an increase in velocity head will take place at the expense of the static head between points 4 and 5 (Fig. 5(b)).

The proportions of ram and fan required for cooling air-head development will depend on the type of vehicle, cooling requirements, system layout, the locations of air inlet and outlet, and vehicle speed. For example, it has been illustrated, in non-dimensional terms [16], for typical vehicle types, how the cooling airflow varies with vehicle speed in the manner shown in Fig. 6, from which the proportions of fan and ram induced airflow can be deduced. It will be realised that, in Fig. 6, diminishing curve-slope implies decreasing ram effect, and, typically, for saloon cars, at high power/high vehicle speeds the predominant proportion of the air-head will be ram-induced, for maximum air-flow rate, whereas, at low vehicle speeds, eg, idling, stop, start, and crawl road traffic conditions, the predominant proportions will be fan-induced for a relatively low percentage of the maximum air-flow rate. At the other end of the range, off-road vehicles, such as, tractors, earth-moving vehicles etc, with frequent forward and reverse operation at low vehicle speed, will be predominantly fan-dependent at the maximum air-flow rate.

4.3 Resistance factors

When evaluating the cooling air-head terms of fan and ram pressure, as defined in the left-hand side of equation (10), consideration must always be taken of installation effects on the development of the air-head. Firstly, from equation (10), it can be seen that the ram dynamic-pressure is the product of the static air pressure coefficient difference between system air inlet (Cp_i) and outlet (Cp_e) and the dynamic head, and, consequently increases as the airspeed (vehicle speed) squared. The pressure coefficient C_p, is obtained by dividing the local surface static pressure (measured with zero internal flow) by the dynamic head. As shown in Fig. 7, pressure coefficients vary over the areas to be used for air inlets and outlets, and since extreme values of pressure are often confined to small areas, the available ram pressure will be based on mean values of C_p for the areas under consideration. In the case where ΔC_p is positive, ie, for front-engined vehicles (with forward-facing inlets), then the ram pressure is subtracted from the system resistance to give the required fan-pressure rise, hence for engine full-load operation in low gear, the fan must supply the greater proportion of the air-head. Similarly, for zero, or, negative ΔC_p values, eg, in the case of rear-engined vehicles (ie, assuming the cooling system is mounted at the rear of the vehicle), the fan must overcome the system resistance, or, the system resistance plus the ram pressure.

As can be seen from Fig. 7, typically for front-engined vehicles, ΔC_p might be expected to be approximately + 1.0. However, in practice, because of vehicle front-end losses, in addition to the effect of the ornamental grille, the actual achieved value is likely to be lower [13]. Also, the effects of head, tail, and cross-winds, must be taken into account, the head and tail-winds being particularly detrimental respectively in the cases of vehicles with negative and positive values of ΔC_p (as measured in the condition of zero natural wind).

As regards the installed fan performance, account must be taken of factors which affect data obtained from performance tests measured in accordance with standard test methods [17],, as typified in Fig. 8. These factors, presented diagrammatically in Fig. 9, are; cowl shape and depth, fan-tip/cowl clearance, fan/cowl penetration, fan/radiator separation distance, and fan/engine separation distance. In order to achieve an accurate assessment of the installed fan performance, and ultimately a prediction of the fan duty point, the effect of the above mentioned installation aspects can be expressed in terms of a 'fan derating coefficient [8 12 18], with which to correct non-installed fan data.

A fan cowl is normally used to provide a more uniform distribution of radiator-face airflow, but to achieve the highest possible fan efficiency, the best cowl shape, and optimum value of fan/cowl penetration, and a minimum value of fan-tip/cowl clearance, are essential. As far as cowl shape is concerned this will depend on whether the radiator is mounted on the suction-side of the fan or downstream of the fan; for the former, more general case, the venturi-type has been found to be an acceptable design, particularly where the fan-tip clearance is 1.5 per cent or less of the fan diameter, giving reductions in airflow (relative to standard performance test values) of between 2 and 6 per cent depending on the separation distance between the fan and radiator. In addition to optimum shape, in order to prevent air recirculation it is essential that the cowl is airtight.

The ideal location of the fan in the cowl depends on a number of installation variables, but the optimum value of the fan penetration into the cowl is typically between 0.5 to 0.66 of the projected depth of the cowl. Reducing the penetration to half the optimum value can give reductions in airflow as high as 10 per cent [18]. General guidance on fan/cowl penetration, and cowl type, in particular relating to both 'suction' and blowing fans, can be found elsewhere [6].

Fan-tip/cowl clearance can appreciably affect fan performance, and as stated above should, ideally, be as small as possible; however, in the case of an engine-driven fan with the fan cowl attached rigidly to the radiator, the clearance can be as high as 5 per cent of the fan diameter. This is necessary to make allowance for movement of the fan relative to the cowl since the fan moves with the engine on its anti-vibration mounts under torque 'wind-up', etc. To permit a small tip-clearance with engine-driven fans, a common solution is to divide the cowl into two parts, one part mounted on the engine surrounding the fan and the other part on the radiator with a flexible member or gaiter joining the two parts.

The optimum value of fan/engine separation distance will depend on whether the engine is upstream or downstream of the fan. It has been suggested that the distance should be held at a maxima of 20 per cent, and 10 per cent of the fan diameter, respectively, for the upstream, and downstream conditions [12]. Although there is no general rule for derating ideal fan-performance data for the above mentioned factors, clearly, the airflow reduction could be at least 15 per cent [18 19].

Because of the influence of air density on fan performance, fan manufacturers' test data is usually corrected to standard atmospheric conditions, where air is taken as having a density of 1.2 kg/m^3, at a dry bulb temperature of 20°C and pressure of approximately 100 kPa, the resulting relative humidity being about 47 per cent. It follows, therefore that the effect of air density on installed fan performance must also be taken into account. Both fan pressure and system pressure loss will vary directly with changes in air density, such as may occur with a change in ambient conditions. Where a change of air density occurs within a system, such as at a hot radiator, the location of the fan relative to the radiator will influence fan performance. The vehicle application will generally dictate the fan location, eg, on-road vehicles normally use a suction fan, and off-road vehicles frequently use blower fans. Blower fans are generally more efficient in terms of power expended for a given mass-flow since they will nearly always operate with lower temperature air as compared to a suction fan. The air entering a suction fan is heated as it passes through the radiator whereas a blower fan, except, perhaps, when drawing air over the engine, will receive air nearly at ambient temperature. Although, as stated above, fan pressure development will vary directly with air density, the air volume flow will remain the same, hence, air-mass flow will be directly proportional to density also.

In the general case of the right-hand side of equation (10), the major factors contributing to the total system resistance can be categorised [13] , eg, (1) the front-end resistance, comprising, air inlet, grille, and radiator ($\Delta P_i + \Delta P_r$), (2) the engine-bay exit path, ΔP_e, and, (3) the engine-bay back pressure due to vehicle motion. This additional factor of engine-bay back pressure, which has been shown to be dependent on vehicle speed and cooling airflow rate, tends to be moderately positive, and extremly sensitive to external effects. It can be considered to relate to radiator/airflow exit area ratio, A_r/A_e [13].

Having examined the factors governing both cooling air-head and total system resistance, it is useful to re-define equation (10) in terms of the other important design parameters of radiator face air velocity, V_r, and vehicle speed, V_o. Such an expression has been developed [14, 20, 21, 22]. which states that V_r is a function of V_o, the total forcing function (air-head), and the total system resistance, thus:-

$$V_r = V_o \sqrt{\frac{(1 - Cp_e) + \Delta P_f/0.5\rho V_o^2}{kp_r + kp_i (A_r/A_i)^2 + kp_e (A_r/A_e)^2 + (A_r/A_e)^2}} \qquad (11)$$

where kp_r = radiator pressure drop coefficient = $\Delta P_r/0.5\rho V_r^2$

kp_i = air inlet pressure drop coefficient (including grille) = $\Delta P_i/0.5\rho V_r^2$

kp_e = air exit pressure drop coefficient = $\Delta P_e/0.5\rho V_r^2$

From equation (11), the required fan pressure rise (ΔP_f) can be calculated, for any vehicle-operating condition, providing values for the loss coefficients, areas, and mean air-outlet pressure drop coefficient are known. V_r is determined via equation (8), once a particular radiator has been chosen from manufacturers' performance data, and V_o is evaluated from information relating to engine speed, overall gear ratio, and tyre rolling radius. Also, kp_r can be determined from radiator performance data.

As already shown in Fig. 7, the value of Cp_e could be typically - 0.2, but as regards actual area ratios, these are often difficult to define; nevertheless, their effect is likely to be negligible in comparison with the loss coefficient terms, particularly, kp_r. Pressure drop coefficients of radiators vary with V_r (ie, with Reynolds number R_e) and core construction, that is, with core density (solidity factor, S_c) and number of water-tube rows (core depth). Typically, the range of values of kp_r could be from 1 to 12 [20, 21]. the lower- to mid-range values being representative of radiators for passenger cars, and the mid- to upper-range values for radiators installed in heavy-commercial and off-road vehicles. The design of the decorative air-inlet grille appears to be the major contribution to the inlet pressure drop coefficient, and the value of kp_i will vary with grille shape and solidity factor, S_g, a good design having a loss coefficient equal to or below 2. In some cases, kp_i, could be as high as 5 [2]. The effect of the air exit conditions on radiator airflow has been found to result in values of kp_e which fall in a range between 1 to 8 [20].

If the fan de-rating factor is taken into account, then in the case of the worst possible cooling-system installation, the total-system loss coefficient could be as high as 50. Some recent work [12], on off-road vehicles, heavy commercial vehicles, and buses* ie, highly fan-dependent vehicles, has indicated system loss coefficients between 15 and 45, with the radiator loss coefficient being approximately 20 to 75 per cent of the total-system resistance. It is probable that in most passenger cars the radiator resistance represents approximately 75 per cent of the total-loss coefficient with a typical total value of 10.

5. FAN SELECTION

5.1 Fan type

After the fan duty requirement has been quantified, then a fan can be selected by extrapolation of known test data using the 'fan laws'. The data can be obtained by carrying out tests on representative fans, from manufacturers' specifications, or from other sources.

NB Equation (11) above assumes all air entering inlet is passing through radiator

As already discussed (and shown in Fig. 8), characterising fan performance is usually a matter of showing the airflow versus pressure, and power consumption, at various fan speeds. This basic information can be presented in another form that has certain advantages when considering the operating point of a fan when installed in the vehicle. Each type of fan operates most efficiently within a prescribed range of volume flow rate Q_f, static pressure head, ΔP_f, and rotational speed, N; this range is described in terms of the specific speed parameter, N_s, as:-

$$N_s = N\left[Q_f^{0.5}/\left(\frac{\Delta P_f}{\rho}\right)^{0.75}\right] \qquad (12)$$

This characteristic is not dimensionless, and for any particular application it is of course necessary to indicate the units employed when deriving it. It is customary when evaluating N_s to use air at standard density. The typical specific speed range covered by each fan type is shown in Fig. 10.

Any one type of fan, or size of fan, may be operated over a given duty range, either by throttling the flow or by changing the impeller speed. For example, one particular duty may be achieved by using different sizes of the same type of fan. However, the noise levels of the fans will be different because of speed and efficiency changes [23]. Therefore, it is necessary, in addition to considering the specific speed, to consider the non-dimensional terms of flow coefficient, Φ, and pressure coefficient, Ψ, which are defined as follows:

$$\Phi = k_1 \left(Q_f/ND_f^3\right) \qquad (13)$$

and

$$\Psi = k_2 \left(\Delta P_f/\rho(ND_f)^2\right) \qquad (14)$$

where D_f = fan diameter

and k_1 and k_2 = constants

Flow and pressure coefficients should be as low as possible for maximum efficiency and minimum noise, and, Fig. 11 typifies the use of the flow and pressure coefficients in conjunction with static efficiency to show the case of two different fans meeting the same duty but with different static efficiency points.

After the designer has tentatively chosen a number of impeller types, all of which satisfy requirements for the air mass flow, static pressure rise, and speed, the final selection will be made on the basis of optimum static efficiency, available space, and impeller cost. Several impellers may be eliminated because either size or shape or both are unacceptable. The relative static efficiencies and initial cost of the remaining impellers of acceptable sizes are evaluated in order to make a final selection. Typically, the efficiencies of current automotive 'propeller-type' fans range approximately from 20 to 50 per cent, with the most efficient operating at close tip-clearance [23]. Sometimes an impeller that operates at a relatively inefficient point on its operating curve (static pressure v air mass flow) is chosen in order to obtain desirable advantages in size; however, this must not be done at the expense of noise.

5.2 Noise considerations

Fans produce two types of noise; vortex noise and rotation noise. Vortex noise is broadband in character, ie, with no discrete frequency components, which for typical axial flow fans used in road vehicles is of relatively low intensity. Rotation noise predominates with axial flow fans and is periodic in character, consisting of the fundamental blade-passage frequency and many higher harmonics.

Vortex noise is a function of the blade shape, fan speed, and duty, and Fig 12 shows that for minimum vortex noise the fan duty point (the intersection of the fan pressure-flow curve and the cooling system resistance curve) must lie somewhere between A and B on the fan pressure rise curve. This means that the fan is then operating on the best part of its efficiency curve (ie, in the axial flow regime shown in Fig. 8). If however, the mass airflow is reduced below a certain limit a sharp drop in pressure (point C) ensues as a result of blade overloading. The air velocity through the impeller is too low and since there is too great a static pressure in the impeller, the air cannot remain attached to the low pressure side of the blade and flow separation occurs, manifesting itself in turbulence and consequent noise. Illustrations (a) and (b) in Fig 12 show the respective air flow patterns around the fan blade.

Rotation noise is basic to all kinds of fans, in that every time a blade passes a given point the air at that point receives an impulse. The intensity of the fundamental blade passage frequency and relative strength of the various harmonics are determined by the shape of the impulse [24]. Therefore blade shape, width, and thickness, can be optimised for minimum rotation noise.

In addition, turbulence in the airstream approaching a fan can affect both the rotation and vortex noise, either as a result of fan blade incidence fluctuations (increase in vortex noise) as illustrated in Fig. 13(a), or by the wake of an obstacle, for example, an inlet guide vane, impinging on the fan blade (increase in rotation noise) as illustrated in Fig. 13(b). Turbulence can be reduced by avoiding sharp edges or bends, and by good grille design, etc [25].

For specific fan performance details, as already mentioned, reference should be made to fan manufacturers' literature. In general for a given duty the axial fan would have the advantage on size; however, its speed to achieve the duty could be double that of a centrifugal fan. With increasing noise legislation the choice of a particular kind of fan may mean that space and cost may have to be sacrificed in order to achieve the cooling system noise targets. This has been demonstrated in recent work, particularly as regards reducing the 'drive-by' noise output from heavy commercial vehicles [26].

5.3 Fan-drive methods

Because the cooling system has to be designed to meet the most demanding requirements likely to be met in service, the radiator and fan have to be dimensioned for the engine operating at full-load, in conjunction with low vehicle speed and in the highest operational ambient temperature. Typically, these conditions exist for only a small percentage of a vehicle's operating time, resulting in overcooling for the rest of the time. With the current emphasis on noise and energy consumption, considerable attention has been paid to fan-drive methods, particularly, methods which can de-activate the fan when not required.

Fan-drive methods employed in automobiles can be placed in the following three categories; (1) the fan rigidly connected to the engine, with the fan having a fixed ratio to the speed of the engine, (2) the fan driven by the engine via a variable drive system, which does not give a fan speed/road speed characteristic which is a fixed ratio of the engine speed/road speed characteristic, and (3) the fan driven entirely separate from the engine, ie, usually an electrically driven fan, whose drive characteristics are essentially distinct from those of the engine and with a transmission system which can be manual or automatic.

Fan-drive methods (2) and (3) have become increasingly popular, with the most common type used in method (2) being the viscous-drive coupling [27 28] whose torque-limiting properties mean that the fan speed increases with engine speed though at a lower rate until the limit of the pre-set torque capacity has been reached in the manner shown in Fig. 14. The characteristics shown are for an air-sensing type which has the additional ability to 'idle' the fan whenever engine cooling demand is low. Popular on small-to-medium passenger vehicles is method (3), where the fan is usually run at constant speed [29].

Fan controlling devices generally fall into two groups, ie, sensing speed or temperature [30] .

5.4 Fan materials

The choice of a material for a fan must involve the anticipated loads, the ambient conditions, the dimensions, and by no means least the number required (ie the unit cost). Like most mechanical components mounted on the engine or chassis, the fan must withstand a variable vibration environment. The source of vibration force inputs can be mechanical or aerodynamic. Although it is desirable to remove the natural frequencies of the fan, fan drive, and mounting system out of normal operating range, it is not always possible. Frequencies which produce high amplitude at the fan blade should receive particular attention. Intermittent loading of the fan blades due to local obstructions disturbing the symmetry of the pressure field surrounding a rotating fan (eg, see Fig. 13) can produce noise and high stresses in the fan structure. Heavy fans with high inertia are undesirable if fan drive problems are to be avoided. From a safety aspect fans are normally designed to burst at 100 per cent overspeed.

Fans can be produced in a variety of ways, eg, cast in steel or aluminium alloy, moulded in synthetic materials, such as nylon and polypropylene, or fabricated from sheet metal. The latter method is the most convenient, from the point of view that a prototype fan can be manufactured very quickly and cheaply. Plastic fans offer reductions in weight, inertia, and production costs, without compromising the blade form design. However, the initial tooling costs are high. Typically, plastic fans are used on cars and light commercial vehicles, whereas pressed-steel and cast-aluminium fans are used on heavy commercial and off-road vehicles.

The chosen material must be capable of operating at the design maximum operating speed of the fan in elevated temperatures (typically 120°C), and not be affected by hydrocarbons, battery acid, and engine cleaning fluids, etc.

6. VALIDATION OF SYSTEM DESIGN

Road measurements of engine cooling system performance involve similar difficulties to those found in other vehicle aerodynamic road measurements, eg, the measurement of vehicle body aerodynamic forces or pressures. Not only should still-air conditions be achieved, but the engine load and speed must be kept constant for long periods to allow the coolant and other temperatures to stabilise. The advantages of a full-scale wind tunnel with a chassis dynamometer installed in permitting the control of conditions are appreciable. Steady-state measurements of cooling system performance and some indication of component durability can be made at constant engine speed, for full or part throttle, and particular gear conditions can be simulated either by matching the tunnel airspeed to that of the dynamometer rolls (ie. the vehicle road speed) with the vehicle in the correct gear, or by deliberately mis-matching air speed and roll speed with the vehicle in some other, more convenient gear ratio. Also, idle and stop conditions (after-boil) can be simulated, and measurements made of temperatures other than that of the coolant, eg, those of the underbonnet air and the various lubricating oils, etc. In order to determine whether the coolant or other temperatures will exceed specified limits, it is not necessary to use any particular ambient temperature,* since the results are obtained in terms of a temperature rise over ambient air temperature (as indicated in Section 2.2).

---

* Although the heat dissipating capability of a cooling system can be established with reasonable accuracy in a wind tunnel which is not temperature controlled, there are increasing demands for tests to be carried out in 'fully climatic' tunnels, in which temperature can be controlled over a wide range (typically - 32°C to 52°C), with the addition of humidity control and solar simulation. Whilst providing facilities for studying cooling system behaviour in critical environmental conditions, taking account, for example, of the actual heat rejection characterisitcs of the engine under these critical conditions, climatic tunnels are of course important for studying vehicle heating and air-conditioning systems, and for studying the thermal behaviour of vehicle components generally.

Figs. 15 and 16, show typical results of system performance measurements made on a heavy commercial vehicle (150 kW engine power output) in the MIRA full-scale tunnel. Fig. 15 shows the variation of installed radiator specific dissipation with vehicle road speed at both constant engine speed (with fan/engine speed ratio = 1.18:1) and constant gear ratio, and Fig. 16, shows the resultant variation of air-temperature to boil with engine speed for various gear ratios.

Although wind tunnel testing is generally accepted as being the most accurate and expedient way of determining cooling system performance, some road measurements are still carried out, and are in fact necessary for the purpose of durability testing and also to obtain a continuous evaluation of likely customer usage. The information obtained from road testing can then serve as a verification of the relationship between the actual operating conditions on the road and wind tunnel tests.

7. CONCLUDING COMMENTS

Within the limitations of this relatively short paper, the factors affecting the selection of a fan for an automotive engine cooling system have been identified. Because of the influence of associated components in the particularly cramped environment in which an automotive-type cooling fan has to operate, a definitive mathematical solution for the prediction of the installed air-side performance is not possible. Nevertheless, if, as indicated in this paper, a rational approach to cooling system design is followed, then the most suitable size and type of fan can be selected and system development time minimised.

8. ACKNOWLEDGEMENT

The author wishes to thank the Director of the Motor Industry Research Association for permission to prepare this paper.

9. REFERENCES

1. Sitkei, G.: "Heat Transfer and Thermal Loading in Internal Combustion Engines". Akademiai Kiado, Budapest, 1974.

2. Emmenthal, K.D and Hucho, W.H.: "A Rational Approach to Automotive Radiator Systems' Design". SAE Paper 740088, 1974

3. Koffman, J.L: "Some Aspects of Cooling System Design for Diesel Engines". Diesel Engine Users Association, 1950.

4. Person, F.W.: "Truck Cooling System Requirements". SAE Paper 99A, 1958.

5. Stratton, D.G., Stringer, R.E., and Taylor, S.R.G: "Engine Cooling System Design and Development". Proc. Auto. Div., I.Mech.E., 180 (Pt2A), No 8, 1965/66

6. General Motors Corporation: "Cooling of Detroit Diesel Engines". Engineering Bulleting No. 28, 1967.

7. Hawes, S.P.: "Improved Passenger Car Cooling System". SAE Paper 760112, 1976.

8. Hebard, P.J., and Tebby, S.W.: "The Design and Optimisation of Engine Cooling Systems for Low Noise and Power Consumption". Proc. FISITA 15th Conference, Paris 1974.

9. Tenkel, F.G.: "Computer Simulation of Automotive Cooling Systems". SAE Paper 740087, 1974.

10. Klinge, E.R.: "Truck Cooling System Airflow". SAE Paper 99C, 1958.

11. Rising, F.G.: "Engine Cooling System Design for Heavy Duty Trucks". SAE Paper 770023, 1977.

12. Costelli, A., Gabriele, P., and Giordanego, D.: "Experimental Analysis of Air Circuit for Engine Cooling Systems". SAE Paper 800033, 1980

13. Schaub, U.W., and Charles, H.N.: "Ram Side Air Effects on the Air Side Cooling System Performance of a Typical North American Passenger Car". SAE Paper 800032, 1980

14. Davenport, C.J., Beard, R.A., and Scott, P.J.: "Optimisation of Vehicle Cooling Systems". SAE Paper 740089, 1974.

15. Paish, M.G., and Stapleford, W.R.: "A Rational Approach to the Aerodynamics of Engine Cooling System Design". Proc. I. Mech.E., Vol 183, 1968/69.

16. Costelli, A., Gabriele, P., and Giordanego, D.: "Experimental Analysis of Engine Cooling Systems". SAE Paper 790397, 1979.

17. British Standards Institution. "Methods of Testing Fans for General Purposes". BS848: Part 1, 1963.

18. Baranski, B.R.: "Designing the Engine Cooling Fan". SAE Paper 740691, 1974.

19. Hay, N., and Taylor, S.R.G.: "The Effects of Vehicle Cooling System Geometry on Fan Performance". I.Mech.E (Fluid Div) Conference on Fan Technology and Practice, London, 1972.

20. Kuchemann, D., and Weber, J.: "Aerodynamics of Propulsion". McGraw Hill, New York, 1953.

21. Paish, M.G., and Stapleford, W.R.: "A Study to Improve the Aerodynamics of Engine Cooling Systems". MIRA Reports 1966/15 and 1968/4.

22. Carr, G.W.: "Reduction of Cooling System Aerodynamic Power Losses in Road Vehicles". MIRA Project Report for Department of Industry, Contract No. K/A72C/332, Annex No. 1, 1975.

23. Mellin, R.C.: "Noise and Performance of Automotive Cooling Fans". SAE Paper 800031, 1980

24. Regier, A.A., and Hubbard, H.H.: "Status of Research on Propeller Noise and its Reduction". Journal ASE, Vol. 25, No.3.

25. Institution of Heating and Ventilating Engineers: "Flow of Fluids in Pipes and Ducts", Extract From I.H.V.E. guide, Section 3, 1965.

26. Tyler, J.: "The TRRL Quiet Heavy Vehicle Project". Proc.I.Mech.E., Vol. 193, No. 23, 1979.

27. Givens, L.: "Fan Clutches; A Must for Heavy Trucks". Automotive Engineering, Vol. 83, No.4, 1975.

28. Pisarski, J.J.: "Fuel Savings and Noise Reduction with the Fan Drive". SAE Paper 801013, 1980.

29. Laise, T.D., Mellin, R.C., and Pryjmak, B.I.: "Electric Cooling Fan with High Ram Airflow - A Fuel Economy Improvement". SAE Paper 790722, 1979.

30. Mascall, D.V.: "The Engine Cooling Fan Installation". Proc.I.Mech.E., Vol. 184, Pt 3A, 1969/70.

Table 1. Vehicle Manufacturers' Typical Cooling Performance Targets

(a) Small Saloon Car (40 kW)

Gear and Engine Conditions	Vehicle Speed, km/h	Engine Speed, rev/min	Target Air-Temperature-to-Boil, °C*
Second, Peak Torque	30.6	2600	25 (34)
Second, Max Power	56.3	4500	28 (36)
Third, Peak Torque	46.7	2600	30 (40)
Third, Max Power	85.3	4800	33 (43)
Fourth, Peak Torque	62.8	2450	38 (46)
Fourth, Max Power	136.5	5200	40 (50)
Idle +	0	1020	40 (50)

*This denotes the European Target, the bracketed figures denote the Worldwide targets, where TC_b = 119°C (100 per cent water and at sea level), at pressure cap setting = 90 kPa.

+ Tested immediately after fourth/max power run.

(b) Heavy Truck (150 kW)

Gear and Engine Conditions	Vehicle Speed, km/h	Engine Speed, rev/min	Target Air-Temperature-to-Boil, °C*
Fourth to Tenth Gear, Peak Torque to Max Power	16 (and higher speeds)	1500-2350	35

*Temperate Climate, Tc_b = 111°C (100 per cent water and at sea level), pressure cap setting = 48 kPa.

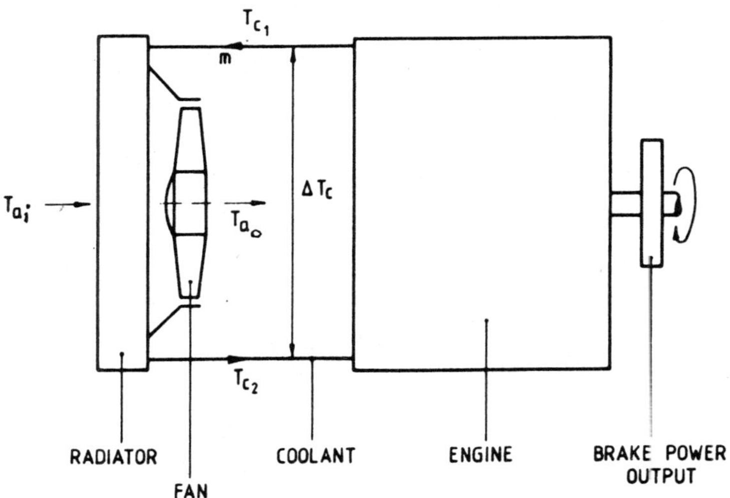

Fig. 1 Diagrammatic Representation of an Engine and its Associated Cooling System.

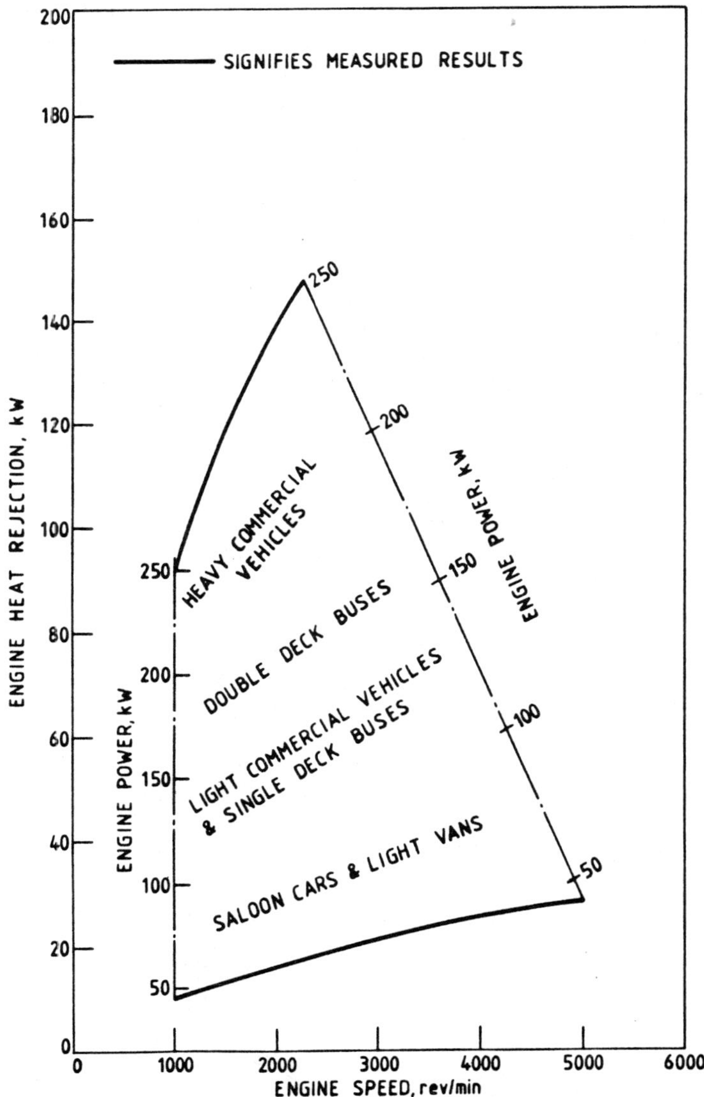

Fig. 2 Typical Heat-Rejection/Speed Envelope for Various Types of Road Vehicle.

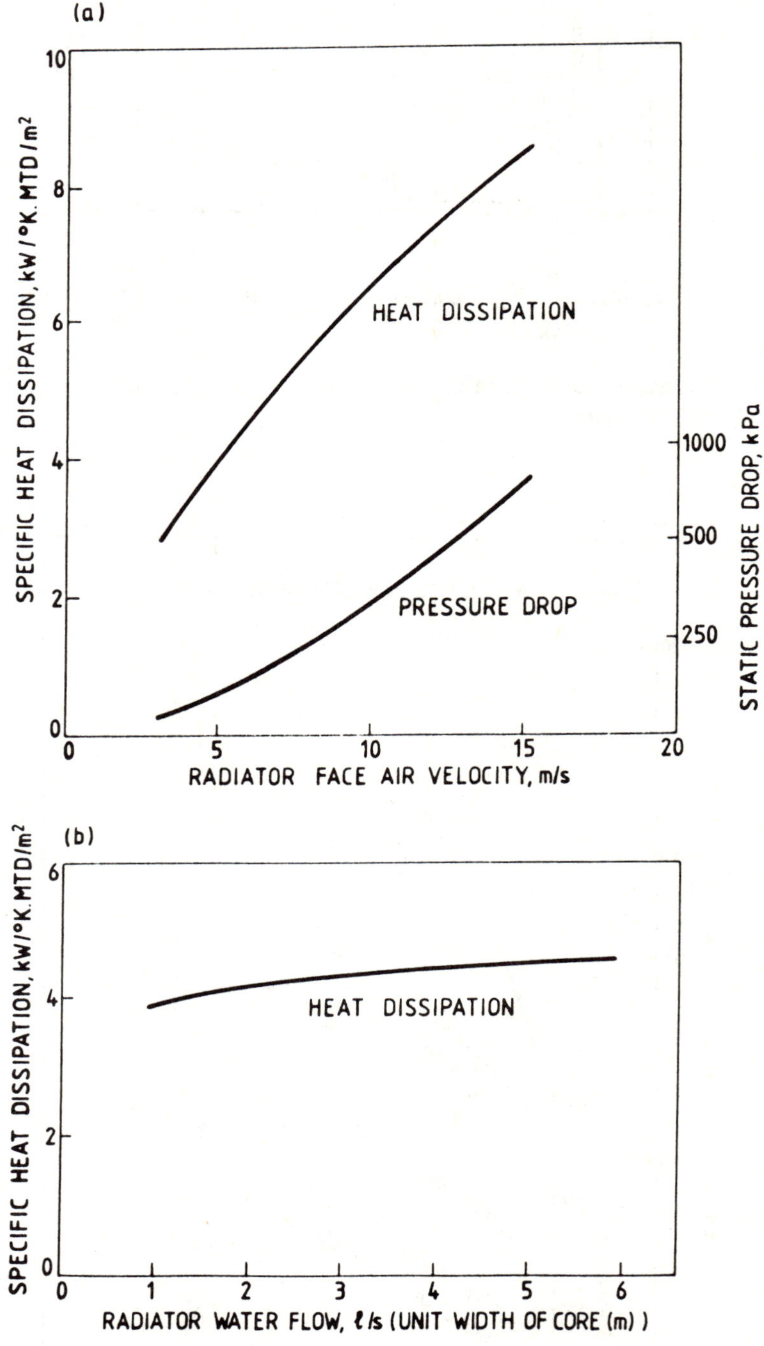

Fig. 3 Typical Form of Radiator Manufacturers' Performance Data.

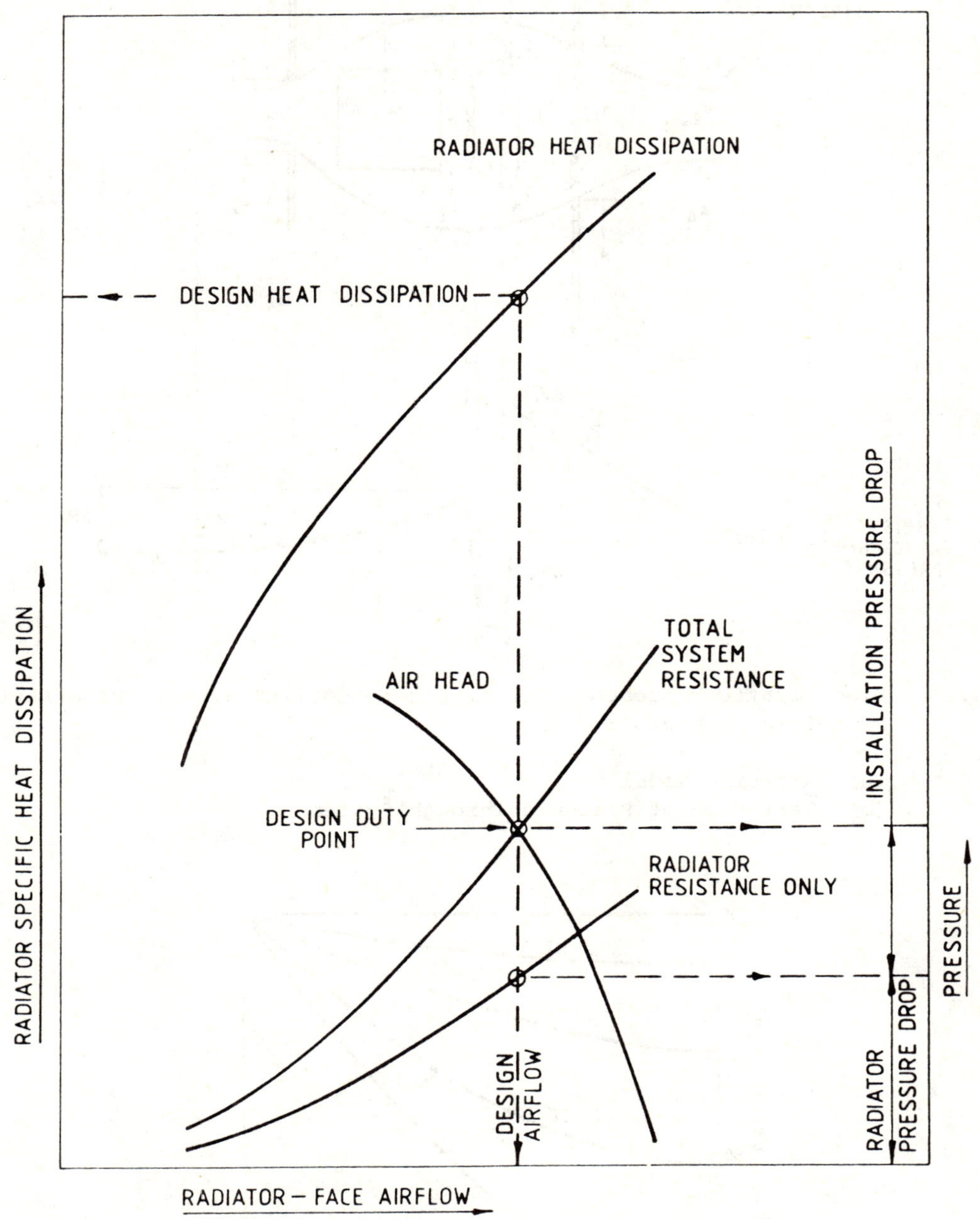

Fig. 4 Relationships Between Radiator Specific Heat Dissipation, Radiator and Total System Resistance, Radiator Airflow, and Air-head at Constant Fan and Vehicle Speeds.

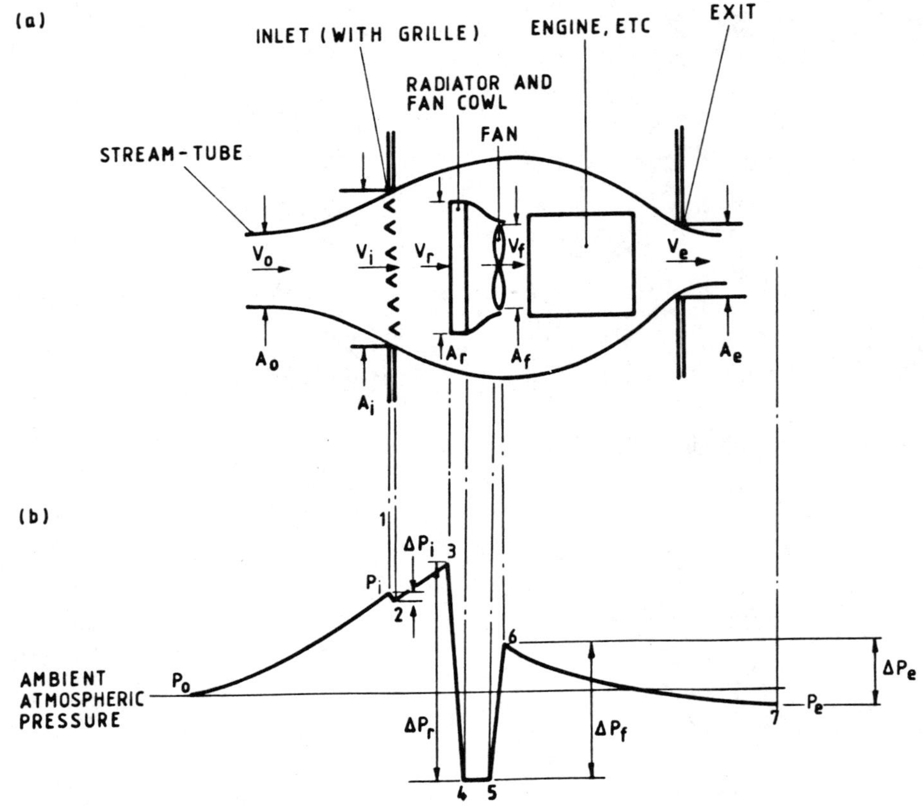

Fig. 5 Diagrammatic Representation of Engine Cooling System Installation Principles, showing:-

(a) Air-side Model
(b) Variation of Pressure Through System

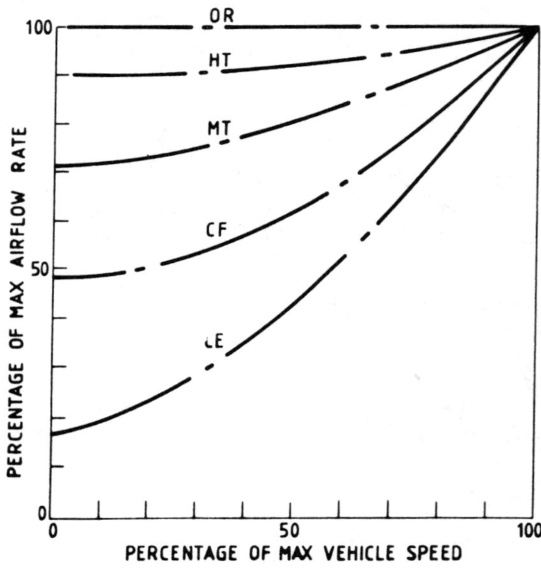

OR = Off-road vehicles
HT = Heavy Duty Trucks
MT = Medium Duty Trucks
CF = Cars with Belt Drive Fans
CE = Cars with Electric Drive Fans

Fig. 6 Non-dimensional Representation of Engine Cooling Air Flow as Function of Vehicle Speed.

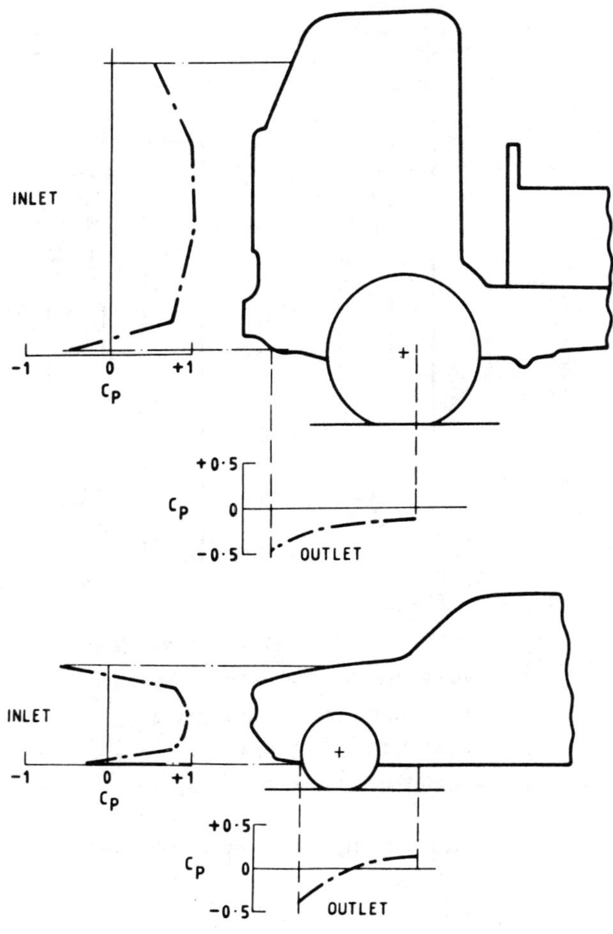

Fig. 7 Typical Variation of Static Pressure Coefficient in Region of Air Inlets and Outlets for a Truck and Car, as measured on Vehicle Centre-line.

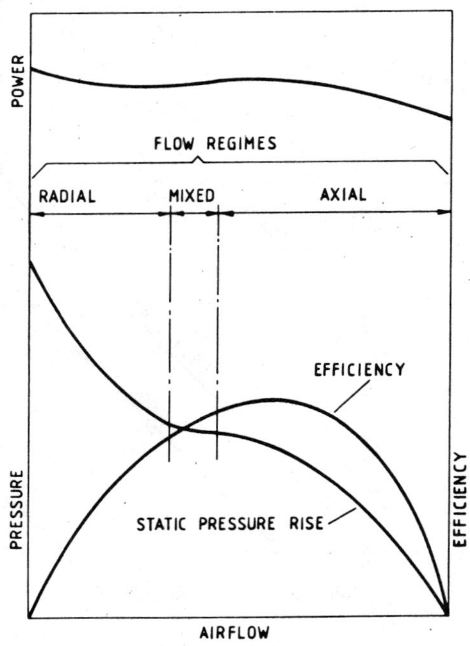

Fig. 8 Typical Form of Constant Speed Fan Performance Characteristics.

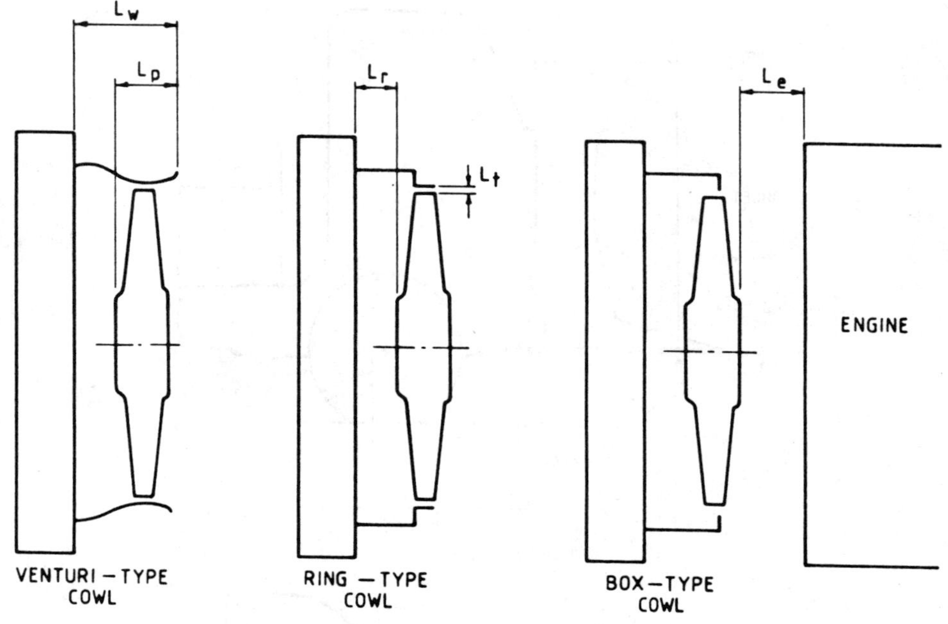

Fig. 9 Factors Affecting Installed Fan Performance

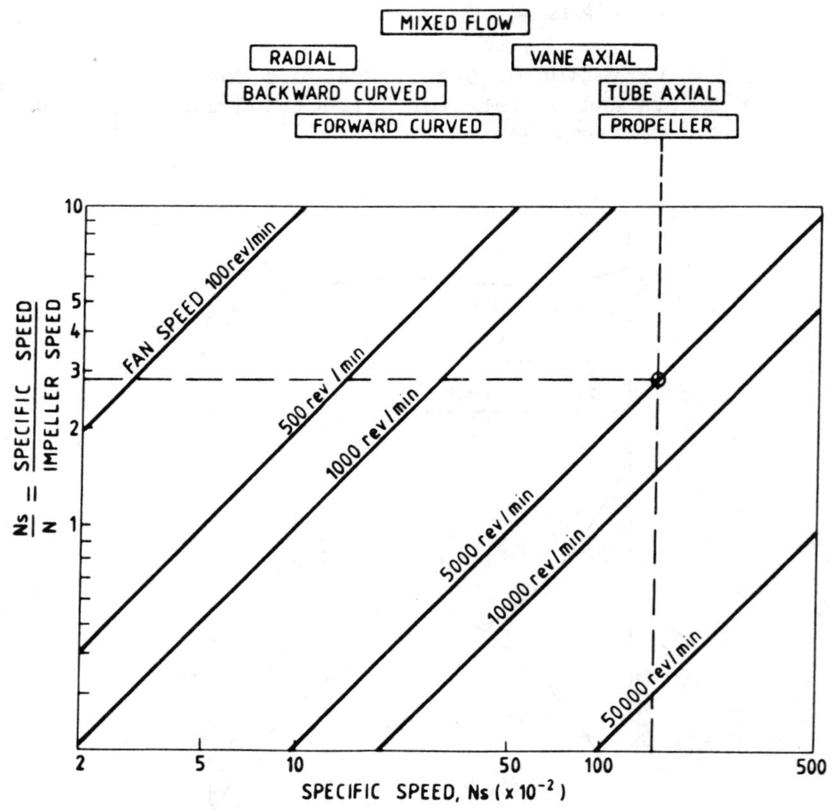

Fig. 10 Range of Specific Speeds Covered by Each Fan Type. Point of Maximum Efficiency Occurs Mid-Range for Each Fan Type.

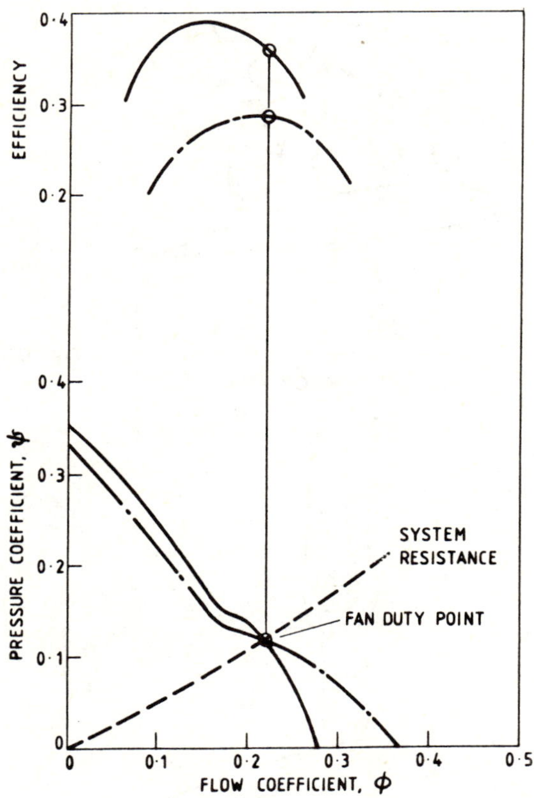

Fig. 11 Dimensionless Performance Comparison of Two Fans Satisfying same Duty Point.

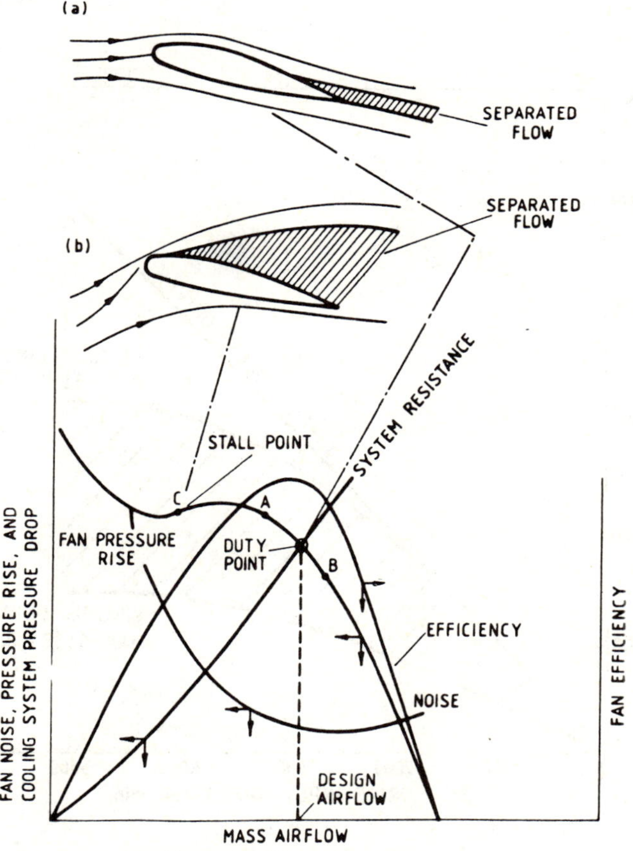

Fig. 12 Relationship between Noise and Fan Performance

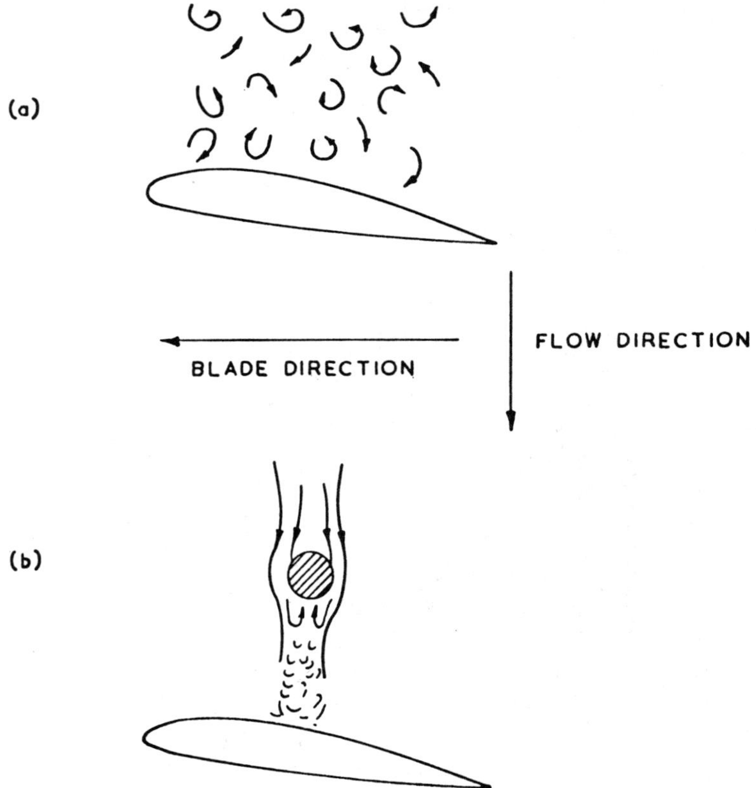

Fig. 13 Diagrammatic Representation of Aerofoil Operating in Turbulent Flow

(a) In Completely Turbulent Air Stream

(b) In Turbulent Wake of Obstacle in Otherwise Smooth Flow

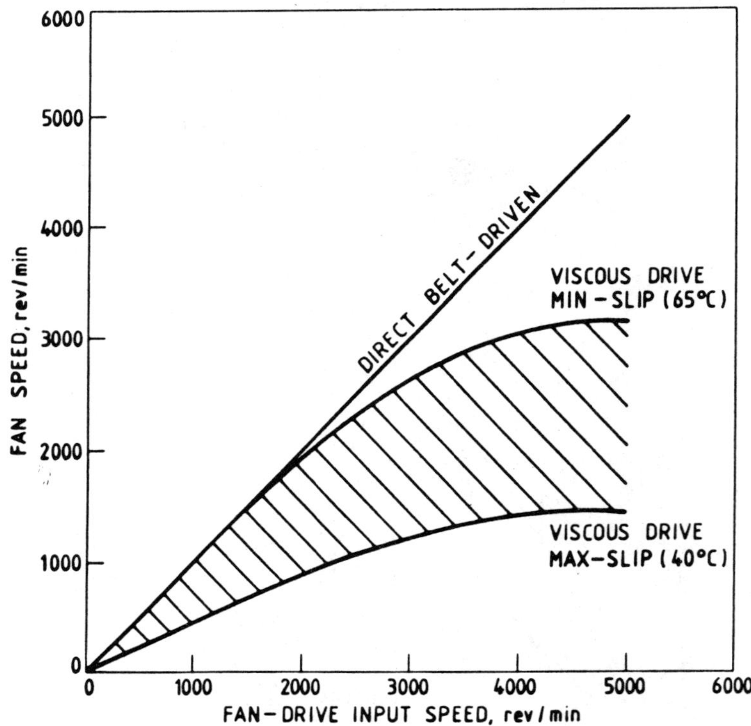

Fig. 14 Viscous Fan-Drive Characteristics (Air-Sensing Type).

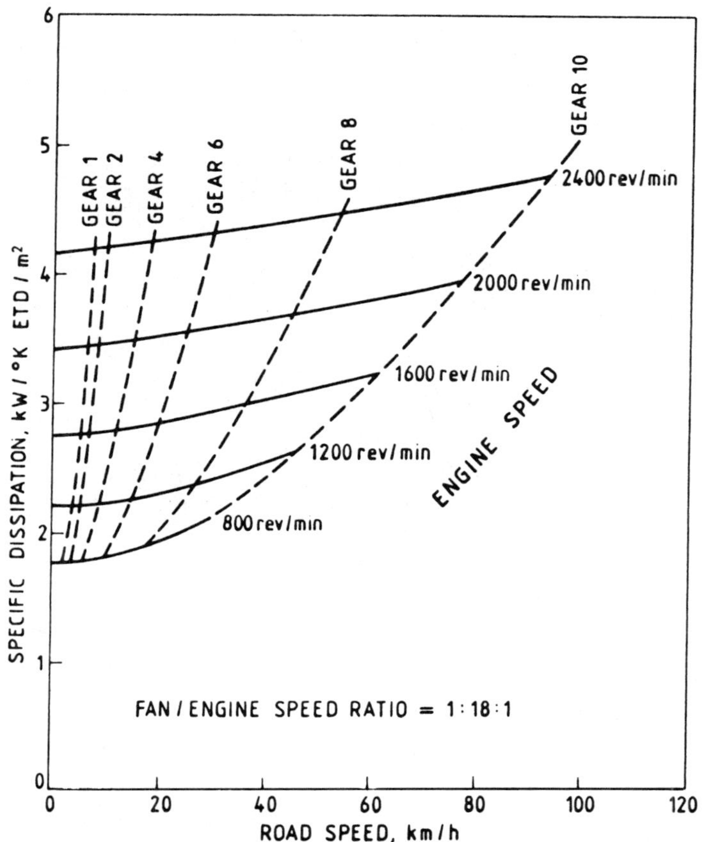

Fig. 15 Variation of Specific Dissipation of Installed Radiator at Constant Engine Speed and Constant Gear Ratio with Vehicle Road Speed. (150 kW Truck).

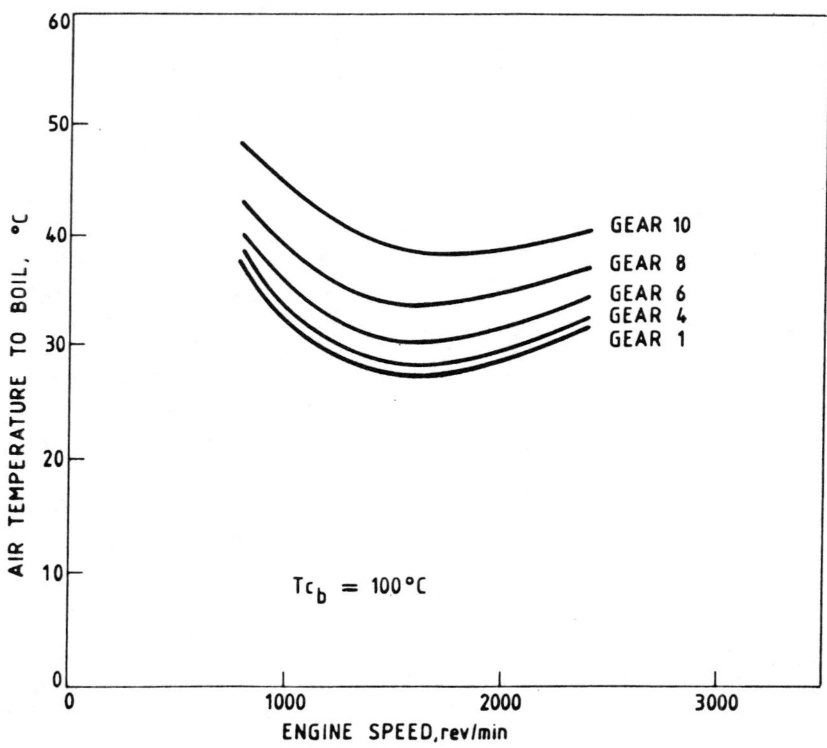

Fig. 16 Variation of Cooling System Air-Temperature-to-Boil, with Engine Speed, for Various Gear Ratios. (150kW Truck).

International Conference on

Fan Design & Applications

Guildford, England: September 7-9, 1982

PAPER F3

DESIGN DEVELOPMENT AND PERFORMANCE OF LARGE AXIAL FLOW FANS IN COOLING TOWERS

K.S. Murthy

Indian Institute of Technology, India

S.J. Basu

Recondo Limited, India

T.S. Rama Murthy and A.K. Rao

Indian Institute of Science, India

Summary

This paper outlines the step-by-step procedure evolved for the design of axial flow fans for cooling tower applications. The design equations have been used to finalise the chord variation, twist variation and number of blades of a 7.315 m diameter fan for an induced draft cooling tower. A computer program was developed to plot the performance curve of the fan. The details of the hollow FRP blade manufacture and the field performance test results have also been given. Comparative performance test results of a conventional aluminium fan have revealed a 26% saving in power consumption by the newly developed fan. The factors affecting the performance have been discussed. The paper also proposes that further investigations have to be carried out to specify the limit of the tip speeds and tip clearances allowable in site installations. It has also been pointed out that development of ultra-high efficiency aerofoils can facilitate the operation of cooling tower fans at much lower tip speeds.

Organised and sponsored by
BHRA Fluid Engineering, Cranfield, Bedford MK43 0AJ, England.

0263 - 421X/82/01 00 - 0001 $5.00
The entire volume can be purchased from
BHRA Fluid Engineering for $82.00

NOMENCLATURE

c	=	blade chord at any radial station
C_L, C_D	=	lift, drag coefficient of fan blade
D	=	Diameter of fan
k_{rl}	=	loss of total head due to blade profile drag
k_{th}	=	theoretical gain of total head at a given radius
N_b	=	number of blades
Q	=	air flow rate
Q_c	=	torque coefficient
R	=	radius of fan
R_s	=	radius of stack at plane of traverse
T_c	=	Thrust coefficient for the fan
$V = VR$	=	air velocity relative to aerofoil
V_t	=	tangential velocity of blade
a	=	lift curve slope of aerofoil per radian
q	=	dynamic head
u	=	axial flow velocity
x	=	r/R - nondimensional radial station
x_b	=	nondimensional hub radius
Z	=	C_L / C_D
ε_s	=	dimensionless swirl at the outlet of fan
λ	=	$u/\Omega r$ - inflow coefficient
σ	=	solidity of fan at radius r
η_a	=	aerodynamic efficiency
ρ	=	air density
ϕ	=	angle of zero lift line of the blade with plane of rotation
ϕ_r	=	angle of local flow with plane of rotation
Ω	=	angular velocity of the fan

1. INTRODUCTION :

The energy crisis all over the world has necessitated researchers to develop and exploit the alternate energy sources like wind, solar, biomass etc. However, it has been realized that before any meaningful and commercial use of the alternate energy sources on a large scale can be made, increasing the installed capacity of power stations and conservation of energy in the industry are the only methods of meeting the challenges of the energy crisis. A recent survey of the power industry has revealed that each KW of power saved by the end user is equivalent to four kilowatts of power saved at the national level.

Cooling towers are necessary in power stations when a recirculating type of cooling system is envisaged. The induced draft cooling towers have axial flow fans installed at the top to induce the required volume of air so as to cool the hot water in the cooling towers and are also used in fertilizer, chemical, sugar, paper, steel and petro-chemical industries. This paper deals with the development of an axial flow fan for use in cooling towers of power stations and fertilizer industries. The design methodology, computer program developed to predict the performance of the fan and the results of a field performance test conducted by the authors have also been given. The cooling tower fan which has been generally treated as a step-child of the industry hitherto, can not tolerate excess power consumption with the increasing operating costs. The efforts of the authors have been to develop a highly efficient axial flow fan so that a substantial saving in power could be achieved.

2. PERFORMANCE REQUIREMENTS :

In India, there are only two major cooling tower manufacturers with foreign collaboration who manufacture the large diameter fans for cooling towers. Recondo Limited has done an intensive survey of the cooling tower fans (Ref. 1) upto diameter of 9.00 Mtrs in the country with a view to the indigeneous design, development and manufacture of large cooling tower fans. Out of the 164 power generating units covered for the survey in India, 74 units have a total number of 1195 fans of diameter of 7.315 operating at an average total power consumption of 66,860 KW i.e. 66.86 MW. At a consumer cost of energy of 0.50 Rs. per KWH, the total operating cost of these fans per year would amount to about 300 million Rupees (roughly 16 million pounds). With an estimated demand of another 1200 fans upto 1987, a saving of 25 % in power consumption of these fans by developing a new design of the fan would reduce the operating cost by 75 million Rupees per year. The survey findings have revealed that 70 percent of the cooling towers in thermal stations and 45 percent of the cooling towers in fertilizer industries use axial flow fans of the diameter of 7.315 Mtr. (24 ft.). Moreover, the cooling tower owners in India started penalising the fan supplier upto Rs.15,000 (800 pounds) per each KW excess of power consumption guaranteed for the duty conditions. With this in view, an indigeneous fan development programme was under taken by the authors. The cooling tower fans handle humid (wet) air and are operated continuously. The cooling tower fan should have high resistance to corrosion and erosion in addition to the requirement of minimum power to handle air at a low pressure drop of the order of 98 Pa (10 mm Water). The fan constructional features should facilitate ease of installation and maintenance.

The cooling tower fan chosen for design development has a design capacity of 9,36,000 m^3/hr at a static pressure rise of 58.84 Pa at a speed of 150 rpm and a diameter of 7.315 Mtr. These design requirements were finalised so as to install the fans in a 210 MW power generating station in Maharashtra, India.

3. DESIGN CONSIDERATIONS :

For an axial flow fan of the aforesaid performance requirements a study of the comparative designs of different types of axial flow fans (Ref. 2) indicated a 25 % hub diameter and has been accordingly chosen.

For the aerodynamic design of the fan blades it has been assumed that :

1. The compressibility effects are ignored
2. The density of air (ρ) is constant
3. The total head rise and the axial flow component remains constant along the length of the blade
4. There is no radial component of flow
5. The preswirl is assumed to be zero i.e. the inflow into the fan is assumed to be axial.

The aerodynamic design of the fan blade consists of determining the details of the section of the fan blade at a moderate number of radial sections and aerofoil selection. The major design parameters for cooling tower fan blade design are : the swirl parameter ε_s at various radial stations and the flow coefficient. The spanwise load distribution on the blade can be estimated on the basis of : (a) constant axial flow velocity and (b) a swirl velocity inversely proportional to the radius giving rise to 'free vortex' flow.

Due to the large variations in relative air velocity and direction it would be necessary to analyze the blades in narrow increments of width dr. By analysing the axial velocity and the tangential velocity (Ref.3), the Fig. 1 shows clearly that there would be no useful work at the centre of the fan axis (i.e. R = 0) as the swirl velocity (V_s) has been assumed to be inversely proportion to the radius. By examining the vectors V_R shown in Fig. 2, in which the mean swirl is used ($V_s/2$) rather than total swirl (V_s), it has been found that no useful work condition exists upto 0.3 R i.e. 1/3 D of the fan. To eliminate the possibility of any reverse flow at the hub of the fan, a hub disc of 0.3 D of fan has been incorporated over the hub. This hub disc diameter would make the effective area for the air flow of the fan to be 91 % of the total area of fan. The cooling tower fan discharges air into the atmosphere at the top of the fan. Due to this there will be an energy loss in the form of the dynamic pressure or velocity pressure loss. The effective work done by the fan can be increased by reducing the outlet velocity of the fan. For this purpose, a diverging section in the form of a diffuser which is sometimes called as a recovery stack is necessary. The optimum performance of these stacks is obtained usually with a semicone angle of 7.5 to 8° of divergence. The divergent section at the exit of the fan reduces the exit velocity and hence the velocity pressure loss. The recovery stack thus converts a portion of velocity pressure to static pressure. The regain of this static pressure would be available as additional suction pressure at the fan inlet which results in reduction of power consumption of the fan. The velocity pressure regain efficiency of 0.80 is usually obtained with 8° semi-cone angle of divergence of the recovery stack.

The total pressure of the fan which is a sum of static pressure and velocity pressure can not be used as a criterion for the relative power required by fans as a fan without recovery stack can have larger velocity pressure (which does no useful work) with a certain static pressure. Hence, a cooling tower fan would be best defined by the static efficiency of the fan instead of total efficiency. Hence, it is obvious that the fan with highest static efficiency will require minimum input power to deliver a given air volume against a given static pressure. The fan development was, for this reason, aimed at, to develop a fan with high static efficiency.

3.1 Aerofoil Selection :

The previous research workers (Ref. 4) have estimated the fan performance by considering the discharge conditions at 0.75 D of the fan to be a true representative of the flow conditions of the complete fan. In this paper the blade configuration was worked at various radii of the fan, for example, 0.35 R, 0.5 R, 0.75 R, 0.90 R, and 1.0 R. The Clark Y aerofoil was chosen as 'good aerofoil' in the early 50s. However, with the advancement of aerodynamic research, the NACA aerofoils have been found to have better aerodynamic characteristics. The aerofoil efficiency can be best defined by the ratio, $Z = C_L/C_D$. The higher the value of C_L/C_D, greater would be the aerofoil efficiency. A comparison of the characteristics of Clark - Y aerofoil and the NACA 6 series aerofoil (chosen for the present fan design) is shown in Fig. 3. The chosen 6 series are low drag aerofoils. In order to have an adequate margin between the operating lift coefficient and C_L max, the operating design C_L is selected as 0.8. The Fig. 3 clearly shows that the maximum C_L/C_D for NACA aerofoil is of the order of 110 whereas that for Clark - Y aerofoil is abour 78. In the design of the fan a Z value of 95 has been used to account for any excess profile drag due to manufacture and skin friction effects.

4. ANALYTICAL CONSIDERATIONS :

The actual design of the fan blade, giving its dimensions and the blade twist along its full length with a provision for fixing it into the hub of the fan, can be finalised by considering the following step-by-step procedure and the equations :

Step - 1 : By knowing the effective area of flow (fan area - hub disc area), and the design air handling capacity of the fan, the air flow velocity (u) can be calculated.

Step - 2 : The pressure head loss coefficient k_{ac} can be calculated (Ref.5) by knowing the design static pressure rise and the flow velocity (u) as given by :

$$k_{ac} = \frac{\Delta H}{1/2 \rho u^2} \qquad \ldots (1)$$

Step - 3 : The inflow coefficient λ can then be found out at various radii by using the equation,

$$\lambda = u/\Omega r \qquad \ldots (2)$$

Step - 4 : The rotor loss coefficient k_{rl} can be found out by the equation,

$$\frac{Z\, k_{rl}}{k_{th}} = \frac{\lambda}{\sin^2 \phi_r} \qquad \ldots (3)$$

The swirl parameter ε_s and the angles of flow direction ϕ_r at various radii of the fan can be found out by an iterative procedure by using the equations,

$$\tan \phi_r = \frac{\lambda}{1 - 0.5\ \varepsilon_s \lambda} \qquad \ldots (4)$$

$$k_{th} = 2/\lambda \varepsilon_s \qquad \ldots (5)$$

and

$$k_{ac} = k_{th} - k_{rl} \qquad \ldots (6)$$

Step - 5 : From the aerofoil characteristics, the design C_L will be known. The chord width required at various radii of the fan blade to meet the design conditions can be found out by solving for C in the equations,

$$\sigma \cdot C_L = 2\, \varepsilon_s \sin \phi_r \qquad \ldots (7)$$

$$c = \frac{2\pi\sigma r}{N_b} \qquad \ldots (8)$$

The design equations given above would give the value of total chord width required at each radii. Theoretically, one wide blade or eight narrow blades will do the same useful work as long as total chord width remains constant. The method of fabrication and ease of manufacture of the blades would also decide the number of blades.

5. ROOT SHOULDER DESIGN :

The clamping of the blades in the hub would require a transition of the aerofoil shape of the blade at the shoulder to a circular shape at the neck. The aerofoil transition to a circular section at .25 D is done by following well known streamlining principles of aerodynamics. The centre line of the root circular portion is to be arranged so that under operating conditions the torque acting is zero at the blade clamp of the hub. This may be achieved by locating the centres of pressure of all the aerofoils along the root axis. Such a root shoulder design would eliminate any chance of excessive untwisting torque load acting on the blade while in operation.

6. DESIGN DEVELOPMENT :

The above design procedure developed by the authors has been applied in the development of the 7.315 Mtr. axial flow fan with the following design operating duty conditions.

Fan diameter	:	7.315 m
Capacity of fan	:	260 m^3/sec.
Static pressure rise	:	58.84 Pa
Fan rotational speed	:	150 rpm
Air density	:	1.1 kg/m^3
Hub diameter	:	1.828 m
Hub disc diameter	:	2.660 m
Net area of flow	:	36.878 m^2

By making use of the equations (1) to (8) and the design procedure outlined above with a design C_L of 0.8 and choosing NACA 6 series aerofoils, fan blade design and the hub design have been finalized to develop the design of a four-blade fan with a chord varying from 0.250 m at the tip to 0.760 m at the shoulder portion. The blade twist angle was zero at the tip and 15° at the shoulder. The 0.760 m shoulder has been aerodynamically smoothly transformed into a 0.160 m circular neck portion to facilitate clamping of the blades.

7. PERFORMANCE ANALYSIS :

The characteristic of the fan can be predicted theoretically so as to plot the performance curve. The performance curves show a plot of variation of flow, pressure and power of the fan at various settings of the blades. The performance curves are generally drawn at a constant density and a certain design tip speed. The tip speeds that are generally used in Indian large cooling tower fans have been found to be varying from 55 m/sec. to 61 m/sec. The tip speed chosen for the design is 57.5 m/sec. The upper limit of the tip speed is mainly from the mechanical and noise consideration of the fan to achieve good compromise between them.

The performance of the fan can be predicted by knowing the blade geometry and the number of blades. By keeping the blade setting and the flow of air constant, the swirl coefficient, ε_s , may be estimated at various radii of fan, by the equations :

$$\varepsilon_s = \frac{2 \cos \phi \quad \tan (\phi / \lambda) - 1}{4/a\, \sigma(x) + \sin \phi} \qquad \ldots (9)$$

where $\sigma(x) = \dfrac{N_b\, c(x)}{2\pi R x}$

$\lambda(x) = u/\Omega Rx$

Knowing ϕ from equation (5), k_{rl} is obtained as

$$k_{rl} = \frac{\lambda \quad k_{th}}{Z \sin^2 \phi} \qquad \ldots (10)$$

The thrust and torque parameters are evaluated from the equations :

$$\frac{d\, T_c}{dx} = 2\, x\, (k_{th} - k_{rl} - \varepsilon_s^2) \qquad \ldots (11)$$

$$\frac{d\, Q_c}{dx} = 4\, x^2\, \varepsilon_s \qquad \ldots (12)$$

where $T_c = T/0.5\, \rho\, u^2 \quad \pi R^2 \qquad \ldots (13)$

$Q_c = Q/0.5\, \rho\, u^2 \quad \pi R^3 \qquad \ldots (14)$

The design duty point performance data can be fount out by evaluating Z in equation (10) outside the integral since the total length of the blade is loaded uniformly at design point. To predict the performance of the fan at off - design points, the C_L value which would be varying at each setting of the blades will have to be evaluated. The value of C_L can be either greater or lesser than the design value at off - design condition depending on the blade setting. It has been found that

$$Z = a_1\, C_L \quad \text{for} \quad C_L \leq C_{Li}$$

$$Z = a_1\, C_{Li} \quad \text{for} \quad C_L \geq C_{Li}$$

For the present study, $C_{Li} = 0.7$ and $a_1 = 95$ have been evaluated from lift slope curve of the aerofoil. Integrating the equations (11) and (12) the torque Q required at the input to the hub of the fan and the thrust T developed by the fan can be evaluated. The aerodynamic efficiency of the fan would then be given by :

$$\eta_a = \frac{T\, u}{Q\, \Omega} \qquad \ldots (15)$$

To plot the complete range of off - design performance characteristics of the fan, a computer program was developed and was written in Fortran-IV. A DEC - 10 computer was used to plot the predicted performance curve of the fan. The stalling range of the fan operation can be easily obtained as the computer output results would show a negative value of thrust for regions of operation beyond acceptable limitations. The values obtained from the above equations would become negative at negative thrust values clearly indicating stall conditions.

8. MANUFACTURING PROCESS :

The larger chord widths were chosen for the four blade design of the 7.315 m, diameter fan with a view to the use of alternative materials and method of manufacture. The fan blades are subjected to :

a) Steady tensile loads due to centrifugal forces

b) Steady bending loads due to aerodynamic thrust and torque forces, and

c) Vibratory bending loads due to cyclic variations in the air loads and other pulse excitations due to the fan support beam structures in the cooling tower.

The most conventional material, cast aluminium, has many disadvantages for large fan blades, for example : the weight of wider chord, and high camber blades would be so high that fixing of the blades in the hub of the fan would pose severe problems as the blade is supported as a cantilever at its root neck portion. The use of efficient aerofoils, hence, becomes unwieldy with cast aluminium material. With the heavier blades the operating /dynamic stresses would also be large, and the heavy weight on the fan gear box bearings would result in a reduction of gear box life. Additionally, since the cooling tower fans operate in a humid environment, aluminium blades could be prone to severe oxidation unless protected.

The Fibre glass Reinforced Plastic (FRP) material has been shown to have a higher strength to weight ratio, and a higher fatigue strength to density ratio when compared with aluminium (Ref.6). FRP is also immune to atmospheric and galvanic corrosion with a negligible attack by fungus. A FRP structure is a composite which has a resin (plastic) matrix reinforced with fibreglass. In India, E-glass is only available and so has been used for the manufacture of blades although S-glass may have superior properties. The FRP blade structure should have low water absorption with strong chemical bonds between the glassfibre and the resin to have a greater fatigue resistance under dynamic conditions. The resin matrix usually is chosen to be commensurate with the properties of the glass. Among the various types of polyester and epoxy resins, it has been established by conducting various tests that epoxy resin is superior with a lower water absorption, lower matrix temperature rise in dynamic conditions and higher bond strength when compared to the polyester resins. Hence, an epoxy resin was chosen as the resin system of the blades. A pressure-bag moulding technique similar to that used for the manufacture of helicopter rotor blades has been developed to manufacture the fan blades which resulted in a hollow blade structure. The details of process optimization (Ref. 7) have been worked out to exploit the advantages of epoxy resins for the FRP Structures. The hollow construction of the blade has made the total weight of the blade lighter by 45 percent when compared to the conventional narrow, cast Aluminium blade for the same size of fan. The structural design is based on the static stresses with a liberal margin of safety. A $\pm$ 45° and 0/90 lay up pattern of the Glass Cloth was used to achieve maximum strength of the blades.

9. FAN PERFORMANCE BY TEST :

The fan performance can be measured in the field in an actual installation, sometimes as a proof of the performance of the fan. A detailed field performance test as has been suggested elsewhere (Ref.8) and also based on the authors experience (Ref.9) with relevance to the Indian conditions has been conducted to check for the design operating point. The newly developed fans were installed in the two induced draft cooling towers of a 210 MW thermal power station. Each cooling tower handles a water flow of 17,000 m^3/hr at a temperature of 38.82°C to cool down to a temperature of 28.33° C at an inlet wet bulb temperature of 25.55° C requiring 18 axial flow fans each 7.315 m diameter.

The fan blades were pitched up to the design blade angle. The air flow velocity above the fan level was measured at five radii producing equal annular areas of the fan recovery stack/cylinder at r_1 = 0.316 Rs, r_2 = 0.548 Rs, r_3 = 0.707 Rs, r_4 = 0.837 Rs and r_5 = 0.948 Rs., Rs being the radius of the fan stack at the plane of measurement. A calibrated vane type anemometer was traversed along four radii of the stack. An average of the 20 readings was taken to give the mean velocity of flow. The area of the flow was taken as the area of the stack at the plane of measurement. By multiplying the average velocity with the area of flow, the air flow was calculated. The static pressure was measured with an inclined tube manometer connected by 6 mm Ø polythene tubing. The polythene tube was inserted, from top of the cooling tower, beneath the fan in the cooling tower at four stations along the diagonals of the cooling tower cell. The static pressure was measured before and after the completion of the flow test. The average of the 8 readings of the static pressure was taken as the operating static pressure of fan. The power consumption at the motor terminals was measured by using calibrated voltmeters, Ammeters and a power factor meter. The test air density was measured by a calibrated psychrometer to measure the wet bulb depression at the fan discharge. The fan speed was assessed from the gear ratio and the motor speed. The motor input power was calculated from the measurements of voltage, current and power factor in the 3-phase electrical input terminals to the motor. The impeller power, which is available as input to the fan from the reduction gear box driven by the electric motor was calculated by knowing the gear box efficiency and the motor efficiency at the operating load.

$$\text{Impeller Power} = (\text{Motor power}) \times (\text{gear box efficiency}) \times (\text{motor efficiency})$$

The aerodynamic efficiency of the fan can be calculated by knowing the air power reduced to the test density values.

$$\text{Air power} = Q \times P_s \text{ Watts}$$

Q - air flow rate ($m^3/sec.$)

P_s - static pressure (Pa)

$$\therefore \quad \text{Aerodynamic efficiency} = \frac{\text{Air power}}{\text{Impeller power}} \times 100$$

The total efficiency of the impeller of fan can be also calculated knowing the total pressure. Substituting total pressure instead of static pressure in the air power equation and taking the ratio of total air power to impeller power, the total aerodynamic efficiency can be calculated.

The field test procedure described was repeated to evaluate the performance of a conventional aluminium fan of the same diameter, to compare/evaluate the new design of the FRP fan developed by the authors, in a nearby cooling tower operating under similar design conditions.

10. RESULTS AND DISCUSSION :

The design procedure outlined in this paper has made it possible to develop a design for the fan. The performance characteristics of the fan as predicted by the computer program are given in Fig. 4. The field performance test result is shown as an operating point on the same performance curve shown in Fig. 4.

It can be seen that the result of the field test is very close to the design point and it can be concluded that the teoretical analysis of the fan performance is accurate.

The analytical methods of prediction of fan performance have not been very convincing since the early 50s. However, with the advancement of computer technology and the availability of more data on the performance of aerofoils, it is possible nowadays, as shown in this paper, to accurately predict the performance of a fan. Such an attempt might eliminate to some extent laboratory tests on the fans especially when there are test uncertainties manufacturing uncertainties, and specification uncertainties involved in the methods of rating fans.

It may be mentioned here that field conditions such as tip clearance, improper pitch setting of the blades,etc. can adversely affect the performance of the fan. A fan tip speed value of the order of 60 m/sec. specified as an upper limit for large cooling tower fans, needs further investigation to clearly explain its use than just to state that it is a good compromise between mechanical and noise considerations. The fan engineers in different countries internationally specify fan tip speed of any value between 55 to 61 m/sec. However, if low noise requirements of the fan become more stringent, then much lower tip speeds in the order of 35 m/sec. may have to be used. To achieve the required duty point of the fan, ultra-high efficiency aerofoils will then have to be developed with Z values of the order of 200.

The inflow conditions in the actual installation of a cooling tower fan are seldom purely axial, as the cooling tower sizes would have to increase substantially to achieve an ideally smooth entry of flow from the packing/fill of the cooling tower. However, cooling tower designers should be aware of the importance of smooth airflow entry into a fan as there would be as high as 50 % energy loss at the entry of the fan with an inefficient and poor design of fan inlet. The newly developed fan has been found to have a value of static efficiency of 71 percent and a total efficiency of 92 percent was successfully achieved. On comparing the power consumption of this newly developed fan with a conventional Aluminium fan, a power saving in the order of 26 percent was achieved. It is hoped that fan designers would concentrate their efforts to further enhance the performance of cooling tower fans. The achievement of higher efficiency would be possible only by more efficient aerofoils and close control of site conditions such as tip clearance, etc. The endeavours of the authors have been to conserve energy by the development of such an efficient fan.

11. CONCLUSIONS :

1. A design methodology has been developed to evolve the design of a cooling tower fan.
2. A computer programme was developed to predict the performance characteristics curve of the axial flow fan.
3. FRP material has been successfully used to manufacture hollow blades for the fan.
4. Field performance tests have shown close agreement with the predicted fan performance curve and operating point.
5. A substantial saving in power has been achieved with the newly developed FRP fan compared to the conventional aluminium fan.

12. ACKNOWLEDGEMENTS :

The authors wish to gratefully acknowledge the whole hearted support of the Board of Directors of Recondo Limited for having encouraged the R and D programme on fans. The authors also acknowledge the Scientists at National Aeronautical Laboratory and Indian Institute of Science, Bangalore, and Indian Institute of Technology, Bombay.

13. REFERENCES :

1. Market Survey for Industrial Cooling Tower Fans - Unpublished Report, Recondo Limited, Bombay.
2. Daly B. B. : Woods Practical Guide to Fan Engineering, England, Essex Telegraph Press Ltd., Colchester, 1979. pp. 132 - 133.
3. Davidson E. M. : Axial Flow Fans and Their Application to Cooling Towers, In : Proc. Annual Meeting, (Monteleone, New Orleans, Jan. 28 Feb : 1, 1968). C.T.I. USA 1968.
4. Greorge W. Forman and Neil W. Kelly : Cooling Tower Fan Performance. Jnl. of Engg. for Power ASME April 1961, pp. 155 - 160.
5. Wallis R. A. : Axial Flow Fans. London, George Newnes Ltd., 1961.
6. Salkind M. J. : The Twin Beam Composite Rotor Blade. Fibre Science and Technology, 8, 1975 pp. 91 - 102.
7. Murthy, K. S. and Halageri, A. S. : Manufacturing Process Optimization for large Axial Flow Fans In : Proc. National Seminar on Fans, Pumps and Compressors (Calcutta, India, Dec. 27, 1981), Calcutta, India Society of Engineers, 1981.
8. Herrmann D. D. : Field Tests of Fan Performance on Induced Draft Cooling Towers In : Proc Annual Meeting, St. Louis, Jan. 16, 1962), C.T.I. USA, 1962.
9. Murthy K. S., Basu, S. J., and Subir Kar : Standardization of Field Testing of Induced Draft Cooling Tower Fan. Draft Proposal of Standard Submitted to Electro-Technical Divn. of Indian Standard Institute, New Delhi, India, Jan. 1982.

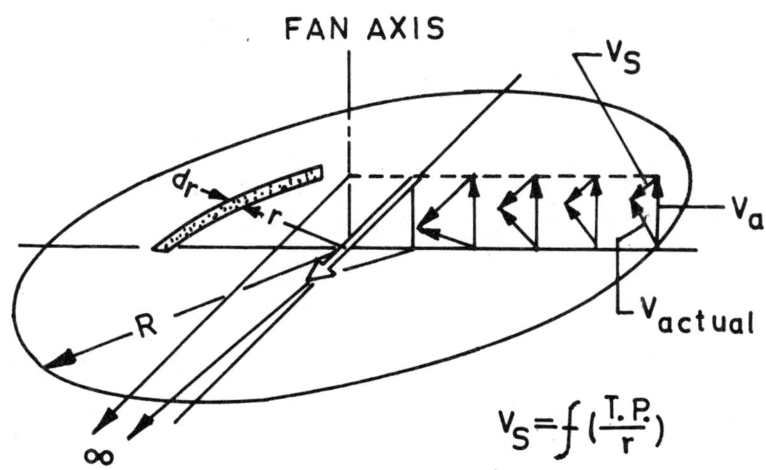

Fig. 1 - Actual discharge velocity

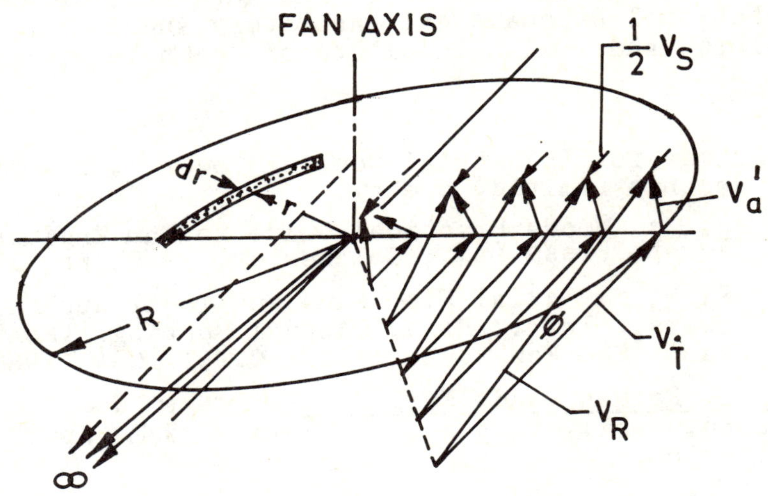

Fig. 2 - Velocity relative to blade

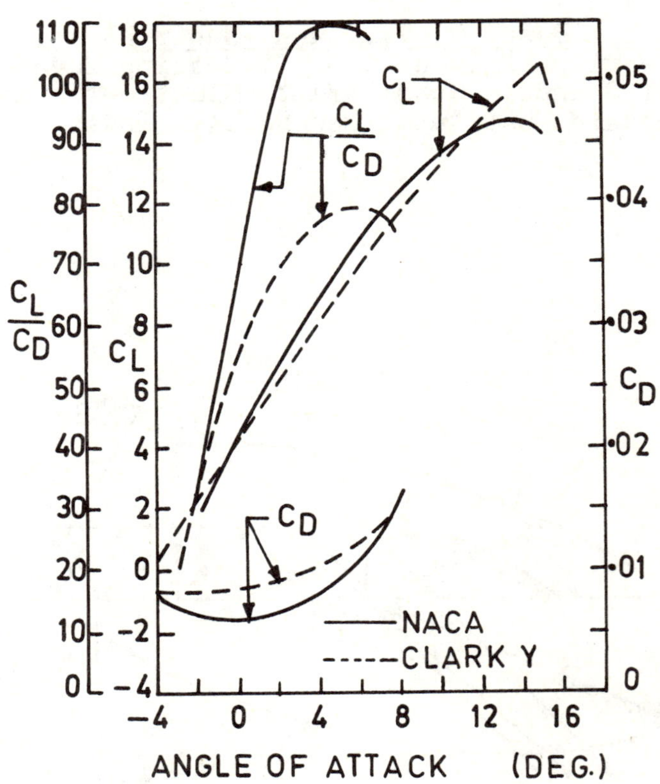

Fig. 3 - Comparision of aerofoil characteristics

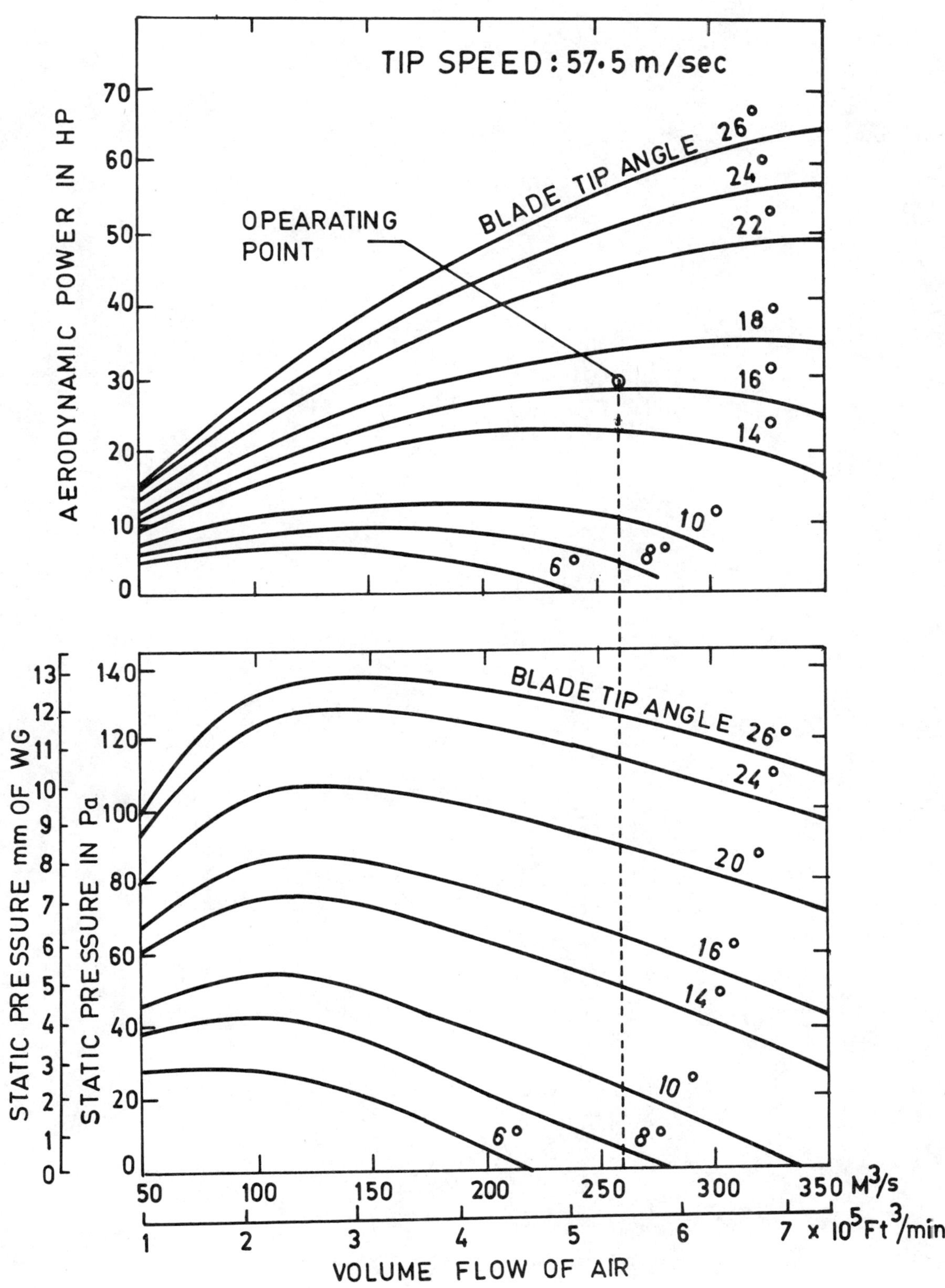

Fig. 4 - Performance curve of 7.315 Mtr. fan.

International Conference on

Fan Design & Applications

Guildford, England: September 7-9, 1982

PAPER G1

HIGH REACTION FANS

A. Smith

Grubb Parsons, U.K.

Summary

Fan reaction is defined as the percentage of the static pressure rise over the rotating blade row to that over the complete stage.

In conventional high performance fans outlet guide vanes are usually employed to redirect the swirl generated by the rotor into the axial direction thereby partially recovering the dynamic head in terms of static pressure and improving further recovery of the axial velocity in the following diffuser. Such fans have reactions of between 60 and 80%.

By suitably changing the blading configuration, reactions of over 100% are possible and fans built to this principle can offer advantages in specific output and mechanical layout without sacrifice in efficiency.

Two high reaction fan designs are examined, the first an early application for gas circulation duties in Bradwell nuclear power station and the second, a recent proposal for "Jet" ventilation of vehicle tunnels. Both aerodynamic and mechanical design aspects of these fans are discussed and test results are given.

Organised and sponsored by
BHRA Fluid Engineering, Cranfield, Bedford MK43 0AJ, England.

0263 - 421X/82/01 00 - 0001 $5.00
The entire volume can be purchased from
BHRA Fluid Engineering for $82.00

NOMENCLATURE

A_1	=	Diffuser inlet area
A_2	=	Diffuser outlet area
C_L	=	Vector mean lift coefficient
Vo	=	Vector mean incidence
U	=	Blade velocity
Um	=	Blade velocity at mean diameter
V_0	=	Absolute outlet velocity from inlet guide vanes
V_1	=	Velocity into moving blades
V_2	=	Velocity leaving moving blades
V_3	=	Absolute outlet velocity
Va	=	Axial velocity
h	=	Blade height
c	=	Mean blade chord
h/c	=	Aspect ratio
ρ	=	Gas density
ΔPs	=	Static pressure rise
ΔP_T	=	Total pressure rise
η_F	=	Friction factor
η_D	=	Diffuser effectiveness
ζ	=	Inlet guide vane stagger angle (to axis)
γ	=	Blade stagger angle (to rotor plane)

1 HISTORICAL BACKGROUND

1956 saw the commissioning of Calder Hall, the first nuclear power station for commercial electricity generation in the world. The reactor was of the graphite moderated, gas cooled type, heat for steam power generation being extracted by circulating the carbon dioxide coolant through the reactor core before passing it through four parallel circuits to the external boilers. Having extracted the heat, a fan was installed in each circuit to return the cool gas to the reactor for reheating.

The fans were of the centrifugal type (Fig 1), each driven from a Ward Leonard controlled variable speed DC motor rated at 1640 kW at 960 RPM. The carbon dioxide coolant was pressurized to 7.9 bar to improve heat transfer and the blower casings had to seal the gas absolutely, at a temperature level of 423 K to prevent radio active contamination should a reactor fuel rod burst. This resulted in the shaft seal being the subject of an extensive development programme, the success of which can be judged by the fact that they have been running at Calder Hall and Annan power stations for up to 25 years without replacement. No such success can be claimed for the fan casings, however, which were of cast steel and split at a horizontal joint to permit the withdrawal of the impeller and shaft.

Some three months of intensive effort was required before the leakage rate on the first unit was brought down to the requisite standard, hand scraping of all joints proving necessary because of restraints in the use of sealing compounds.

The site problems of the first blower prompted the building of special erection and test beds for the remainder at Heaton Works but only by exercising extreme care with joints, was it found possible to meet nuclear gas tightness standards.

The sealing difficulties of the early Calder centrifugal machines proved sufficiently daunting that axial flow fans were adopted for the second generation of nuclear power stations to be built at Bradwell (Fig 2), and Latina. Although ratings were considerably higher than the Calder fans, the change was essentially made to reduce casing sizes, the rotor being overhung from a single seal and bearing cartridge assembly (Fig 3) to limit sealing to one vertical circular flange in the outer casing.

In the interests of compactness and mechanical integrity Parsons adopted a new type of high reaction fan design in which the more usual outlet guide vane arrangement was rejected in favour of inlet guides. In these the flow was accelerated into the oncoming path of the rotor blading, blade angles being chosen to produce an axial discharge into the diffuser. The flow distribution into the moving blades could consequently be improved within the space restraints of the pressurized casing over a conventional axial entry while avoiding swirl, which detracts from diffuser pressure recovery where flare on the outer fairing is restricted.

Because the rotor blading had to compensate for the acceleration (and pressure drop) through the inlet guide vanes, the stage reaction, defined as the pressure rise over the rotor to that of the stage, exceeded 100%. High reaction fans have other applications, however, and the second example selected for this paper is concerned with air momentum transfer in road tunnels, the "jet" ventilation concept.

2 ROAD TUNNEL VENTILATION

Transverse ventilation had been the favoured method of ventilation for long road tunnels for many years, fresh air being supplied through grills from a full length duct usually below the roadway and extracted through overhead or side vents into an extract duct. The expense of providing these ducts had resulted in the more recent adoption of semi traverse ventilation, where the roadway itself is allowed to act as the extract duct. The development of one way traffic systems, however, has favoured the adoption of twin bore tunnels where full advantage can be taken of the piston effect of the traffic to induce longitudinal air flows. To augment this induced ventilation under adverse wind or traffic conditions "jet" fans are installed along the tunnel roof which transfer their momentum into the direction of traffic flow when required. Economies in the civil engineering can consequently be effected by totally eliminating the supply (and extract) ducts and, with them, the surface ventilating fans.

A second and perhaps a less convincing argument is that overall savings in tunnel running costs can result because of the ease of switching off unnecessary fan units under favourable traffic conditions. This, however, must be examined against the inefficiency of momentum transfer from such fans to the main stream as compared with conventional ventilation methods. Results from European tunnels, nevertheless, would appear to justify the economic claims for jet ventilation and, with the accumulation of operating experience, the length of tunnels ventilated by this technique is being considerably extended. It is the development of a high reaction fan for such ventilation duties that forms the second and principle subject of this paper.

3 AXIAL FAN DESIGN CONSIDERATIONS

Fans are characterized by having a low rotor compression so that unlike compressors, their overall performance becomes critically dependent on intake and discharge losses. For example, the additional losses of a poor intake can not only reduce the static pressure before the fan, but can distort the velocity distribution into the rotor and thereby reduce the compression performance. The job of the diffuser, on the other hand, is to recover the dynamic head generated at inlet plus any residual swirl from the rotor. Failure to do this introduces an unacceptable penalty because of the relative magnitude of these heads as compared with the rotor compression head.

Where there is no external diameter restriction on the diffuser, the swirl introduced by rotor work can be accommodated by care in design (Ref 1), but where outlet diameter is restricted, it becomes necessary to provide the requisite increase in flow area by tapering the inner diffuser fairing. Should rotor swirl be present in this situation then the free vortex which develops downstream raises the tangential velocity component close to the inner wall destabilizes the boundary layer and results in flow separation and high losses.

To avoid such diffusion problems, conventional high performance axial fans draw the gas axially into the moving blades and subsequently discharge the flow through outlet guide vanes (OGV). The swirl produced by the rotor work can consequently be recovered in the form of static pressure rise by redirecting the flow into the axial direction before entering the diffuser (Fig 4). The resulting reaction of such a design, expressed as a percentage of the static pressure rise over the rotor blading to that of the stage, is usually between 60 and 80%.

When designing for high specific output, a disadvantage of the conventional (OGV) design arrangement is the difficulty of effective removal of the high rotor swirl because of aerodynamic loading restraints on the guide vanes themselves.

In high reaction fans, the flow into the rotor is pre swirled by inlet guide vanes (IGV) into the path of the moving blades (Fig 5); moreover, by suitably matching the inlet guide vane with rotor blade angles, a swirl free discharge can be obtained. Scope for the recovery of the outlet dynamic head in a diffuser of limited diameter is consequently improved. Apart from possible mechanical design improvements an added aerodynamic advantage of this arrangement is that the inlet acceleration through the guide vanes improves the flow distribution into the moving blades so that the influence of upstream maldistribution of the flow is reduced.

Since pre-swirling the flow is essentially an acceleration process, boundary layer conditions on the surfaces of the IGV's are considerably more stable than in OGV's where diffusion predominates. Fan output is consequently no longer restricted by aerodynamic loading in the vanes, but only by the inlet Mach number into the moving blades, a limit which is unlikely to be approached in fan applications.

Because of the fall in pressure through the inlet guide vanes the pressure rise through the blades must exceed that of the stage so that the resultant reaction of this design exceeds 100% and hence has been termed the high reaction design principle. Compared with the OGV form of design however, the small entropy gain through the guide vanes of the IGV arrangement is more than compensated by the improvement in pressure recovery in the moving blades which results from the improvement in Reynolds number and the transfer of swirl recovery into the efficient rotor blading.

4 THE BRADWELL FAN

4.1 The Design:

Each of the six gas circulating fans for the Bradwell nuclear power station was designed for a flow of 35.8 m^3/s with a rise of 53 KPa. The absolute pressure and temperature of the carbon dioxide at inlet correspond to 9.65 bar at 447 K with a design shaft power input of 2.35 MW at 3000 RPM and a maximum of 3.3 MW at 3300 RPM.

The high working pressure resulted in the adoption of a cylindrical external casing, the diameter of which was kept to a minimum by the adoption of an inward reversed flow type of inlet (Fig 2). With the circuit resistance being essentially constant, the setting angles of both the inlet guide vanes and rotor blades were fixed, the frequency to the induction motor drive being changed to vary the speed to suit reactor output.

Both guide vanes and rotor were overhung from the inlet assembly so that a single flanged joint resulted with the rotating parts being supported on outboard palms sitting on horizontal rails on the machine centre line when the flanged joint was broken (Fig 2). By removing the flange nuts, oil pipes and motor jack shaft, the rotor assembly could subsequently be axially withdrawn for maintenance.

Because of the external pressure casing diameter restriction, the diffuser employed a cylindrical outer tube with a short finishing taper and a conical inner fairing. The taper towards the outlet of the inner cone was, however, increased to maintain a near constant equivalent cone angle (Ref 2).

Thirty five untwisted inlet guide vanes were used, the twelve rotor blades being designed for constant radial work so that the resultant axial swirl at outlet varied from 9.9° at the inner diameter of 0.838 m to + 0.6° at the mean, and - 5.3° at the outer diameter of 1.27 m. The blade twist conformed to the then popular "near vortex" form of design which had been shown experimentally in compressors to behave as effectively as a full vortex design with the economic advantage of untwisted guide vanes. A mean reaction of 110% resulted. To prevent leakage the fan was provided with a radial shaft gas seal, the oil supply to which was kept 0.3 bar above the gas pressure under all conditions, by a differential pressure control valve.

The shaft itself was supported by an inboard journal bearing and a combined outboard bearing and tilting pad thrust, these being accommodated in a cartridge assembly flange bolted to the external casing.

A direct two pole, squirrel cage induction motor drive was employed, the duty of all six fans being controlled simultaneously by the supply frequency from a single variable speed turbo generator.

4.2 Test Rig:

A full scale prototype circulator was tested on air in a closed circuit loop which could be pressurized to a level of 3 bars absolute. Input powers were consequently reduced to 750 kW at the design flow conditions which also suited the rating of the cross flow, gilled tube cooler incorporated in the circuit.

The flow rate was measured by an annular orifice, a model of which had been calibrated on steam using scaled ductwork in both the approach and discharge lengths. This particular form of orifice was chosen following its recommendation by the Royal Aircraft Establishment because of its insensitivity to swirl and was positioned downstream from the cooler so that flow maldistribution from a louvred throttle valve at the circulator outlet could be minimized.

Because of the possible use of the pressure casing at site, pressure tappings and instrument holes were confined to positions immediately upstream and downstream from the inlet and outlet flanges. Pressure rises were therefore confined to overall static values.

Input was measured by torquemeter, electrical input and a differential temperature arrangement of twelve iron constantan thermocouples disposed in grids across the inlet and outlet duct planes. This thermocouple arrangement amplified the electro motive force for potentiometer measurement by a factor of twelve to reduce inaccuracies resulting from the low temperature rise across the circulator.

To establish the influence of guide and blade angle on performance, three sets of blading were manufactured, the guide vane settings varying by 10°, and the rotor by 3° above and below design so that nine combinations of guide vanes and rotor settings could be tested, the profiles and twist remaining unaltered in each case.

4.3 Results:

Results at the design guide vane and blade angles are presented in the conventional non-dimensional form of overall static pressure rise coefficient ($\Delta Ps/\frac{1}{2}\rho Um^2$) against a flow coefficient (Va/Um), from which the closeness to the design duty can be judged (Fig 6).

High overall efficiencies were precluded by high inlet duct losses (Fig 6), a value of 79.3% being achieved at the design flow coefficient of 0.308, peaking at 82% at a flow coefficient of 0.28. Assuming a diffuser recovery of 0.5 of the axial dynamic head at outlet from the blading, total efficiency of the stage was deduced to be 89% an acceptable value in view of the restriction imposed on the aspect ratio (h/c = 1.15) by high gas bending loads. Subsequent traverses on a model of the Oldbury circulator inlet also suggested that the inlet guide vane acceleration was insufficient to totally suppress radial maldistribution of the flow into the blading from the 180° flow reversal before the guide vanes.

To establish the variation in duty with IGV and blade angle, four sets of results have been taken from the nine IGV and blade angle combinations tested, and compared with the performance at the design settings (Fig 7).

The characteristic curves show that duty is considerably more sensitive to the 6° change in blade stagger angle than the 20° change in IGV angle. It should be noted, however, that the accelerating flow regime of the guide vanes improves boundary layer stability so that their working range can be extended to $\pm$ 40° with suitable profiled vanes. The adoption of IGV angle control to vary duty therefore becomes a viable alternative to more usual methods of varying blade angles, especially if there is the requirement for this to be done while the machine is running.

5 THE JET FAN

5.1 The Design:

Jet fans are employed to augment the flow induced by the piston effect of the traffic in longitudinally ventilated tunnels. In a fan producing swirl at outlet only the axial velocity component provides axial momentum to improve ventilation (Ref 3); not to attempt to recover the swirl component is therefore wasteful since this energy has to be provided by the motor. It has been suggested that swirl may help to improve mixing and therefore dilute exhaust gases, but chemical tracer evidence (Ref 4) would suggest that the vorticity generated by the traffic flow is more than sufficient to promote effective mixing. In recognition of these ventilation considerations, a 600 mm prototype jet fan has been designed and built for a flow of 16.5 m^3/s and a discharge velocity of 30 m/s.

A high reaction single stage axial flow fan with inlet guide vanes was employed, the vane and blade angles being chosen to provide a swirl free discharge. The inherent stiffness of the guide vanes was also used to provide a robust mounting inside the external cylindrical casing from which to bolt the moving assembly (Fig 8).

The guide vanes were formed from rolled mild steel plates, their chord being progressively increased with radius by axially extending the leading edges for additional stiffening. The curved vane trailing edges, however, were progressively trimmed back from inside to outside diameter to provide a free vortex variation in outlet angle.

Leading edges of the vanes were rounded to prevent local flow separation and the trailing edges sharpened to reduce downstream mixing losses. Primarily chosen to eliminate three dimensional pressings, this trailing edge trimming technique also reduces noise by providing additional distance for wake coallescence towards the outer diameter where blade speeds are highest.

The ten moving blades were cast in LM4 aluminium, their profile being chosen to minimize two dimensional losses, and twist conforming to contant circulation principles to minimize induced drag and three dimensional loss. Unlike the Bradwell design with its constant IGV outlet angle, an axial discharge at all radii was consequently a feature of the design duty.

The rotor was keyed to a cylindrical shaft extension on the motor, which partially counterbalanced the weight of the upstream two pole flange mounted motor about the IGV assembly. The motor was, therefore, both accessible and cooled by the incoming air to the fan.

The 400 mm diameter fan rotor was spun from mild steel, the rim of which was spherically shaped, both for strength and to enable the blade setting angles to be changed without introducing excessive clearances and loss. The control hub was subsequently welded into the contre of the spun disc before being bored and slotted to suit the key on the motor shaft extension.

Recesses were machined in the rim to receive the cylindrical blade stubs, each blade being held by a single high tensile bolt. The blade stubs were drilled and tapped to receive helicoil spring inserts which served to distribute the tensile load along the bolt and to prevent bolt seizure to the aluminimum. Dished washers of known rate were also fitted between bolt head and rim to prevent the blades slackening under centrifugal load enabling them to be accurately pretensioned by prescribed angular movement rather than by the less accurate torque spanner method.

Resetting the blade angles to change the fan duty was facilitated by a gauge using the vertical downstream casing flange as a datum. The primary purpose of this flange, however, was to preserve the circularity of the casing close to the blade tips where clearances were small.

The diffuser was constructed from a cylindrical outer member, with a truncated inner cone to give an outlet to inlet area ratio of 1.75. The equivalent cone angle (Ref 2) varied from 7° at inlet to 3° at outlet to prevent flow separation with the growth of the boundary layer along the walls.

5.2 The Test Rig:

Since jet fans are designed to produce kinetic instead of pressure head there is no overall total to static pressure rise. To test such a fan over its full characteristic, it therefore becomes necessary to employ a booster fan to overcome the resistance of the test circuit and the internal resistance of the fan under test.

Since that part of the pressure volume characteristic lying above the duty point was of greatest interest, revealing as it does, the surge or working limit of the fan, a simple tubular test rig of the same diameter as the fan was chosen for the tests (Fig 9).

With the dynamic head of the flow through the duct-work being sufficient to produce an acceptable pressure drop for measurement purposes, it was decided to calibrate the intake ducting by pitot static traverse rather than suffer the additional losses introduced by a venturi or orifice. The inlet configuration of the ductwork consequently took the form of the recommendations given in BS 848, Part 1, Section 27.4 (a).

The test ductwork was constructed from rolled 3 mm plate, the 600 mm bore of which was identical to that of the fan casing. Essentially the rig consisted of the inlet metering section followed by a straight through silencer before the flow entered the fan (Fig 10). The outlet diffuser was followed by a second silencer before the flow entered the final discharge length which terminated in a disc type throttle valve. Such a valve, mounted on a threaded central supporting rod, is both easy to make and causes little upstream flow disturbance.

Inlet flow calibration was made by total head and static pressure traverses across two diameters at 90° to one another, in the same plane as four metering tappings. Kiel and Prandtl type instruments were used for these traverses (Fig 11) both of which had been calibrated over the working Reynolds number range in a special tunnel, additional readings being taken in the boundary layer zone close to the walls. To determine the discharge coefficient of the inlet, the velocities derived from the pressure differentials were subsequently mass weighted and compared with the isentropic flow based on the pressure drop from atmosphere to the static tappings in the metering plane.

Initially, the radial velocity distributions at the two orthogonally disposed transversing positions were found to be in poor agreement and this was traced, with the visual aid of flow streamers, to separation on the upper lip of the inlet cone. This difficulty was subsequently rectified by raising the centre line of the test rig from 0.77 to 1.5 diameters above the floor and adding a streamlined wooden fairing before the inlet cone. Re-traverses showed the modifications to be totally effective, the velocity profiles across the traversing plane being almost identical and the mass weighted flow being within 0.5% of the theoretical value.

Flow angles in the annulus immediately behind the blading as well as at the diffuser outlet, were measured by a calibrated arrow head yaw meter which was nulled into the direction of flow using a differentially connected manometer, the flow angles being read from a protractor attached to the instrument. Total head and static pressure traverses were subsequently made in the traversing planes by rotating the Kiel and ball static probes to the appropriate angle at each radial station. Such instruments had been chosen because of their sensitivity to incidence, particularly since it was not possible to null them in the pitch plane, (Fig 11). Both instruments had again been previously calibrated, and were read on a micromanometer capable of discriminating to $\pm$ 0.02 mm of water gauge. Rotational speed was measured by a Hasler revolution counter from the end of the motor shaft.

The two wattmeter method was adopted for measuring the motor power input, class AL potential and current transformers being used in conjunction with precision grade meters which had been calibrated before test.

Acoustic surveys were made with a 12.5 mm condenser microphone coupled through a Bruel and Kjaer level meter to a Magra IV precision tape recorder. Calibration was made by a pistonphone.

The test procedure at each blade angle consisted of taking sets of measurements by closing the throttle valve in convenient steps, the valve always being re-opened to its full extent for the final reading to check for flow hysteresis following surge. The procedure was subsequently repeated for stagger angles of 2° above and 3° below the design blade stagger angle of 39.7^{o}.

5.3 Results:

Since the total to static pressure rise over a jet fan is designed to be zero, both pressure rise and efficiency have been based on the rise from atmosphere to the total head mass weighted value behind the moving blades. Performance characteristics have consequently been expressed non-dimensionally in terms of total head pressure rise coefficient ($\Delta P_T/\frac{1}{2}\rho Um^2$) and efficiency against flow coefficient (Va/Um). To facilitate comparison, the design prediction has been superimposed on the test values at the design stagger angle of 37.7^{o} (Fig 12).

It will be noted that at the design blade flow coefficient (Va/Um) of 0.69, the test characteristic falls slightly below prediction, which can be partly explained by the inlet guide vane angles being 3° lower than cascade tests would have predicted. The ratio of mean blade clearance to annulus height of 0.03 was also somewhat higher than design because of ovality in the casing.

Of considerable interest, was the late surge, the onset of which could be detected by change in note at a flow coefficient of 0.42 instead of the predicted 0.54 to give a greater working range (turn down) than expected. A possible reason for this is in the interaction of the guide vane wakes with the boundary layer of the moving blades which tends to induce earlier transition from a laminar to turbulent condition and consequently improves stability at high incidences, a result which would be similar to that of increasing the blading Reynolds number.

The design point total efficiency, based on shaft input, was 77% as compared with the prediction of 79%, the peak value being 84% at a flow coefficient of 0.53. Since intake losses are included in these figures, there would appear to be little doubt that design efficiencies can be comfortably met, especially with more favourable blade clearances.

The test characteristics at blade stagger angles of 34.7^{o} and 39.7^{o} have been added to the design stagger angle results of 37.7^{o} (Fig 13) to illustrate the sensitivity of performance to blade angle. This amounts to an overall rise of 15% in flow and 33% in pressure rise for a 5° increase in blade angle. It is, however, of importance to note that efficiencies are not seriously affected, the general level being 80% with a peak at the higher stagger of 85%.

As a check on the blading performance, the vector mean lift coefficients at all three blade settings were derived from the shaft power input and the measured mean outlet angle from the inlet guide vanes (Fig 14). The consistency of the results is good at all three angles with a mean incidence shift from design of under 1°.

Measurement of the overall diffuser performance was complicated by the wakes from the struts supporting the inner cone. To minimize this difficulty, a downstream traversing plane was chosen before plane of the three struts, the upstream position being that behind the blading (Fig 8). For convenience, the method of expressing diffuser efficiency has been reduced to two factors, a geometric effectiveness based on the area ratio $[1-(A_1/A_2)^2]$ and aerodynamic friction (η_F), the product of which is deemed to give the overall effectiveness (η_D) (Fig 15). η_D is consequently expressed as the ratio of the static pressure recovery to the inlet kinetic head. All results are derived from the mass weighted values of total and dynamic head at the two traversing positions.

By plotting aerodynamic friction factor (η_F) against mean inlet swirl angle to the axis (Fig 15) it will be noted that the aerodynamic performance suffers progressively with swirl, the factor falling from 0.73 to 0.55 with a swirl change from 2° to 16°. This illustrates the importance of keeping inlet swirl to a minimum in diffusers of this type.

Acoustic measurements on the fan were confined to the design blade setting angles. Levels at the 3 m, 45° off-axis positions at inlet and outlet corresponded to NR 95 and 98 respectively without silencing and NR 82 and 84 with silencing. Expressed in "A" weighted terms, the mean levels at these positions corresponded to 98 dbA without and 84 dbA with silencing.

6 CONCLUSIONS

6.1 The high reaction design arrangement employing inlet guide vanes offers an aerodynamically acceptable and mechnically attractive design for special purpose fans and particularly for those requiring a high specific output.

6.2 Stage efficiencies on the Bradwell gas circulator under the unfavourable aspect ratio and clearances were estimated to be 89% and, at more favourable aspect ratios and clearances 90% could be comfortably exceeded.

6.3 The sensitivity of performance to moving blade setting angles has been demonstrated on the Bradwell machine. Variation in the stationary inlet guide vane angles also offers an effective way of regulating the duty. Although performance is less sensitive to IGV than blade angle changes, their wide working incidence range compensates for this lack of sensitivity to make this a mechanically attractive alternative to feathering blades while in motion.

6.4 The retention of swirl in jet fans to promote mixing in one way vehicle tunnels appears unnecessary and wasteful. From current experimental evidence it is recommended that such fans should be designed to inhibit swirl.

6.5 Jet fans designed on the high reaction principle can offer lower power inputs by virtue of their stage efficiency of approximately 80% and absence of outlet swirl.

6.6 The aerodynamic duty of the jet fan prototype fell marginally below design primarily because of excessive blade clearances.

6.7 Surge margin on the jet fan was better than prediction and has been attributed to early boundary layer transition resulting from IGV wake interaction with the moving blades.

6.8 Duty of the jet fan is sensitive to blade setting angles, a change of 5° increasing the flow by approximately 15% at the design resistance level.

6.9 The performance of a diffuser constructed from a cylindrical outer tube and conical inner fairing is sensitive to inlet swirl, probably because of the downstream formation of a free vortex which increases the swirl component at the inner diameter.

6.10 If care is taken in the spacing between inlet guide vanes and moving blades, acceptable noise levels result (98 dbA at the 3 m, 45° off-axis position). This can be readily reduced to 84 dbA using tubular straight through silencers.

7 ACKNOWLEDGEMENTS

Much of the testing and anaylsis of this paper was conducted by graduate apprentices during their engineering training at Grubb Parsons. Special mention must be made of Messrs G Cruddas, A Fredriskson, G M Horseman and A Levitt in this respect.

The author also wish to thank Grubb Parsons Limited for permission to publish this paper and also to Messrs A Abraham, B Dixon and G R Grew for their contributions in the design and manufacture of the jet fan prototype.

8 REFERENCES

1 Hawley R and Smith A : "Fans for the Tyne and Wear Metro" Transactions of North East Coast Institution of Engineers and Shipbuilders." Vol 96 p 105 Dec 1980.

2 Howell A R : "The Present Basis of Axial Flow Compressor Design (Part 1) - Cascade Theory and Performance" ARC R & M No 2095 (P 7 and Appendix VI June 1942).

3 Pinter R : "Possibilities for the Reduction of Energy Consumption on Ventilation Systems with Jet Fans". Fourth International Symposium on the Aerodynamics and Ventilation of Vehicle Tunnels, P 279 BHRA Fluid Engineering, Cranfield March 1982

4 Gotaas Y: "The Study of the Dispersion of Road Tunnel Air by Tracer Experiments". Fourth International Symposium on the Aerodynamics and Ventilation of Vehicle Tunnels, P 233 BHRA Fluid Engineering, Cranfield March 1982

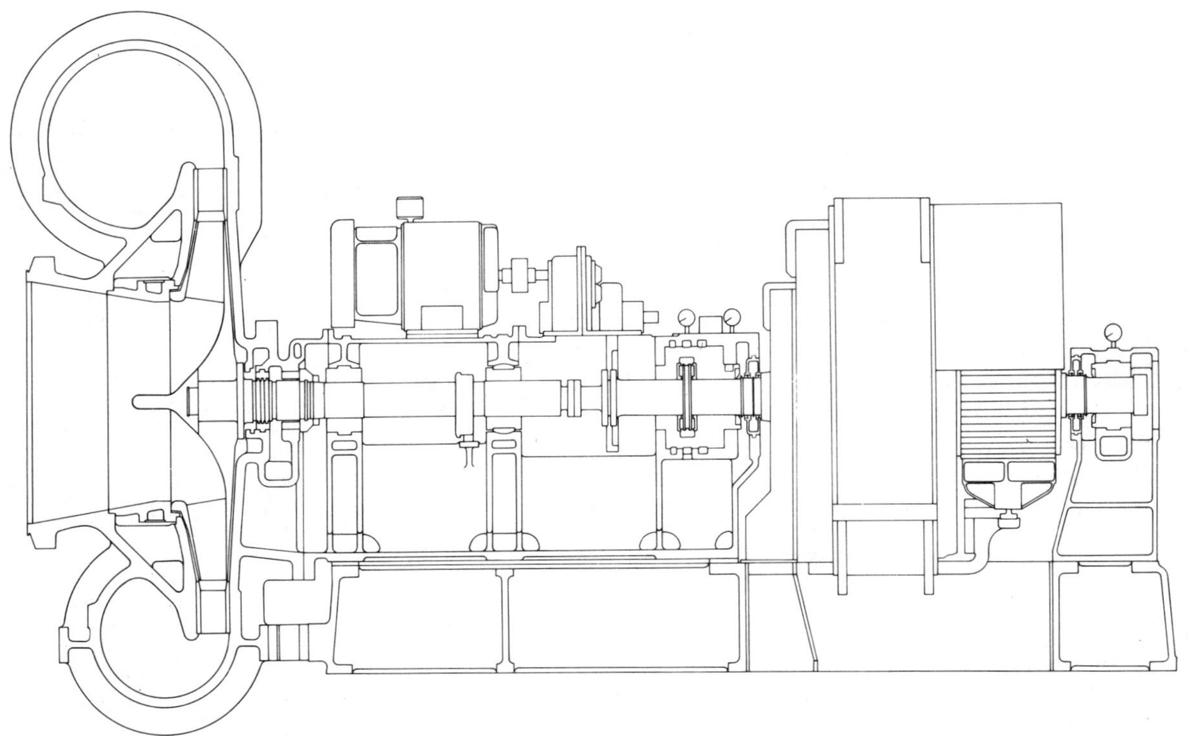

Fig. 1   Calder Hall gas circulating fan.

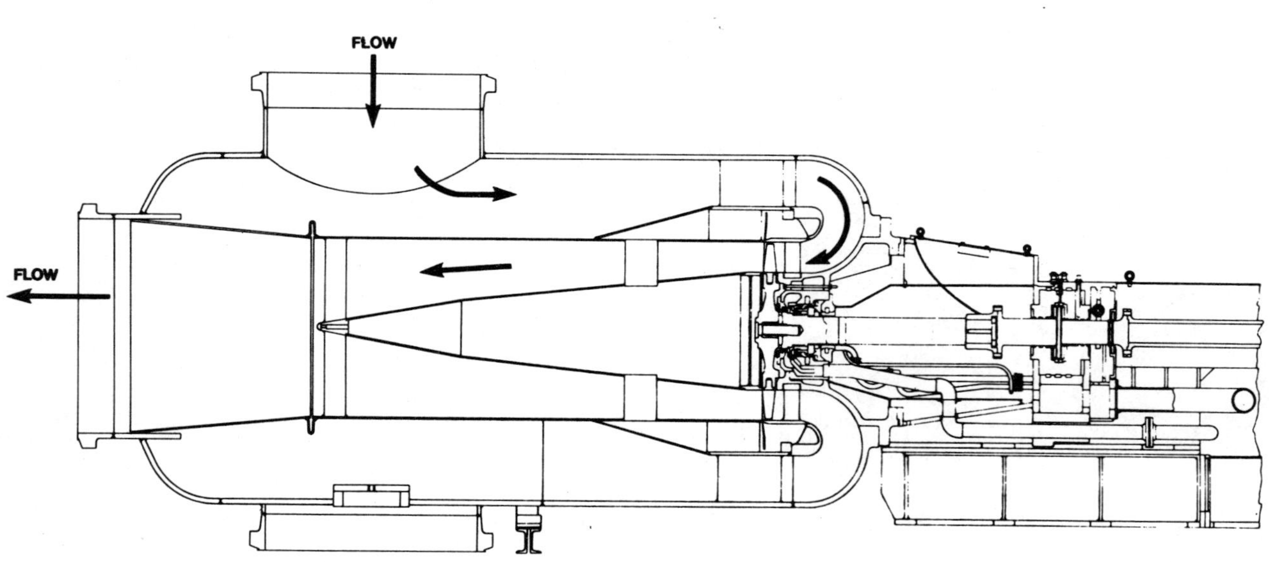

Fig. 2   Bradwell nuclear gas circulator.

Fig. 3 Bradwell rotor assembly

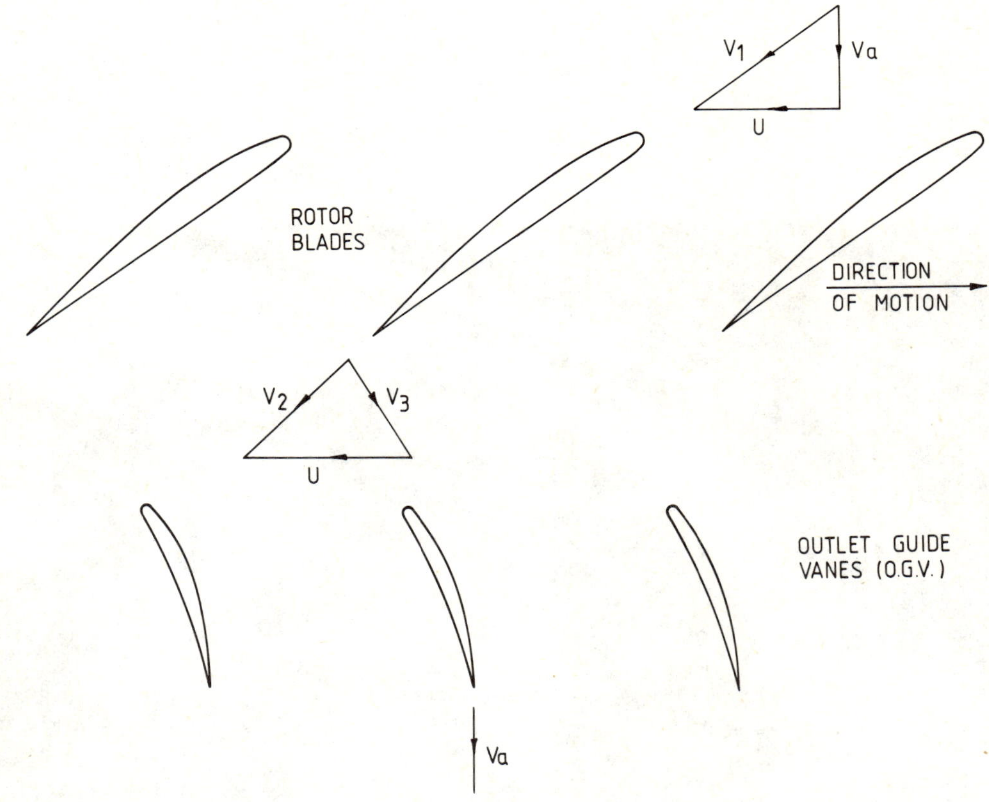

Fig. 4 Conventional low reaction fan principle.

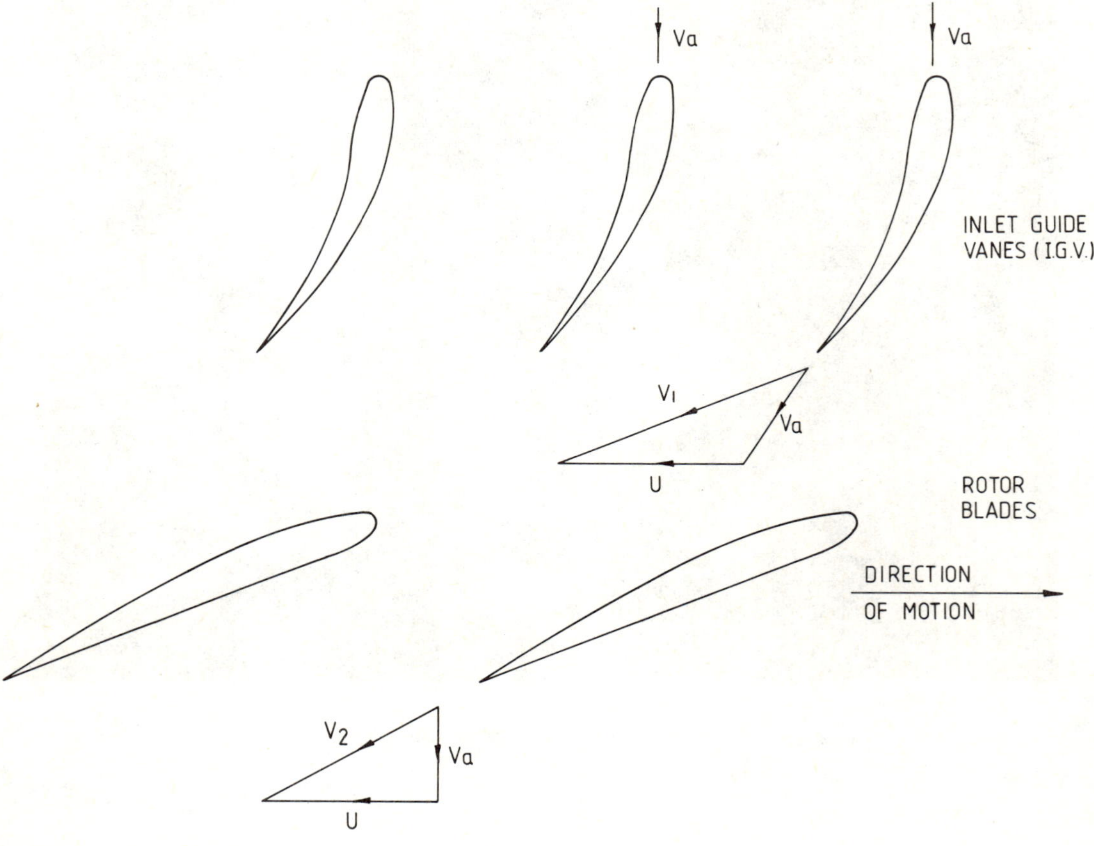

Fig. 5 High reaction fan principle.

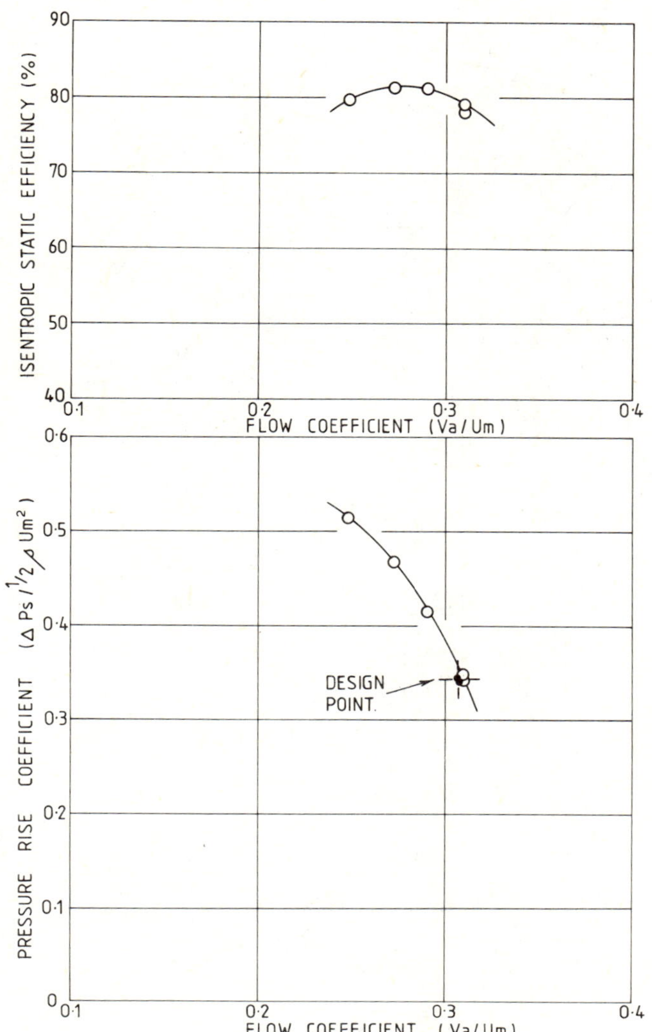

Fig. 6 Bradwell design performance.

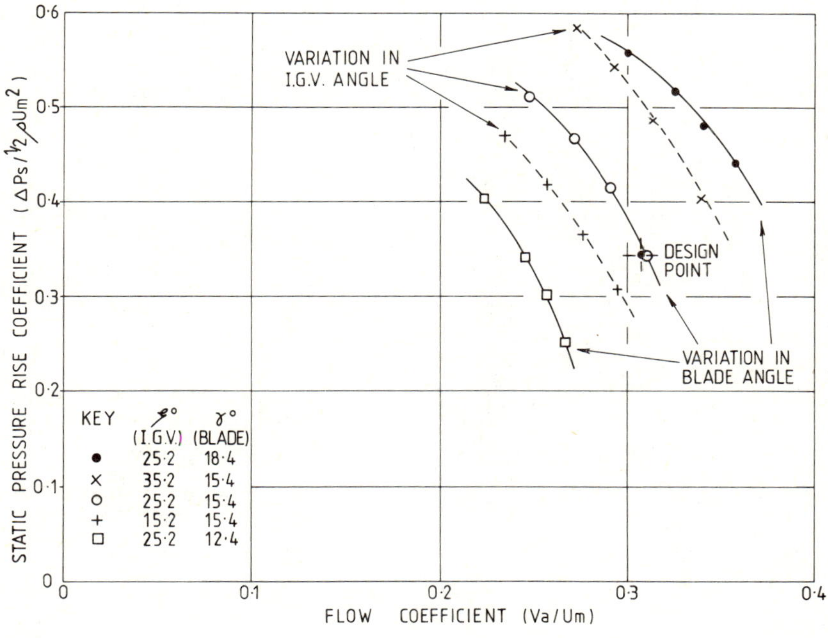

Fig. 7 Influence of inlet guide vane and blade angle on Bradwell circulator duty.

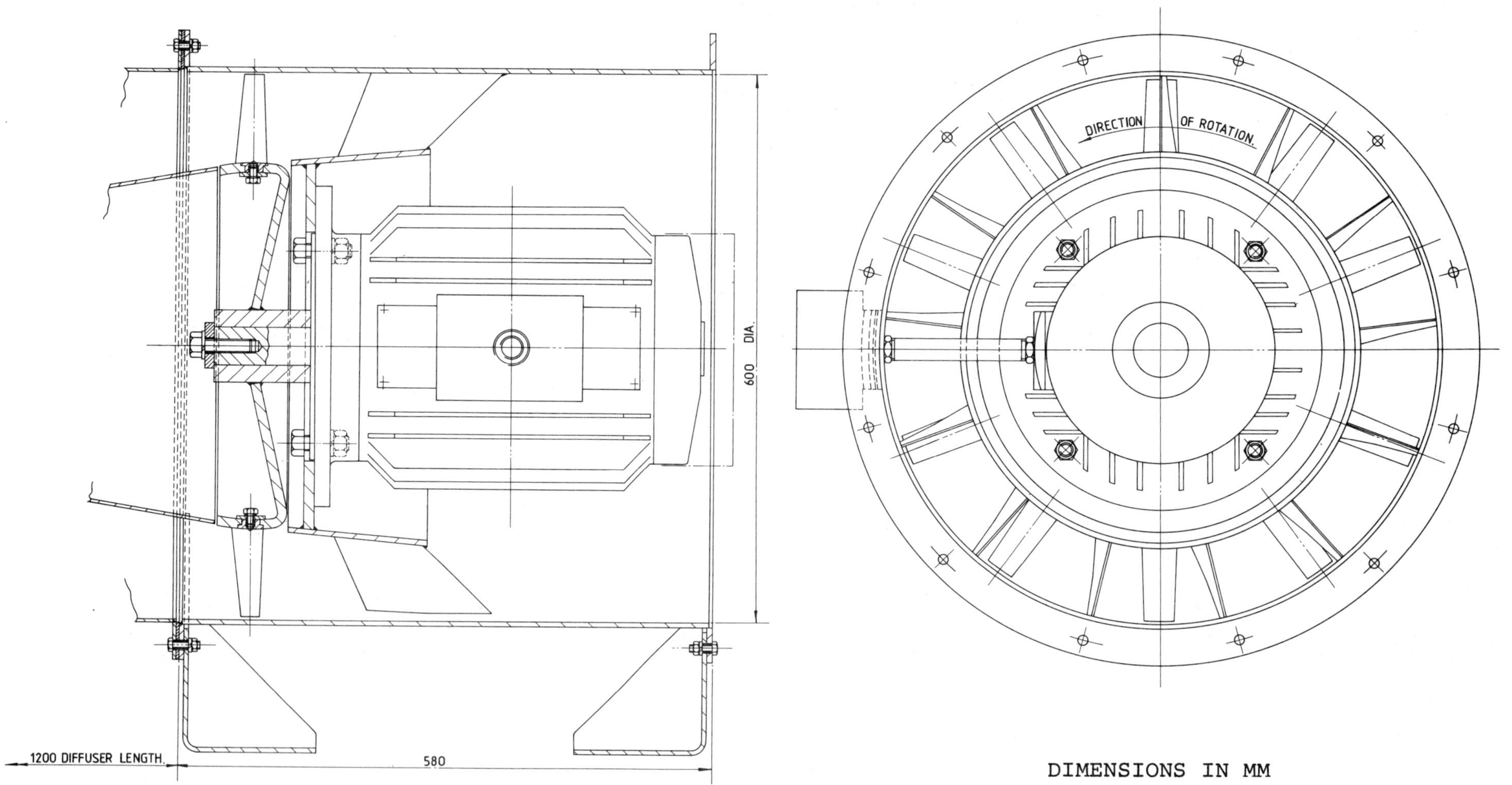

Fig. 8 Jet fan cross section.

Fig. 9 Jet fan under test.

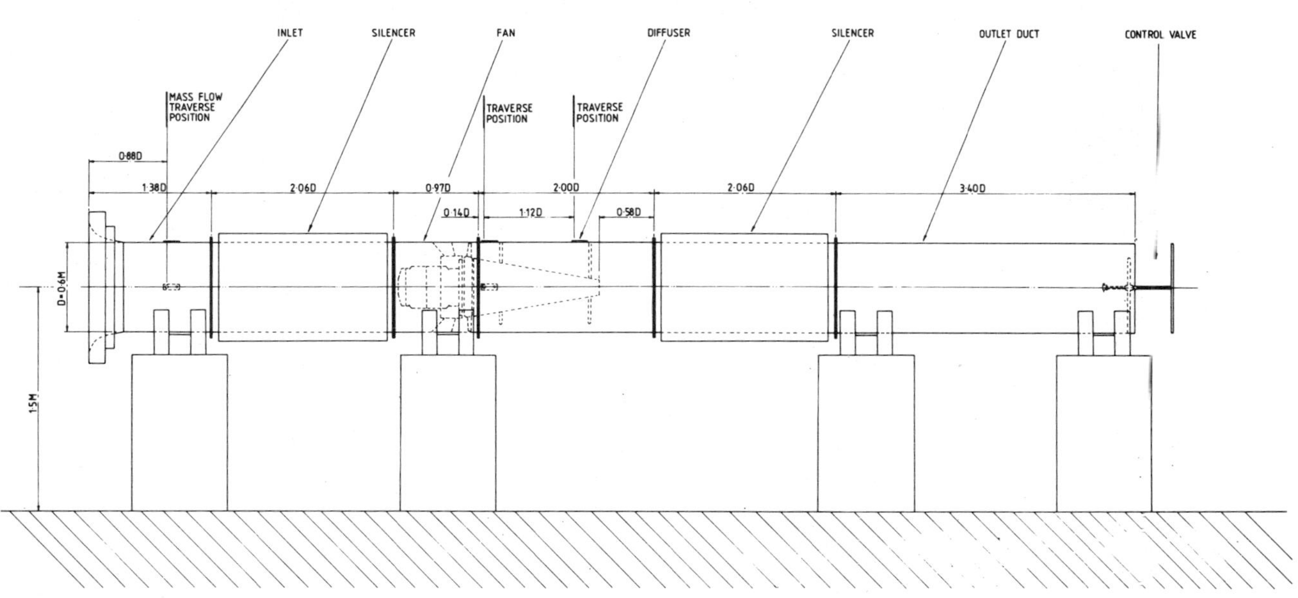

Fig. 10 Jet fan test rig.

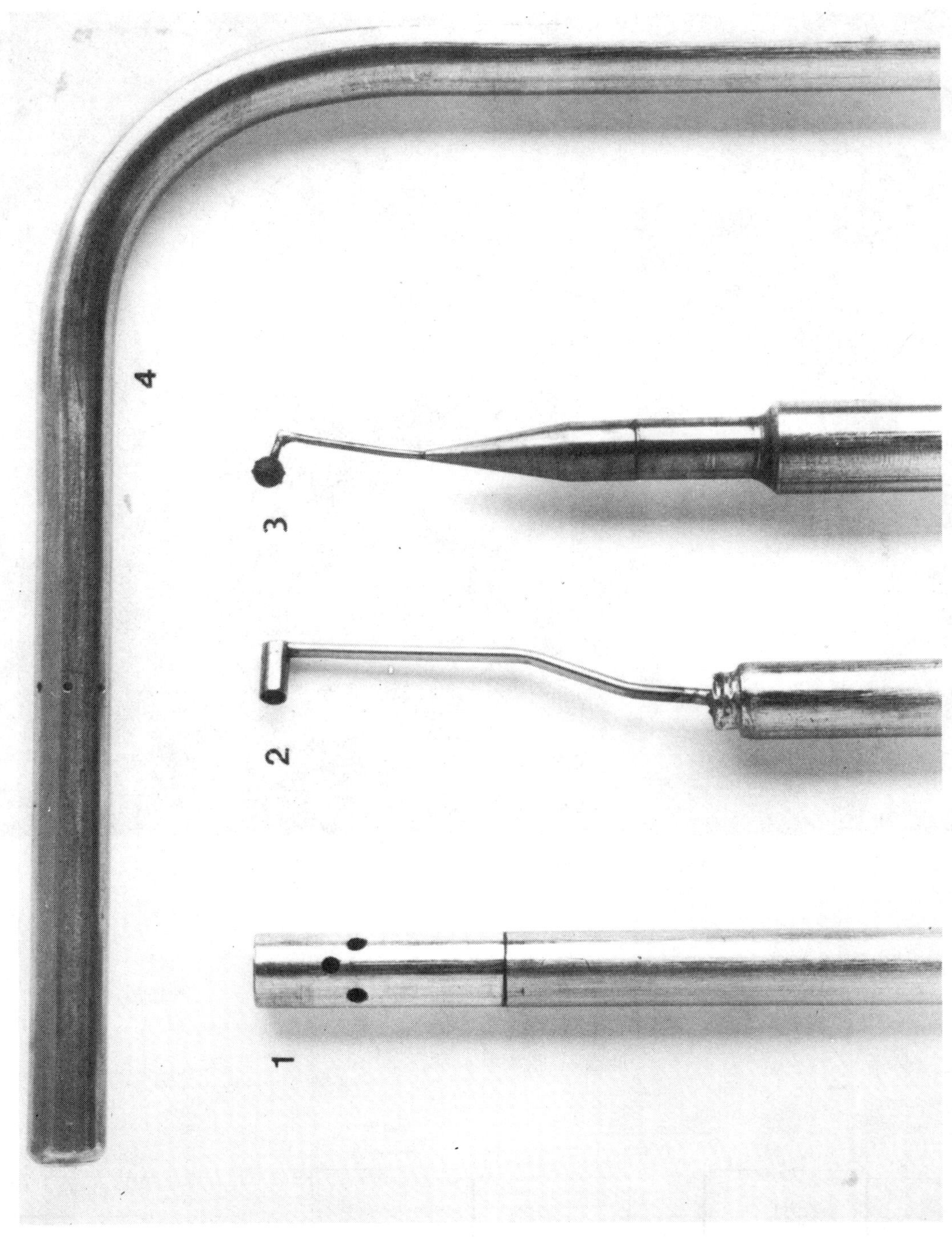

(1) Yaw Angle Probe

(2) Kiel-Type Total Pressure Probe

(3) Spherical Head Static Pressure Probe
(Employed at Upstream Diffuser Traverse Plane)

(4) Prandtl-Type Static Pressure Probe
(Employed at Downstream Diffuser Traverse Plane)

Fig. 11 Traversing probes.

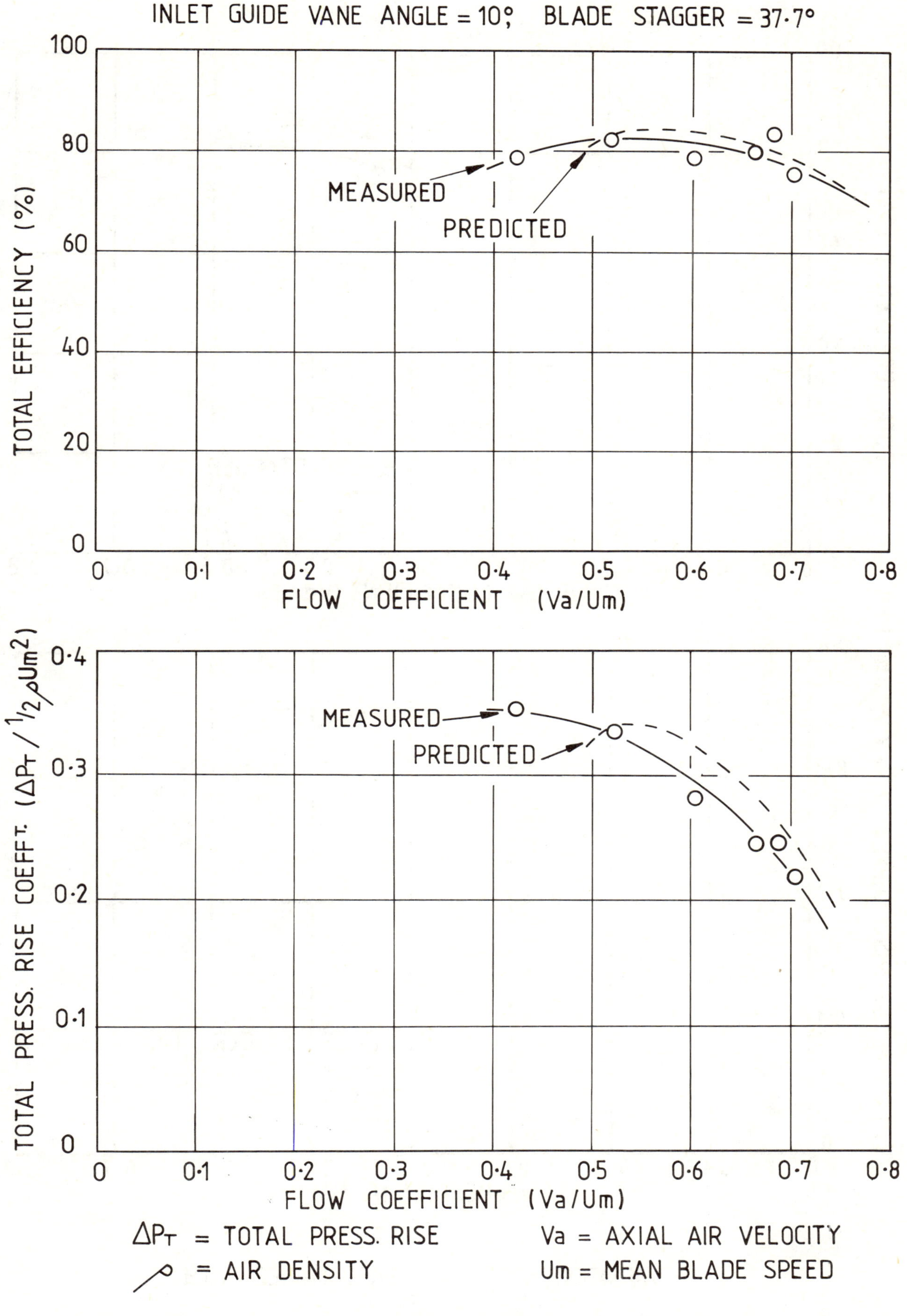

COMPARISON OF JET FAN PERFORMANCE WITH PREDICTION

FIG. 12

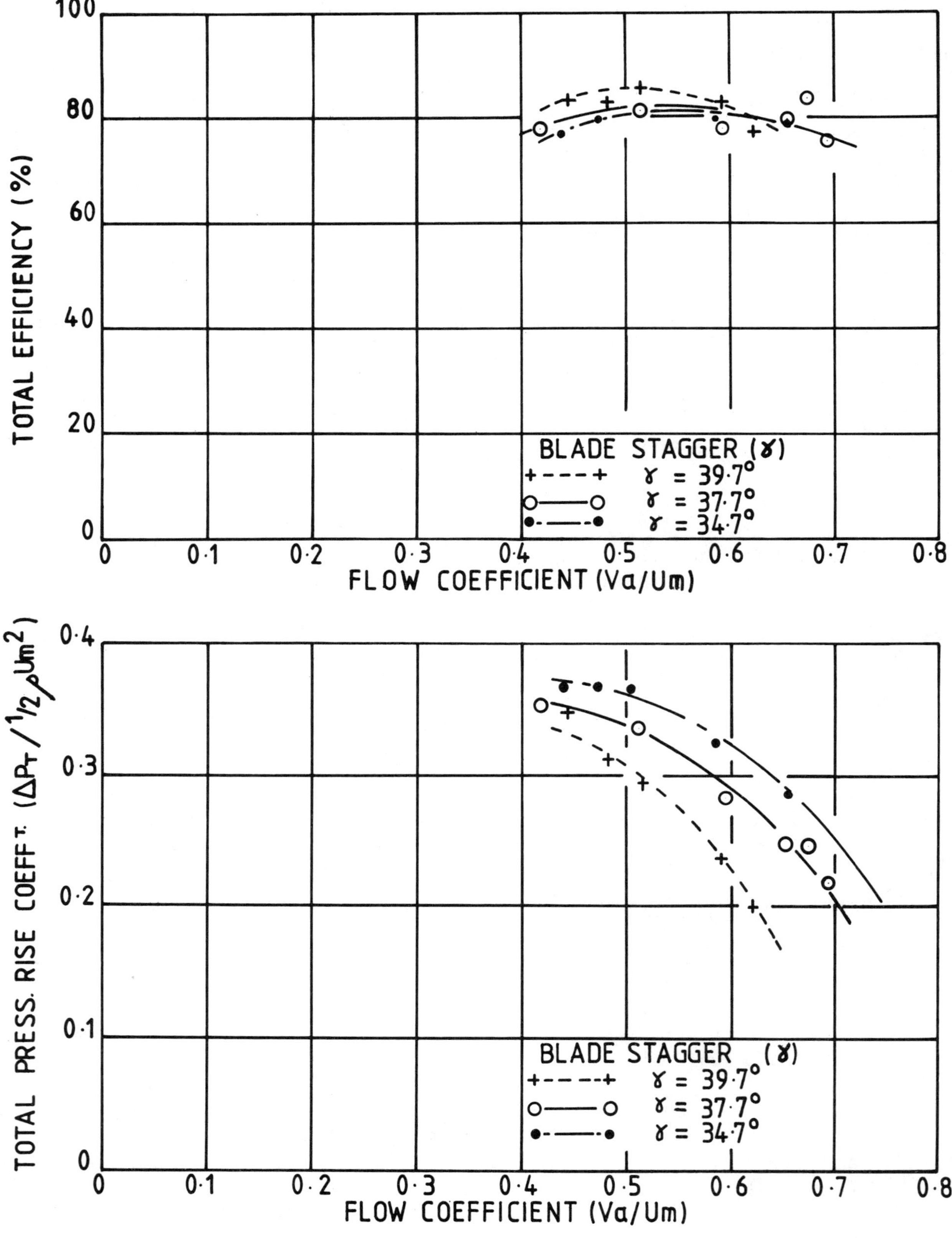

VARIATION IN JET FAN PERFORMANCE WITH BLADE STAGGER ANGLE.

FIG. 13

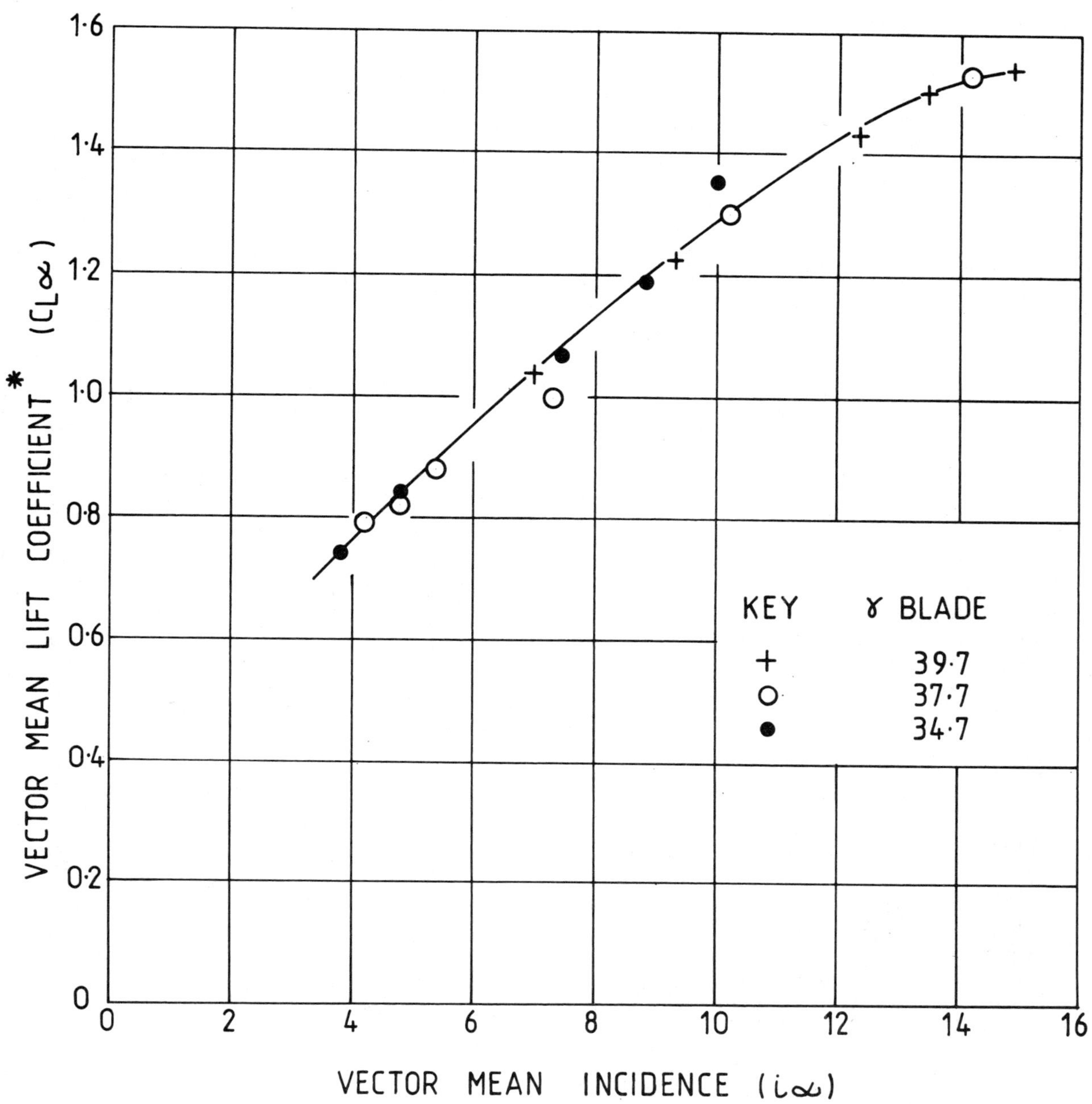

* VECTOR MEAN LIFT COEFFICIENT DERIVED FROM FLOW, MEAN BLADE SPEED SHAFT. POWER INPUT AND MEAN I.G.V. OUTLET ANGLE

VARIATION IN LIFT COEFFICIENT WITH VECTOR MEAN INCIDENCE IN JET FAN

FIG. 14

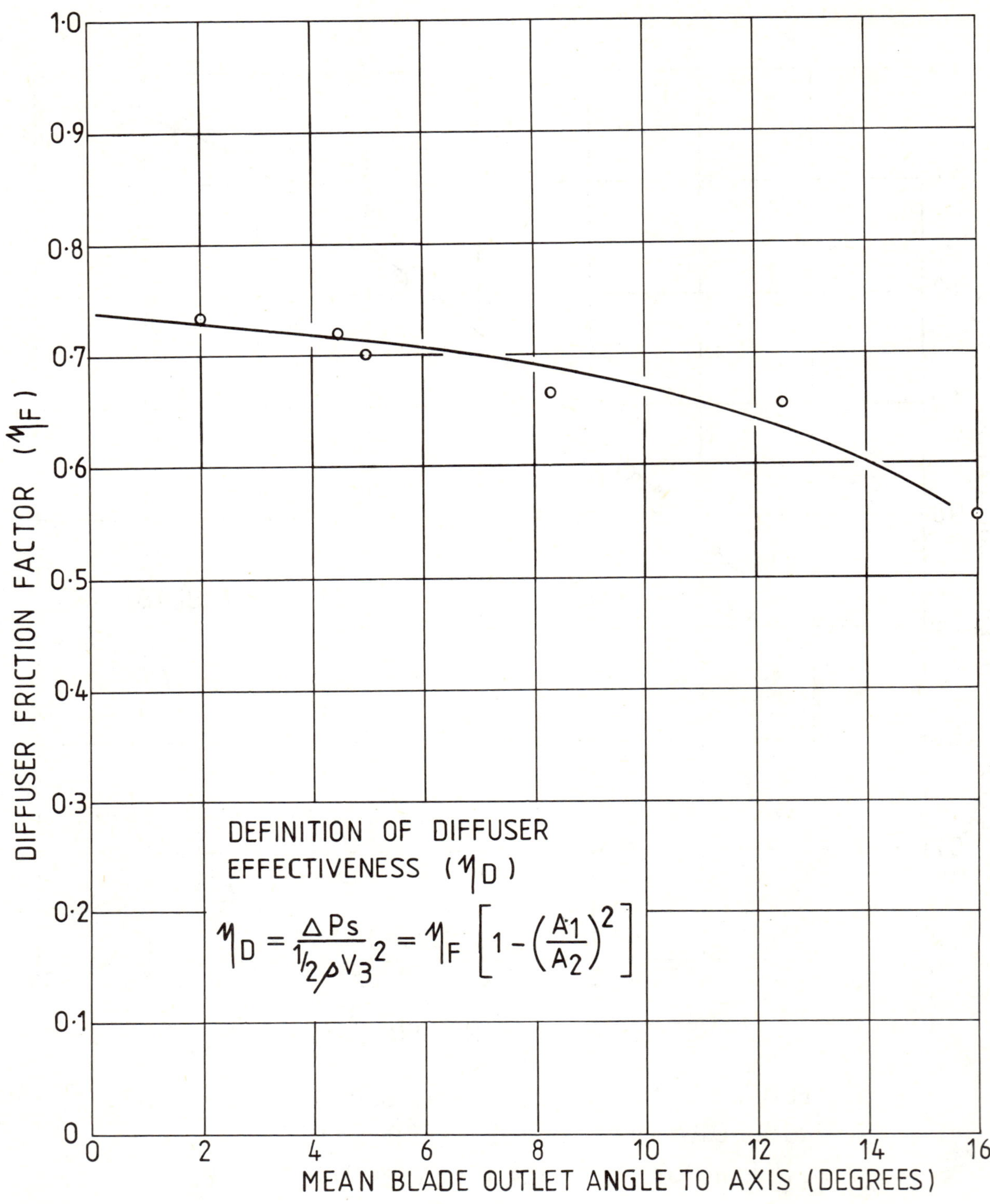

VARIATION IN DIFFUSER FRICTION FACTOR
WITH INLET SWIRL ANGLE.

FIG. 15

International Conference on

Fan Design & Applications

Guildford, England: September 7-9, 1982

PAPER G2

THE INTRODUCTION OF LEADING EDGE SLATS TO IMPROVE THE OFF - DESIGN PERFORMANCE OF AXIAL FLOW FANS

M. M. Javur

Karnataka Regional Engineering College, India

K. S. Murthy and S. Kar

Indian Institute of Technology, India

Summary

This paper deals with investigations carried out on a fixed pitch axial flow fan. The leading edge slats, for the first time in the field of fan engineering have been introduced on the blades to improve the off-design performance of the fan. Theoretical analysis has been made to optimize the dimensions and gap of the leading edge slats. Experimental verification of the optimization procedure for slats has been made by conducting fan performance tests in a tunnel. The flow visualisation test results have been shown in the form of photographs to confirm the boundary layer control effects achieved by leading edge slats. The introduction of the optimum size of leading edge slats have widened the stable operating range of the fan by 27% with a 10% increase in overall efficiency. The power consumption at best efficiency point has been reduced by 4% with the introduction of leading edge slats on the tested fan blades. The introduction of leading edge slats on fan blades has been shown to improve the off-design performance of axial flow fans.

Organised and sponsored by
BHRA Fluid Engineering, Cranfield, Bedford MK43 0AJ, England.

0263 - 421X/82/01 00 - 0001 $5.00
The entire volume can be purchased from
BHRA Fluid Engineering for $82.00

NOMENCLATURE

C	=	Chordwidth of the blade
C_m	=	Meridional velocity
H	=	Pressure head
P_{sht}	=	Shaft power = motor input power x η_m
P_{tot}	=	Air power developed by the fan = $2 \cdot 723 \times 10^{-3} x Q x \Delta P_{tot}$
U_o	=	Velocity at inlet section
U_2	=	Peripheral velocity at the tip of the impeller
$U(x)$	=	Velocity in the central stream at any distance
x	=	Distance of velocity profile from inlet
a	=	Half of the width of gap
g	=	Acceleration due to gravity
h	=	Gap of the slat at the leading edge
l_s	=	Length of the slat
ΔP_{tot}	=	Total pressure developed by fan in mm of water column
δ_1	=	Displacement thickness
ε	=	Dimensionless inlet length
η	=	Overall efficiency of fan = P_{tot} / P_{sht}
η_m	=	Motor efficiency
ν	=	Kinematic viscosity of air
ϕ	=	$\frac{C_m}{U_2}$ = flow coefficient
ψ	=	$\frac{2gh}{U_2^2}$ = pressure coefficient.

1. INTRODUCTION :

The axial flow fan may be described as a fan in which the flow of air is substantially parallel to the axis of the impeller. The shape of the impeller passages depends on the basic parameters like air volume flow, pressure and speed. The design of a fan is commonly made to operate the fan at a duty point which is usually the best efficiency point. However, it is also necessary for the fan to operate smoothly over a large range of volume flow rate and pressure depending on the demand of the system in which the fan is installed. The axial flow fans when operated below a certain minimum and above a certain maximum air flow, show unstable behaviour. Hence, for any fan, depending on the design, there would be a best efficiency operating point and a stable range of operation. The performance of the fan beyond the stable range of operation is known as stalling condition. At near zero-flow conditions and maximum flow conditions, the air flow direction ceases to be axial resulting in an erratic operational behaviour of the fan.

The stall conditions exists in all types of axial flow machines like axial flow compressors, axial flow pumps and fans.

2. MOTIVATION :

In aeronautical industry, the improvements in air craft performance have been obtained by increasing wing loadings while maintaining acceptable landing and take-off speeds. This is usually obtained by changing the geometry and the aerodynamic characteristics of the wing section aerofoils by the introduction of movable elements at the trailing edge and also the leading edge of the wing. These elements called as high-lift devices have been used to improve the maximum lift coefficient of the wings without changing the aircraft characteristics for the cruising and high-speed flight conditions. The elements on the trailing edge are known as flaps and those fixed on leading edge are called slats (Ref. 1). The leading edge slats are relatively smaller aerofoils mounted ahead of the leading edge of the wing to assist the turning of the air around the leading edge at high angles of attack, primarily, to delay the stalling. The fixed leading edge slat investigation was first reported by Weik and Sanders (Ref. 2) and a maximum lift coefficient of 1.64 was achieved at lower Reynolds numbers. In aircrafts, leading edge slats are generally used on wings as one of the methods of boundary layer control which result in high lift.

The boundary layer control achieved by leading edge slats can be explained with reference to Fig. 1 as follows : The boundary layer formed on the slat A is carried into the main stream flow of air before separation occurs giving rise to the formation of a new boundary layer on the wing from point B onwards. Under certain conditions, this newly formed boundary layer will reach the trailing edge C of the wing without separation even at larger angles of incidence producing higher lift coefficients. The polar diagram for a wing section (Ref. 3) shown in Fig. 2, has been plotted to show the comparative lift coefficient values with and without slat. The gain in lift coefficient can be seen to be quite large. The aerofoil shown in Fig. 2, without a slat has a maximum lift coefficient of about 1.3 at an angle of incidence of 14.5^{o}. For the same aerofoil with a leading edge slat, a maximum lift coefficient of 1.75 was achieved at a comparatively much higher angle of incidence of 26°. It is thus clear that a 34 °/. increase of maximum lift coefficient even at a higher angle of incidence of 26° which is a 80 °/. increase in the aerofoil angle of incidence, may be achieved with slats.

The unstable region of operation of the axial flow machines has been a topic of interest for the research workers during the past two decades. Several investigators : for example, Tanaka and Muraka (Ref. 4) Toyokura et al (Ref. 5), Panamareff(Ref. 6), Koch (Ref. 7) and

Scheer (Ref.8) have reported the work on unstable performance of axial flow compressors and pumps. Comparatively very few researchers, for example, Tomio (Ref.9) and Venkatrayulu et al (Ref.10) have suggested methods of improving performance of axial flow fans. In this paper, the authors have investigated the use of slats on the leading edges of the axial flow fan blades with a view to improve the lift coefficient, similar to that in aircraft wings, of the fan blades and increase the stable operating range of the fan with a net gain in the efficiency of the fan. To the best of the information available with the authors, there has not been any earlier research work reported on the boundary layer control by leading edge slats to improve the performance of the axial flow fans. The theoretical analysis developed to estimate the size of the slat and the gap of the slat has been given. Experimental programme conducted to verify the correctness of the method of estimation of slat gap size as well as flow visualization tests conducted to study the effect of slat on the boundary layer control of the axial flow fan blades are also enumerated.

3. ANALYTICAL INVESTIGATIONS :

The slat and the blade move together with no relative velocity between them maintaining a certain gap between each other. The flow situation in the gap of the slat would then be analogous to that of the inlet section of a channel i.e. flow between two fixed plates. To calculate the size of the gap, a theoretical approach suggested by Schlichting (Ref.11) was further modified by the authors as an attempt to develop an optimization procedure to predict the size of the gap between the slat and the blade. The analytical procedure developed would facilitate calculation of the gap of the slat for the fan without the necessity of doing any 'trial and error' type of experiments to improve the fan performance by use of slats. The gap calculation is based on the velocity distribution in the gap and the critical Reynolds number for the flow between the flat plate walls. Hence the inlet flow at the leading edge of the blade is always maintained laminar as the critical Reynolds number chosen to finally calculate the gap is 2300. The velocity distribution for the case of a laminar flow was analyzed by Schlichting (Ref.11) and is shown in Fig. 3.

The equation of continuity :

$$\int_{y=0}^{a} u \, dy = U_0 a \qquad \ldots (1)$$

The displacement thickness, δ_1, of the boundary layer is given by,

$$\delta_1 = \frac{1}{U} \int_0^a (U - u) dy \qquad \ldots (2)$$

The velocity distribution can be written as :

$$U(x) = U_0 \frac{a}{a - \delta_1} = U_0 \left[1 + \frac{\delta_1}{a} + \left(\frac{\delta_1}{a} \right)^2 + \ldots \right] \qquad \ldots (3)$$

Near the inlet section, the boundary layer develops in the same way as on flat plate at zero angle of incidence.

$$\frac{\delta_1}{a} = 1.72 \sqrt{\frac{\nu x}{a^2 U_0}} \qquad \ldots (4)$$

or

$$\frac{\delta_1}{a} = 1.72 \, \varepsilon$$

$$\text{where} \quad \varepsilon = \sqrt{\frac{\nu x}{a^2 u_o}} \qquad \ldots (5)$$

The gap of the slat, 2a, can be calculated by knowing the value of x which is same as that of the length of the slat, l_s. The recommended lengths of the slat suggested by Reigels (Ref.12) vary from 0.16 to 0.25 times the chord width. Therefore,

$$l_s = 0.16 \text{ to } 0.25\ C, \qquad \ldots (6)$$

where C is the chord width.

From equation (5),

$$100\ \varepsilon^2 = 100\ \frac{\nu\ l_s}{a^2\ U_o} = Y \qquad \ldots (7)$$

Knowing velocity U_o and l_s, the value of Y can be calculated for different assumed values of a, remembering that h = 2a.

The velocity profile can be found out by knowing Y,

$$U_{max} = Y\ U_o \qquad \ldots (8)$$

and

$$U_{av} = \frac{U_{max_1} + U_{max_x}}{2} \qquad \ldots (9)$$

where U_{max_1} is the maximum velocity at entry

U_{max_x} is the maximum velocity at any distance x

The value of h, the gap between the slat and blade can be found out at critical conditions of laminar flow where Reynolds number would be 2300, from the equation,

$$h = \frac{R_{cr}\ .\ \nu}{U} \qquad \ldots (10)$$

The value of h(= 2a), calculated by equation (10) should be twice that of the a and also same as that assumed in equation (7). If the equation (7) does not get satisfied by substituting equation (10) into it, the interative procedure of calculation should be made till both the values of h are same.

By using equations (6), (7) and (10) the length of the slat and the gap of the slat can be calculated if the design delivery of the axial flow fan is known from the fan manufacturer.

4. SAMPLE ANALYSIS :

The analytical equations developed have been applied to predict the size and gap of the slats for a four-blade fixed pitch axial flow fan of 0.610 m diameter. The fan manufacturer guaranteed a design flow of 118.3 m^3/ min.

Air Flow = 118.3 m^3/min. = 1.97 m^3/sec.

Area of flow = 0.297 m^2

Velocity of air = U_o = 6.63 m/sec.

The fan blades were made of cambered steel plates with a chord length of 0.285 m at the mean diameter. The slat length is taken as 0.25 C.

$$l_s = 0.07125 \text{ m}$$

$$\therefore \quad l_s = 71.25 \text{ mm}$$

A slat length of 72 mm and velocity of 6.63 m/sec. will be used in the calculation of gap of the slat.

As a first approximation, in the iterative procedure, a gap value of 2a = 5 mm may be assumed. The air is assumed to have a value of $\nu = 16 \times 10^{-6}$ m²/sec. From equation (7)

$$Y = \frac{100 \cdot (16 \cdot 10^{-6}) \cdot 72\ (10^{-3})}{(2.5^2)\ (10^{-6}) \cdot (6.63)}$$

$$Y = 2.79$$

From Fig. 3, the velocity profile corresponding to the value of Y = 2.79, can be read off as $U_{max}/U_o = 1.3$

$$\therefore \quad U_{max} = 1.3\ U_o$$

The average of the maximum values of velocity at entry, U_{max_1} and at l_s, $U_{max_{ls}}$ can be given by

$$U_{av} = \frac{U_o + 1.3\ U_o}{2}$$

$$= 1.15\ U_o = 1.15 \times 6.63 = 7.62$$

$$\therefore \quad U_{av} = 7.62 \text{ m/sec.}$$

From equation (10)

$$h = \frac{2300 \times 16(10^{-6})}{7.62}$$

$$\therefore \quad h = 4.83 \text{ mm.}$$

The calculated value of the gap height is quite close to that of the assumed value of 5 mm and hence would require no further iteration.

Theoretically, it has been predicted that for 0.610 m dia fan at a guaranteed design air delivery of 7100 m³/hr, the slat dimensions are:

$$l_s = 72 \text{ mm}$$

$$h = 5 \text{ mm}$$

5. EXPERIMENTAL PROGRAMME :

To study the improvements in the off-design performance of the axial flow fan of 0.610 m diameter, by the introduction of the slats at the leading edges of the blades, tunnel tests have been conducted. The optimum size and gap of the slats predicted theoretically have been used to verify the proposed method. The experiments were conducted with slat gaps of 2, 5 and 10 mm, although theoretically 5 mm gap of the slats gives the best performance. The flow visualization tests were also conducted to observe the effect of leading edge slats on the boundary layer control. The experimental tunnel was made as per the Indian standard specifications for axial flow fans (Ref.12). The dynamic pressure and static pressure measurements were made at sections A and B respectively as shown in Fig. 4. The fan was fitted in the discharge duct which in turn was connected to the inlet duct of 0.615 m diameter. The motor was directly coupled to the fan. A stroboscope was used to to measure the speed of the fan. The speed of the fan was maintained at 650 rpm with the aid of the voltage stabiliser. The motor was connected to the 3-phase mains through a voltage stabiliser. The motor was calibrated to evaluate the efficiency of the motor (η_m) at various loads. The efficiency of motor is required to be known to find out P_{sht} which is given by, P_{sht} = Input power x η_m . Close mesh screens were used to increase the resistance in the duct and flow could be completely shut-off by closing the duct with a disc at the same position of the screens.

5.1 Instrumentation :

The barometric pressure and room temperature were recorded by calibrated meters. The specific weight of air for each test run was found out from these values. The static pressure measurements were made both at stations A and B through four holes at each station around the circumference of the duct. A 10° inclined tube manometer with a manometric liquid of specific gravity 0.659 was used to measure the static pressures. The static pressure measurements at station B were made to calculate the total pressure of the fan.

A calibrated vane-type anemometer was used to measure the velocity of flow of air. The anemometer was traversed through suggested locations as per the standard (Ref.12) to calculate the average velocity of air and hence flow.

The power input to the motor was measured by two calibrated watt meters. The motor was calibrated by the mechanical method - spring and rope at various loads to calculate the efficiency of the motor.

5.2 Tunnel test procedure :

The fan was tested for its complete operating characteristics in the tunnel without any slats. The tunnel diameter was 0.615 m and the area of flow was based on this tunnel diameter of 0.615 m and not on fan diameter of 0.610 m. The barometric pressure, temperature, speed of fan, static pressures and power were all recorded for each setting of the resistance in the screens. The performance curve of the fan was then studied by providing slats at the leading edge of the blades with varying gaps of 2, 5 and 10 mm and lengths of 0.18 C and 0.25 C. The experiments were conducted with different gaps to evaluate the performance of the fan.

5.3 Flow visualisation test :

The flow visualisation test to study the effect of introduction of slats on the boundary layer was conducted in a wind tunnel. The wind tunnel has a capability of achieving maximum velocity of 50 m/sec. at full opening. The velocity could be controlled by a damper mechanism. One of the four blades of the fan was removed and was fixed on a vertical stand in the wind tunnel. Initially the blade was kept in such a way that its chord is parallel to air flow from the wind tunnel. At this position the angle of attack is zero. The vertical stand could be rotated on a base plate to vary the angle of attack. The angle of attack of the blade was varied from 0 to 15°.

The surface of the blade was smeared with a thin film of titanium dioxide powder and kerosene before placing it on the wind tunnel test stand. When the test run was completed, on the surface of the blade, wherever the flow separation occured the white titanium dioxide powder remained intact as air flow did not touch the surface of the blade due to separation. In this manner a quantitative estimation of boundary layer separation was made. The test procedure was repeated after introducing slats on the leading edge of the blade to study the boundary layer control achieved. The slat gap was maintained at 5 mm and the flow visualisation test was conducted at test section velocity of 6.6 m/sec.

6. RESULTS AND DISCUSSION :

The theoretical equations developed for the estimation of the gap and size of the slat have been used to optimize the dimension and the gap height of the slat for a 0.61 m diameter fan. The experimental performance curve of the fan without slats on blades is shown in Fig. 5. It can be seen that the stable operating condition exists only above 0.17 value of flow coefficient. The actual stable discharge range of the fan was found to start from 3431 m^3/hr. The overall efficiency was calculated by knowing fan air power, motor shaft power and motor efficiency at each opening of the test tunnel. The maximum efficiency of the fan was found to be 58.37 %. The pressure fluctuations were observed at tested flow coefficients less than 0.17 which is the lowest limit of the stable operating range. The fan blades were fitted with slats of length 0.18 C and 0.25 C and the gap was varied in each case from 2 mm to 10 mm. The performance curves of the fan with 0.18 C and 0.25 C slat lengths are shown in Fig. 6 and Fig. 7 respectively for various gap heights. It can be seen that at a 5 mm slat gap the overall efficiency of the fan has increased upto 68.51 % and 68.79 % with 0.18 C and 0.25 C slats respectively from a no-slat overall efficiency of 58.37 %. The stable operating range of the fan has gone up by 27 % in the case of 0.18 C slat and by 20 % in the case of 0.25 C slat at the same gap height of 5 mm. The power consumed at the fan motor terminals with the introduction of slats is 4 percent less than that of the fan without slats. It is clear from these experimental results that the length of the slat whether it is 0.18 C or 0.25 C is not a critical parameter as long as it is within the limits set by equation (6). The gap of the slat is very critical and it has been shown that the optimization procedure for slat gap determination suggested in this paper is quite accurate. The flow visualisation test results are given in plates 1 to 3. The fan blades were made of cambered steel plates and one of the blades has been tested at various angles of attack with and without slat. The photographs of visualisation test as shown in plate 1 and plate 2 have been taken after removing the slats. Plate 1 shows the flow situation on the suction side of the blade at zero angle of attack. It can be seen that the boundary layer separation has been completely eliminated with the introduction of the slat. Plate 2 shows the results of the wind tunnel test with an angle of attack of 5°. The boundary layer separation at 5° angle of attack has also been completely eliminated by the introduction of slat. Plate 3 shows a comparison of flow situation on the blade at 15° angle of attack. It is clear from the pictures that the boundary layer control by the slat has been satisfactory even at a high value of angle of attack of 15°. The flow visualisation test results have confirmed the advantage of using optimum gap of slats on the leading edge of the blades. The flow pattern clearly shows the control of the boundary layer achieved by introduction of engineered slats on the leading edges, which explains the reason for the improvement of the fan performance at off-design conditions as well as the best efficiency point.

7. CONCLUSIONS :

1. The introduction of leading edge slats on fan blades has been shown to be an effective method of boundary layer control for fan blade performance improvements.
2. Equations and optimization procedure have been suggested to introduce the slats on the leading edge of the fan blades.
3. The flow visualisation test results have confirmed the boundary layer control achieved by the slats.
4. The tunnel experiments conducted on the fan have shown 10 % increase in the efficiency of the fan and a maximum value of 27 % increase in the stable operating range of the fan resulting in a 4 % reduction in power by the introduction of engineered slats on the leading edges of blades.

8. ACKNOWLEDGEMENTS :

The authors of the paper would like to gratefully acknowledge the suggestion of the organising committee of the conference to change the original title of the paper to its present form. The referee's comments which have contributed to improve the presentation of the paper are also acknowledged.

9. REFERENCES :

1. Abbott, I.H. and Albert E. Von, Doenhoff : Theory of Wing Sections, New York, Dover Publications, 1959, pp.189 - 230.

2. Weik, F.E., and Sanders, R. : Wind tunnel Tests of a Wing with Fixed Auxiliary airfoils having Various Chords and Profiles. NACA Report No.472, 1933.

3. Javur M. M. : Investigation on Axial Flow Pumps and Fans, Ph.D. Thesis, Department of Mech. Engg., Indian Institute of Technology, Bombay, India, 1980.

4. Tanaka, S. and Muraka, S. : On the Partial Flow Rate Performance of Axial Flow Compressors and Rotating Stall (3rd Report on the devices for improving the unstable performances), Bulletin of JSME., 18, 125, Nov. 1975. pp. 1277-1284.

5. Toyokura, T., Kitamura, N., and Kida, K. : Studies on the Improvement of High Specific Speed Pump Performances at Low Flow Rates. Bulletin of JSME 8, 29, Feb. 1965, pp. 78-88.

6. Panamareff, A. J. : Axial Flow Compressors for Gas Turbines Trans. ASME 70 , 1948, pp. 295-306.

7. Koch, C. C. : NASA Report CR 54592, 1970, U.S.A.

8. Scheer, W. : Investigations and Observations on the Working Method of Axial Flow Pumps with Particular Reference to the Past Load Range, Brennstofft-Warme-Kraft, (in German), 11, 1959, pp.503-511.

9. Tomio, I. : Characteristics and Flow Conditions of Forced Vortex Impeller of an Axial Flow Fan, Bulletin of JSME, 13 , 10, June 1970, pp.773- 780.

10. Venkatrayulu, N., Prithviraj, D. and Narayana Murthy, R. G. : Some experimental investigations on the improvement of Off-design performance of a single stage axial flow fan. in : Proc. 3rd International Symposium on Air Breathing Engines, Munich, Germany, 1976, Paper 76-106.

11. Schlichting, H. : Boundary Layer Theory, New York McGraw Hill Book Co., 1960, pp. 168-169.

12. I S : 3588 - 1976 : Specification for Electrical Axial Flow Fans, Indian Standards Institute, New Delhi, India, 1976.

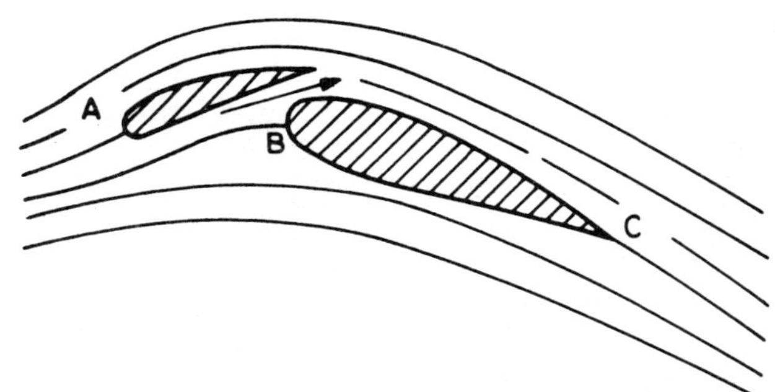

FIG. 1. BOUNDARY LAYER CONTROL BY SLAT

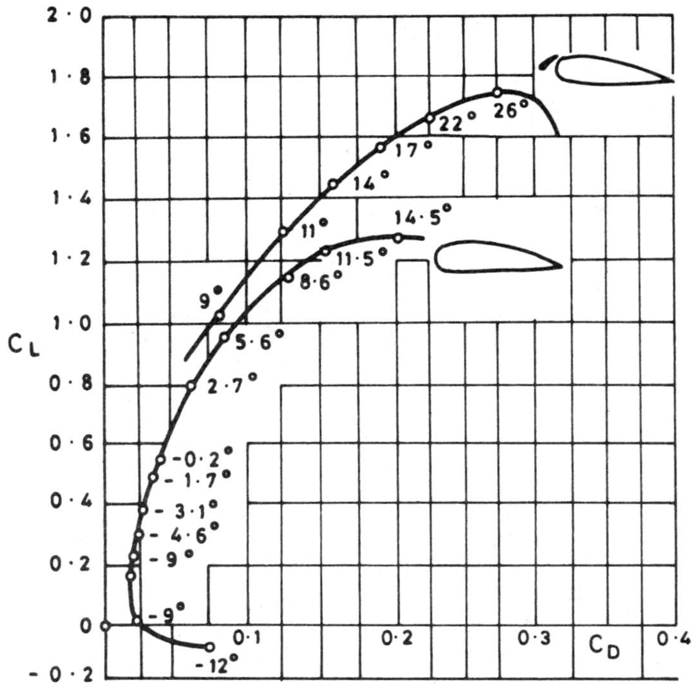

FIG.2. POLAR-DIAGRAM OF A WING SECTION

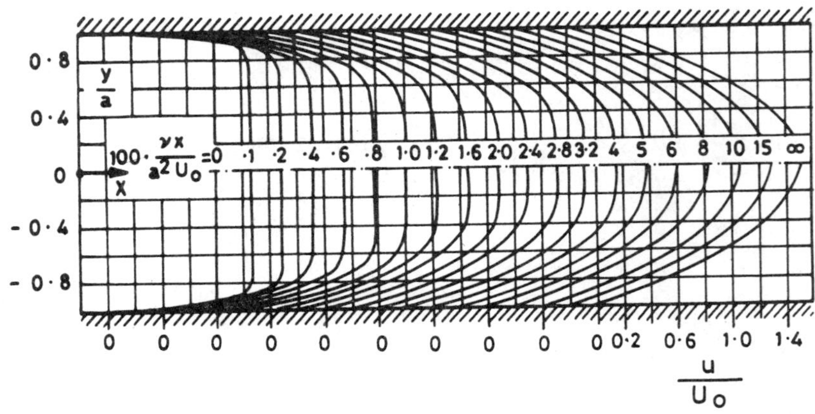

FIG. 3. VELOCITY DISTRIBUTION FOR LAMINAR FLOW IN THE INLET SECTION OF A CHANNEL

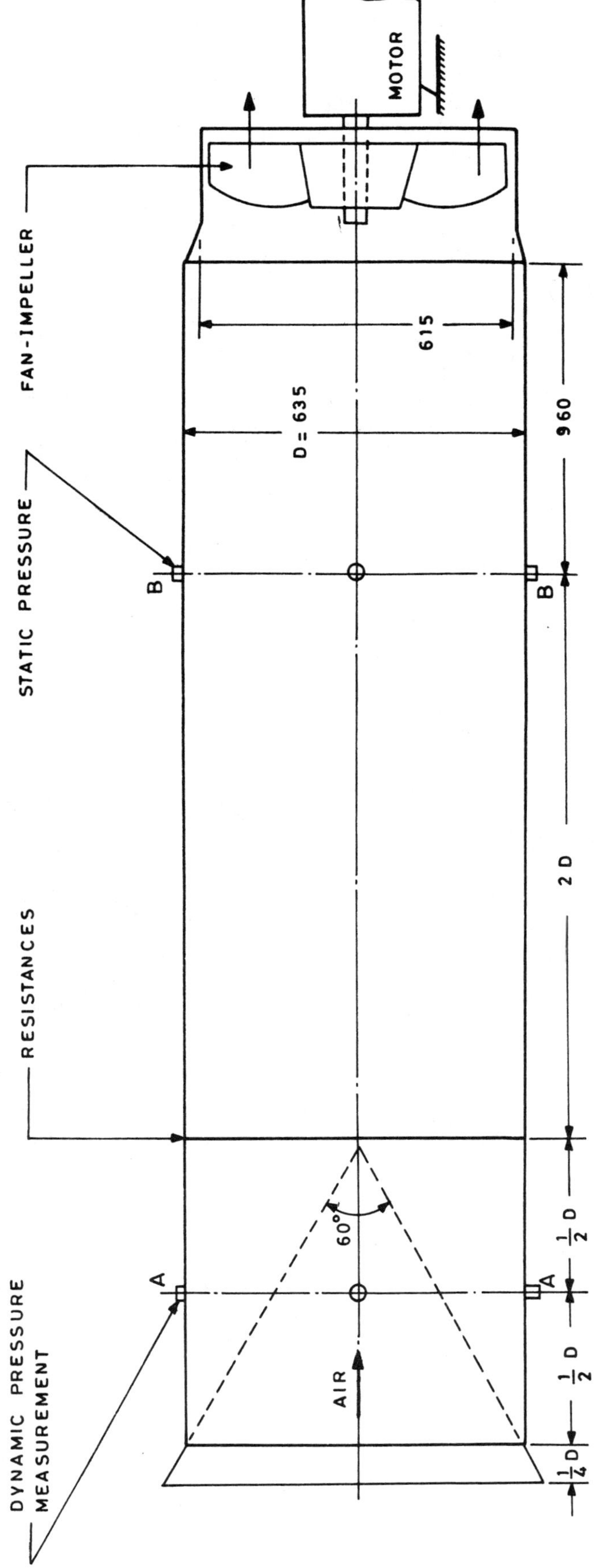

FIG. 4. FAN TEST TUNNEL DETAILS

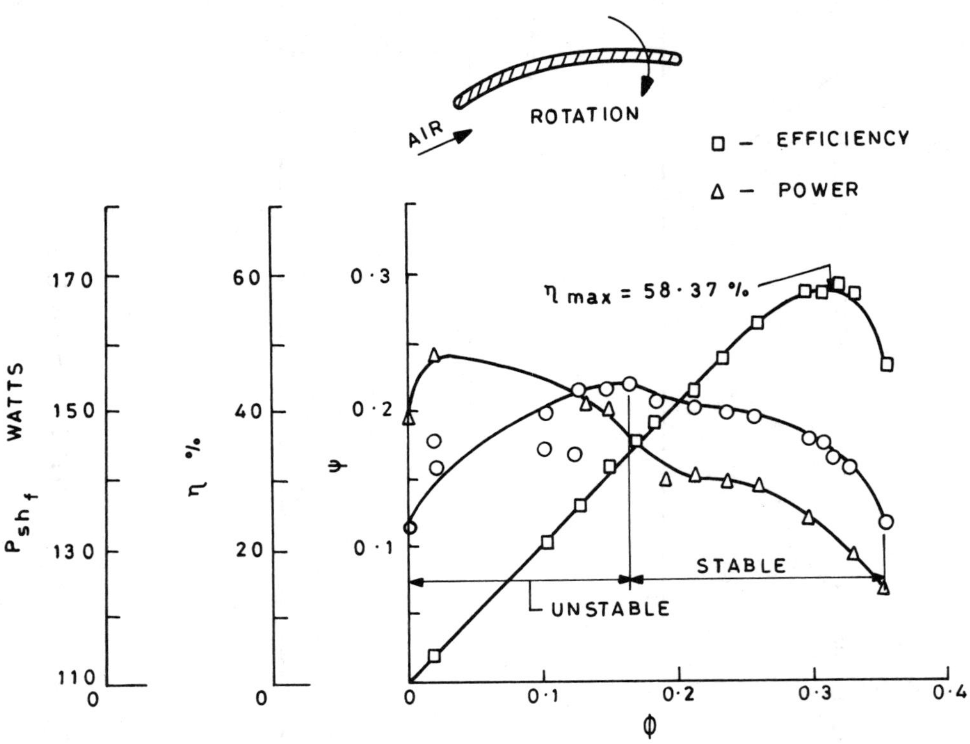

FIG. 5. PERFORMANCE CURVE OF FAN

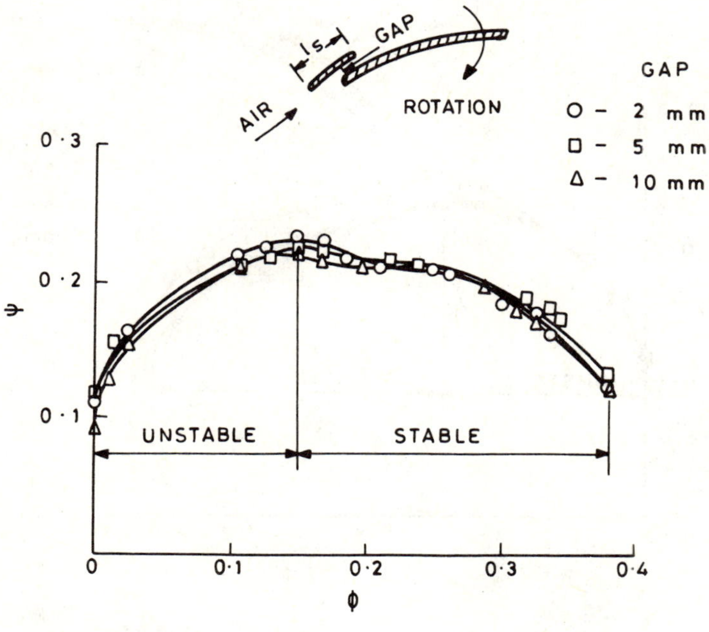

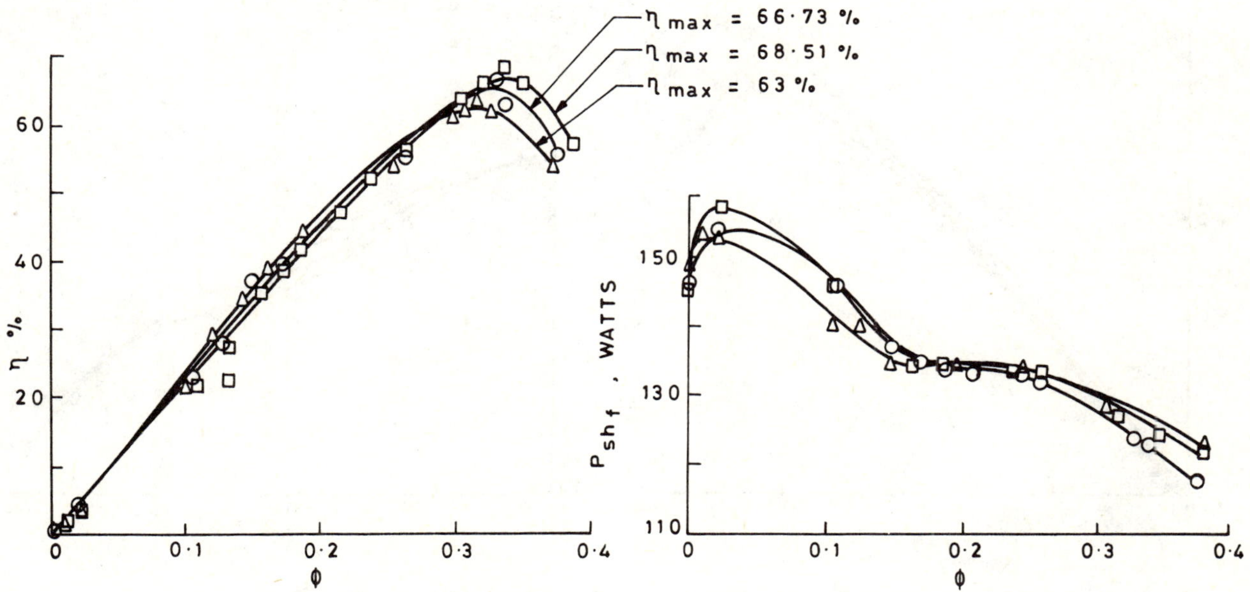

FIG. 6. PERFORMANCE CURVES OF FAN WITH 0·18 C SLAT LENGTH

STEEL BLADES WITH SLATS

$l_s = 0 \cdot 25C$

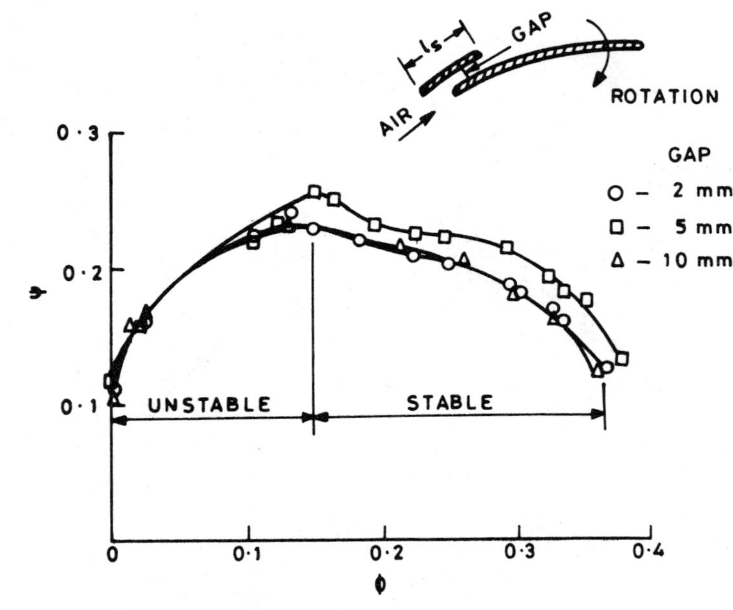

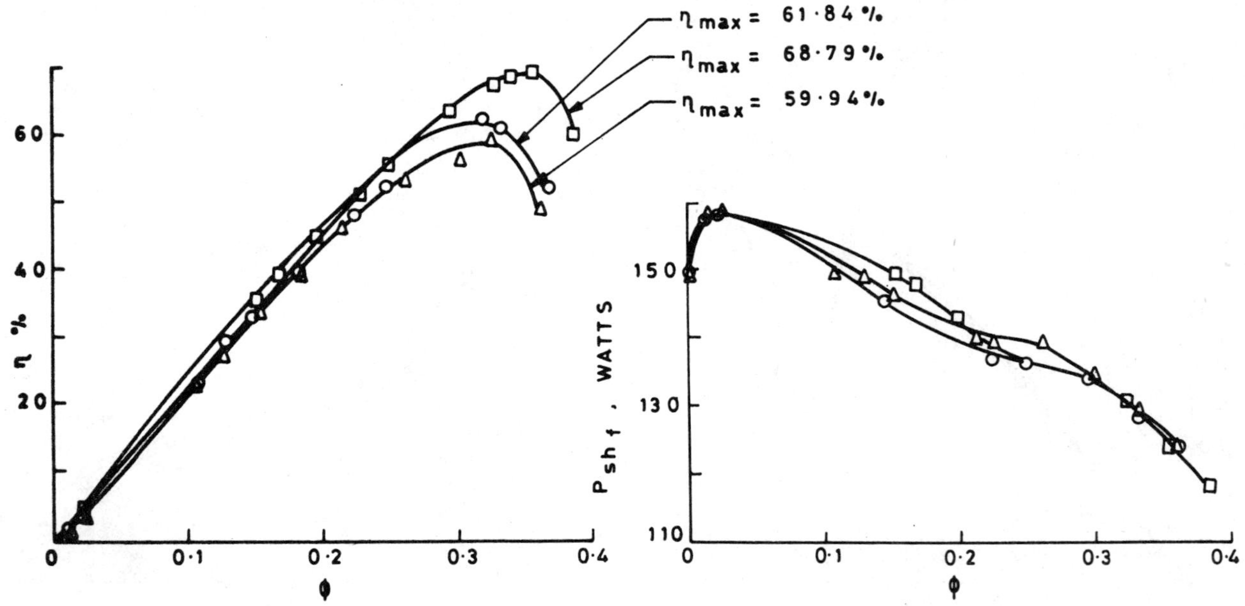

FIG. 7. PERFORMANCE CURVES OF FAN WITH 0·25 C SLAT LENGTH

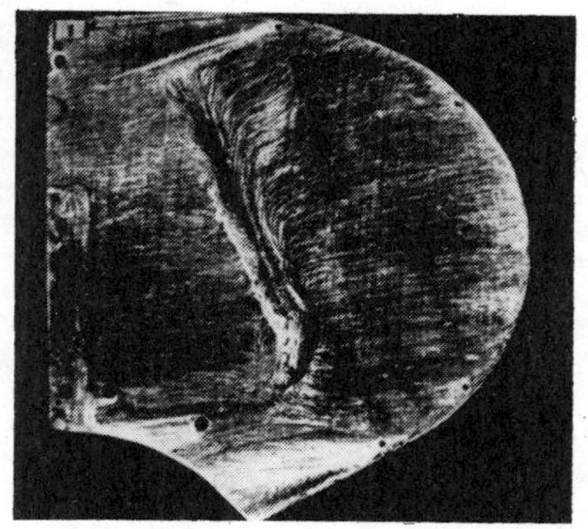

WITHOUT SLAT

WITH SLAT

PLATE.1. FLOW VISUALISATION AT 0°

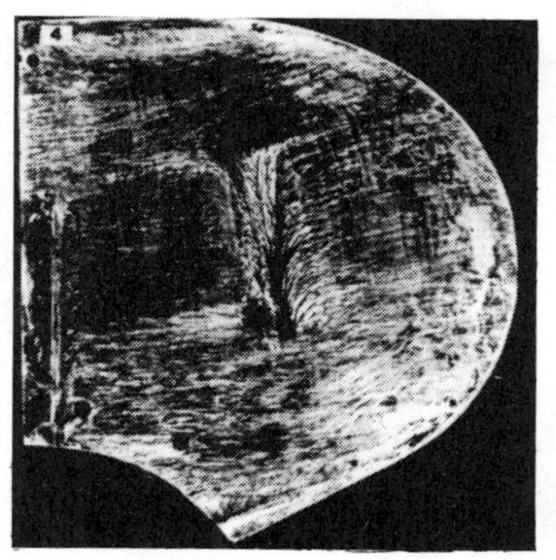

WITHOUT SLAT

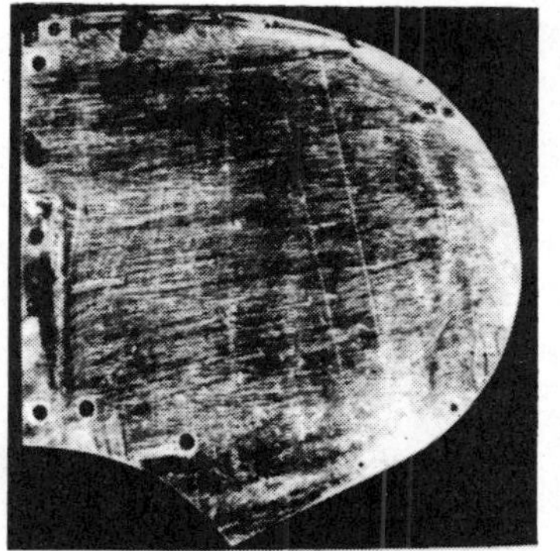

WITH SLAT

PLATE.2. FLOW VISUALISATION AT 5°

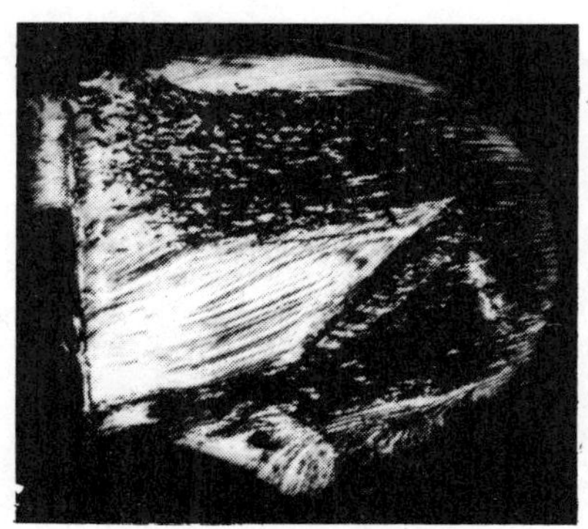

WITHOUT SLAT

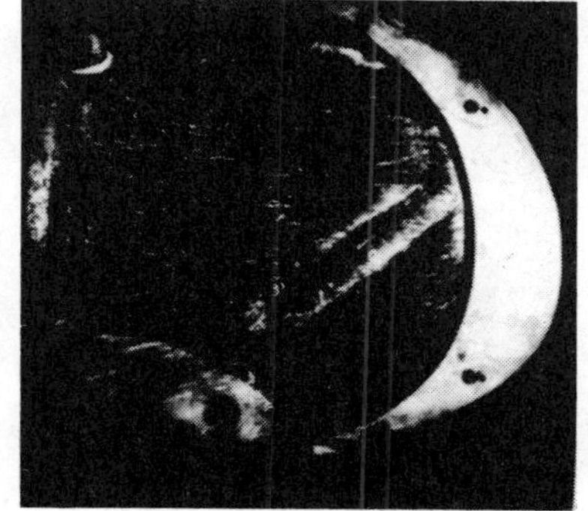

WITH SLAT
(slat shown in position)

PLATE.3. FLOW VISUALISATION AT 15°

International Conference on

Fan Design & Applications

Guildford, England: September 7-9, 1982

PAPER G3

TURBULENCE AND STABILITY IN CENTRIFUGAL FAN IMPELLERS

H. Bard

Flakt Industri AB, Sweden

Summary

This paper presents the development of a hot wire anemometer technique for the measurement of turbulence intensity and flow velocity in centrifugal fan impellers during fan operation. Test data from critical areas in various impellers, operating at the commonly used part of the fan characteristic, are presented together with fan capacity data and efficiency. The investigations have been concentrated at the flow patterns in the vicinity of the impeller shroud. It is demonstrated that there is a pronounced velocity gradient and a considerable increase in turbulence intensity within this area.

Organised and sponsored by
BHRA Fluid Engineering, Cranfield, Bedford MK43 0AJ, England.

0263 - 421X/82/01 00 - 0001 $5.00
The entire volume can be purchased from
BHRA Fluid Engineering for $82.00

NOMENCLATURE

A = area

b = impeller width

D = impeller diameter

d = inlet diameter

p_t = total pressure

q = volumetric flow

s = overlap at inlet clearance

t = inlet clearance

u = velocity component

$\overline{u}$ = time average value

u' = deviation from $\overline{u}$

U_2 = impeller tip speed

x, y, z = coordinates

ρ = density

τ = turbulence intensity = $\frac{u'}{\overline{u}}$

$\varphi = \frac{4q}{\pi D^2 U_2}$

$\psi = \frac{2 \cdot p_t}{U_2^2 \cdot \rho}$

1. INTRODUCTION

Fan performance in general is strongly affected by the inlet flow conditions. A comparison of performance data for various types of centrifugal fans indicates that local concentrations of flow losses in the impeller inlet area occur also for close to ideal inlet conditions. Certain areas having a predominant influence on fan performance and efficiency, there is a pronounced need for experimental data on the flow conditions in the rotating blade channels. The hot wire anemometer technique proves to be a most useful tool in the study of rotating flow phenomena, where relative flow measurements and turbulence levels are of interest.

The velocity gradient in the impeller inlet area affects the conditions within the boundary layer at the shroud, see fig 1. Also the effects of secondary air injected at the impeller shroud have vital importance. Boundary layer separation causes instability and reduced fan efficiency. Centrifugal fans therefore often have a conservative inlet design with respect to expansion rate.

There are various possible effects of the inlet clearance on the flow characteristics in the impeller. A proper supply of secondary air into the boundary layer would tend to stabilize a lean velocity profile. On the other hand recirculation represents a throttling of the open blade channel area. This is accentuated for narrow impellers.

For manufacturing reasons most centrifugal fans have an unsealed inlet clearance design and accordingly recirculation. Increased turbulence levels are likely to occur close to the impeller shroud, even without recirculation, and even if a stabilizing effect is taken into account, recirculation naturally involves a loss of energy. The influence from the inlet clearance being concentrated to the regions close to the impeller shroud makes the impeller width an important design parameter. Aerodaynamic design criteria, optimum for a sealed inlet clearance, are not necessarily ideal for an inlet design with recirculation.

2. EXPERIMENTAL PROGRAM

2.1 Instrumentation

A subminiature hot wire probe, commercially available from DISA, was attached to a specially designed probe support, used for traverses in the blade channels. The specially developed constant temperature hot wire anemometer was built into a hollow shaft connected with the impeller hub, see fig 2. The preamplified signals were transmitted from the rotating members by means of a slip ring unit.

DC and AC signals were displayed on a digital voltmeter. A one-dimensional probe being used, indicated values of turbulence intensity and flow velocity refer to blade-conformal streamlines. This restriction has been considered acceptable since the intention of the measurements is to provide relative values for various configurations rather than absolute values for the three dimensional flow.

The flow rate and pressure head were measured according to the AMCA 210-74 test code. A digital wattmeter was used for power readings, and the fan motor was calibrated with respect to efficiency.

2.2 Test fan

Two impellers with backward curved blades, identical except for width, were selected for testing. The basic geometry is shown in fig 3.

The test probe was inserted into the blade passages immediately downstream of the impeller shroud radius, and was traversed from the highly turbulent area close to the shroud into the low turbulence regions outside the boundary layer, see fig 2.

The inlet section was machined in order to provide a symmetrical and reproduceable inlet clearance. The fan motor and impeller being mounted on a slide, and the alignment of the impeller and inlet being carefully adjusted, made the axial position of the inlet clearance variable under fully controlled conditions during fan operation. All tests were performed with two sizes of inlet clearance: 0.3 mm and 2.5 mm.

The width of the casing was adjusted to fit the width of the impellers.

2.3 Test procedure

Parameters investigated for each impeller and inlet geometry were

a) Turbulence and velocity at peak efficiency as a function of distance from the impeller shroud.

b) Turbulence and velocity in the boundary layer at various loads.

c) Turbulence, velocity and fan performance for various sizes and positions of the inlet clearance.

d) Fan capacity.

3. RESULTS

The expansion rate in the impeller inlet is expressed in terms of the area ratio A_1/A_0, as defined in fig 4. The first test series being presented was performed with an impeller designed for only small velocity gradients in the inlet section. For the second test series an impeller was selected with a substantial retardation of flow velocity in the inlet area. The size and position of the inlet clearance on each was varied as described above.

3.1 Impeller with area ratio A_1/A_0 = 1.107 and inlet clearance t = 0.3 mm

3.1.1 Turbulence and velocity at peak efficiency

As was suggested earlier a clearly defined boundary layer is indicated within a distance of approximately 25 mm from the shroud. There is a considerable increase in turbulence intensity within this area, see fig 5. Close to the shroud are measured τ -values in the range of 10-15 %, while the level in the central parts of the blade channel is 5 %. At larger distances than 25 mm from the shroud are measured almost constant velocities and turbulence intensities. Variations in the axial position of the inlet clearance had only minor influence on the flow conditions.

Even though the concentration of turbulence in the vicinity of the shroud is pronounced, the absolute values remain at a comparatively low level. Turbulence intensities in the range of 2-8 % are to be considered normal in a turbomachine.

3.1.2 Turbulence and velocity at various loads

Throttling of the fan results in increased pressure gradients along the streamlines. The boundary layers grow and tend to separate at a certain point of instability at the fan characteristic. The position of this point is closely connected to the width of the recommended range of operation for the fan. It therefore is important to determine variations in the stable range of operation when parameters like impeller width and size and position of the inlet clearance are varied. Stability over a wide range of duty points often is preferable to a narrow band of peak efficiencies.

The inlet clearance of this test series being small makes the variation in overlap between the stationary fan inlet and the cylindrical section of the shroud the only variable. The amount of recirculated, secondary air is negligible. Since the flow conditions within the boundary layer is of major interest, these tests were performed at distances 2, 7 and 12 mm from the shroud. Test data at 2 mm distance are shown in fig 6.

Boundary layer separation occurs at a clearly defined point $\varphi = \varphi_b = 0.06$. This point can be identified on an oscilloscope as a steep peak value in the turbulence signal. A narrow band analysis of the turbulence frequency spectrum shows that flow separation, or break-down, occurs with a frequency 55-85 % of the frequency of rotation. This is within the expected range of a rotating stall. Rotating stall occurs due to the mismatch between impeller geometry and the direction of air into the blade passages for flows lower than design flow.

Going back to fig 6 we observe the small influence of φ on the velocity u for $\varphi > \varphi_b$. For $\varphi < \varphi_b$ an increase in velocity results and the turbulence decreased as the inlet overlap s is increased from 0 to 30 mm.

Tests performed at 7 mm and 12 mm distance from the shroud are presented in fig 7 and 8. The φ_b value is clearly defined, but at these larger distances there is no tendency for increased τ values or instability at large flow rates. Velocities are continuously increasing as a function of φ .

The tests show that even at this very moderate expansion rate, instability occur close to the shroud, not only for small flow rates and large pressure gradients, but also for the larger flow rates.

3.2 Impeller with area ratio $A_1/A_0 = 1.677$ and inlet clearance $t = 0.3$ mm

In comparison with the impeller in section 3.1 the expansion rate in the inlet has increased by a factor of 52 %. In order to create equal flow conditions the width of the fan casing was increased by the same amount. The position of the impeller backplate was unchanged in position relative to the backplate of the casing.

3.2.1 Turbulence and velocity at peak efficiency

The velocity and turbulence profiles are shown in fig 9. The increased expansion rate causes further growth in turbulence intensity close to the shroud. The velocity profile within the boundary layer is leaner due to the rapid expansion, but the boundary layer thickness still remains 25 mm, which indicates a maintained high peak efficiency. Even at this rate of expansion the velocity profile remains stable within the boundary layer at the shroud also without recirculation. However, there is a pronounced decrease in velocity in comparison with fig 6. The centrifugal forces evidently stabilizes the potential wake region and allows for increased expansion rates.

3.2.2 Turbulence and velocity at various loads

Boundary layer separation (breakdown) occurs at $\varphi_b = 0.09$, see fig 10. The increase in φ_b-value corresponds to the increase in impeller width ($\varphi_b = 1.515 \cdot 0.06 = 0.091$). Accordingly the increased expansion rate has not affected the lower limit of stable fan operation. The turbulence intensity close to the shroud increases rapidly as a function of flow rate, while flow velocities are decreasing. The corresponding trends, although not so accentuated, are measured also at 12 mm distance from the shroud, see fig 11.

3.3 Impeller with area ratio $A_1/A_0 = 1.107$ and inlet clearance $t = 2.5$ mm

3.3.1 Turbulence and velocity at peak efficiency

A boundary layer of similar type as in fig 6 is developed, and the boundary layer thickness is unaffected by the secondary air, see fig 12. The turbulence intensity close to the shroud has increased, which indicates reduced fan efficiency. This trend is accentuated by the internal energy represented by the recirculation itself.

The turbulence level outside the boundary layer is unaffected by the inlet clearance A variation in inlet overlap from 10 to 25 mm has no significant effect on the flow parameters.

3.3.2 Turbulence and velocity at various loads

The τ-values have increased over the whole range of operation, and there is a slight decrease in φ_b-value, see fig 13. In spite of the inlet clearance and recirculation of air, velocities are low in the vicinity of the shroud.

Going back to fig 7 we observe an increased velocity, specially at large overlaps when the clearance increases from 0.3 mm to larger values in the curved section of the shroud. When the corresponding increase in clearance is introduced in the cylindrical section there is no significant change in velocity close to the shroud. The increased τ-values, due to recirculation, are reflected also in the values measured at 12 mm distance, while flow velocities are maintained at the levels measured for $t = 0.3$ mm, see fig 14.

3.4 Impeller with area ratio $A_1/A_0 = 1.667$ and inlet clearance $t = 2.5$ mm

3.4.1 Turbulence and velocity at peak efficiency

The increased inlet clearance has made no significant change in the velocity or turbulence profiles at the shroud. The boundary layer thickness remains constant, and the performance is stable. The hot wire traverse data are shown in fig 15.

3.4.2 Turbulence and velocity at various loads

The shape of the φ-u and $\varphi - \tau$ curve is practically unaffected by the inlet clearance, see fig 16 and 17. Recirculation has a delaying effect on breakdown which earlier was stated also for the narrow impeller.

3.5 Performance data for the tested impellers

A comparison of performance data for the two tested impellers with no recirculation proves the total fan efficiency to remain at a high level also for the wide impeller, see fig 18. The increased τ-values within the boundary layer is compensated by the increased areas of low turbulence flow.

For the impellers with recirculation a considerable decrease in efficiency will be noted for the narrow impeller. The wider impeller practically retains the high level of efficiency measured for the non-recirculation design.

4. CONCLUSIONS

The turbulence and velocity profiles in the rotating blade passages of a centrifugal fan impeller can be studied by means of a specially designed hot wire anemometer device. The tests show that a pronounced boundary layer is developed at the impeller shroud, and that there is a strong concentration of flow losses within this area of the impeller.

It also is shown that the boundary layer thickness remains practically unchanged within a considerable range of inlet velocity decelerations. With no recirculation through the inlet clearance there is a pronounced increase in turbulence intensity within the boundary layer at large expansion rates. This counteracts the increase in fan efficiency which otherwise would be implied due to an increase in low turbulence areas outside the boundary layer.

For the case with recirculation the level of turbulence intensity within the boundary layer remains practically uninfluenced by the rate of expansion in the impeller inlet. The amount of recirculated air is uninfluenced by the selected impeller width and the boundary layer thickness remains unchanged in comparison with the non-recirculation case.

Therefore, for a fan design with a normal unsealed inlet clearance the impeller width is of design importance, not only with respect to fan capacity but also fan efficiency.

A point of initial instability, identified by means of a narrow band analysis as a rotating stall, is indicated for all tested impellers. The rate of inlet expansion does, within reasonable limits, not influence the position of this point on the fan characteristic. The strongest fluctuation in flow velocity under stalled conditions is found within the boundary layer. Recirculation has a slightly delaying effect on instability.

The introduction of the hot wire anemometer technique for the measurement of flow parameters inside a rotating fan impeller improves the understanding of the flow phenomena in the blade channels. This paper should be looked upon as the initial step on an alternative way of finding improved designs for future centrifugal fans.

5. REFERENCES

1. Balje O.E.: Loss and flow path studies on centrifugal compressors. Journal of Engineering for Power, July 1970.

2. Hönmann W: Zum Problem der optimalen Laufradbreite bei Radialventilatoren. Heiz.-Lüft-Haustechn. 12, 1961.

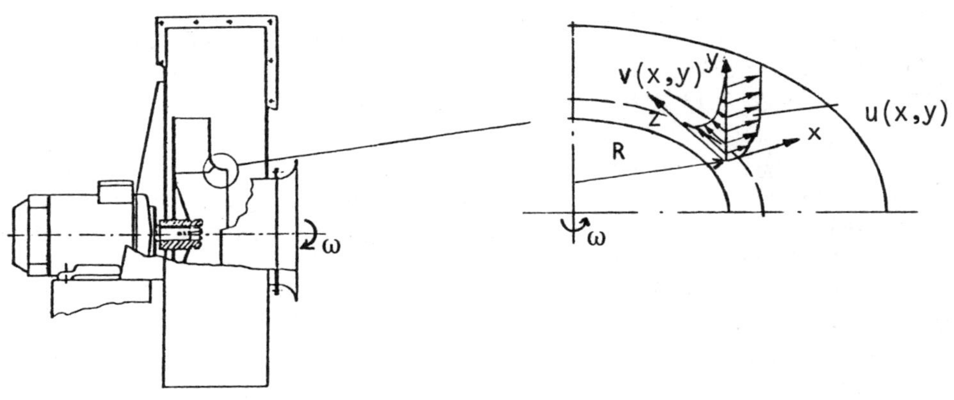

Fig 1: Boundary layer at impeller shroud

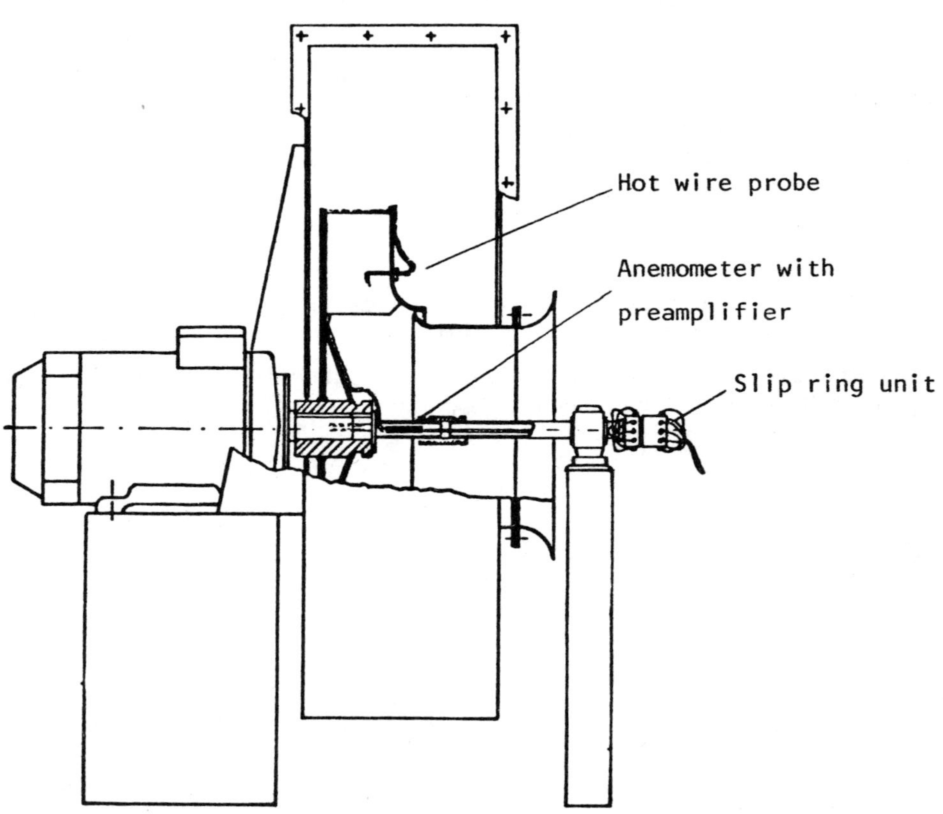

Fig 2: Hot wire anemometer arrangement

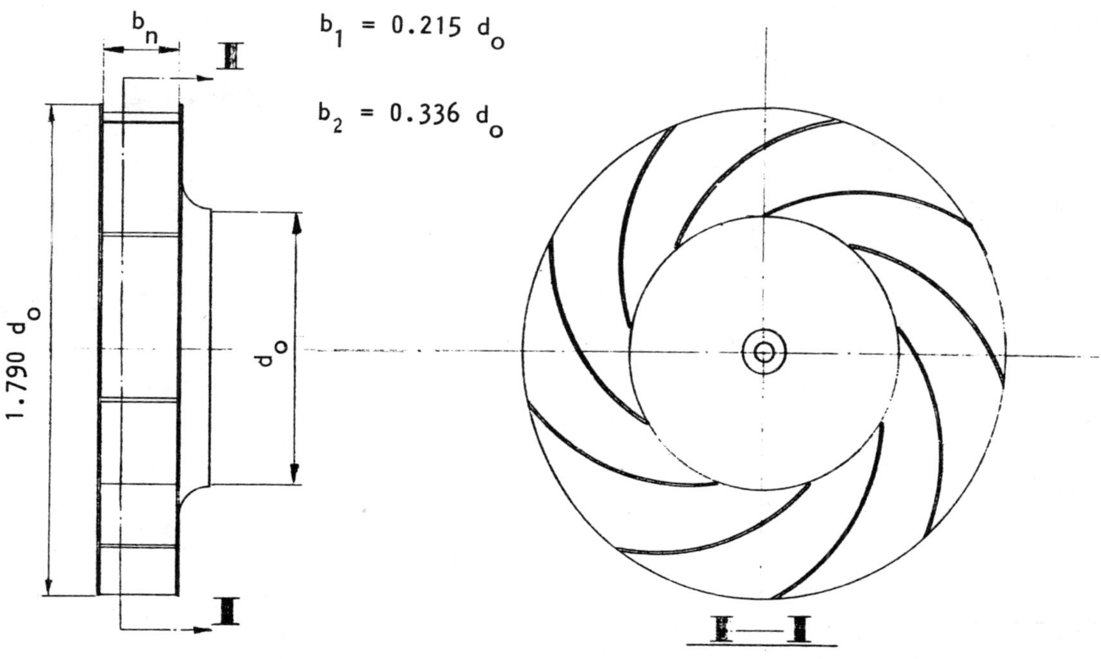

Fig 3: Geometry of the tested impellers

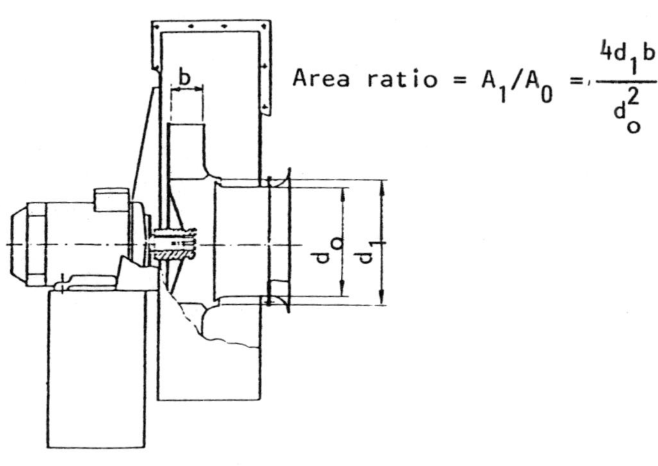

Fig 4: Definition of area ratio

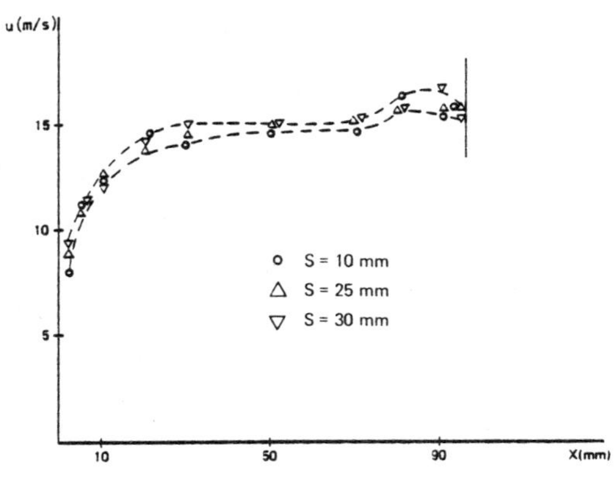

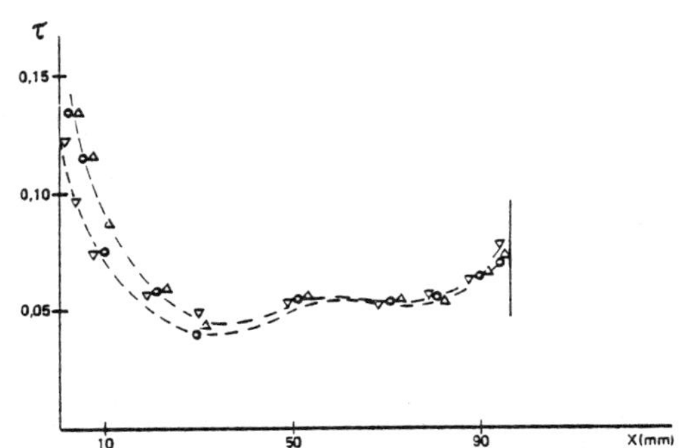

Fig 5: Traverse at peak efficiency
A_1/A_0 = 1.107; t = 0.3 mm

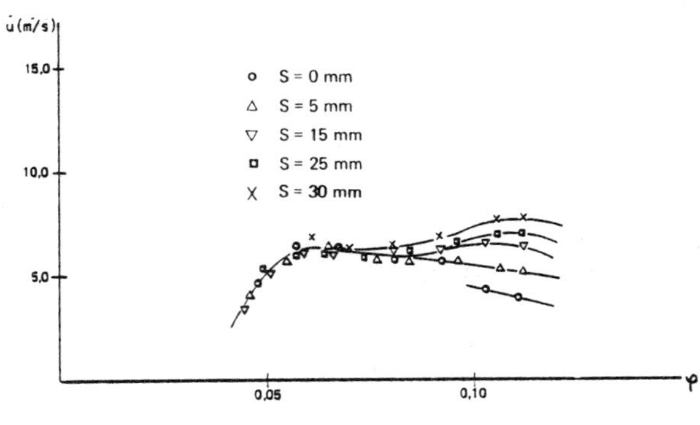

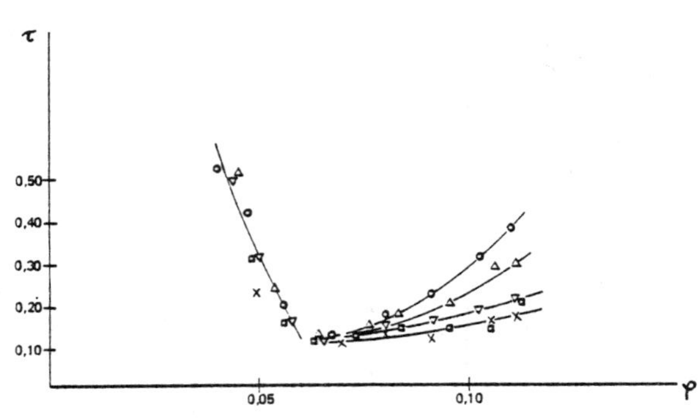

Fig 6: Flow conditions 2 mm from the shroud
A_1/A_0 = 1.107; t = 0.3 mm

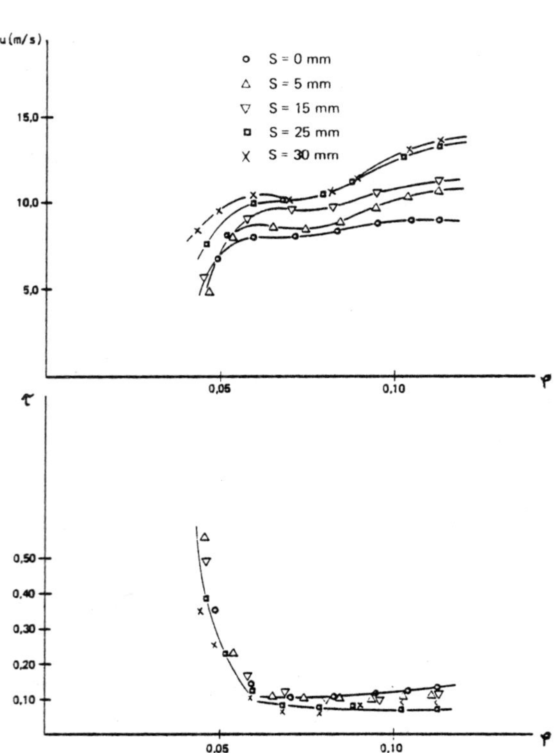

Fig 7: Flow conditions 7 mm from the shroud
A_1/A_0 = 1.107; t = 0.3 mm

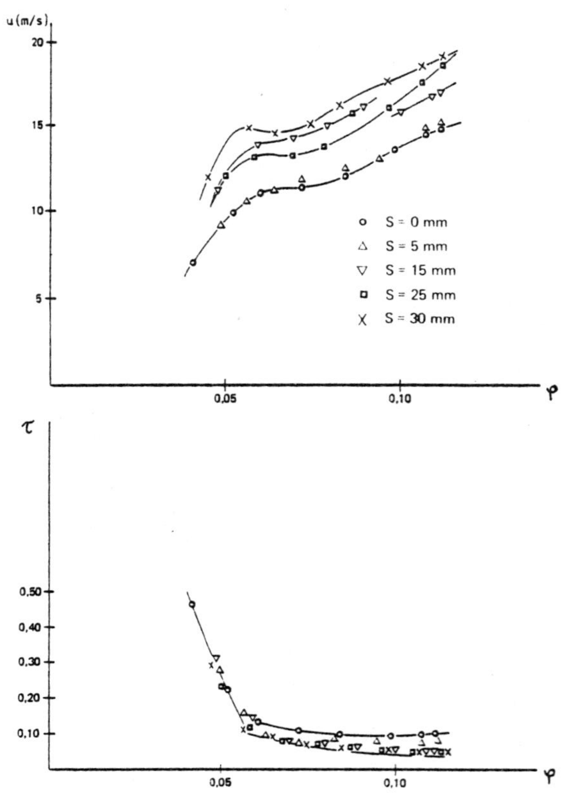

Fig 8: Flow conditions 12 mm from the shroud
A_1/A_0 = 1.107; t = 0.3 mm

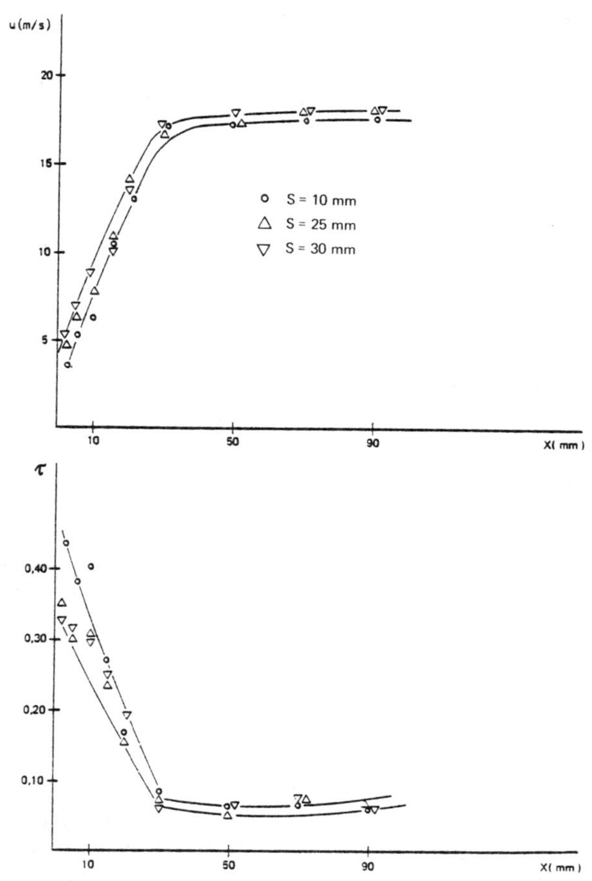

Fig 9: Traverse at peak efficiency
A_1/A_0 = 1.677; t = 0.3 mm

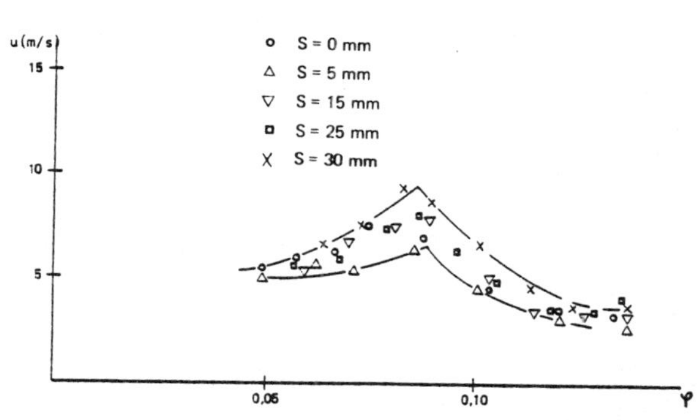

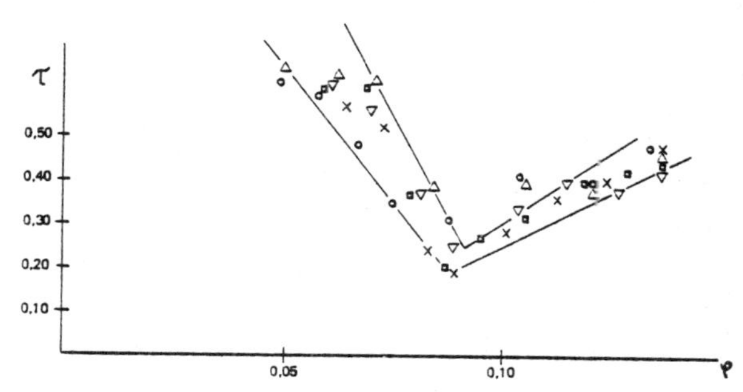

Fig 10: Flow conditions 2 mm from the shroud
A_1/A_0 = 1.667; t = 0.3 mm

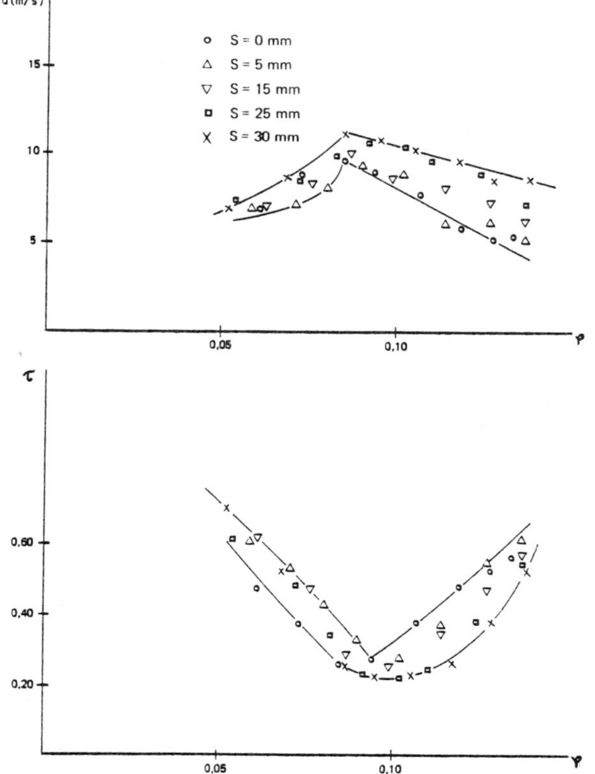

Fig 11: Flow conditions 12 mm from the shroud
A_1/A_0 = 1.667; t = 0.3 mm

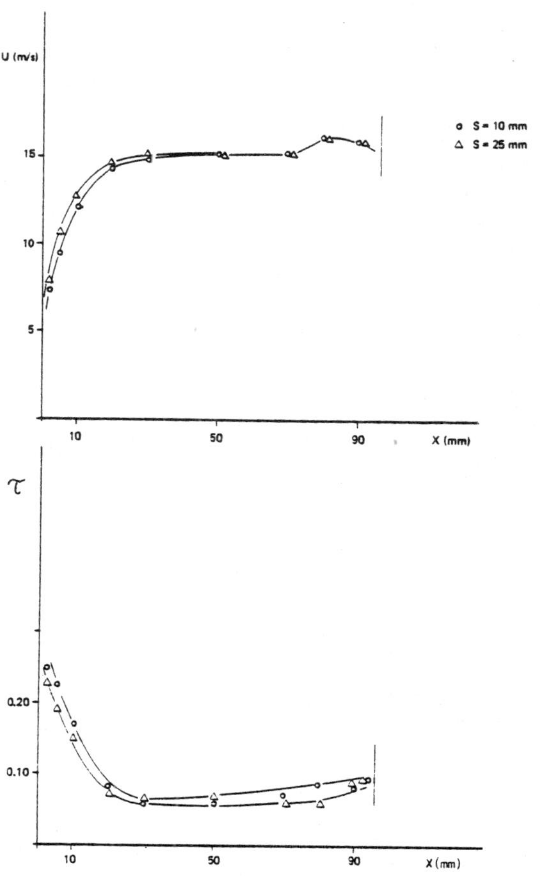

Fig 12: Traverse at peak efficiency
A_1/A_0 = 1.107; t = 2.5 mm

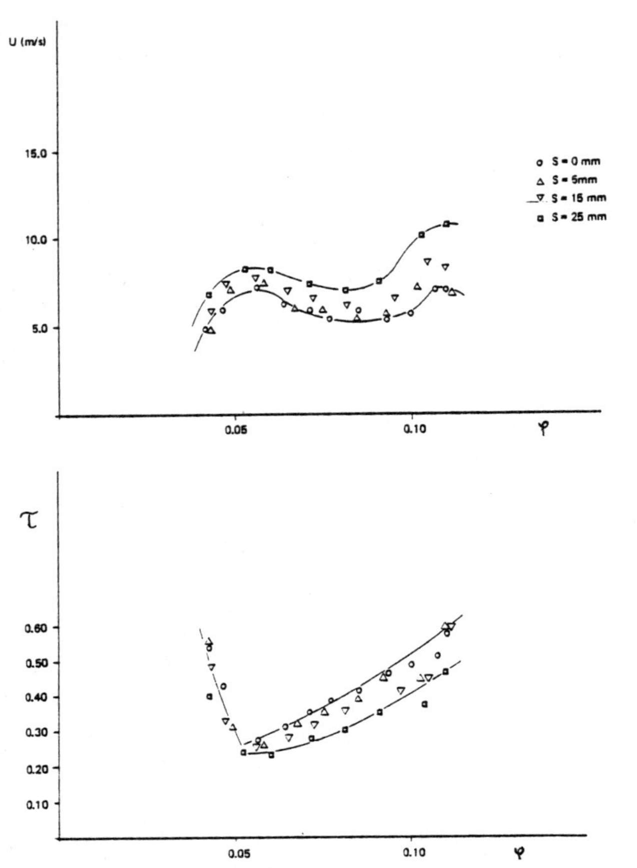

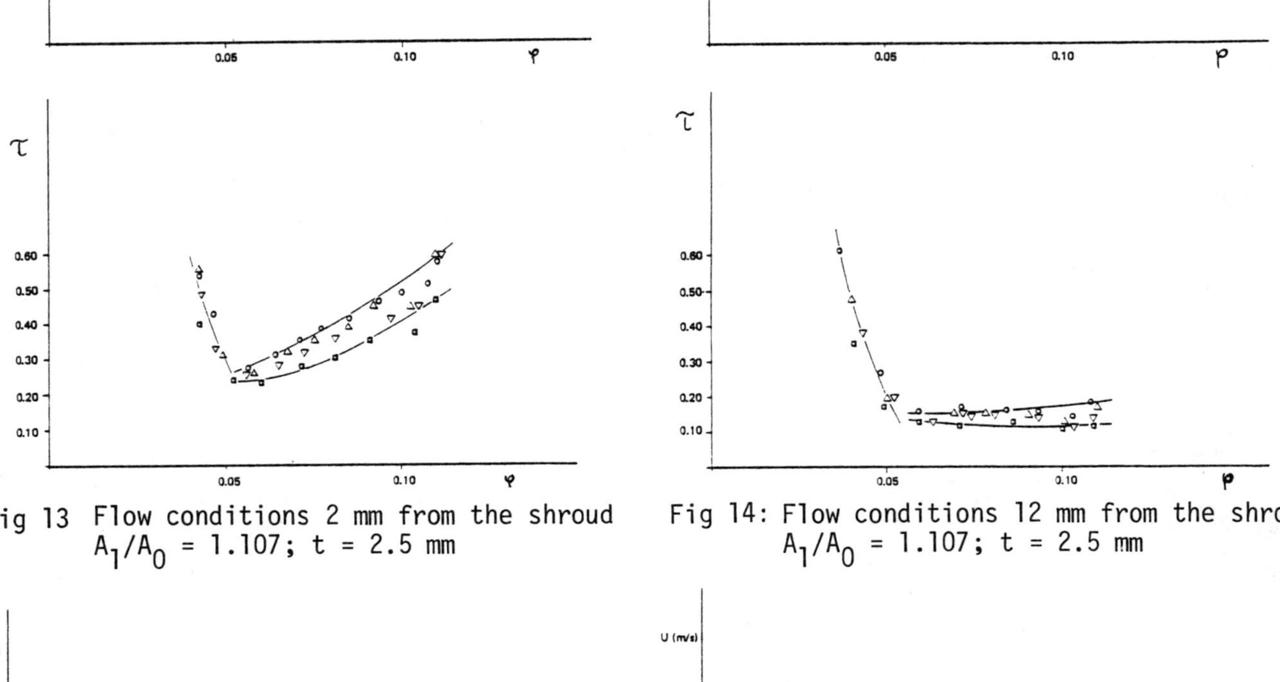

Fig 13 Flow conditions 2 mm from the shroud
A_1/A_0 = 1.107; t = 2.5 mm

Fig 14: Flow conditions 12 mm from the shroud
A_1/A_0 = 1.107; t = 2.5 mm

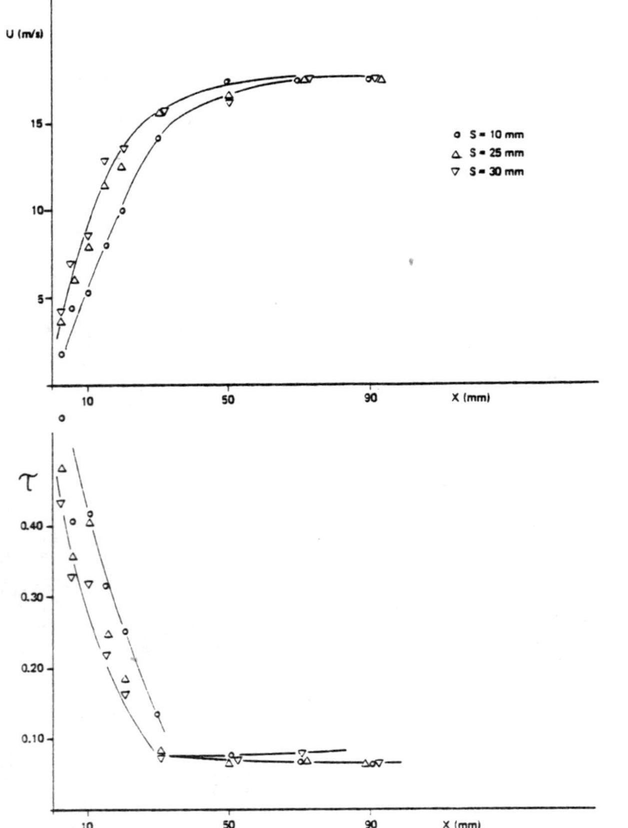

Fig 15: Traverse at peak efficiency
A_1/A_0 = 1.667; t = 2.5 mm

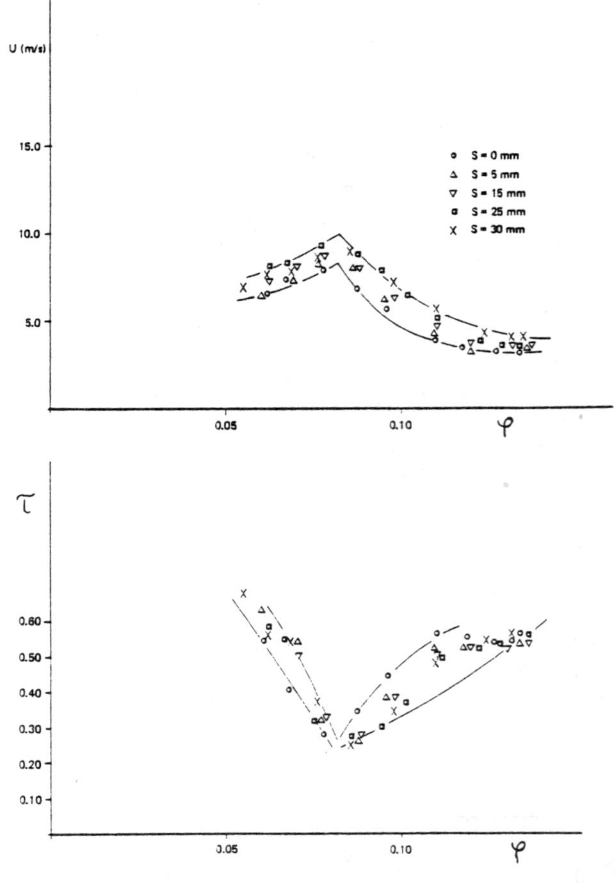

Fig 16: Flow conditions 2 mm from the shroud
A_1/A_0 = 1.667; t = 2.5 mm

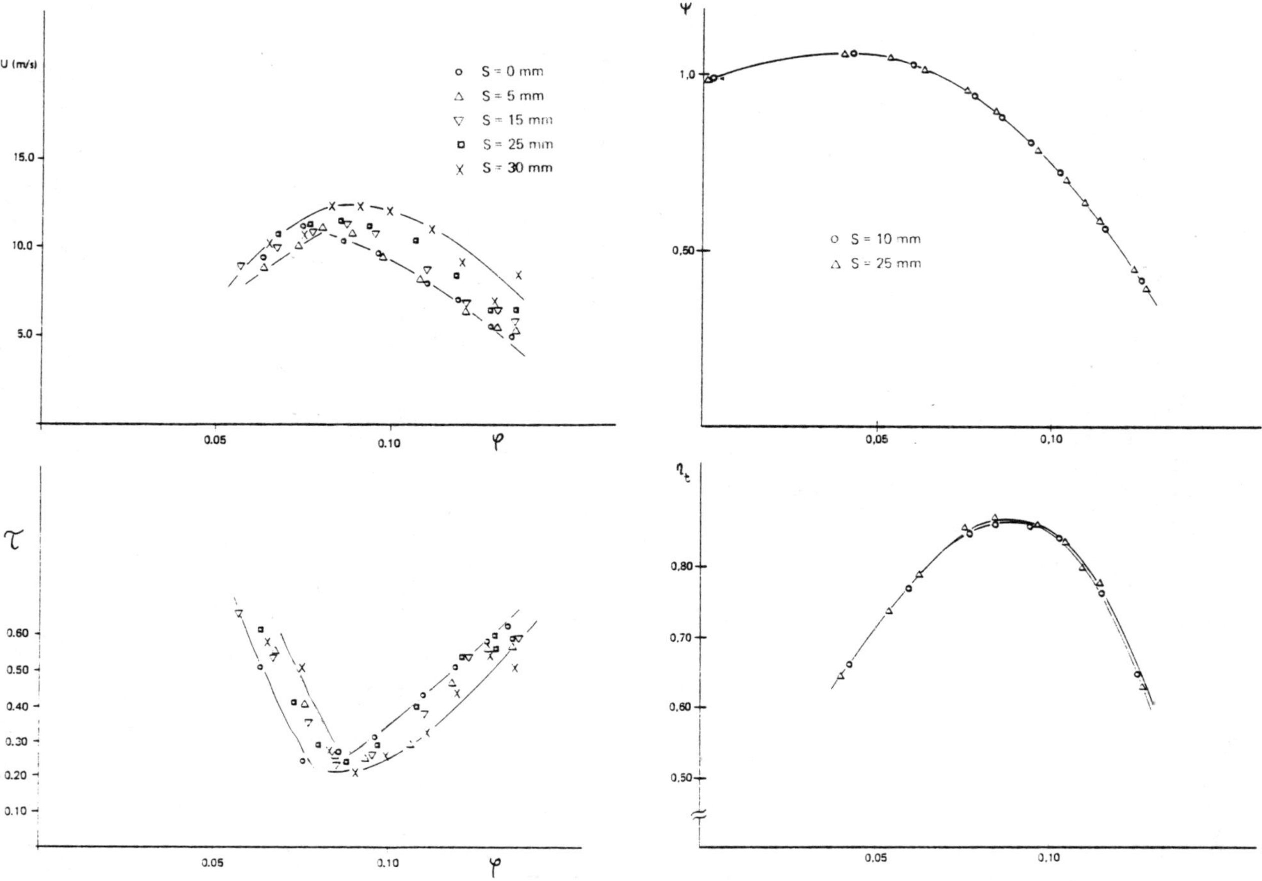

Fig 17: Flow conditions 12 mm from the shroud
A_1/A_0 = 1.667; t = 2.5 mm

Fig 18a: Fan performance.
A_1/A_0 = 1.107; t = 0.3 mm

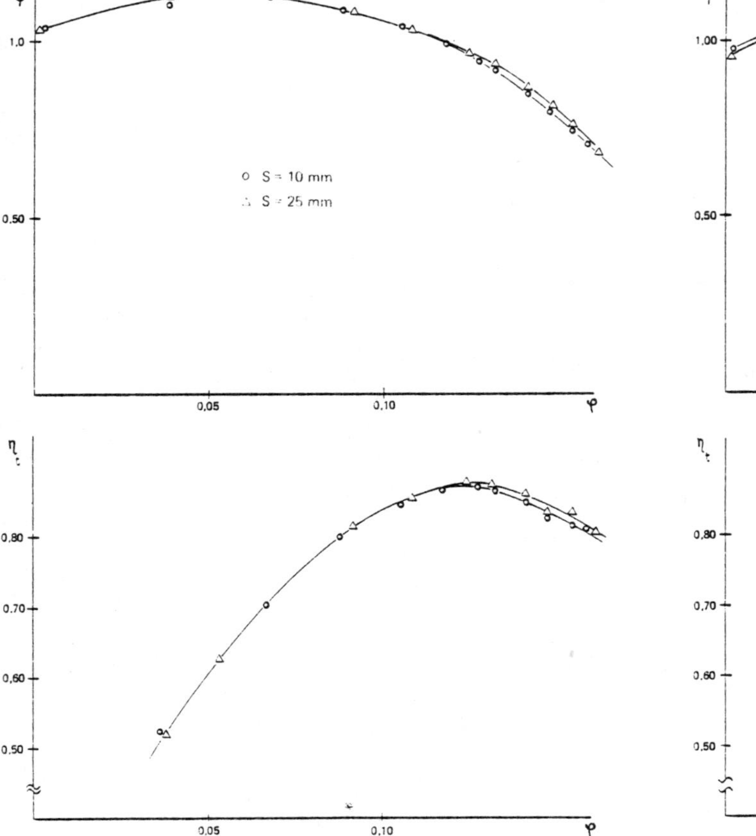

Fig 18b: Fan performance.
A_1/A_0 = 1.667; t = 0.3 mm

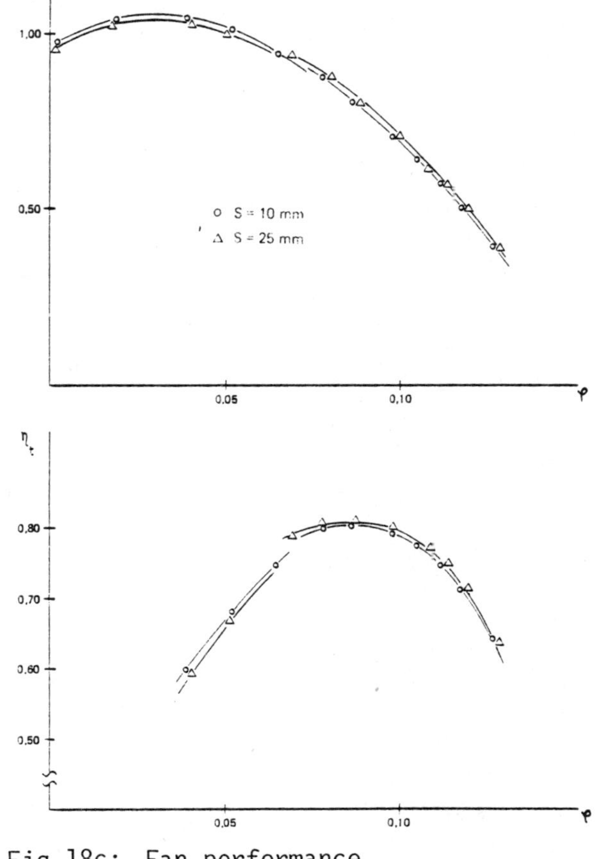

Fig 18c: Fan performance.
A_1/A_0 = 1.107; t = 2.5 mm

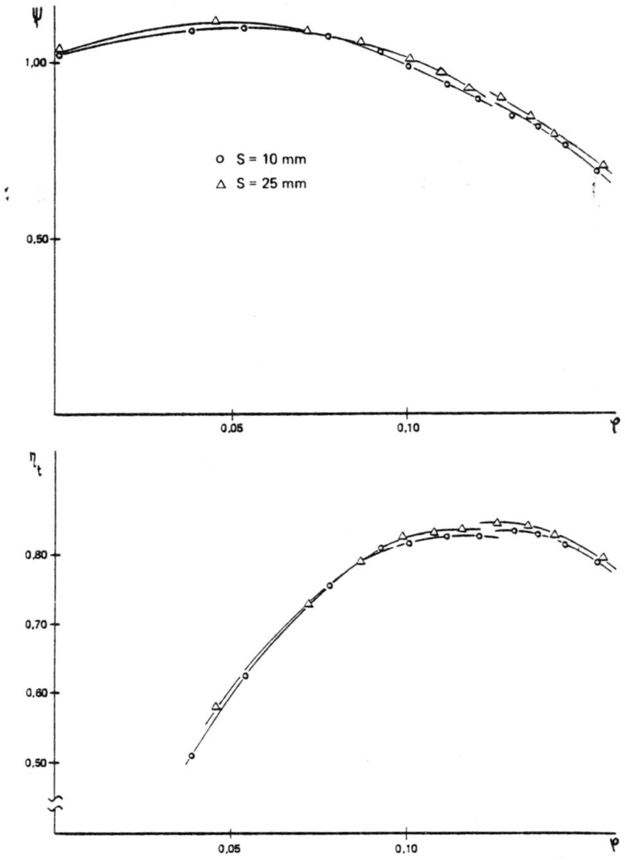

Fig 18d: Fan performance. A_1/A_0 = 1.667; t = 2.5 mm

International Conference on

Fan Design & Applications

Guildford, England: September 7-9, 1982

PAPER H1

ANECHOIC TERMINATIONS FOR IN-DUCT FAN NOISE MEASUREMENT

A.N. Bolton and E.J. Margetts

National Engineering Laboratory, East Kilbride, U.K.

Summary

The requirement for anechoically terminating a fan test duct during an assessment of in-duct fan sound power is briefly examined and the general acoustic features of terminations are described. A simple design procedure is presented and comparisons are made between the measured and predicted performances of a number of different terminator units. The design and the measured performance of a novel termination using a stepped, parallel-sided expansion section are presented.

Organised and sponsored by
BHRA Fluid Engineering, Cranfield, Bedford MK43 0AJ, England.

0263 - 421X/82/01 00 - 0001 $5.00
The entire volume can be purchased from
BHRA Fluid Engineering for $82.00

NOMENCLATURE

A	Amplitude of incident sound pressure	-
B	Amplitude of reflected sound pressure	-
b	Shape factor (see equation 10)	m^{-1}
c	Speed of sound	$m\ s^{-1}$
d	Duct diameter	m
f	Frequency	Hz
j	Imaginary constant ($\sqrt{-1}$)	-
k	Wave number	rad/m
Lp	Sound pressure level	dB re 20×10^{-6} N/m^2
l	Length of exponential expansion	m
M	Area expansion ratio	-
m	Expansion flare constant	-
p	Sound pressure	N/m^2
R_T	Reflection coefficient specified by ISO or BSI standard	-
r	Reflection coefficient	-
S	Cross-section area	m^2
U	Acoustic volume velocity	$m\ s^{-1}$
x	Axial coordinate	m
x_o	Parameter equal to length of apex of truncated cone (see Fig. 7)	m
Z	Acoustic impedance	$N\ s\ m^{-3}$
θ	Angle (see equation (10)	rad
λ	Wave length	-
ρ	Density of air	kg/m^3
ω	Angular frequency	$rad\ s^{-1}$

SUBSCRIPTS

b	Refers to body section
i	incident
r	reflected
t	Refers to taper section or transmitted

1. INTRODUCTION

Proposed British (Ref. 1) and International (Ref. 2) Standards for in-duct fan noise measurement require that the test duct must be anechoically terminated. There are two reasons for this. One is that if the duct is not properly terminated but stops abruptly, sound from the fan travels along the duct and is reflected from the open end. Within the duct the resulting sound pressure is a combination of the direct and reflected noise. At some positions in the duct the pressures will reinforce while at others they will cancel so that the in-duct sound pressure level becomes a function of axial position within the duct. The second reason is that recent work (Refs 3-5) has shown that the sound generated by a fan is a function of the acoustic loading on the fan. In order that comparative assessments of fan sound pressure can be taken it is necessary that a standard loading be specified. This standard loading is most readily and conveniently specified to be anechoic.

An anechoic termination, as its name suggests, should, ideally, be completely reflection free. In practical terms this is virtually impossible and the standards specify the acceptable maximum reflection coefficients as a function of frequency. The actual values are given in Table 1.

The ISO Standard presents a number of alternative designs of anechoic terminator. Some of them have only been tested on small diameter test ducts and no indications are given as to how they may be scaled up or how practical designs may be derived. In relation to the cost of conventional fan test ducting an anechoic terminator is a relatively expensive item and a manufacturer will wish to be fairly certain that any particular design will be satisfactory before embarking on its construction. At the same time, the capital cost is not sufficient to justify the considerable effort which would be required to develop an accurate predictive design method. It is the purpose of this paper to present a relatively simple analysis of terminator performance and, by comparing the predicted and measured performances of a number of terminators, to provide guidelines for the design of further units.

2. ASSESSMENT OF TERMINATOR PERFORMANCE

The performance of a terminator is determined using the test arrangement as indicated in Fig. 1. A loudspeaker is connected at the opposite end of the test duct to the termination and a pure tone signal applied to the loudspeaker. It is shown in most acoustical text books that sound waves incident from the loudspeaker and reflected from the terminator will form an interference pattern in the duct with maximum and minimum sound pressures occurring at regular half wave length positions along the duct. The reflection coefficient, r, which is the ratio of the reflected sound pressure to the incident sound pressure is determined from the ratio

$$r = \frac{p_{max} - p_{min}}{p_{max} + p_{min}} \qquad (1)$$

where p_{max} and p_{min} are the maximum and minimum sound pressures.

Experimentally, a microphone is traversed along the length of the duct and the maximum and minimum sound pressure levels Lp_{max} and Lp_{min} noted, the reflection coefficient then being

$$r = \frac{10^{\left(\frac{Lp_{max} - Lp_{min}}{20}\right)} - 1}{10^{\left(\frac{Lp_{max} + Lp_{min}}{20}\right)} + 1}. \qquad (2)$$

To determine the performance of the termination measurements are taken at frequencies corresponding to the centre frequencies of the ⅓-octave bands from 50 Hz up to the band containing the cut-on frequency, f_o, of the first cross mode, where $f_o = 0.586\ (c/d)$, c being the speed of sound and d the test duct diameter.

The maximum values of reflection coefficient allowed by the standards permit differences between Lp_{max} and Lp_{min} as given in Table 1.

Since measurements of fan noise will be made in ⅓-octave bandwidths the variation of sound pressure level with axial distance along the duct will be substantially less than that shown in the final column of that table for the non-tonal components during an actual fan noise test.

3. TERMINATOR DESIGN

3.1 General Features

There are three basic elements in an anechoic termination - an expansion section, a body, and a throttle plate in the case of an outlet side termination or a further set of tapers in the case of an inlet side termination.

The role of the expansion section is to provide a gradual change in the duct impedance, acting as an acoustical transformer between the test duct and the terminator body. When a sound wave is incident on a boundary where there is a sudden change of impedance part of the sound energy is reflected. The acoustic impedance, Z_a, at any surface is the ratio of the acoustic pressure, p, averaged over that surface to the effective volume velocity, U, through it, so that $Z_a = p/U$. Consider Fig. 2. At the boundary between the ducts the sum of the pressures must be equal so that

$$p_i + p_r = p_t \tag{3}$$

and the volume velocities must be conserved so that

$$U_i - U_r = U_t. \tag{4}$$

Since Z = p/U, equation (4) becomes

$$\frac{p_i}{Z_1} - \frac{p_r}{Z_1} = \frac{p_t}{Z_2} \tag{5}$$

and after manipulation of equations (3) and (5) we obtain

$$r = \frac{p_r}{p_i} = \frac{Z_2 - Z_1}{Z_2 + Z_1}. \tag{6}$$

It is necessary to have a transformation section with impedance Z_2 such that the reflection coefficient does not exceed the specified limits.

The role of the terminator body is to absorb or to radiate away from the test duct much of the acoustic energy transmitted to it through the transformation section so that very little is reflected back into the test duct. Since the expansion section has to be an efficient transmitter of sound energy from the test duct to the terminator it follows that it will also be efficient in the reverse direction.

Each element of the terminator is important in the overall design of a successful unit. Referring to Fig. 3, if r_t represents the reflection coefficient of the taper section, r_b the reflection coefficient of the body section then in order to meet the target reflection coefficient R_T we have the following relationship

$$r_t + (1 - r_t)^2 r_b = R_T. \tag{7}$$

Since the reflection coefficient is equal to the ratio of the reflected pressure to incident pressure, the difference in sound pressure levels between incident and reflected sound is equal to 10 log(r^2) dB. In order to achieve a certain overall value for R_T there are many combinations of r_t and r_b as illustrated in Table 2.

While there can be a trade off between the performance of the different components the design of a good taper section is more critical than that of the body in that, once manufactured, modifications can generally be made to improve the performance of the body whereas the tapers cannot readily be modified.

In the following sections brief studies are made of the design features of the expansion sections, the body and the backplate or inlet duct transition.

3.2 Expansion Sections

3.2.1 A simple theory

While there can be an infinite number of transformation profiles linking the test duct to the terminator body probably the most commonly used are conical, exponential or catenoidal in shape. Fig. 4 compares the ideal transmission performance of three such expansions of the same overall dimensions. Above a certain frequency (known as the cut-off frequency) the catenoidal expansion is clearly superior to the other two and the conical section is the least efficient. Since the catenoidal and exponential expansions are geometrically very similar towards the mouth (large diameter end) of the expansion the difference in acoustic performance must be largely accounted for by the rate of change of cross-sectional area at the start, or throat, of the flare. In the catenoidal horn, the rate of change of area at the origin of the expansion is zero and the initial expansion is very gradual. The shape of exponential and catenoidal horns can be described by the equations

$$S(x) = So \exp(mx) \qquad (8)$$

and

$$\text{and } S(x) = So \cosh^2(mx) \qquad (9)$$

respectively, the parameter, m, being a measure of the rate of the flare of the expansion.

For an infinitely long transformation the cut-off frequency, ω_o, is given by $\omega_o = 0.5\ mc$. Fig. 5 is a graphical representation of equation (8) for an exponential expansion. When x is large the catenoidal and exponential profiles are indistinguishable.

When the expansion sections are of finite length their transmission properties depart from those indicated in Fig. 4. The mathematical modelling is rather difficult but fortunately there exists a text book solution for the case of an exponential connector between two infinitely long circular ducts of different diameters which can be used as a basis for deriving the reflection coefficient of the expansion. Olson (Ref. 6) presents the expression for the throat specific acoustic impedance of an exponential connector looking from the small to the large end as

$$\frac{Z(\omega)}{\rho c} = \frac{\cos(bl + \theta) + j\sin(bl)}{\cos(bl - \theta) + j\sin(bl)} \qquad (10)$$

where $b = 0.5(4\omega^2/c^2 - m^2)^{\frac{1}{2}}$,

$\theta = \tan^{-1}(m/2b)$, and

l = length of expansion.

The reflection coefficient, r, of the expansion can then be obtained from the relationship

$$r = \left|\frac{Z/(\rho c - 1)}{Z/(\rho c + 1)}\right|. \qquad (11)$$

Plots showing the resultant reflection coefficient spectra as a function of several lengths and expansion rates are shown in Fig. 6. A study of many similar plots for a range of length and expansion rates shows that in order to meet the ISO limits on reflection coefficient the flare expansion constant has to be less than 0.7 per metre implying that fairly long, gradually-expanding transitions will be necessary.

For all but a few special cases it will be prohibitively expensive to build truly exponential expansion sections and in general practice, transitions would be approximated by a series of conical sections. The acoustic impedance at the throat of an infinitely long cone is given by

$$\frac{Z}{\rho c} = \frac{1 + j\lambda/2\pi x_0}{1 + (\lambda/2\pi x_0)^2} \tag{12}$$

and the reflection coefficient can be derived from equation (11).

The variation of reflection coefficient with the parameter x_0 and frequency is plotted in Fig. 7. To meet the ISO reflection criteria a value of x_0 of greater than about 2 m is required and, to allow for reflections from the terminator body and to account for effects of finite length, a substantial margin of safety should be incorporated.

In the following paragraphs four different horn shapes which have been tested at NEL are analysed with reference to the foregoing theory.

3.2.2 Analysis of experimental data on four expansions

Sketches of the four different expansions are presented in Figs 8 and 9 with some further details summarised in Table 3. Expansion 1 was constructed as per the suggestion in BS 848 : Part II : 1963 (Ref. 7). Expansions 2-4 were designed at NEL as approximations to catenoidal sections. In the catenoidal expansions some of the final conical sections were lined with a tapering noise absorbing polyurethane foam. Part of the reason for lining the sections was to minimise the change in impedance at the junction between the expansion and the terminator body. The initial conical pieces in each catenoidal expansion had 'x_0 values' in excess of 9 m. Each of these expansions was tested using the method indicated in Section 2 as part of a termination using the body and throttle backplate described in Section 3.3.

While the performance of the terminator body has never been independently assessed it is thought that its reflection coefficient is relatively good and not likely to be dominant. Plots of the measured reflection coefficient are presented in Fig. 10 superimposed on curves derived from equations (10) and (11) assuming a perfect exponential expansion. There is sufficiently good agreement between the measured and predicted data to conclude that an expansion design method based on the foregoing is likely to lead to a satisfactory performance. The three catenoidal expansions were probably rather conservatively designed in that the flare constants were in the range 0.3-0.42 per metre and the overall reflection coefficients were well within the specified limits. It is likely that a constant of around 0.5 per metre could be used provided that the attenuation of terminator body is good. In approximating the expansion with conical sections it would appear prudent to design the first section so that the 'x_0 value' was in excess of 8 m, and not to have any section with an 'x_0 value' below 2 m.

3.3 Terminator Body

The role of the terminator body is to absorb most of the sound energy transmitted to it by the transformation section allowing only a small fraction to be reflected back towards the source. Referring to Table 2 and bearing in mind the performance of the expansion sections it would appear that, in general, the differences between the sound pressure levels incident into and reflected from the terminator body have to be in excess of 15 dB at 50 Hz and 25 dB at frequencies at and above 125 Hz. In the case of an inlet duct termination, sound will effectively pass through the body to the inlet of any duct work where it will be reflected back through the termination. The body is thus required to act as a silencer giving transmission losses of around 8 dB at 50 Hz and about 13 dB at frequencies in excess of 125 Hz. Attenuations of this order of magnitude can be achieved by commercially manufactured silencers of length equal to two diameters. For the case of an outlet side terminator fitted with a backplate faced with an acoustically absorbent lining, the transmission loss need not be so great, provided the absorbent material on the backplate is effective.

A drawing of a successful terminator body designed at NEL is shown in Fig. 9. It has a 1330 mm diameter section and is 1730 mm long. There is a fixed lining of 150 mm thick polyurethane foam glued onto the walls. In addition a further 100 mm thick layer can be inserted and held in position using steel bands when smaller diameter test ducts are being used.

3.4 Throttle Backplate

An anechoic termination used on the discharge side of a fan is generally fitted with a throttle device. The requirements of a throttle are that it should control the flow and not generate noise which could interfere with the sound being measured. Probably the most convenient form of throttle is a plate at the end of the terminator body which can be moved axially to vary the annular discharge area. Such a unit is shown in Fig. 9. The plate should be covered by at least 150 mm thickness of noise absorbent lining. Experience has shown that the plate must be quite rigid. Otherwise, it tends to flex and can develop a severe vibration instability. If a high pressure fan is being tested, a considerable axial load can be developed across the backplate and bearings must be adequately sized.

3.5 Inlet Termination

An anechoic termination on an inlet duct is generally fitted to ducting on both sides. On the side remote from the test duct the expansion sections need not have as high a transmission coefficient as on the test duct side and it should be possible to have a much steeper flare on that side.

4. A POSSIBLE ALTERNATIVE FORM OF EXPANSION SECTION

The proposed BS and ISO Standards permit a transition from a duct equal to the fan diameter to duct sizes 44 per cent larger and 16 per cent smaller. Daly (Ref. 8) has shown that within the range of Standard R20 series ducting from 160-1400 mm diameter it is necessary to use only five diameters of test duct and possibly only two terminator bodies. It would thus be necessary to have 'only' five sets of expansion pieces with each duplicated if a wide range of fan sizes were to be tested in a ducted inlet/ducted outlet configuration. With each expansion having four or possibly five elements, the amount of ductwork required clearly becomes very substantial as there is no commonality between any of the conical sections. Some further work possibly using the analysis procedure outlined in this paper might allow a few common elements to be derived but substantial savings are unlikely.

In an attempt to reduce the stock of ducting required, a novel design of expansion section has been considered and tested at NEL. Behind the design was a radical change in operational concept. The conventional philosophy is to have the expansion sections gently tapering so as to minimise sound pressure reflections. In the NEL design the expansion sections are formed from short lengths of stepped, parallel-sided ducting as shown in Fig. 11. Reflections occur at each change in cross section and the anechoic effect is achieved by having the lengths of the sections adjusted so that the reflected waves cancel out. The theoretical background is quite simply illustrated using the example shown in the sketch in Fig. 12. It is assumed that there is no reflected wave of amplitude, B_4. At the boundaries between Sections 1-3 the equations for equality of acoustic pressure and continuity of volume velocity yield the pairs of equations

$$\left.\begin{aligned} A_1 + B_1 &= A_2 + B_2 \\ A_1 - B_1 &= M_1(A_2 - B_2) \end{aligned}\right\} \qquad (13)$$

$$A_2 \exp(-ikl_2) + B_2 \exp(ikl_2) = A_3 + B_3$$
$$A_2 \exp(-ikl_2) - B_2 \exp(ikl_2) = M_3(A_3 - B_3) \quad (14)$$

$$A_3 \exp(-ikl_3) + B_3 \exp(ikl_3) = A_4$$
$$A_3 \exp(-ikl_3) - B_3 \exp(ikl_3) = M_4 A_4. \quad (15)$$

Manipulation of the last pair of equations will enable the variable A_4 to be eliminated and B_3 expressed as a function of A_3. A_3 can then be eliminated by manipulating the second pair of equations leaving B_2 expressed as a function of A_2. Then finally, by manipulating the first pair of equations to eliminate A_2, the ratio B_1/A_1 can be obtained as a function only of section lengths and expansion ratios. Appendix A describes this procedure in more detail. It is clear how the calculation may be extended to a much larger number of expansion sections. A number of units have been tested both at NEL and at Woods of Colchester Ltd and the measured reflection coefficient spectra are plotted on Fig. 13. While some of the reflection coefficients are greater than the specified limits, the performance is sufficiently encouraging to justify further development work. If this stepped type of expander can be used, the overall length of the terminator is shortened and the stock of expansion ducting is dramatically reduced.

5. A SIMPLE DESIGN PROCEDURE

The following design steps should enable a termination with a conventional, gradually increasing expansion section to be constructed to meet the reflection coefficient criteria of the proposed new standards.

For a specific size of test duct:

a Select the diameter of the terminator body. Experimentally it has been shown that a 1300 mm diameter body proved satisfactory for test duct diameters of 600-900 mm. It is likely that a diameter of about 1000 mm would be suitable for test ducts below 400 mm diameter and a 2500 mm diameter body should be adequate for a test duct diameter of 2000 mm.

b Choose a flare constant m of not greater than 0.5 per metre and, assuming a catenoidal expansion, draw the profile to scale.

c Approximate the profile using conical sections ensuring that for the first section the x_o value is greater than 8 m and for no section is x_o less than 2 m.

d Make the terminator body 1½-2 diameters long.

e Line the terminator body with at least 150 mm thick acoustically absorbent material and taper the lining into one or two of the adjacent conical expansion sections.

f Line the backplate, if fitted, with at least 150 mm thick acoustically absorbent material.

g If, on test, the unit fails to meet the reflection criteria, increase the amount of absorption in the terminator body, for example by the inclusion of wedges mounted on the backplate, by an increase in lining thickness or by an increase in length.

6. CONCLUSIONS

This paper has examined some of the features of anechoic terminator design. The role of the profile of the expansion sections in the transformation from the test duct to the terminator body has been studied using a simple theory. The agreement between the reflection coefficients predicted by the theoretical model and those experimentally measured on four different terminator units was shown to be sufficiently good as to confirm the use of the model as a design tool. Based on the analysis of experimental work a design procedure has been outlined.

The design and the theoretical basis of a novel form of terminator expansion involving a stepped, parallel-sided expansion section has been presented. The experimentally measured performance is very close to meeting the reflection coefficient criteria specified by proposed International and British Standards and the potential savings in the required stocks of test ductwork justify its further development.

7. ACKNOWLEDGEMENTS

This paper is contributed with the permission of the Director of the National Engineering Laboratory, Department of Industry. It is Crown copyright. The work was supported by the Mechanical Engineering and Machine Tools Requirements Board. The directors of Woods of Colchester Ltd are thanked for their permission to present some of their experimental data.

8. REFERENCES

1. British Standards Institution.: "Draft revision of BS 848. Fans for general purposes. Part 2: Methods of noise testing". (Draft in preparation.)

2. International Standards Organisation.: "ISO/DIS 5136. Acoustics - determination of sound power levels with flow from or by fans radiated into a duct: in-duct method". ISO/TC43/SC1/WG3 N59, revised June 1980.

3. Cremer, L.L.: "The treatment of fans as black boxes". Journal of Sound and Vibration, 16, 1, 1971, pp 1-15.

4. Baade, P.K.: "Effects of acoustic loading on axial flow fan noise generation". Noise Control Engineering, 8, 1, 1977.

5. Deeprose, W.M.: Implications for the fan industry of the new BS 848 Standards. Int. Conf. on Fan Design and Application, Guildford, England. September 1982. BHRA, Bedford.

6. Olson, H.F.: "Elements of acoustical engineering". 2nd edition. New York: Van Nostrand Co Inc., 1947, pp 109-111.

7. British Standards Institution.: "Fans for general purposes. Part 2: Methods of noise testing". BS 848 : Part II : 1963.

8. Daly, B.B.: Submission to BSI Committee MEE/185/-/2, 17 June 1981.

APPENDIX A

SOLUTION TO EQUATIONS FOR THE REFLECTION COEFFICIENT OF A SIX-SECTION STEPPED EXPANSION

As an example this appendix considers the terminator expansion shown in Fig. 11 which is called a six-section expansion. Using the notation as given in Fig. 12, the area exapnsion ratio of the first section is written as M_2 and is equal to $(0.71/0.61)^2$, while the area expansion ratio of the terminator body is written as M_8 and is equal to $(1.33/1.25)^2$. The length of the first stepped section is written as l_2 andis 0.24 m long (while the last section is denoted l_7 and is 0.32 m long). In the notation adopted here the subscript 1 refers to the test duct with the subscript 8 referring to the terminator body.

At the first junction the equations for pressure and velocity are

$$\left.\begin{aligned} A_1 + B_1 &= A_2 + B_2 \\ A_1 - B_1 &= M_2(A_2 - B_2) \end{aligned}\right\} \qquad \text{(A1)}$$

At the junction of the fifth and sixth sections the equations are

$$\left.\begin{aligned} A_6 \exp(-ikl_6) + B_6 \exp(ikl_6) &= A_7 + B_7 \\ A_6 \exp(-ikl_6) - B_6 \exp(ikl_6) &= M_7(A_7 - B_7) \end{aligned}\right\} \quad \text{(A2)}$$

and at the entrance to the terminator body they are

$$\left.\begin{aligned} A_7 \exp(-ikl_7) + B_7 \exp(ikl_7) &= A_8 \\ A_7 \exp(-ikl_7) - B_7 \exp(ikl_7) &= M_8 A_8. \end{aligned}\right\} \quad \text{(A3)}$$

Manipulation of the last pair of equations yields

$$B_7 = A_7 \left\{\frac{1 - M_8}{1 + M_8}\right\} \exp(-2ikl_7)$$

$$= A_7 C_7 \quad \text{(A4)}$$

where

$$C_7 = \left\{\frac{1 - M_8}{1 + M_8}\right\} \exp(-2ikl_7). \quad \text{(A5)}$$

Substituting this value for B_7 into equations A2 results in

$$B_6 = \frac{A_6\{(1 + C_7) - M_7(1 - C_7)\}}{(1 + C_7) + M_7(1 - C_7)} \exp(-2ikl_6)$$

$$= A_6 C_6 \quad \text{(A6)}$$

where

$$C_6 = \frac{(1 + C_7) - M_7(1 - C_7)}{(1 + C_7) + M_7(1 - C_7)} \exp(-2ikl_6). \quad \text{(A7)}$$

Working through the equations yields of the form of A6 and A7 for all subsequent sections. Thus

$$B_5 = A_5 C_5 \text{ with } C_5 = \frac{(1 + C_6) - M_6(1 - C_6)}{(1 + C_6) + M_6(1 - C_6)} \exp(-2ikl_5) \quad \text{(A8)}$$

$$B_4 = A_4 C_4 \text{ with } C_4 = \frac{(1 + C_5) - M_5(1 - C_5)}{(1 + C_5) + M_5(1 - C_5)} \exp(-2ikl_4) \quad \text{(A9)}$$

$$B_3 = A_3 C_3 \text{ with } C_3 = \frac{(1 + C_4) - M_4(1 - C_4)}{(1 + C_4) + M_4(1 - C_4)} \exp(-2ikl_3) \quad \text{(A10)}$$

$$B_2 = A_2 C_2 \text{ with } C_2 = \frac{(1 + C_3) - M_3(1 - C_3)}{(1 + C_3) + M_3(1 - C_3)} \exp(-2ikl_2) \quad \text{(A11)}$$

$$B_1 = A_1 C_1 \text{ with } C_1 = \frac{(1 + C_2) - M_2(1 - C_2)}{(1 + C_2) + M_2(1 - C_2)} \quad \text{(A12)}$$

The reflection coefficient of the expansion is the absolute value of $B_1/A_1 = C_1$. It is found by successively solving equation A5, then A7, then A8, A9, A10, A11 and finally A12. It must be noted that all the equations are complex, that is, they have both real and imaginary parts.

The forms of equations A7-A12 makes it clear how any number of expansion sections may be dealt with.

TABLE 1

ANECHOIC TERMINATION REFLECTION COEFFICIENT CRITERIA

$\frac{1}{3}$-octave band centre frequency	Maximum pressure reflection coefficient	$Lp_{max} - Lp_{min}$ dB
50	0.4	7.4
63	0.35	6.4
80	0.30	5.4
100	0.25	4.4
>125	0.15	2.6

TABLE 2

VARIATION OF COMPONENT REFLECTION COEFFICIENTS

Reflection coefficient			Sound pressure level difference into and out of terminator body
Target R_T	Taper section r_t	Body section r_b	10 $\log(r_b^2)$
0.4	0.3	0.2	14
	0.2	0.3	10
	0.1	0.37	8.5
0.15	0.1	0.06	24
	0.075	0.09	21
	0.05	0.17	15.5

T A B L E 3

DETAILS OF EXPANSION SECTIONS

Terminator	1	2	3	4
First section				
Section length, mm	300	800	1 000	1 400
Initial and final diameter, mm	610/760	355/380	500/530	710/817
Parameter 'x_o', mm	1 220	11 360	16 700	9 300
Lining	No	No	No	No
Second section				
Section length, mm	250	800	1 400	
Initial and final diameter, mm	760/910	380/430	530/650	
Parameter, x_o, mm	1 267	2 020	5 700	
Lining	No	No	No	
Third section				
Section length, mm	210	1 600	1 400	1 200
Initial and final diameter, mm	910/1070	430/700	650/900	817/1062
Parameter, x_o, mm	1 194	2 540	3 640	4 000
Lining	No	Yes	Yes	Yes
Initial thickness, mm		5	5	5
Final thickness, mm		113	77	119
Fourth section				
Section length, mm	200	2 000	1 200	1 000
Initial and final diameter, mm	1070/1220	700/1330	900/1330	1062/1330
Parameter, x_o, mm	1 427	2 200	2 510	3 960
Lining	No	Yes	Yes	Yes
Initial thickness, mm		113	77	119
Final thickness, mm		250	250	250
Approximation to exponential - overall value of m mm^{-1}	1 500	420	300	350

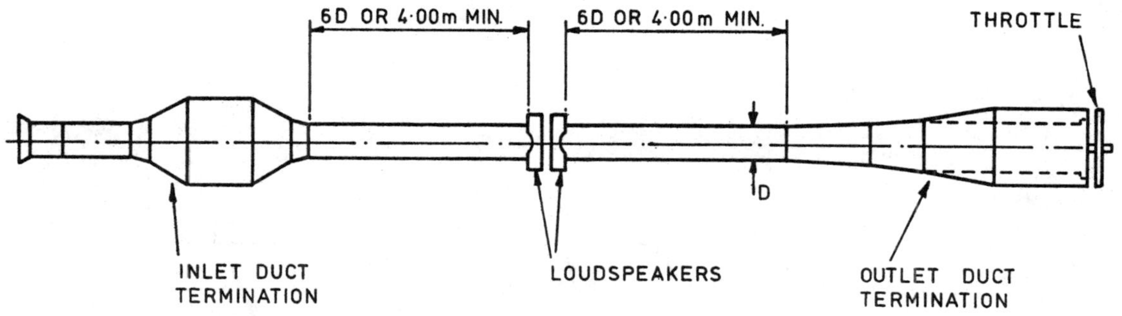

1 Test arrangement for measurements of reflection coefficients

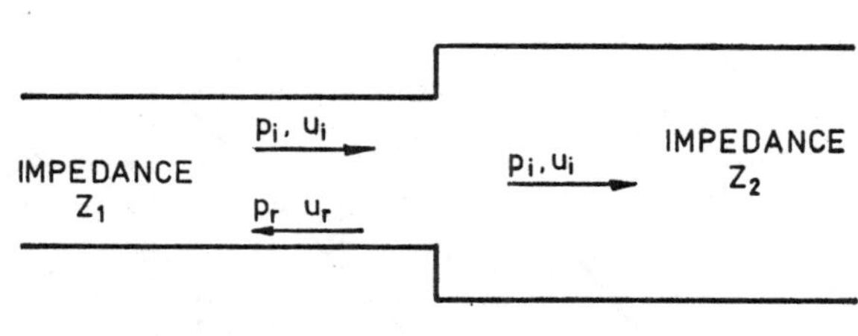

2 Sound transmission at a boundary

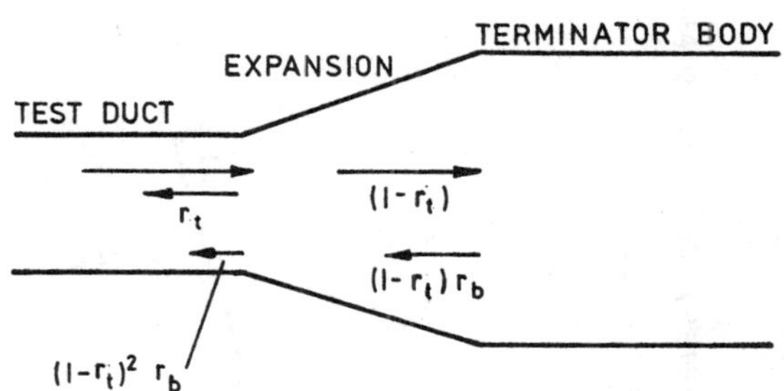

3 Overall reflection coefficient of taper section and body

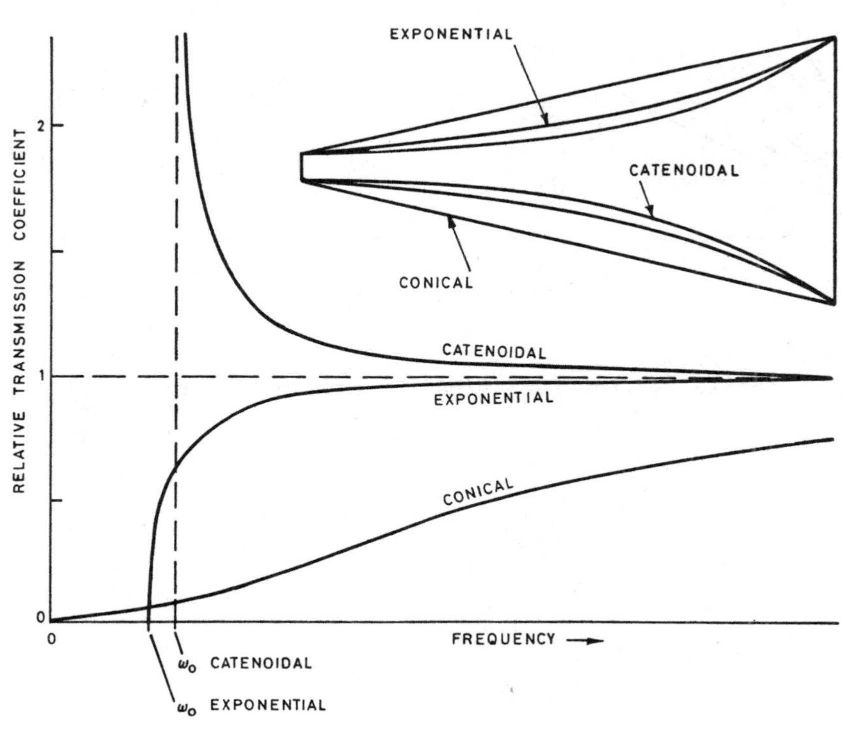

4 Comparative transmission performance of three expansion sections

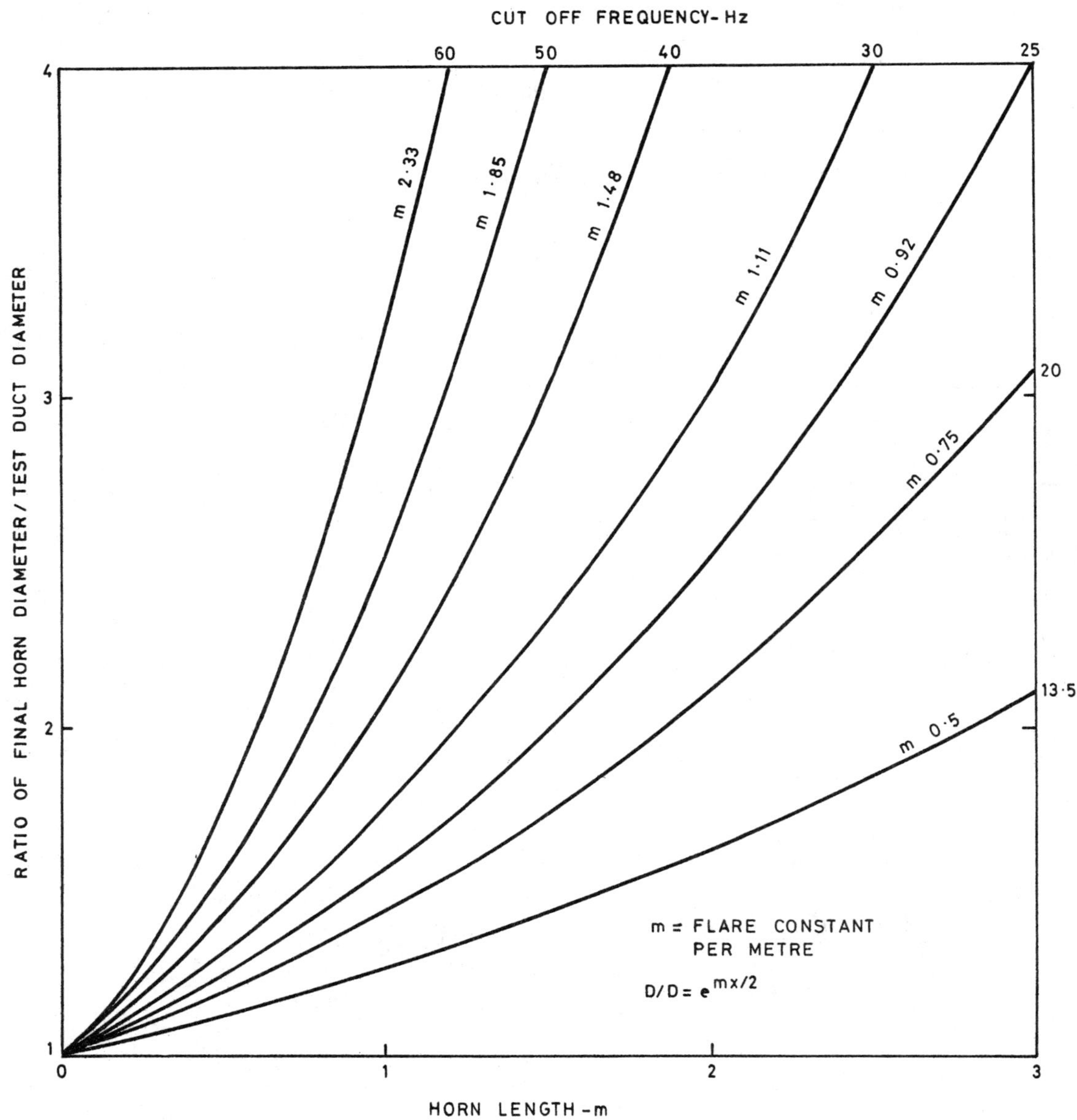

5 Profile of transformation for exponential expansion

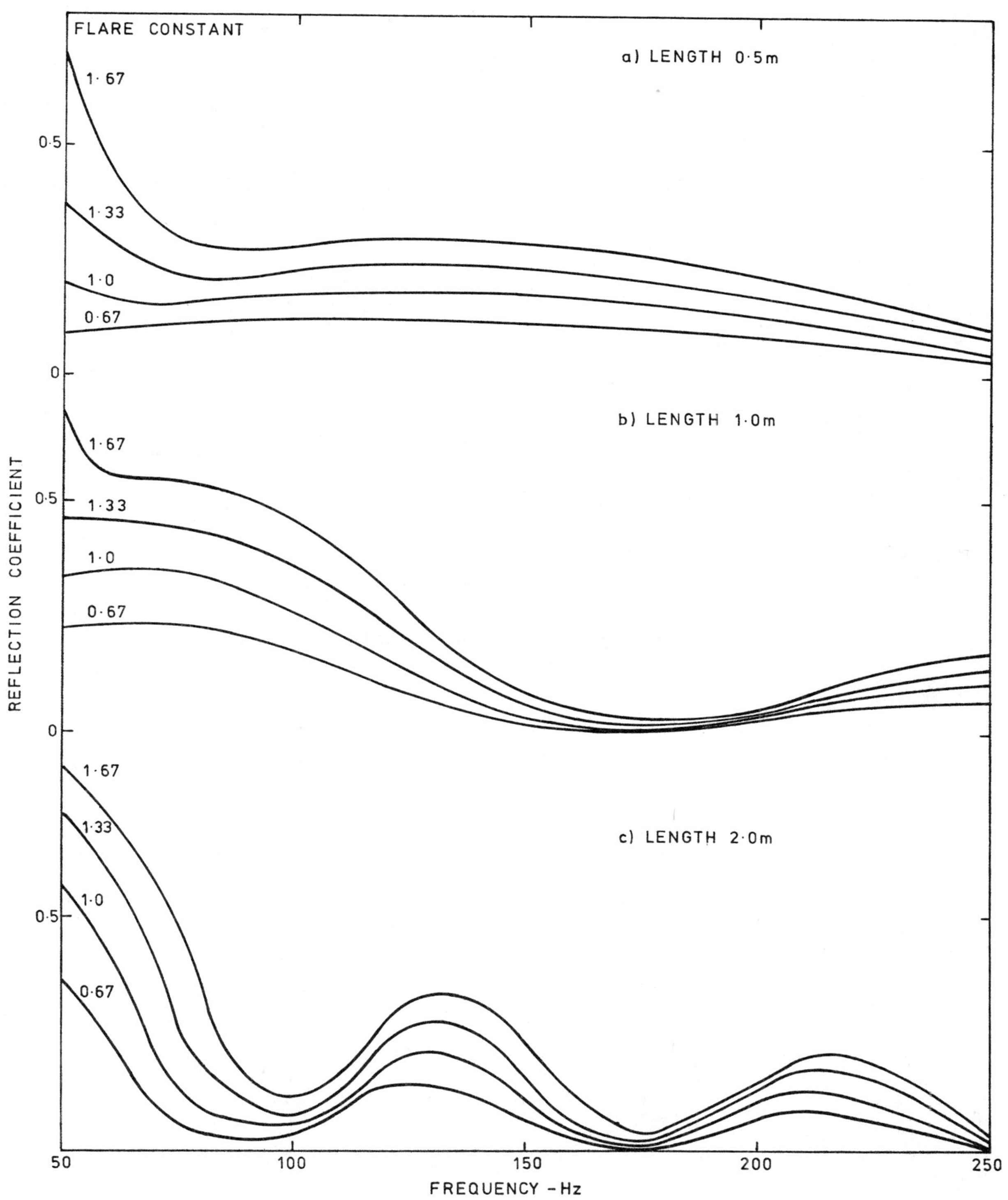

6 Reflection coefficient spectra as functions of length and expansion rate

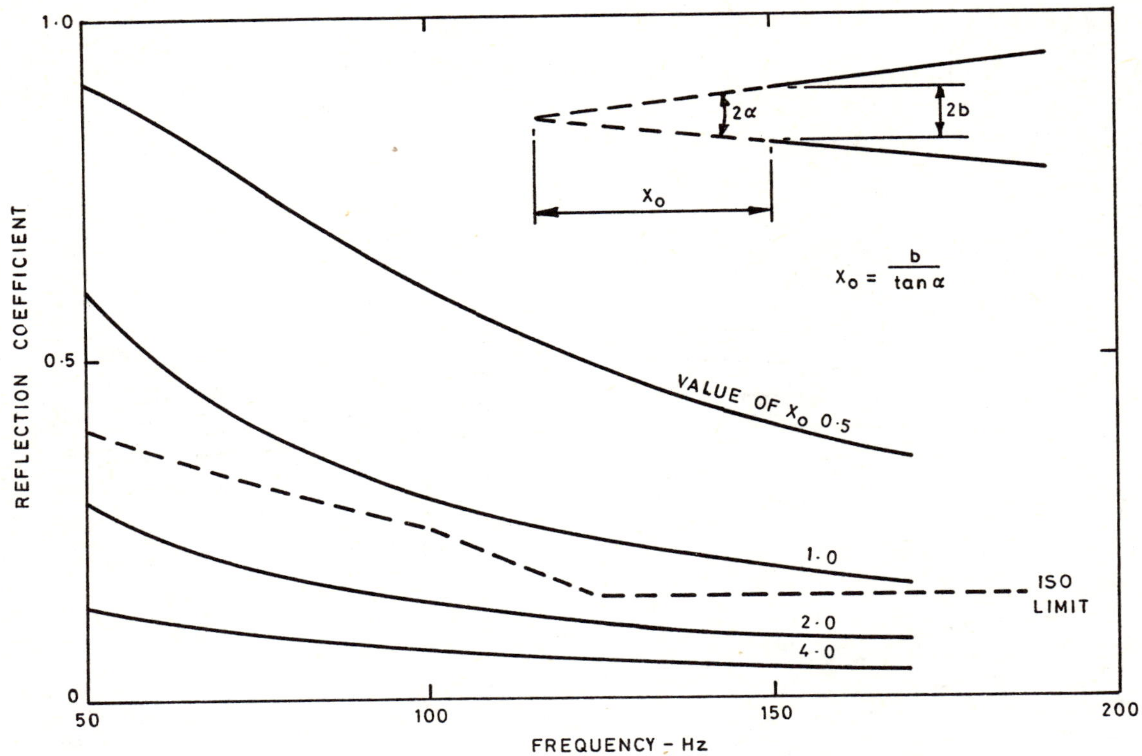

7 Reflection coefficient of an infinite conical horn

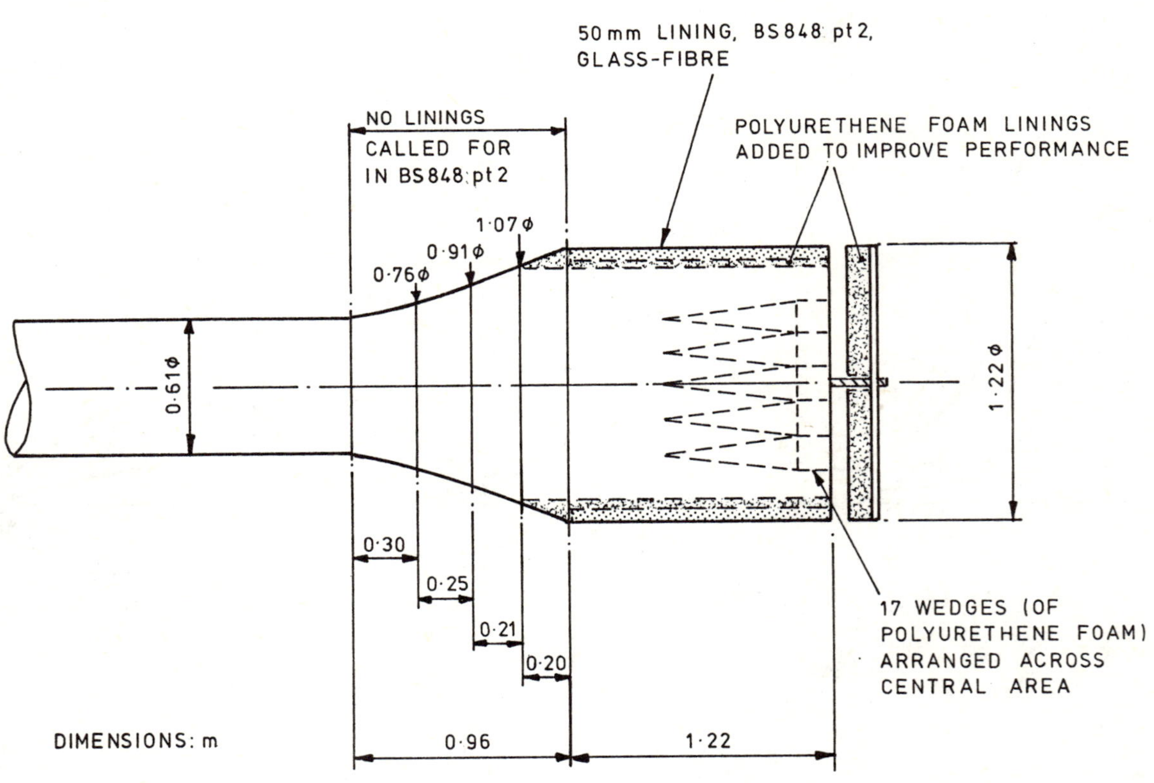

8 Sketch of a BS 848 Part II 1963 type of terminator

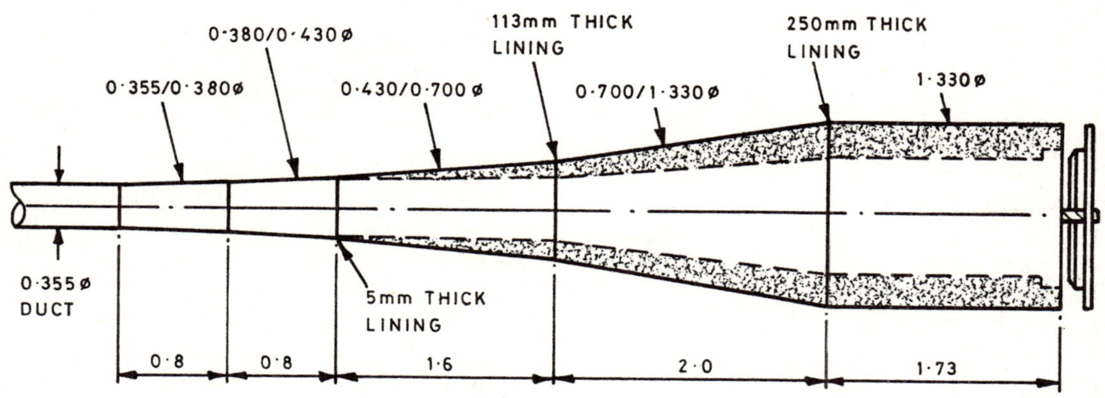

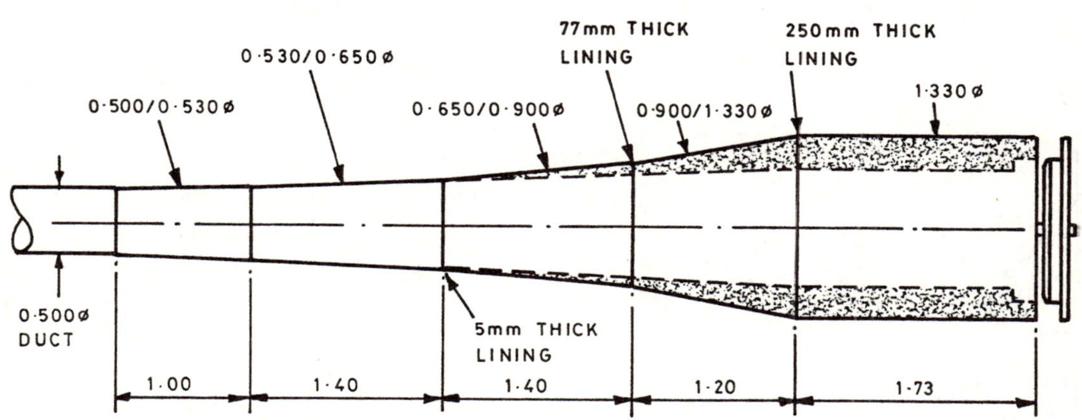

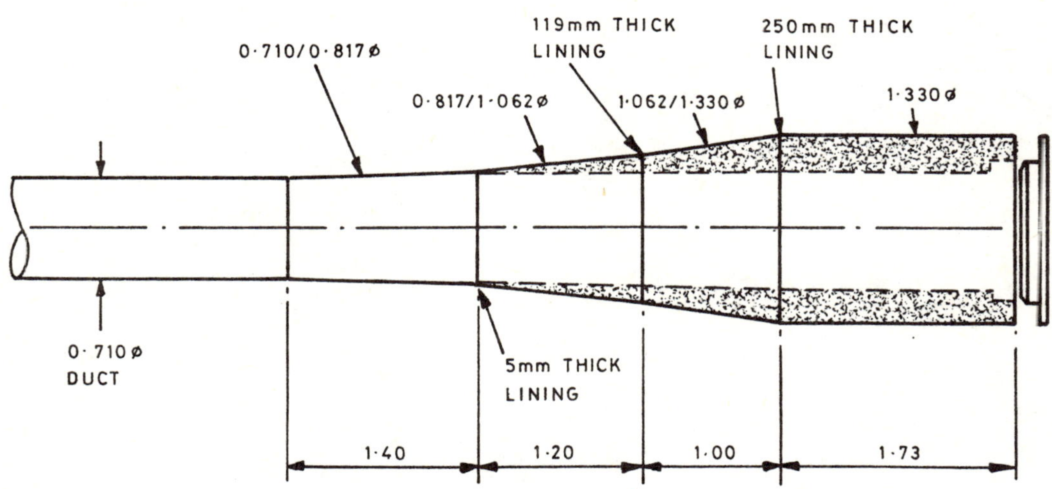

9 Sketches of three catenoidal designs of terminator

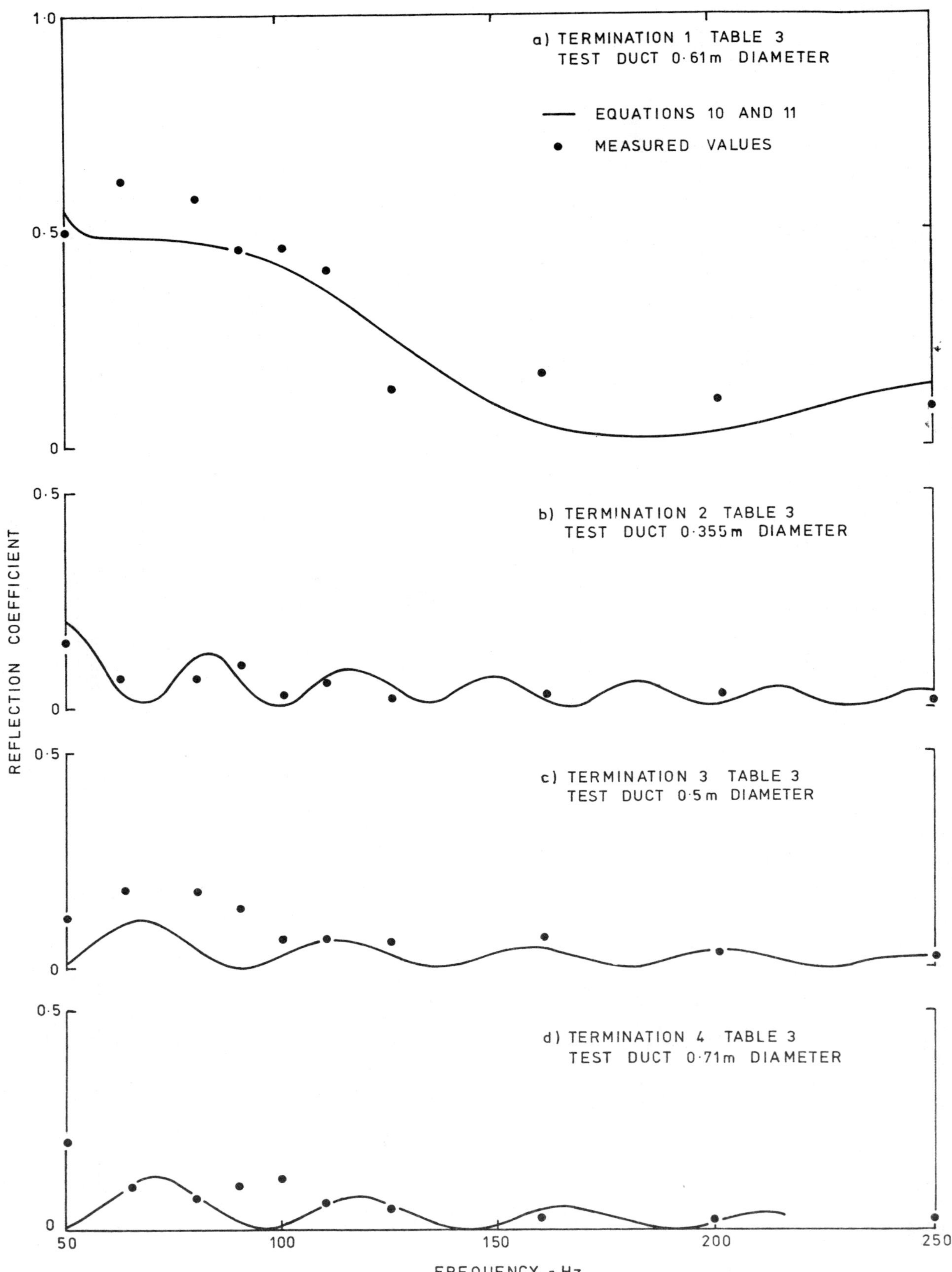

10 Comparison of measured and estimated terminator performance

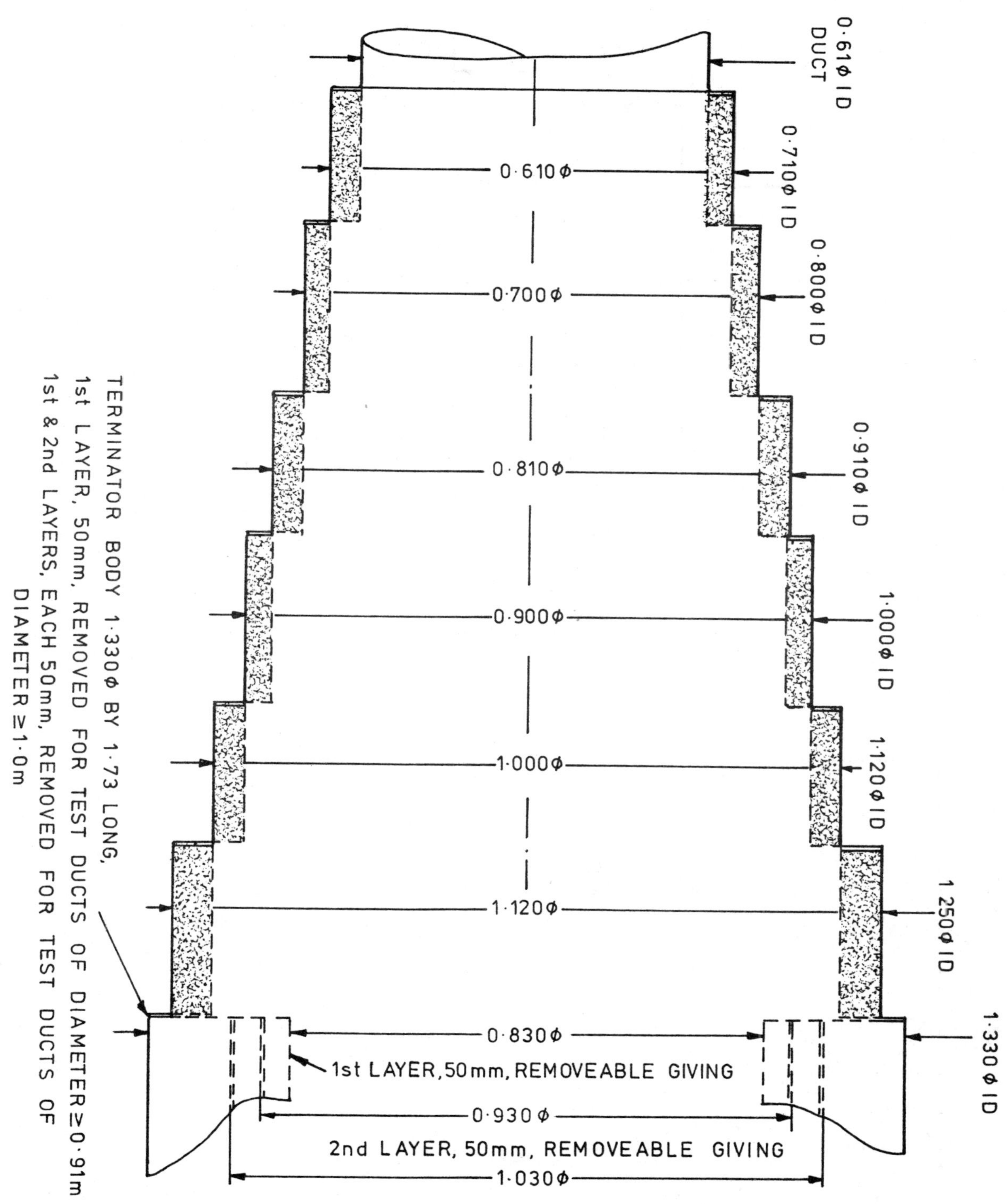

11 Sketch of stepped termination

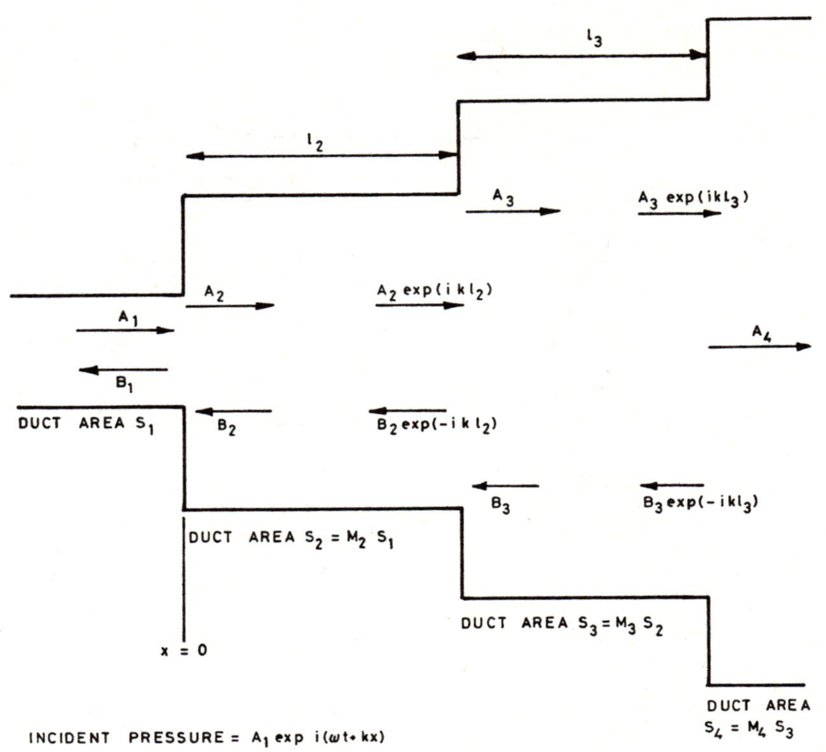

12 Notation for transmission through a stepped expansion

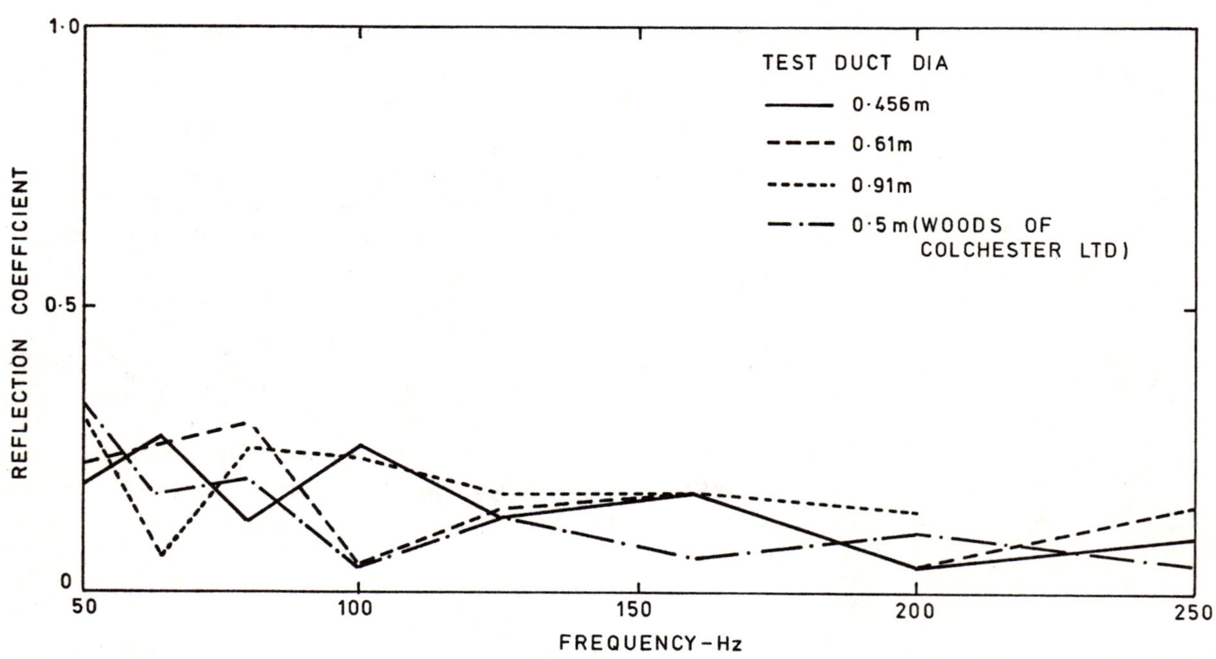

13 Measured performance of terminator with stepped transitions.

International Conference on

Fan Design & Applications

Guildford, England: September 7-9, 1982

PAPER H2

MEASURING THE NATURAL FREQUENCIES OF COMPONENT PARTS OF INDUSTRIAL FAN IMPELLERS

D. Banyay and L. Gutzwiller

Robinson Industries Inc., U.S.A.

Summary

A method of measuring the natural frequencies of component parts of fan impellers is presented. Even with a relatively low investment in testing equipment, accurate and useful results may be determined using simple test procedures. The data gathered is valuable in preventing failure due to vibratory stress and fatigue or in solving problems that already exist in field installations.

Organised and sponsored by
BHRA Fluid Engineering, Cranfield, Bedford MK43 0AJ, England.

0263 - 421X/82/01 00 - 0001 $5.00
The entire volume can be purchased from
BHRA Fluid Engineering for $82.00

NOMENCLATURE

Hz ... frequency, cycles per second

R.P.M. ... rotational speed, revolutions per minute

O.D. ... outside diameter

I.D. ... inside diameter

mV ... millivolts

S_B ... accelerometer sensitivity, $^{mV}/(m/S^2)$

m ... meters

S ... seconds

S_A ... hammer sensitivity, $^{mV}/_{N}$

N ... newtons

X ... analyzer transfer function, $\frac{mV\ (Channel\ B)}{mV\ (Channel\ A)}$

C ... compliance, $^{m}/N$

MEASURING THE NATURAL FREQUENCIES OF COMPONENT
PARTS OF INDUSTRIAL FAN IMPELLERS

1. INTRODUCTION

The various components of large industrial fan impellers have natural resonant frequencies. It is possible to excite the components at their natural frequencies during operation. A simple example is the blade of an axial flow impeller vibrating as a cantilever beam in its first mode as shown in Fig. 1 (Ref. 1).

This type of vibration is not to be confused with unbalance, bearing vibration measurements, or foundation vibration. It is related to the shaft "critical speed" or shaft natural frequency since the impeller/shaft as a system has natural frequencies that one should avoid exciting (See Fig. 2). Similarly, other systems in the impeller (such as the blade of an axial flow unit) have natural frequencies that one must also avoid exciting during operation of the fan in its system.

We all know the result of operating a fan at a rotating speed that coincides with the shaft natural frequency--very high amplitude vibration, extreme sensitivity to unbalance, damage to the bearings and/or shaft and if damping is insufficient, the possibility of catastrophic failure. In the same way any single component of a fan impeller; that is the blade, shroud (rim) or webplate (center disk); is subject to unexpected failure if excited at its natural frequency by some outside force. The purpose of this paper is to explain how one may avoid such failures using the following approach:

1. Measure the vibration response of the various components.
2. Anticipate the frequencies that will be excited in the field installation.
3. Change the natural frequency of some component (if necessary) to avoid excitation in the field.

As a result of our repairing and/or rebuilding both centrifugal and axial flow impellers manufactured by many different suppliers, we have seen a number of impeller failures that resulted from resonant frequency problems.

2. CASE HISTORIES

2.1 Large axial impeller

A 98 inch diameter (2489 mm) axial flow impeller from a large sintering plant developed cracks in the blades (See Fig. 3). In one case, the blade tore completely loose from the center hub! A vibration analysis showed that there were three important natural frequencies associated with blade movement. Mode #1 is the cantilever bending motion of the blade and occurs at a frequency of 92 Hz (See Fig. 4). The second blade bending mode has a node just beyond the heavier cross-section portion of the blade and occurs at 124 Hz (See Fig. 5). The third mode has two nodes as shown (See Fig. 6) and occurs at 372 Hz. The impeller had 10 blades and had been operating in a speed range from 680 to 750 R.P.M. This means the blade passage frequency was 113 to 125 cycles per second which includes the 2nd. mode bending frequency of 124 Hz. The failure occured at the node shown in mode #2. Analysis shows this to be a highly stressed area and with the high number of fairly large amplitude stress reversals at the 2nd. mode natural frequency, cracking and failure occured. Redesigning the blade with a stiffer cross-section raised the natural frequency of the primary mode to 152 Hz and solved the problem.

2.2 Double inlet centrifugal impeller

A 72 inch diameter (1829 mm) double inlet impeller with eight radial blades, flat shroud (rims), and a "spider" type center webplate was shipped to us for repair. One of the shrouds was cracked completely through and several of the blades had developed cracks near the shroud weld (See Fig. 7). Vibration testing showed two significant mode shapes (See Figs. 8 through 9).

The operating speed of the fan impeller was 1180 R.P.M. which gave a blade passage frequency of 157.3 Hz -- almost exactly matching the second bending mode frequency. During rebuilding, the wheel was modified by tapering the shroud and adding stiffener bars. The resulting impeller had only one large response peak at 120 Hz (See Fig. 11). This rebuilt wheel has now been in service for over six years with no further evidence of cracking or failure.

2.3 Variable speed operation

Variable speed operation increases the need to investigate the natural frequencies of various components. Two 70 inch diameter (1778 mm) centrifugal fans in a fixed speed application operated satisfactorily for 18 months. The customer changed to a variable speed drive and suddenly cracks developed in the blades of both units (See Figs. 12 and 13).

One impeller even lost a blade completely causing bearing and shaft damage. Using an FM tape recorder for data collection, a vibration analysis was run on the second wheel in the field. Analyzing the data on the real time analyzer back in the laboratory, we discovered several large vibration peaks that could coincide with the blade passage frequency as the fan operating speed varied over the usual operating range (See Fig. 14). Of primary concern was the response of the shroud at 152 Hz. This was the primary shroud panel movement with associated blade bending that could have been an important contributor to the failure. The customer was cautioned to limit the operating speed of his remaining impeller to either 800-850 R.P.M. or 925 - 1000 R.P.M., which result in blade passage frequency ranges of 160 to 170 Hz and 185 to 200 Hz, while a new impeller was being built. Modified with stiffening members (See Fig. 15), the new wheel was tested and the shroud was found to have no significant peaks below 440 Hz, well above any possible blade passage frequency (See Fig. 16). This rebuilt wheel has continued to operate without failures throughout the range of variable speed operation.

3. IMPORTANT FREQUENCIES EXCITED DURING OPERATION

Measuring the natural frequencies of the component parts is only part of the job of preventing failures. In addition, one must anticipate which frequencies are likely to be excited in the field.

3.1 The operating speed is always of concern. However, most component parts have natural frequencies well above the operating frequency of industrial fans.

3.2 The blade passage frequency is the primary concern as a source of excitation. This is the number of blades times the operating frequency. As seen in the three cases I have discussed, this frequency quite often falls in a range that might coincide with the natural frequency of component parts.

3.3 Rotating stall sometimes occurs in a system with a partially closed inlet damper. Typically, this results in a vibration frequency of approximately two thirds of the operating speed.

3.4 Duct induced vibration due to stack length, duct work turns, unusual fan inlet configuration, etc. may be a source of low frequency excitation.

3.5 Mechanical drive vibration caused by bearing problems, coupling misalignment, the drive gearbox, bent fan or motor shaft and a miriad of other sources can result in exciting a fan component's natural frequency.

3.6 The shaft torsional frequency can be an important exciting force especially in variable speed drive applications.

Hopefully, the importance of considering the natural frequencies of impeller parts is clear. Now how do you determine these frequencies? What investment in equipment is necessary? The following discussion will explain our test method in more detail.

4. TESTING METHOD

4.1 Impulse excitation vs. swept-sine excitation

This procedure utilizes the impact (or impulse) excitation technique. This is relatively new compared to the more common swept-sine wave excitation and offers several advantages (See Fig. 17).

First, it is much more convenient because the amount of time and equipment is greatly reduced (See Fig. 18). The impulse hammer excites a wide frequency range in one strike; whereas, the swept-sine wave technique involves a slow process of spanning through that range. The shaker and corresponding hardware used in the swept-sine wave procedure is all replaced with a simple impulse hammer with integral force transducer and a small power supply. Therefore, both the setup time and actual test time are much less (See Fig. 19).

Another advantage is the excitation point can be easily moved anywhere on the fan impeller. This provides a much better response for each individual component and allows the mechanical compliance (displacement per unit input force) to be determined anywhere on the impeller. The shaker can usually only be attached to one point with its output oriented in one direction. If that particular point happens to be a node for a frequency of interest, then the full response at the frequency will be distorted.

The main limitation with impact testing is that its accuracy is very sensitive to structural non-linearities. However, this is of minimal concern with integrally welded fan impellers (Ref. 2).

4.2 Impeller preparation

All that is required to prepare the impeller for testing is to hang it with the shaft centerline in a horizontal plane (See Fig. 20). The impeller/shaft assembly should be centerhung between two crane straps (or other "soft" supports) so that there is no contact between the fan and the floor or any other obstruction. If the fan is an overhung arrangement, a dummy shaft should be temporarily fixed to the non-drive side to support it or it can be mounted as illustrated in Fig. 21. The test area should be free of ambient noise and vibration.

4.3 Measurement point locations

The following locations are normally tested on a centrifugal fan (See Fig. 22):

1. Shroud O.D. -- midway between two blades
2. Shroud I.D. -- midway between two blades
3. Web O.D. -- midway between two blades
4. Blade tip -- midway between web and shroud
5. Blade inlet -- midway between hub and shroud
6. Torsional Mode

On axial fans, only the blades are tested. However, several points on the blade are tested which allows mode shapes to be determined at each resonance (See Fig. 1).

Test points on impellers other than the centrifugal and axial fans described above must be determined individually.

4.4 Test setup and equipment

Fig. 19 is a photograph of a typical test setup showing the blade tip of a centrifugal fan being tested. This shows all the necessary instrumentation for the procedure, which is discussed below.

Instrumentation: (All included in photo, see Fig. 19)

A. Dual-channel spectrum analyzer---Hewlett Packard #3582A

This is a fast Fourier transform real time analyzer which accepts the incoming signals and processes them into the frequency domain almost instantaneously.

B. Four-channel FM tape recorder --- Dallas Instruments #T-5

This is an option which facilitates field testing without the need for having the spectrum analyzer along. The test data can be conveniently recorded for future analysis in the lab and is permanently stored on cassette tapes.

C. Impact hammer ------------------- PCB Piezotronics #086A03 (including force transducer and integral amplifier)

A single strike with this hammer excites the test specimen over a wide frequency range and the output of the force transducer is proportional to the impact force.

D. Power supply for impact hammer -- PCB Piezontronics #480A

E. Accelerometer ------------------- Bruel & Kjaer #4366

The output signal is proportional to the acceleration of the test specimen.

F. Charge amplifier for accelerometer -- Bruel & Kjaer #2635

G. X-Y Plotter --------------------- Hewlett-Packard #7015B

This can provide a hard copy of the data displayed on the spectrum analyzer screen.

4.5 Test procedure

For each location, the accelerometer is attached to one face of the metal and the hammer strikes the other face, directly opposite the accelerometer (See Fig. 23). Where hollow sections occur or where tight clearances prohibit access to both faces of the metal, the hammer can strike directly beside the accelerometer with minimal effort on accuracy. However, it should be noted that the phase will be shifted 180°.

The most convenient way to effectively mount the accelerometer is with beeswax. This is reliable under most circumstances up to 1 KHz. The use of a magnet is not recommended because there is a risk of the accelerometer jumping upon hammer impact. Under conditions where beeswax is not suitable, such as extreme temperatures or high frequency measurements, other methods such as a screw stud or an adhesive are required.

The hammer should strike each location several times (typically four or eight), allowing time between each hit for the vibration to dissipate. By taking several averages, the effects of ambient vibrations and non-linearities are greatly reduced.

The output of the impact hammer's integral force transducer is connected to Channel A of the spectrum analyzer. This channel is equipped with a threshold signal trigger. The output of the accelerometer is connected to Channel B of the analyzer, through the charge amplifier.

Now a graph of transfer function vs. frequency is displayed on the screen of the spectrum analyzer. An example plot is shown on Fig. 11. The transfer function is essentially the input of Channel B (acceleration) divided by the input of Channel A (force) (Ref. 3). This can be used to calculate compliance vs. frequency as shown in the appendix.

Another useful function on the analyzer is the phase angle. This is used in two ways. First, for each test point a phase shift indicates a resonant frequency. Theoretically, this is a shift from 180° to 0° (or vice versa) with a value of 90° right at the frequency. An example of a phase vs. frequency plot is shown on Fig. 24.

A second use of the phase relationship is determining mode shapes. This is done by moving the accelerometer to different points on the impeller and having the excitation point constant, or varying the excitation point and leaving the accelerometer at one location. By comparing the phase values of all the points at each resonance, a mode shape can be defined. Fig. 1 shows an example mode shape of an axial impeller blade.

5. ANALYSIS OF TEST RESULTS

Once the measurements are finished, the results are ready to be analyzed. The first thing to look for is resonance at or near one of the fan operationally induced excitation frequencies described above. In order to decide whether a particular component is a problem, a predetermined limit must be set of maximum compliance within a specified percent of an excitation frequency. If this limit is not exceeded, then there is sufficient damping to prevent any problems. However, if this vibration does exceed the limit, then something must be done to correct it.

6. CORRECTIVE ACTION

Normally, an extra brace or liner bar is added which increases the stiffness and, consequently, raises the natural frequency of the component. The mode shape determination described above is quite useful in deciding the type and location of stiffener required. Fig. 15 shows a stiffener ring added to the shroud of a centrifugal impeller.

A less desirable solution is to redesign the impeller with different material thicknesses or cross-section. However, this is expensive and time consuming and could result in other problems, such as increased stresses or excessive rotational moment of inertia.

7. CONCLUSION

The test method described here is a reliable, convenient, and inexpensive way to determine impeller natural frequencies and avoid unnecessary, wasteful, and potentially dangerous failures.

ACKNOWLEDGEMENTS

1. Meldrum, Jay, Field Engineer, Structural Kinematics Division of F. Joseph Lamb Co.

2. Blyzinskyk, John, Staff Engineer, Hewlett-Packard Limited

REFERENCES

1. Reynolds, D.D.: "Test Procedures for the Static and Dynamic Testing of Propeller Fans". ASHRAE Research Project RP266: "Propeller Fan Dynamics", June 29, 1981.

2. Halvorsen, W.G. and Brown, D.L.: "Impulse Technique for Structural Frequency Response Testing". Sound and Vibration, November, 1977, pp. 8-21.

3. Hewlett-Packard Publication No. 5952-8773, "Understanding the HP3582A Spectrum Analyzer".

APPENDIX

Calculation of compliance vs. frequency from test data:

Compliance, $C = \frac{m}{N}$

$$m = \frac{(mV_B)}{(2\pi Hz)^2 \; (S_B)}$$

mV_B = millivolts of Channel B (accelerometer)

S_B = accelerometer sensitivity, $\frac{mV_B}{m/s^2}$

$$N = \frac{(mV_A)}{(S_A)}$$

mV_A = millivolts of Channel A (hammer)

S_A = hammer sensitivity, $\frac{mV_A}{N}$

$$C = \frac{(mV_B) \; (S_A)}{(2\pi Hz)^2 \; (S_B) \; (mV_A)}$$

$$X = \frac{mV_B}{mV_A}$$

$$S_A = 2.00 \; \frac{mV}{N}$$

$$S_B = 10 \; \frac{mV}{m/sec^2}$$

$$C = \frac{X}{197.2 \; (Hz)^2}$$

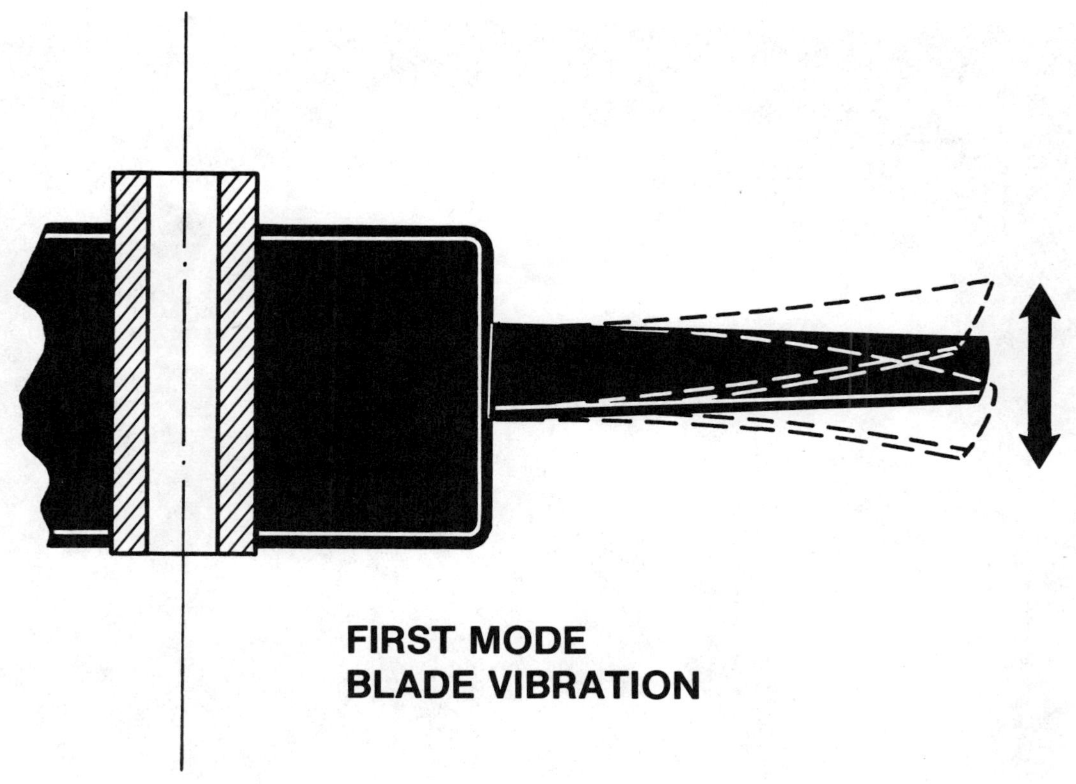

1. Axial flow impeller blade vibrating in first mode

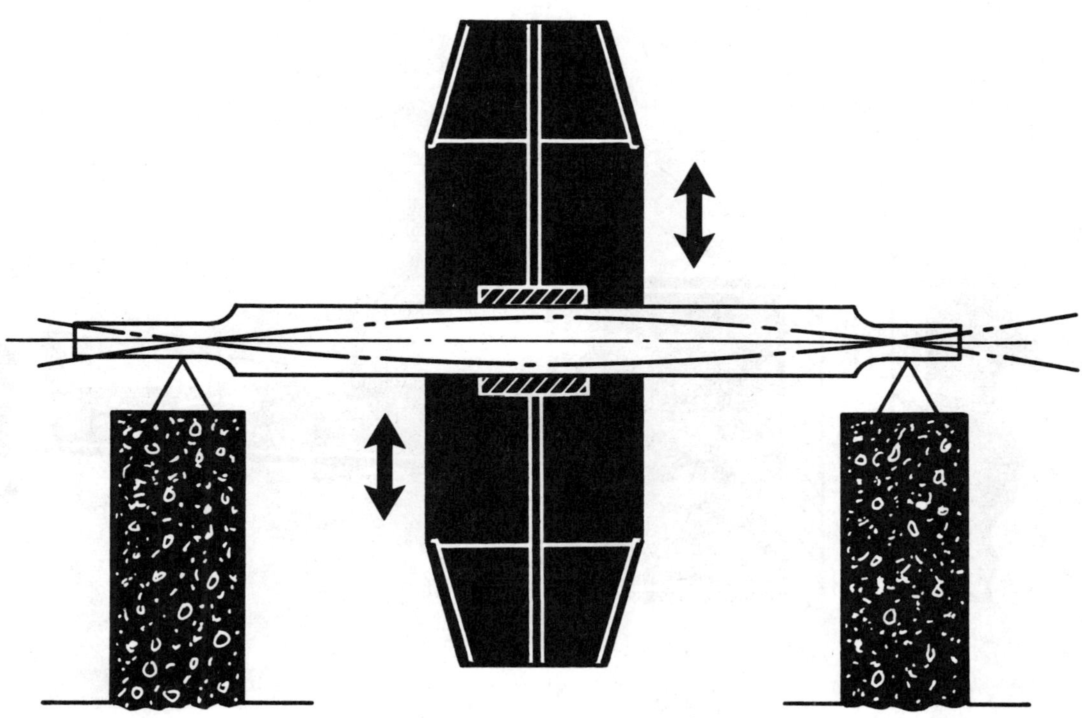

2. Fan impeller/shaft vibrating at its natural frequency

3. 98 inch diameter axial flow impeller

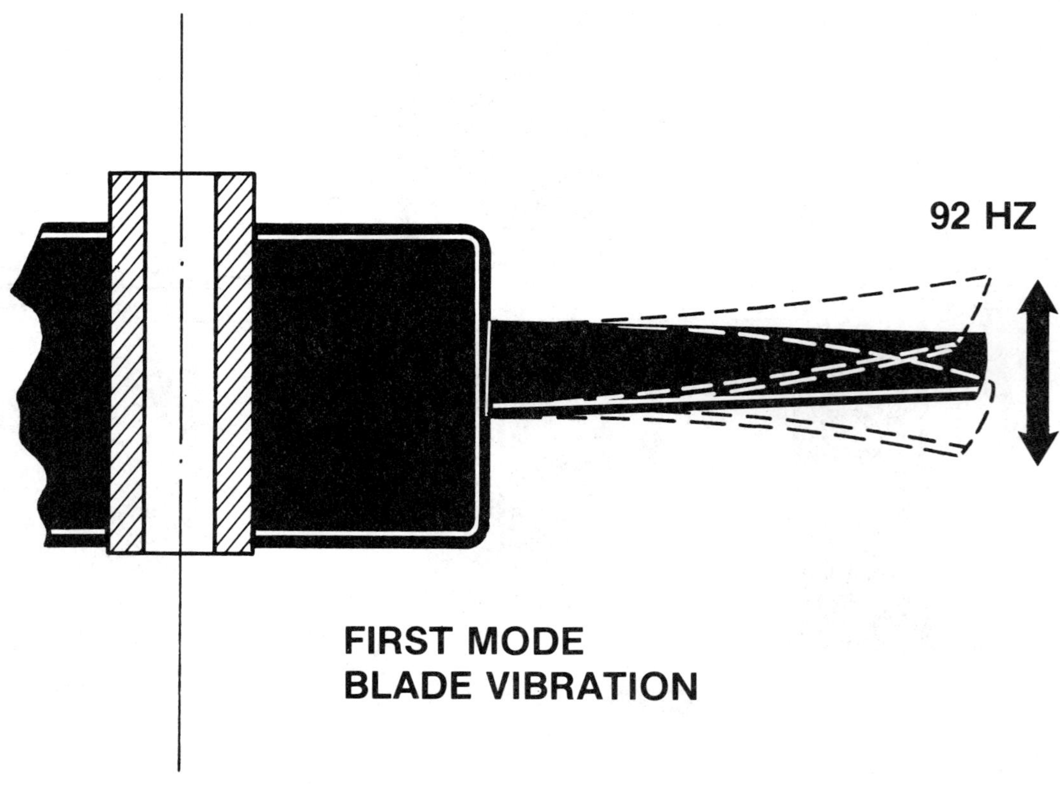

4. Mode #1 vibration of 98 inch axial impeller blade

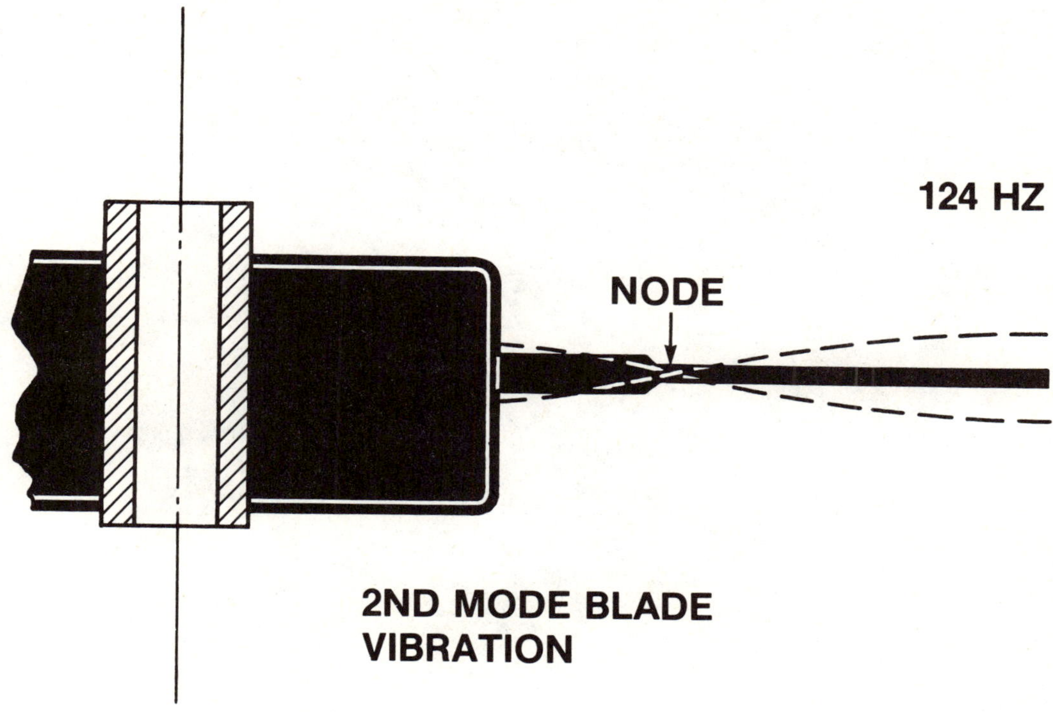

5. Mode #2 vibration of 98 inch axial impeller blade

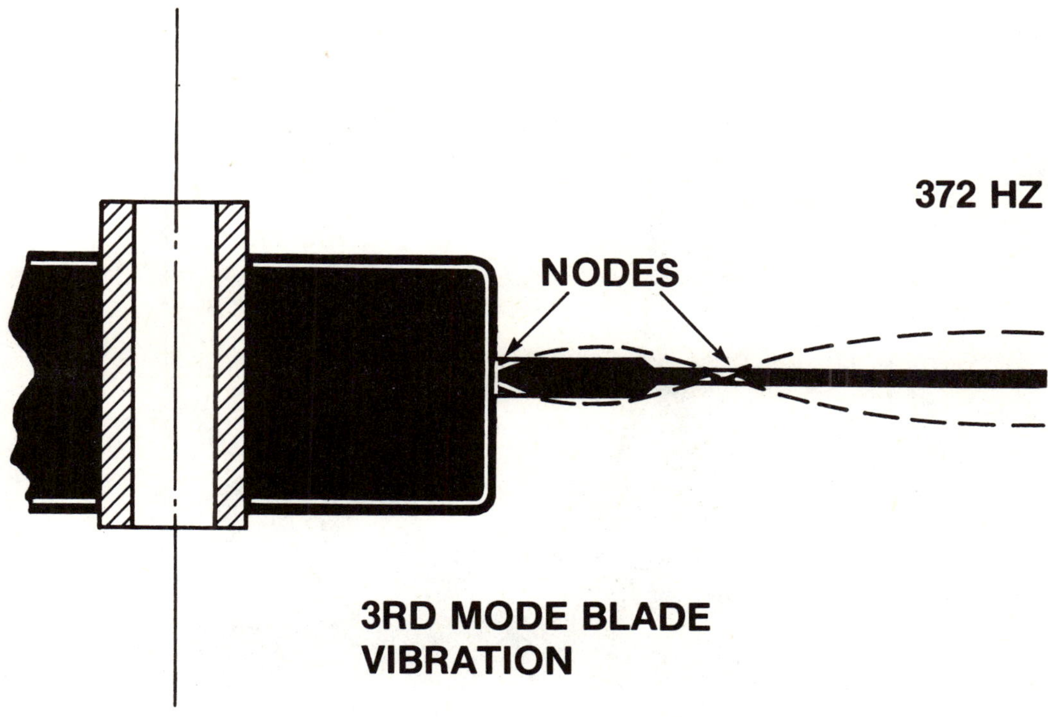

6. Mode #3 vibration of 98 inch axial impeller blade

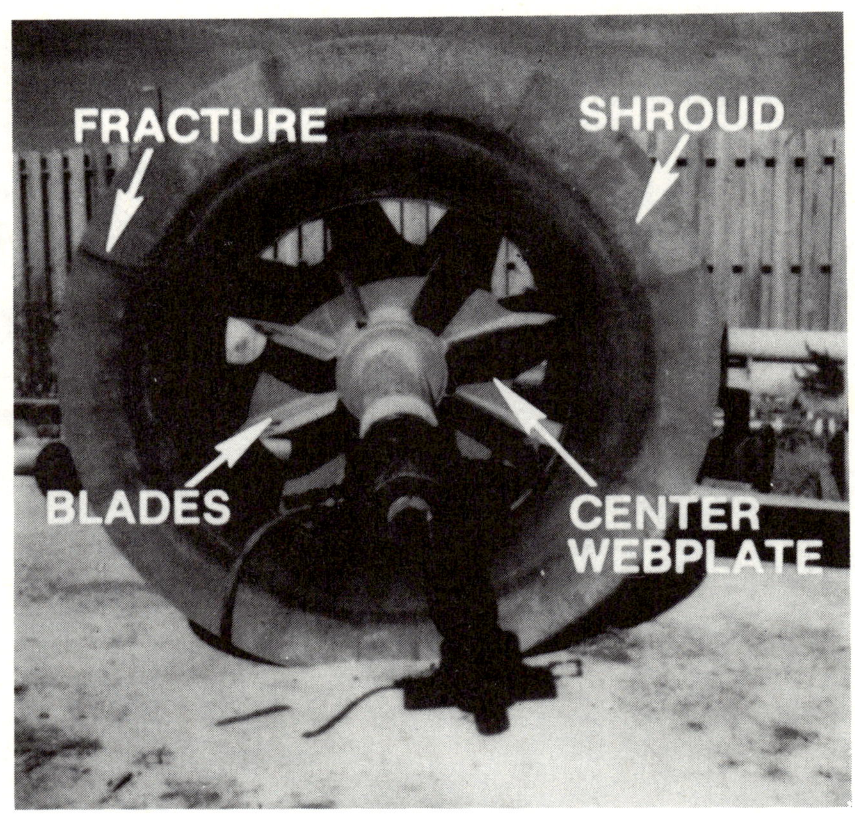

7. 72 inch diameter centrifugal impeller showing cracking

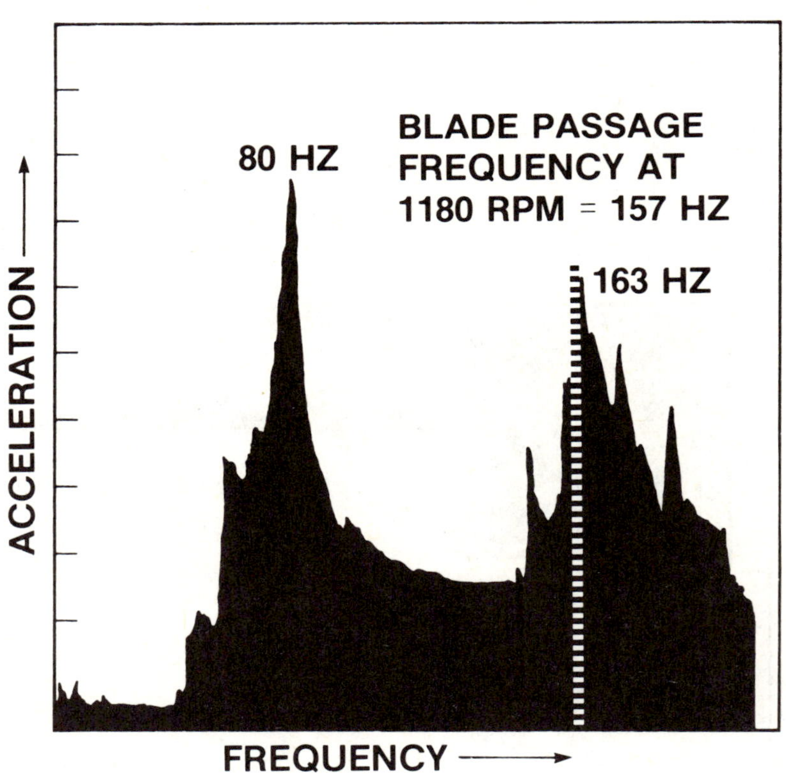

8. Frequency vs. acceleration plot for 72 inch diameter impeller shroud

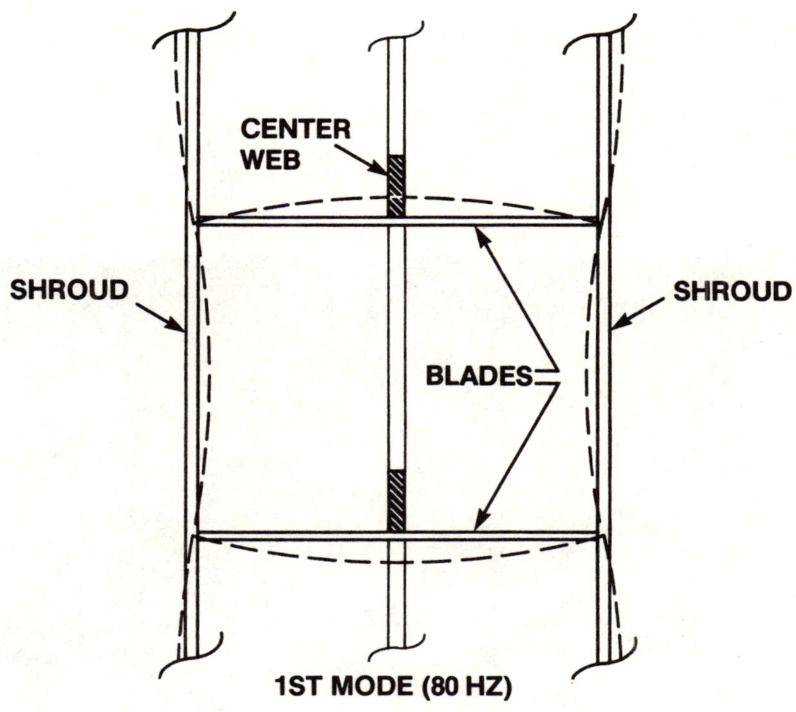

8.5 Mode #1 vibration of 72 inch centrifugal impeller shroud

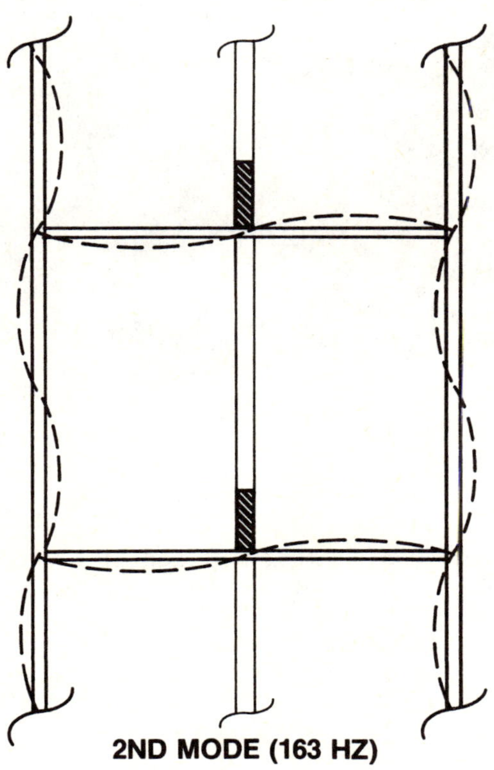

9. Mode #2 vibration of 72 inch centrifugal impeller shroud

10. 72 inch diameter centrifugal impeller showing modifications

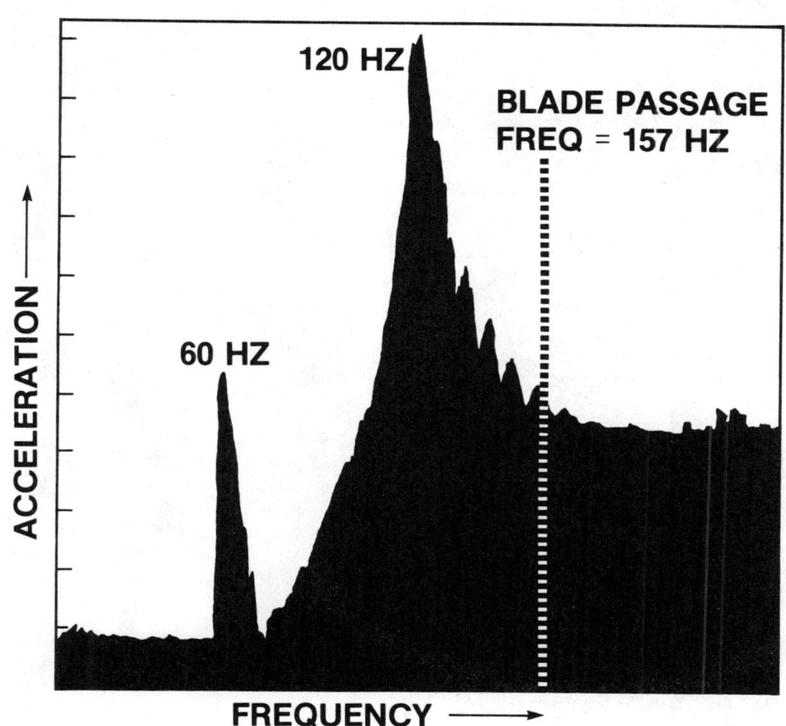

11. Frequency vs. acceleration plot of modified 72 inch centrifugal impeller shroud

12. 70 inch diameter centrifugal impeller operating at variable speed (View #1)

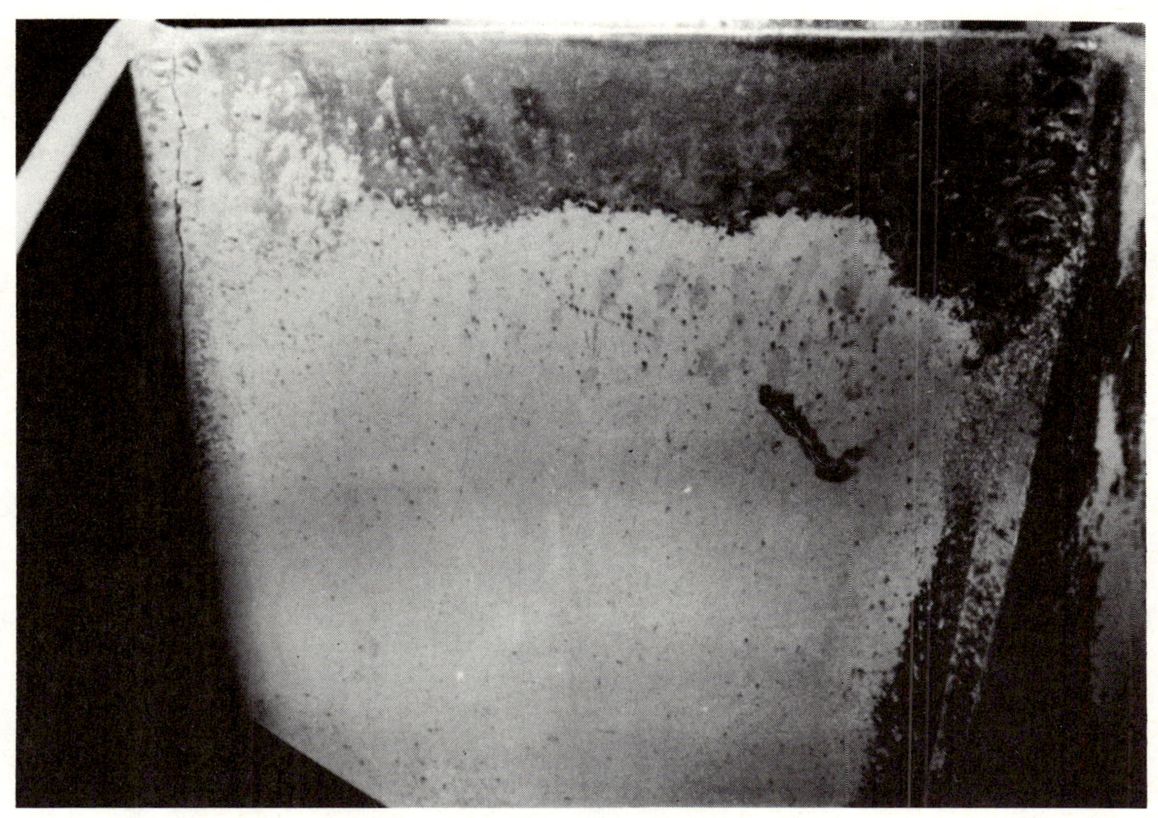

13. 70 inch diameter centrifugal impeller operating at variable speed (View #2)

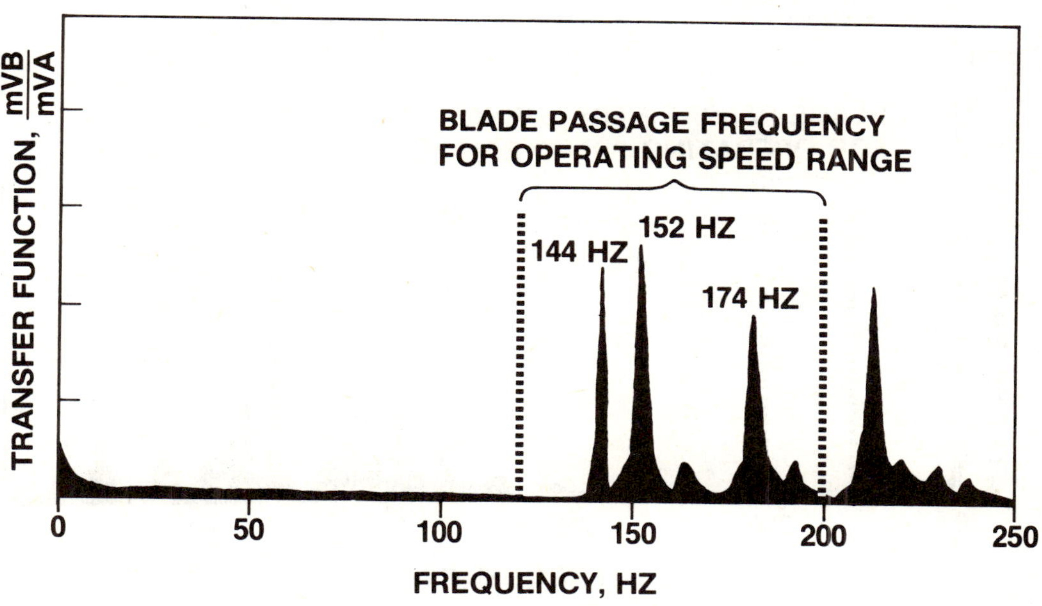

14. Frequency vs. transfer function plot of original 70 inch centrifugal impeller shroud

15. Rebuilt 70 inch diameter impeller

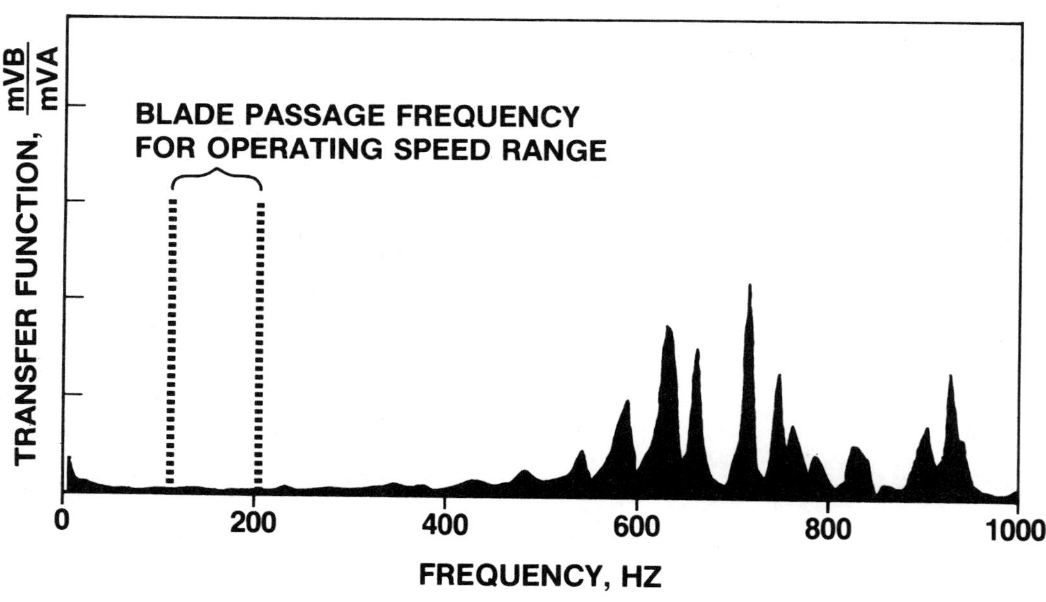

16. Frequency vs. transfer function plot of rebuilt 70 inch centrifugal impeller

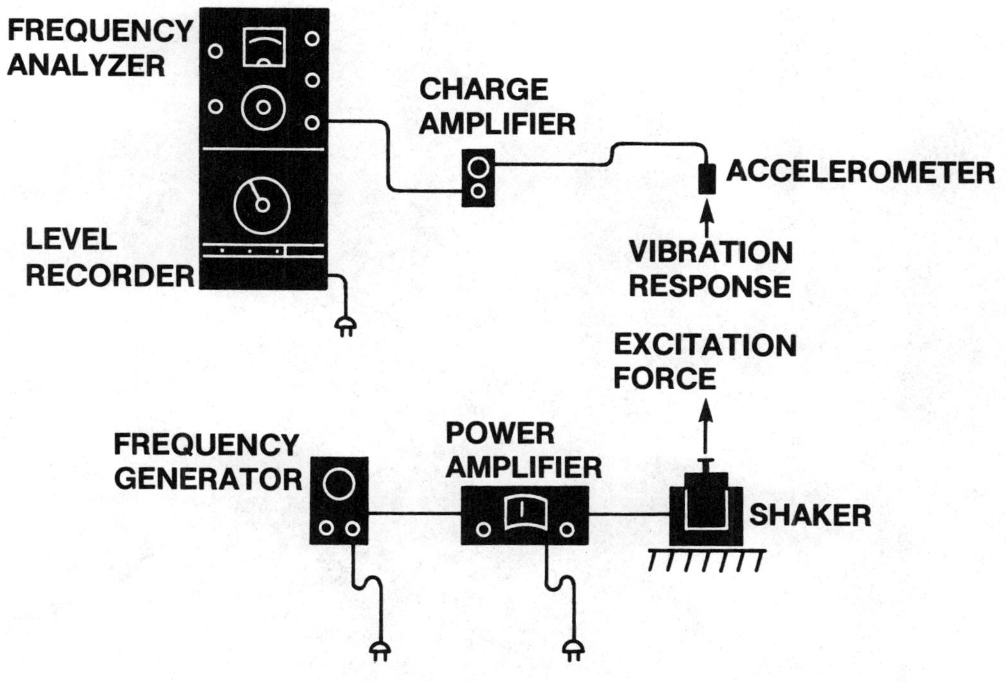

17. Swept-sine excitation test setup

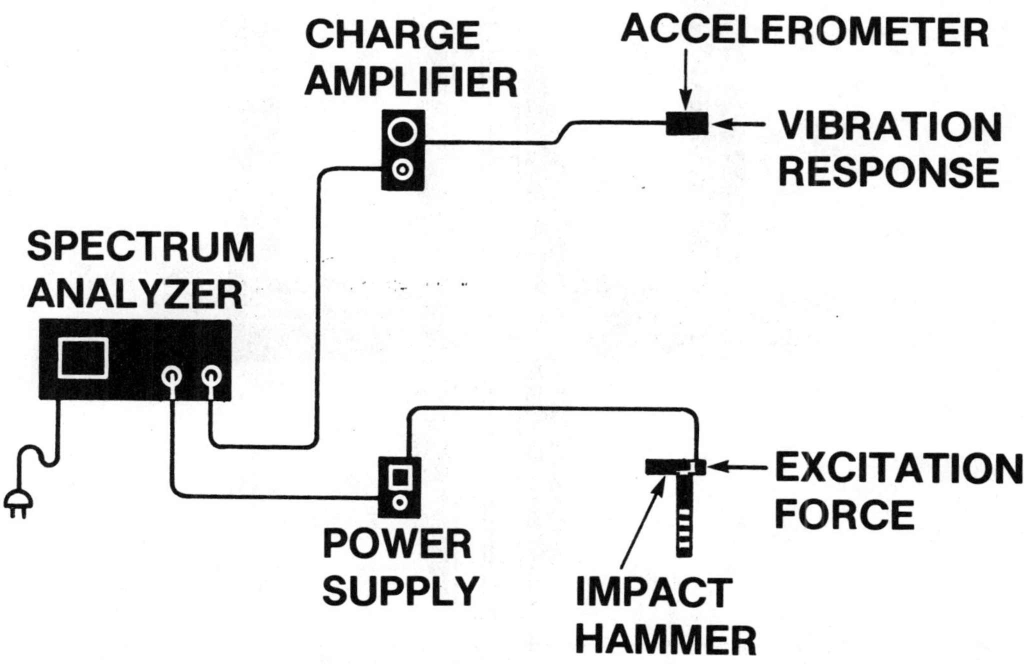

18. Impact testing setup

19. Example of an impact test setup

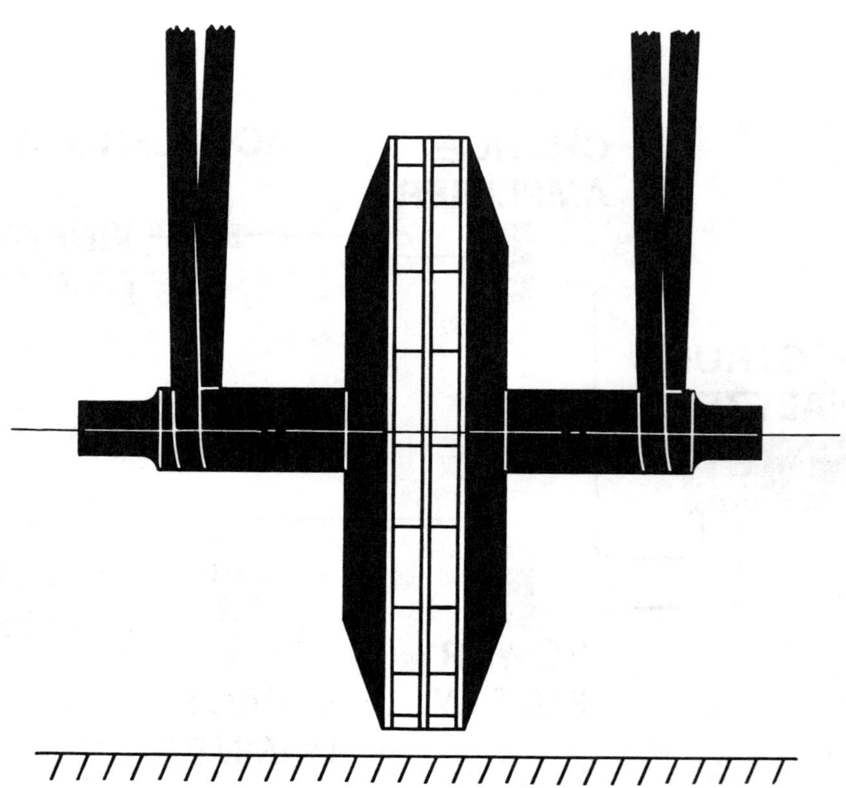

20. Centerhung rotor suspension

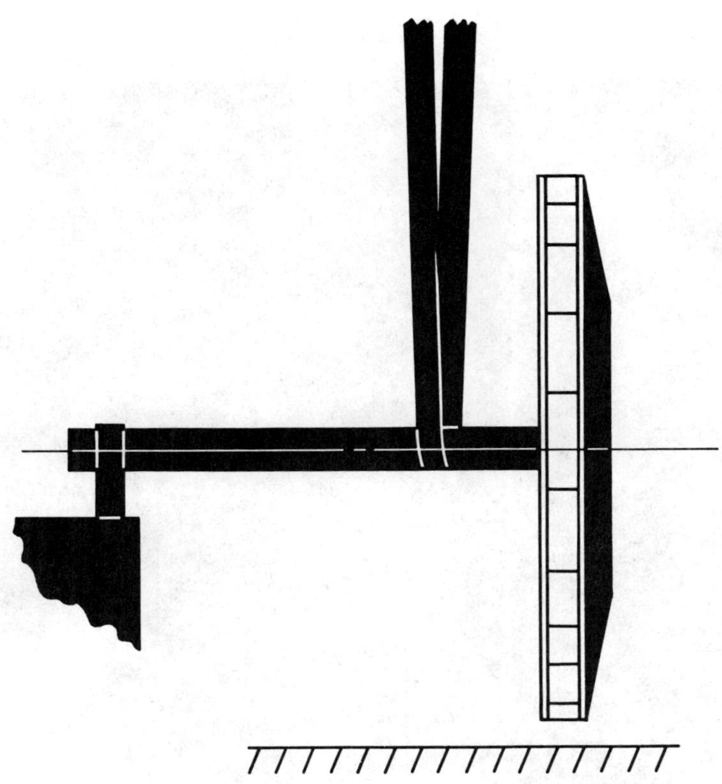

21. Overhung rotor suspension

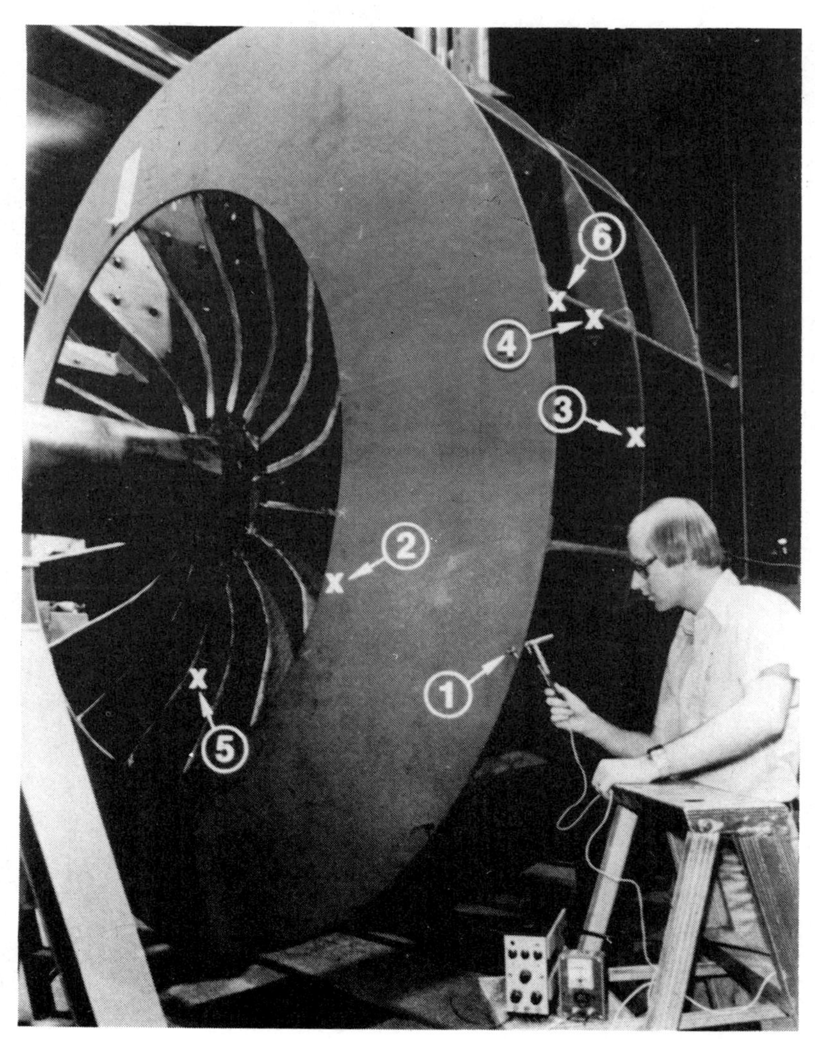

22. Centrifugal impeller test locations

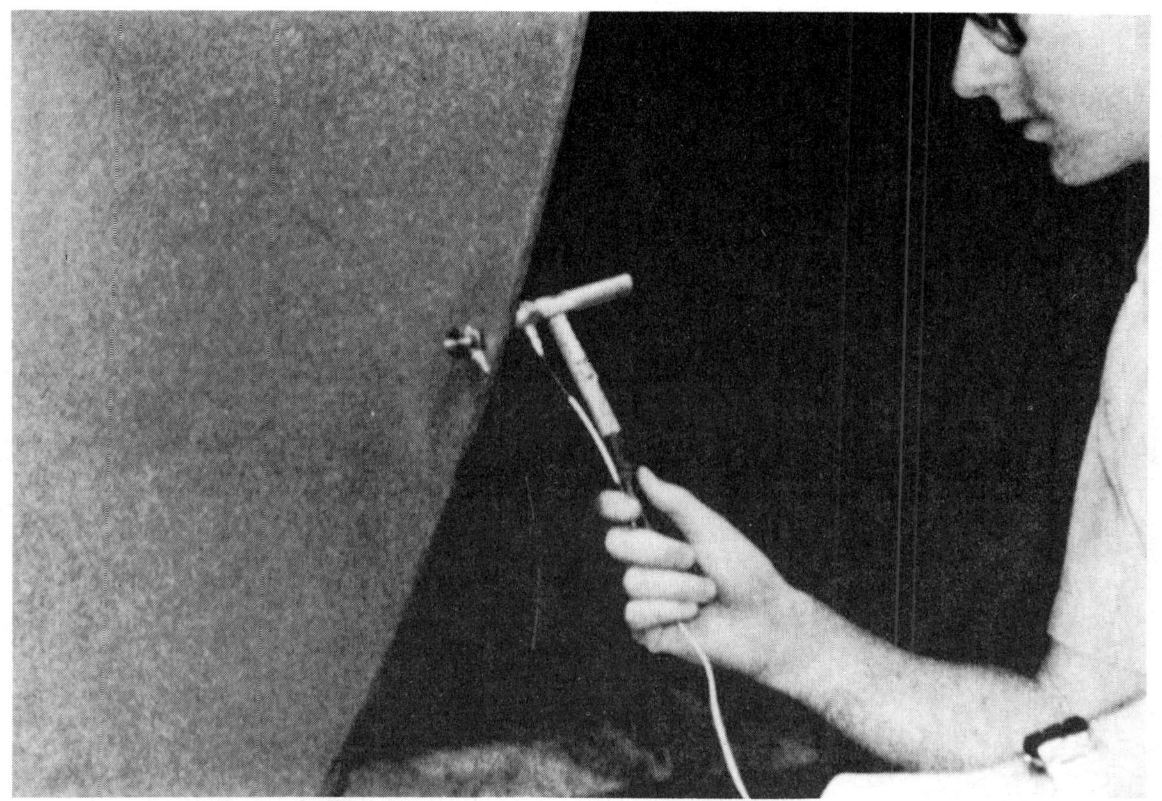

23. Accelerometer and hammer orientation

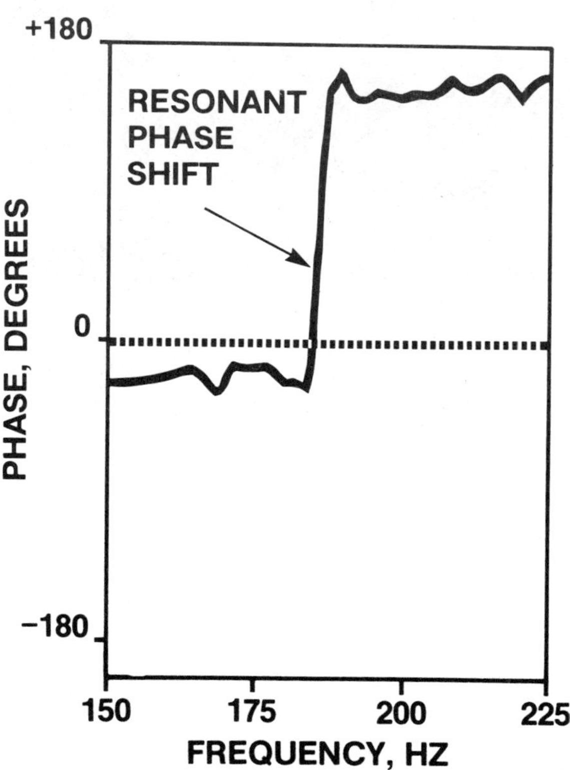

24. Frequency vs. phase angle plot

25. Example of a finished impeller

International Conference on

Fan Design & Applications

Guildford, England: September 7-9, 1982

PAPER J1

THE EFFECT OF ROTOR AND CASING DESIGN ON CROSS-FLOW FAN PERFORMANCE

D. J. Allen

Keith Blackman Ltd., U. K.

Summary

The cross-flow fan has been known of, in various forms, since the late nineteenth century. Nevertheless, it remains a neglected study, with the majority of experimental and theoretical contributions only adding to the mystique of the machine design.

This present work defines important geometric design parameters and combines extensive experimental work with a guiding semi-theoretical approach, to successfully determine their effect on all aspects of cross-flow fan performance.

The aerodynamic performance is principally dependent on five primary parameters and and optimal fan design will be simple, with a rotor and casing geometry within the established design limits.

Organised and sponsored by
BHRA Fluid Engineering, Cranfield, Bedford MK43 0AJ, England.

0263 - 421X/82/01 00 - 0001 $5.00
The entire volume can be purchased from
BHRA Fluid Engineering for $82.00

THE EFFECT OF ROTOR AND CASING DESIGN ON CROSS-FLOW FAN PERFORMANCE

1. Introduction.

Cross-flow fan rotors are similar in construction to forward-curved centrifugal blowers, but they differ fundamentally from all other fan types in that the flow traverses the same blade row twice.

A blade passage experiences continually varying flow conditions through a single impeller revolution: it is this complex flow field which enables a small diameter cross-flow fan unit to achieve a high aerodynamic performance.

The fan casing geometry is important in determining the proportions of the suction and discharge arcs and for governing the aerodynamic performance. Geometric parameters will be defined and extensive experimentation has determined the effect of these variable on performance, including noise and stability of operation.

Figure 1 illustrates essential features of the cross-flow fan.

2. Theoretical Aspects.

Since 1937, when the mechanism of the internal vortex was recognised, a number of models have been proposed to explain cross-flow fan behaviour. The most elementary approach is to treat the internal flow field as a Rankine vortex; a potential vortex with a forced vortex core. The velocities in the potential region are given by equation 1.

$$Vr = \text{constant} \qquad - 1$$

Early experimental work however, proved that the fan peripheral velocities were not adequately predicted by this model. MOORE (5), in 1972, attempted to determine the 'ideal' performance of a cross-flow fan by using the conventional velocity triangle technique of equating total head to the change in angular momentum between suction and discharge. In this analysis the flow field is considered to be a potential vortex of variable strength and the throughflow to be adequately represented by four points on a 'mean-line' through the rotor. Figure 2 shows a typically irregular peripheral velocity distribution and illustrates that any 'mean-line' approach can be subject to great error by over-simplifying the flow-field and ignoring the circumferential variation of a number of parameters.

The velocity triangle analysis may be considerably enhanced by treating the suction and discharge arcs as a number of segments and applying representative values to each segment. Boundary conditions may be determined experimentally and the total head calculated by numerical integration of values within paired suction and discharge segments. The author has developed a model using this approach,which is discussed in Section 3.

Other analyses have assumed Laplacian fields within the rotor, satisfying a boundary condition on the inner blade edge. However, experimental work, traversing the impeller interior, has implied non-Laplacian conditions and a high-order drift in velocities close to the blade.

3. Development Of A Flow Model.

This model is an extension to the conventional velocity triangle technique; and the impeller internal flow-field is largely disregarded. Experimentally determined external boundary conditions are used to fix the main flow-field parameters in the stream function.

The flow-field is considered to consist of two zones:-

i) A forced-vortex type 'recirculation zone' where the radius of the vortex is the distance between the core and the vortex wall leading edge.

ii) A potential field.

Radial and tangential velocities close to the outer suction periphery have been measured for one fan type and for three flow conditions. The flow model matches these observed values closely by manipulating the constituent parameters of a combined hydro-dynamic field. This field is determined by superposition of a free vortex and a line sink; the stream function, S, is expressed in cartesian co-ordinates as:

$$S = \frac{\Gamma}{2\pi} \ln \sqrt{x^2 + y^2} - \frac{M}{2\pi} \operatorname{Tan}^{-1} \frac{y}{x} \qquad -2$$

This stream function is applied only to calculate flow velocities along the outer suction periphery of the impeller, all other values are derived.

The forced vortex zone is considered as circular in cross-section and the flow-field is not evaluated nor energy transfer assumed to take place within this zone. Expansion of the vortex will involve an increased portion of the rotor in the rotational zone.

The principal fan design parameters are E, B, δ, d, D, H and β_2 - see Figure 3. From these inputs, geometrical variables may be determined and the effect of altering the primary design parameters may be theoretically evaluated. The duct depth term, H, is required to facilitate design of an optimum rear wall profile. No account is taken of the inner blade angle, β_1 or rear wall clearance E_2; the effect of altering these must be established experimentally.

3.1 Determining Peripheral Conditions.

Suction flow-field velocities are determined by differentiating the stream function in equation 2 and transforming these values into equivalent fan peripheral velocities. Discharge flow conditions are determined using an 'equivalent sector' hypothesis applied between the inflow and outflow arcs. This hypothesis states that each sector around the suction arc has an equivalent on the discharge arc; where the radial velocities are related by the ratio of the arc lengths.

$$V_{r2i} = V_{r1i} \frac{\lambda_s}{\lambda_d} \qquad -3$$

This condition implies accelerated throughflow where $\lambda_s > \lambda_d$ and also allows for a reversal of streamline curvature; both these phenomena have been observed experimentally.

From velocity triangles :-

$$V_{t2i} = u + \operatorname{Tan} \left\{ \frac{\pi}{2} - \beta_2 \right\} V_{r2i} \qquad -4$$

Volumetric flowrate is calculated by summing the radial velocities over the suction arc:

$$Q = \lambda_s \sum V_{r1i} \qquad -5$$

Total head is determined as the difference between suction and discharge moments of momentum :-

$$P_t = \frac{\rho u}{N} \sum_{i=1}^{n} \left(V_{t2i} - V_{t1i} \right) \qquad -6$$

N = No sectors around each arc. Equation 6 was found to converge rapidly and give only a small computation error.

3.2 Rear Wall Design.

In conventional centrifugal fan design, the ideal volute areas at each cross-section are determined by the principal of 'constant mean velocity'. Although the rear wall is not involved in the energy transfer, the importance of good design should not be underestimated. A poor profile wall will lead to high viscous losses within the volute and, in the case of cross-flow fans, increased flow instability. By incorporating a design parameter, H a rear wall may be designed using the 'constant mean velocity' across each radial cross-section to suit all combinations of the primary casing parameters.

This tailored rear wall design was fundamental to the research program, and the manufacture of a variety of wall shapes was made possible using a laminated module, described in Section 4.

The radial distance at each cross-section, i, is given by :-

$$D_i = Q_i / \bar{V} \qquad -7$$

Figures 4, 5, 6 and 7 show some of the predicted performance trends using this model. Figures 8 and 9 show a typical variation in the predicted suction fields which cause these changes in performance. Figure 10 illustrates the changes in the designed rear wall shape with varying duct depths.

4. Experimental Investigations.

The rotor unit used in the experimental program enabled the fan diameter to be varied from 95mm to 195mm. The rotor length was 320mm, the blade chord 20mm and the blade number 24.

The rear wall module was laminated, formed from 1.5mm gauge wooden slats. Shapes were followed from a template and the stepped finish smoothed by covering with plastic film.

The vortex wall module was made of perspex and slotted vertically to allow movement and location of the tongue support platform. This platform enabled horizontal movement of the vortex wall and the attachment of various vortex wall types.

The discharge ducting consisted of a length of wooden ducting placed downstream of the vortex wall. The duct depth and shape could be altered to allow a smooth transition in profile from the vortex wall module. The discharge ducting could be attached to an outlet plenum chamber and a diffuser unit. The fan had no inlet ducting.

Static pressures were measured using a ring of four tappings placed approximately 3 fan diameters downstream of the vortex wall leading edge.

Volumetric flowrate was determined in two ways: a) using an outlet plenum chamber with two orifice plates (lower flowrates) and b) using an outlet pitot traverse. These two test methods were frequently checked against each other and found to give consistently good results.

Shaft power was determined using a calibrated full-bridge strain-gauge circuit, with the two strain-gauges mounted diametrically opposite each other on an unobstructed length of shaft supported between two bearings.

Acoustic tests were performed using a Bruel and Kjaer sound power source and calibrated loudspeaker unit. Sound pressure levels were measured using a Bruel and Kjaer sound level meter with single octave filter. Measurements were made in a 'diffuse-field' test chamber.

5. Cross-Flow Fan Aerodynamic Performance Experimental Results.

The design parameters investigated were represented under two group headings;

'primary' variables, which fundamentally determine the aerodynamic performance and 'novel' design variables, which modify the flow conditions imposed by the primary casing. Primary design variables, for example, determine the vortex location, dictate the rear wall shape and lengths of the suction and discharge arcs. Novel designs include such items as flow guides, vortex stabilisers and modifications incorporated to suppress noise.

The results demonstrate that aerodynamic performance is largely determined by a few basic geometrical parameters. Vortex wall'shape' terms are neglected because elaborate profiles generally do little more than relocate the vortex opposite some point on their leading face. Figure 3 has already illustrated the primary design variables, Figure 11 illustrates some of the design variants considered novel. As stated, for each primary design variation a new rear wall shape was used. These analytically designed rear walls were consistently proven to give better results than conventional rear wall designs, typically: Figure 12.

All aerodynamic results are expressed in non-dimensional form (Ø, Ψ and η), implying that one set of performance curves may represent a homologous series of fans. BUSH (2) determined that cross-flow fans obey conventional fan laws and that scaling is valid within certain limits. The effect of Reynolds number on fan performance has been researched by HAINES and HOLGATE (4) and their results indicate that cross-flow fans satisfy the accepted similarity relationships. All performance tests done by the author were on a small impeller running at a relatively low tip speed; under these conditions the peak pressure was generally low (150-200 Pa) and the Reynolds number, based on diameter of the order of 0.82×10^5. At these low pressures it is unlikely that all stall zones will be highlighted. Nevertheless, it is considered that the test fan size and Reynolds number was representative of the average marketable unit, but it should be appreciated that larger fans may exhibit stall in operation.

BUSH (2) also notes the existence of three-dimensional effects, a phenomenon well understood in cross-flow fan technology. So, although it may be assumed that cross-flow fans are scaleable on diameter, there are limiting slenderness ratios; with vortex breakdown in slim rotors and secondary flow effects in short rotors.

5.1 Influence Of Primary Design On Aerodynamic Performance.

5.1.1 Vortex Wall Clearance, E.

The vortex wall clearance is defined as the distance between the vortex wall leading edge and the point on the impeller outer periphery in the plane of the unsloped vortex wall. This is physically consistent with locating the vortex.

As the vortex wall clearance is increased the vortex grows; but there is no simple relationship between momentum loss and forced vortex volume, it also depends on angular velocity and the shape of the recirculation zone. The angular velocity of the vortex decreases with increasing cross-sectional-area and the momentum loss in a non-circular vortex is increased because a rotating element of fluid has to alter its radius during each revolution, see MOORE (5).

Figure 13 indicates that there is no strong relationship between vortex wall clearance and static efficiency, although wider clearances have improved efficiencies to the left of 50% Ø max. The pressure characteristic remains fairly constant in the range $1\% \leq E/D \leq 3\%$, but increasing the clearance further reduces the peak pressure development.

5.1.2 Vortex Wall Declineation, B.

This value proves to be the most important single parameter in cross-flow fan casing design. Its importance stems from the increased suction arc velocities formed by lengthening the inlet over a spatially critical sector close to the vortex centre.

Figure 14 clearly illustrates the improved performance possible by increasing the vortex wall declineation over the range $0.0\% \leq B/D \leq 28.2\%$. In general, the efficiencies also increase with increased B. This experimentally determined trend was well predicted by the theoretical treatise outlined in section 3.

5.1.3. Rear Wall Circumferential Angle, δ

This value is important because it defines the extent of the inflow and outflow arcs, for a fixed vortex wall location. The ratio λ_s/λ_d is itself unimportant because it disregards the uneven velocity distribution governed by an eccentrically located vortex, which determines that the majority of throughflow and energy transfer takes place over an arc close to the vortex wall. The spatial distribution of velocities is of considerably more importance that the circumferential length of an arc.

There is a zone of random, low energy flow close to the rear wall edge, caused by the impossibly rapid changes in flow angle demanded by an abrupt suction/discharge arc division. This will inevitably result in blade stall and a reduction in efficiency.

The results in Figure 15 indicate an optimum δ value of 20°, which is in agreement with results reported by HAINES and HOLGATE (4).

Apart from this optimum value, the trend is for a reduced pressure development with increasing rear wall angle.

There exists no obvious trend between δ and aerodynamic efficiency, although peak efficiency occurs with the minimum length of suction arc. Peak efficiency is reduced about 3% by altering the rear wall angle from 60° to 20°.

5.1.4 Diameter ratio d/D.

PORTER and MARKLAND (6) suggest a lower limiting ratio of 70% and an upper limit of 85%. They state that satisfactory operation occurs and that the influence of diameter ratio is small within these limits.

However, the results of Figure 16 indicate a strong dependence of performance on diameter ratio, in that performance increases, within limits, with increasing chord/diameter ratio. Performance is acceptable in the range $70\% \leq d/D \leq 80\%$ but optimal in the range $72\% \leq d/D \leq 70\%$. With the experimental rotor used by the author, decreasing the diameter ratio also increases the slenderness ratio, L/D. It is likely, therefore, that the severely reduced performance withd/D = 68% could be at least partially attributed to transverse vortex breakdown.

Fan efficiency is increased in proportion to the improvement in pressure development.

5.1.5 Duct Height, H.

Duct height is a design criterion governed by preferred dimensions for ventilation ductwork.

Although it is possible that the rear wall breadth may be influential in determining the vortex location, the rear wall is essentially a flow channeller and not involved in the energy transfer. Thus, for correctly designed rear wall profiles the performance would not be expected to vary greatly with H.

The results, Figure 17, depict an optimum H/D of about 80%, although performance varies little over the range $70\% \leq H/D \leq 85\%$. Decreasing the duct height below this range causes a severe reduction in performance, increasing the duct height reduces Ø max. An optimum efficiency with H/D = 80% is indicated.

5.1.6 Rear Wall Clearance, E_2.

It is important to adopt a rear wall clearance which will result in minimum pressure loss. With an ideal fluid the rear wall gap should be no more than a running clearance, but in reality large losses can result from the turbulent zone around the rear wall edge.

The results illustrated in Figure 18 show that there is usually a broad range of clearances where the pressure loss is tolerable, and that too wide clearances are preferable to too narrow clearances. The optimum E_2/D ratio exhibits a complex dependence on many rotor and casing design parameters and it is recommended that minimum-loss values be determined for each type tested - it should generally be in the range $5\% \leq E_2/D \leq 15\%$.

5.2 Novel Designs.

For a commentary on the effects of other design modifications, refer to Table 1.

6. Cross-flow Fan Noise.

Developing a theoretical analysis of cross-flow fan performance and the associated velocity distributions also enabled the author to develop a parallel analysis of cross-flow fan noise. Details of this analysis are beyond the scope of this paper, but the results predicted observed trends moderately well; although sound power values were many orders of magnitude higher than those determined experimentally. Four sound production mechanisms were isolated as fundamentally important, fuller details are available through ALLEN (1).

The noise analysis, as well as helping clarify observed trends, also yielded three main results:

i) The irregular peripheral velocity distribution, inherent with cross-flow fans and responsible for the high performance achievable, causes increased discrete and broad-band noise. Also, the high fluid velocities close to the tongue increase the tongue interaction sound production mechanisms. For the same aerodynamic power (pressure x flowrate), cross-flow fans are likely to be noisier than an equivalent centrifugal fan.

ii) No universal speed exponant exists, but there will be a considerable variation in the value of the exponent with frequency and aspects of fan design. Experimentally, the speed/sound power exponent, α, was found to be approximately 7.29 for an unshrouded rotor and of this order for a cased rotor :

$$PWL = \text{constant} \times Ma^{\alpha} \qquad - 8$$
$$\alpha = 7.29 \pm 0.53$$

iii) Obscuring the suction arc close to the vortex wall will reduce fan noise considerably; but this reduction in noise level will be accompanied by a large reduction in aerodynamic performance, see Figure 19.

Experimental Results.

6.1 Influence Of Primary Design On Radiated Sound Power.

Primary variables were altered as indicated in each sub-section. As far as possible 'optimum' aerodynamic designs were tested, although d/D was kept as 79.5% unless otherwise stated. The tabulated noise levels were taken at the flowrate corresponding with maximum efficiency.

6.1.1 Vortex Wall Clearance, E.

Most researchers have concluded that increasing the vortex wall clearance leads to a monotonic decrease in noise level. The results from a large number of tests are tabulated in Table 2 and show that, for an unshrouded impeller, the dependence upon tongue clearance may be expressed as simple power relationships, with a strong frequency dependence.

Where:

$$PWL \propto f(st) \cdot (E/D)^{\gamma} \qquad (1 \leq E/D\% \leq 5)$$
$$(st = \text{freq.} / BPF)$$

STROUHAL BAND	γ
St including 0.5	-0.63
St including 1.0	-0.88
St including 2.0 & 3.0	-1.56
St including 4.0	-1.61
St including 8.0	-0.30

Table 2.

Table 3 illustrates that the same trend is also observed with a cased impeller and that sound power is reduced with increased vortex wall clearance, for strouhal bands 2, 3, 4 and 5.

6.1.2 Vortex Wall Declineation, B.

This geometric parameter has been shown to strongly influence the aerodynamic performance of a fan. For an uncased impeller the sound power output increases weakly with increasing B; this trend is broadly supported by the experimental results on cased impellers with plane walls and shaped vortex walls. The results, Table 4, are a little inconclusive, but show that in strouhal bands 2, 3 and 4 at least sound power level increases with B.

6.1.3 Rear Wall Circumferential Angle, δ .

Generally, the quietest operating condition occurs with δ = 40°, although the rear wall angle has an irregular influence on radiated sound power. In strouhal bands 1, 5, 6 and 7 adopting a rear wall angle of 0° or 20° gives the noisiest operation, and this trend is commensurate with increased aerodynamic performance. Overall, the noisiest operation occurs with δ = 60°, see Table 5.

6.1.4 Diameter Ratio, d/D.

For an uncased impeller the radiated sound power was found to increase steeply with an increase in the chord/diameter ratio. This trend is obscured by running the impeller cased, although the figures in Table 6 depict an irregular relationship broadly implying increased sound power with increasing d/D, for strouhal bands 2, 3, 4 and 5.

The apparent weakness of the relationship may be attributed to some feature of the experimental rig affecting sound radiation properties.

6.1.5 Duct Height, H.

Closed boundaries are essential to a fan if the impeller is to deliver air at a useful pressure in one direction. The rear wall acts only as a flow guide and any dependence of noise upon its profile will be related to boundary turbulence or the internal impedence of the enclosed volume.

The measured sound powers exhibit an irregular dependence on flowrate and frequency, but the results in Table 7 indicate the following relationship:

H/D : PWL 70% > 60% > 80% > 90%

6.1.6 Rear Wall Clearance, E_2.

The results in Table 8 support the assertion that improved performance generally increases the radiated sound power. Overall, the noisiest operation corresponds to the optimum rear wall clearance. An increased clearance reduces the noise level.

6.2 Influence Of Novel Designs On Radiated Sound Power.

A number of design modifications were tried with the aim of reducing the overall sound power levels; these were mostlyineffective and often resulted in either reduced performance or unstable operation. However, three modifications were moderately effective and these are discussed briefly.

6.2.1 Perforated Vortex Wall.

This consisted of a plane vortex wall 5mm thick, with a castellated leading edge and 5mm diameter holes drilled on a regular pattern giving about 30% free area.

There was observed to be a consistent reduction in noise level of between 1-5dB, with no reduction in performance, but a small decrease in the operational stability, particularly at lower flowrates.

6.2.2 Sloping The Vortex Wall.

Radial sloping of the vortex wall may reduce noise levels by 1-4dB in some frequency bands, but this is accompanied by a reduction in peak efficiency of 8% and a reduced pressure development. Longitudional sloping of the vortex wall was found to be an ineffective noise reduction design modification.

Sloping the vortex wall radially or longitudionally decreases the flow field stability and causes an increased low frequency pulsation of the vortex core.

6.2.3 Irregular Blade Spacing.

Irregular motor manufacture has the effect of altering conditions between blades and disrupting flow processes temporally governed by the blade passing frequency. In all strouhal bands but the fundamental harmonic, noise is significantly decreased by radially offsetting the blades. The magnitude of noise reduction is greatest at the lower multiple harmonics, decreasing towards the higher frequency bands. The noise reduction may be up to 10dB using a random, radial offset of the blades - the mean blade offset was 20% of blade chord.

7. Cross-flow Fan Operational Stability.

Improving operational stability has long been an objective of cross-flow fan researchers. ECK(3), for example, called his profiled vortex wall a 'vortex stabiliser', and BUSH (2) stressed the need to quantify levels of stability when dealing with cross-flow fans.

TRAMPOSCH (7), in 1964, observed a small oscillatory motion of the forced vortex along the impeller inner periphergy and YAMAFUJI (8) found that even with speed-invariant rotation, the internal flow pattern varied cyclically. Both these observations illustrate the intrinsic instability of cross-flow fan operation.

Cross-flow fans, due to the eccentrically located vortex have a range of relative flow angles distributed around the suction arc: it is likely, therefore, that they will always operate with some stalled blade passages, and that these will probably occur towards the vortex wall or rear wall boundaries of the suction arc. This stall will reduce throughflow and lower fan efficiency, but it is not considered the main mechanism of cross-flow fan instability.

Pressure around the forced vortex can only be guaranteed uniform when there is a free vortex velocity distribution, the existence of an eccentrically located vortex eliminates this possibility. Thus, in mose cases, the forced vortex must have an uneven peripheral pressure distribution, which may cause the vortex to migrate between quasi-stable locations. It is this mechanism which the author considers primarily responsible for the operational instability experienced with cross-flow fans.

Experimental Results.

Pulsation analyses were made on a large number of fans, in both the frequency-domain (Fourier transform) and the time-domain (Δ P/P%). The principal conclusions are listed below:

i) Nearly all fans experience a low-frequency 'drift' of pressures and volume flows, with a peak-to-peak time interval of the order of 100 seconds. For a particular fan type the pulsations are not random, but generally exhibit one or two discrete pulsation frequencies below 50% of the rotor RPS.

ii) Unshrouded impellers exhibit severely reduced performance and operate very unstably with well-defined discrete pulsation frequencies superimposed on a lower frequency drift of values. It is considered unlikely that an uncased rotor could ever form part of a useful machine.

iii) Operational stability is always enhanced by avoiding operation at extreme flowrates.

iv) Generally, a fan which gives good aerodynamic performance will also operate stably.

v) Adopting an ECK-type vortex stabiliser does reduce the level of instability, but only at the expense of a vastly reduced aerodynamic performance, see Figure 19.

8. Conclusion.

The best cross-flow fan design consists of a plain vortex wall, unobstructed suction arc, plain discharge and a rear wall with a profile tailored to suit the primary design variables.

A number of novel fan designs have been tested but these have mostly been unable to prove their worth, and frequently to affect the performance adversely. The casing should be kept simple and the geometry of this within the established design limits, Table 1.

Cross-flow fans can offer advantages in size and shape, and also for systems requiring a uniform airflow over a rectangular cross-section. Unless there are fairly severe constraints on noise and operational stability, there is no reason why cross-flow fans should not supplant centrifugal exhausting fans for size-restricted, non-intermittent applications where a total efficiency of between 45% and 55% is acceptable.

SYMBOLS.

B	Vortex wall declineation
d	Inner impeller diameter
D	Outer impeller diameter
D_i	Volute depth
E	Vortex wall clearance
E_2	Rear wall clearance
H	Duct height
i, n, N	Integer number
L	Impeller width
M	Sink strength constant
P_t	Pressure, total
Q	Volume flow
r	General radius
S	Stream function
u	Fan tip speed
V	General velocity
$\bar{V}$	Mean velocity through volute
x, y	Cartesian co-ordinates

α, γ	General exponents
β	Blade angle
δ	Rear wall circumferential angle
η	Efficiency of energy transfer
Γ	Circulation constant
λ_s, λ_d	Suction arc length, discharge arc length
ψ	Pressure co-efficient; $P / \frac{1}{2} \rho u^2$.
ϕ	Flow co-efficient; Q / LDu.
ρ	Fluid density

Subscripts.

1	Relating to suction outer diameter
2	Relating to discharge outer diameter
i, n	Integer counter
r	Radial
t	Tangential

REFERENCES.

1 Allen, D.J. "The effect of rotor and casing geometry on the performance of cross-flow fans". PhD Thesis, University Of Durham 1981.

2 Bush, E.H. "Cross-flow fans". Conference on fan technology and practice; I.Mech.E. London (18-19 April 1972)

3 Eck, B. U.S. Patent 2942 773 (1960)

4 Haines, P & Holgate, M.J. "Scaling of cross-flow fans - an experimental comparison". Proc. I.Mech.E., Stirling (1977)

5 Porter, A.M. "A study of the cross-flow fan". Journal of Mechanical Engineering Science 12(6) pp 421-431 (1970)

6 Tramposch, H. "Cross-flow fan". Paper: A.S.M.E. No. 64-WA FE-26 (1964)

7 Yamafuji, K. "Studies on the flow of cross-flow impellers: (experimental study (1))". Bulletin J.S.M.E. 18 No. 123 (1975).

SUPPLEMENT.

All internal and external flow field traverses were performed on a large, experimental cross-flow fan rig. This fan was 0.625m diameter, 1.0m blade width and operated at 800 rpm. Further details of this fan design are available through reference (4).

Fuller analytical and experimental results are available through reference (1).

TABLE 1.

PARAMETER	COMMENTARY
Vortex way declineation, B	The most important geometrical parameter. B/D should be kept in the range $25\% \leq B/D \leq 30\%$ for optimum performance. Quieter operation occurs below this value, but at the expense of aerodynamic performance.
Diameter ratio, d/D	Ideally, $72\% \leq d/D \leq 76\%$ but acceptable in the range $70\% \leq d/D \leq 80\%$.
Vortex wall clearance, E	From performance considerations and ease of manufacture, E/D should be of the order of 3%.
Rear wall angle, δ	This should be 20°, efficiency is not compromised if δ is increased to 40°.
Duct height, H	The performance varies little in the range $70\% \leq H/D \leq 85\%$.
Rear wall clearance, E_2	The optimum clearance is a function of other design parameters, if this cannot be determined experimentally it should be kept in the range $9\% \leq E_2/D \leq 12.5\%$.
Tongue thickness, T	This will generally be determined by stiffness or material thickness considerations. A flat wall is recommended with a thickness in the range $0.8\% \leq T/D \leq 4\%$.
Irregular blade spacing	Can reduce noise levels by up to 10 dB, but also reduces the fan pressure development.
Slotted endplates	20% free-area endplates reduce the pressure development by 30% and causes a stall zone around the mid-characteristic. Efficiencies are similarly affected.
Sloping the vortex wall	Radial sloping can reduce the noise level by 1-4 dB. This, however, is accompanied by a significant decrease in pressure development and high amplitude vortex pulsations. Longitudianal sloping was ineffective in reducing noise levels.
Wedge on vortex wall	To be effective the wedge must act as a flow guide to the high radial velocities along the vortex wall upper surface. This is rarely the case; generally the wedge acts as a blockage to the throughflow and reduces fan performance.
Perforated vortex wall	Can reduce noise by 1-5 dB, with no apparent reduction in performance. Slightly worsened flow stability to the left of the characteristic.
Outlet diffuser	A uniform velocity profile is required for successful diffusion. With cross-flow fans this profile is irregular and a diffuser placed close to the outlet is ineffective.
ECK-wall	This can reduce the inherent fan instability, but only at the expense of

	performance and efficiency which will be reduced by at least 20% of peak.
Vortex walls with arcuate damper	The vortex wall should cover the minimum possible arc; blocking throughflow in an arc close to the vortex wall, on the suction side, will drastically reduce the throughflow and pressure development.
Outer blade angle, β_2	Because of the variable peripheral flow pattern, no optimum blade angle exists. However, losses may be reduced if angles are matched over a portion of the suction arc with maximum throughflow. This gives $25° \leq \beta_2 \leq 35°$, with increased pressure development at the lower blade angle.
Inner blade angle, β_1	The simple Rankine vortex model requires $\beta_1 = 90°$. Whilst this model is imperfect there is no reason why this angle should not be approximately radial.
Blade number, Z	$24 \leq Z \leq 36$
Internal blades or bodies	The optimum location of such elaborations is a function of flowrate and basic fan design. There is little evidence to suggest that internal bodies do locate or stabilise the vortex, but such additions would greatly complicate fan manufacture and they would be unlikely to find commercial acceptance.
Inlet guide vanes	The optimum vane settings would vary around the fan periphery, they could only be found experimentally and would be a function of flowrate and fan design. Improved operational stability has been claimed with a single recirculation-vane located towards the rear wall.
Tongue shaping	These can only vary the vortex location or act as a guide vane, usually they only obstruct the throughflow. The best design is a simple, flat wall
S-shape, U-shape and L-shape throughflow	Care must be taken to ensure the casing design is not compromised to suit the ducting - and that the suction flow field is not distorted by proximity to the casing.
Single rotor fans (in series)	See U.K. Patent 816, 689. Care must be taken to ensure flow is not distorted on entry to successive rotors. There may well be stability problems when running a fan in series.
Twin-rotor fans (parallel)	This design has the advantage of a simplified casing. If the external inlet flow is not symmetrical about the duct C/L, there may be problems with windmilling .

Generally, the simplest casing design is the best to use. Complicated modifications only decrease the commercial competitiveness of the fans, usually without improving the performance.

E/D% \ St	1	2	3	4	5	6	7
1	90	92	82	77	72	64	61
3	91	88	79	73	71	65	62
5	87	90	76	70	69	62	60
7	84	74	69	68	67	62	60

Table 3 - results in dB, PWL

B/D% \ St	1	2	3	4	5	6	7
0	82	81	74	70	65	60	55
6	83	82	75	70	67	61	56
12	90	86	78	70	69	63	59
28	84	83	84	77	69	60	53

Table 4 - results in dB, PWL

δ° \ St	1	2	3	4	5	6	7
0	86	79	88	78	72	63	56
20	92	88	82	81	68	62	54
40	80	80	83	80	72	62	54
60	81	90	85	84	71	60	53

Table 5 - results in dB, PWL

d/D% \ St	1	2	3	4	5	6	7
68	80	90	94	82	69	61	54
72.4	84	92	86	81	68	60	53
75.8	82	99	88	81	69	60	55
79.5	93	88	84	81	69	60	54

Table 6 - results in dB, PWL

H/D% \ St	1	2	3	4	5	6	7
60	87	88	89	85	72	63	56
70	88	96	85	87	74	65	60
80	93	89	85	82	70	61	55
90	86	84	85	76	70	61	52

Table 7 - results in dB, PWL

$E_2/D\%$ \ St	1	2	3	4	5	6	7
5	85	84	76	73	66	60	54
10	84	80	81	74	66	61	55
15	88	81	71	70	65	60	54
20	81	85	78	70	65	60	54

Table 8 - results in dB, PWL

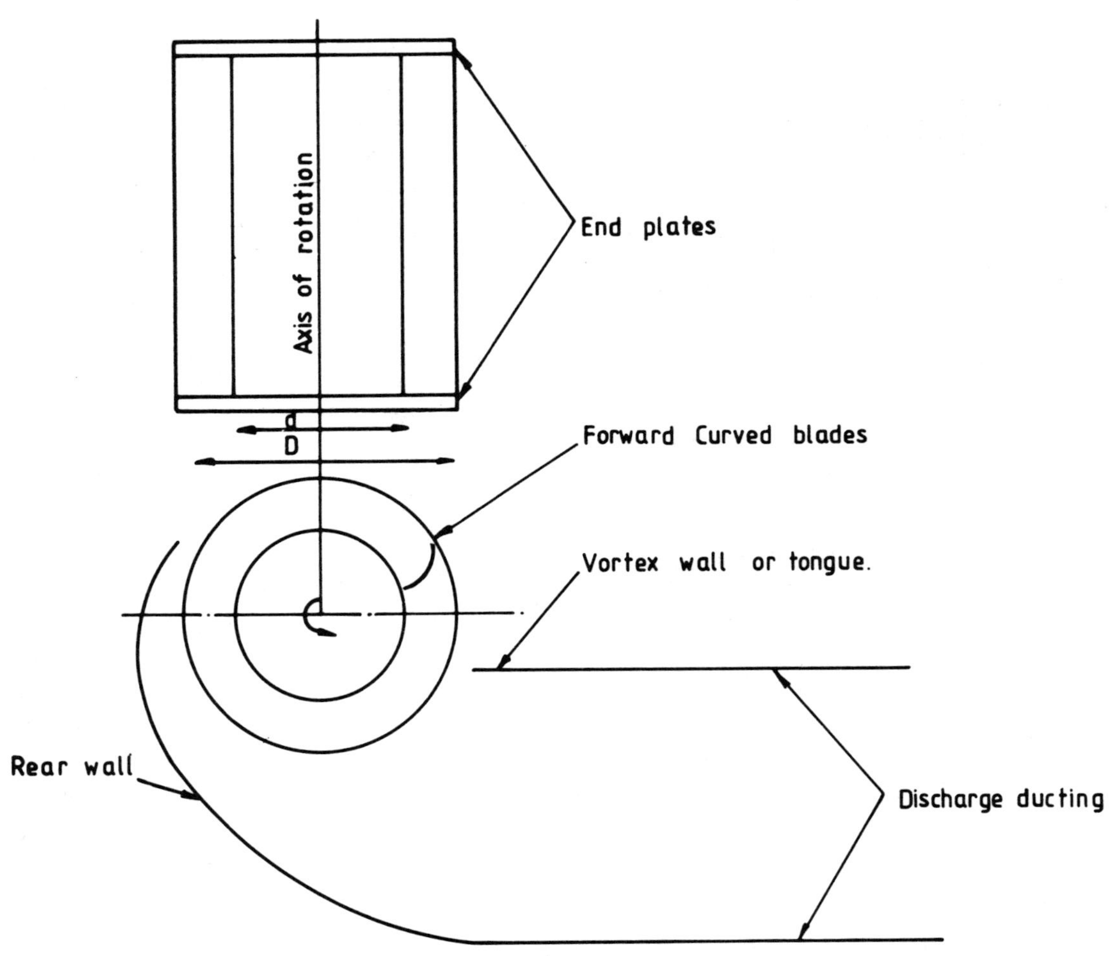

Fig. 1. GENERAL ARRANGEMENT OF A CROSS-FLOW FAN.

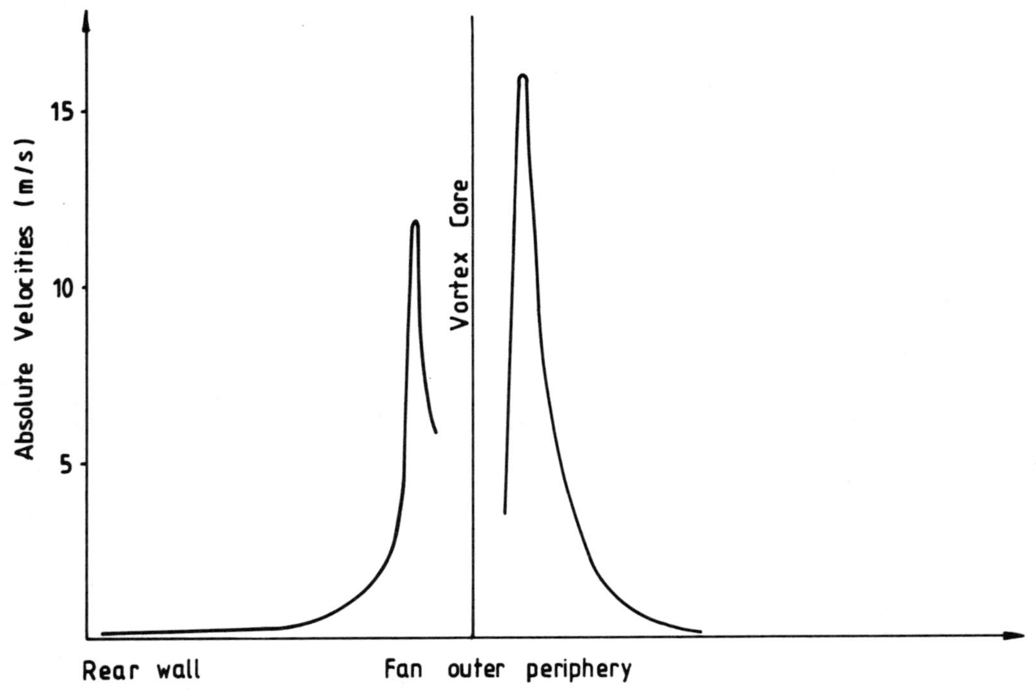

Fig. 2.

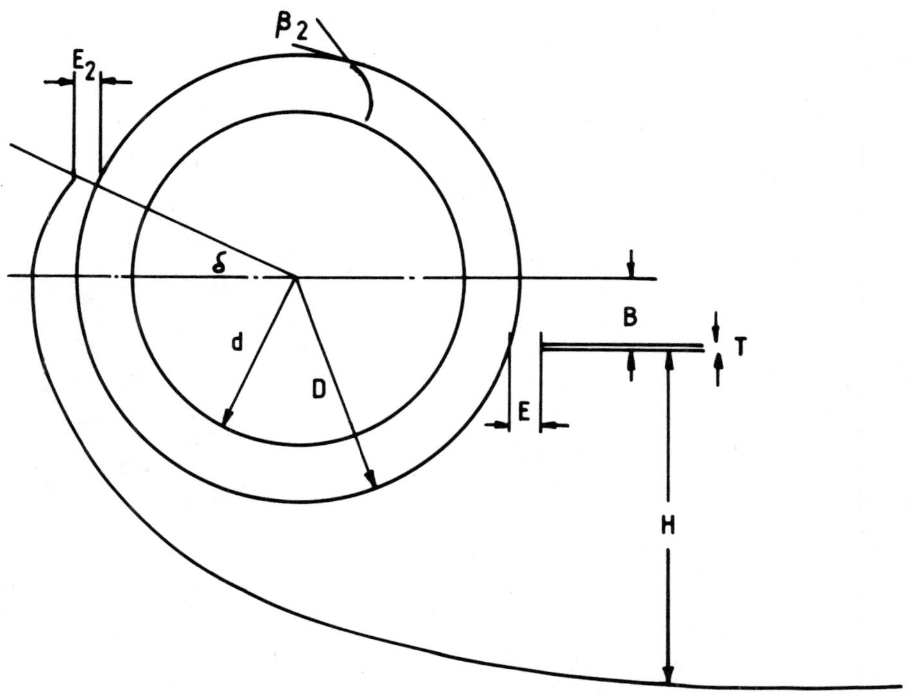

Fig. 3. PRIMARY VARIABLES

E/D
1 %
3 %
5 %
7 %
$\lambda = \frac{Watts}{\rho u^3 D^2}$
ψ
6·0
5·0
4·0
3·0
2·0
1·0
ψ_t
ψ_{st}
λ
10
5
·2
·4
·6
·8
1·0
1·2
1·4
1·6
φ

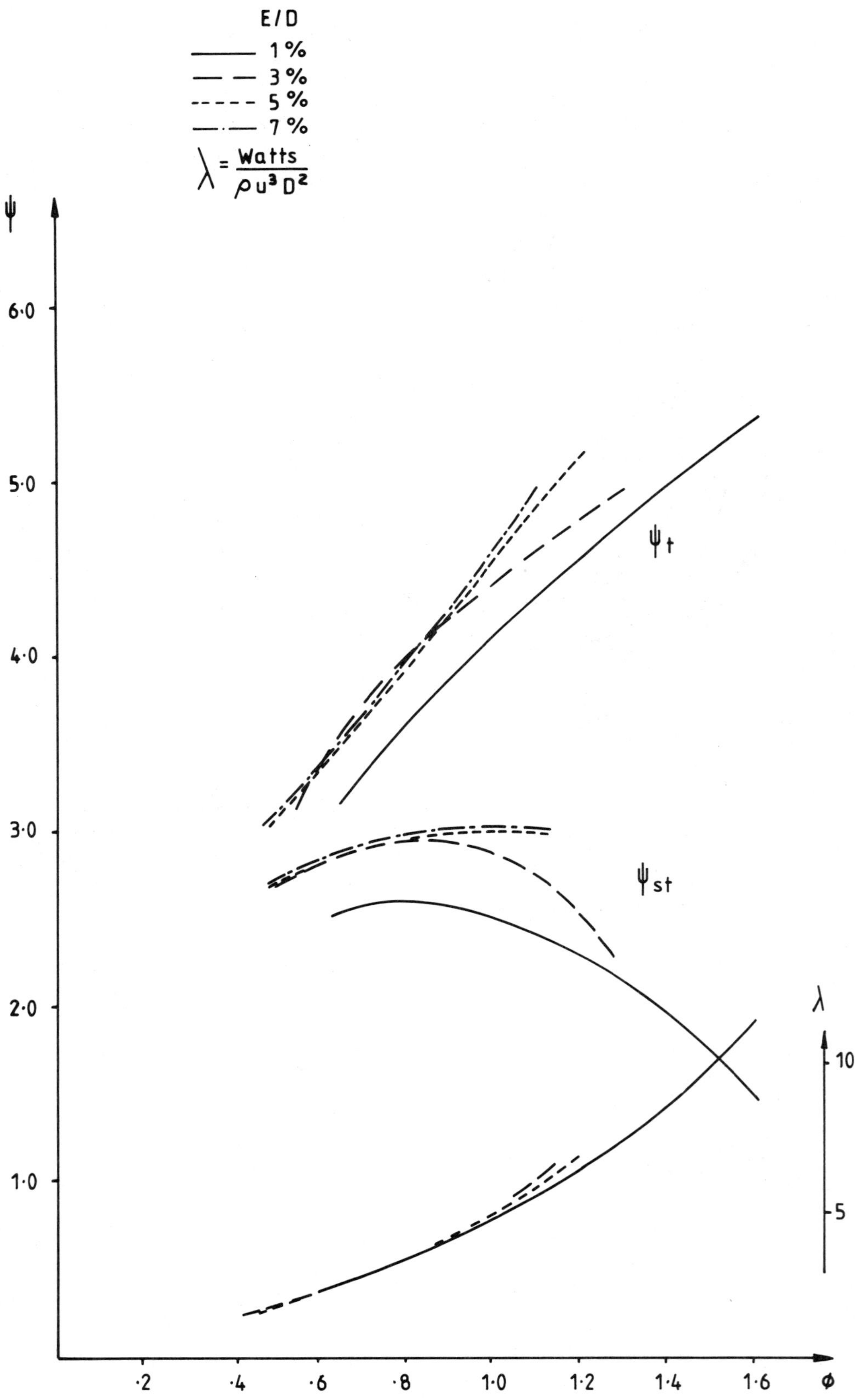

Fig. 4.

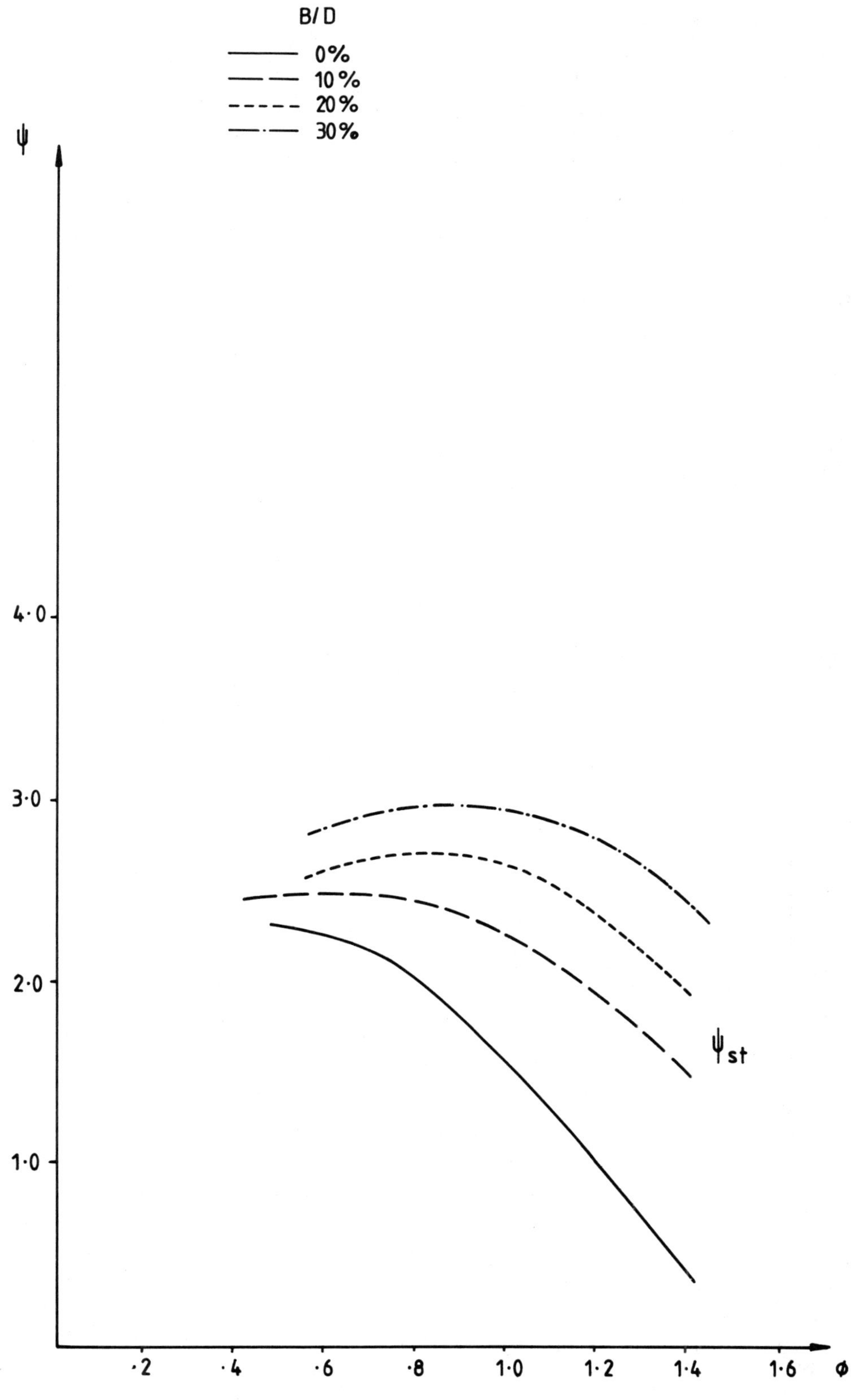

Fig. 5.

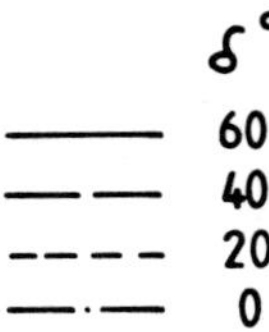

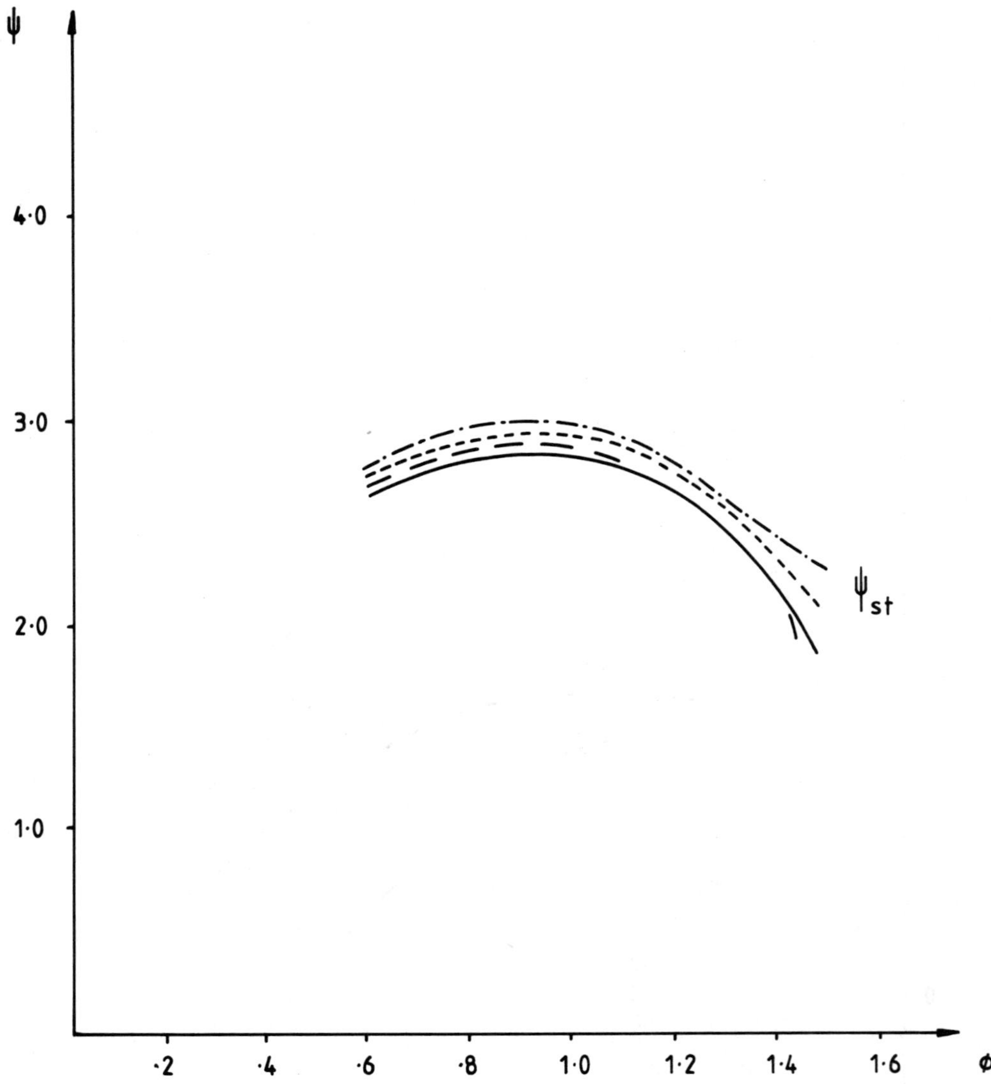

Fig. 6.

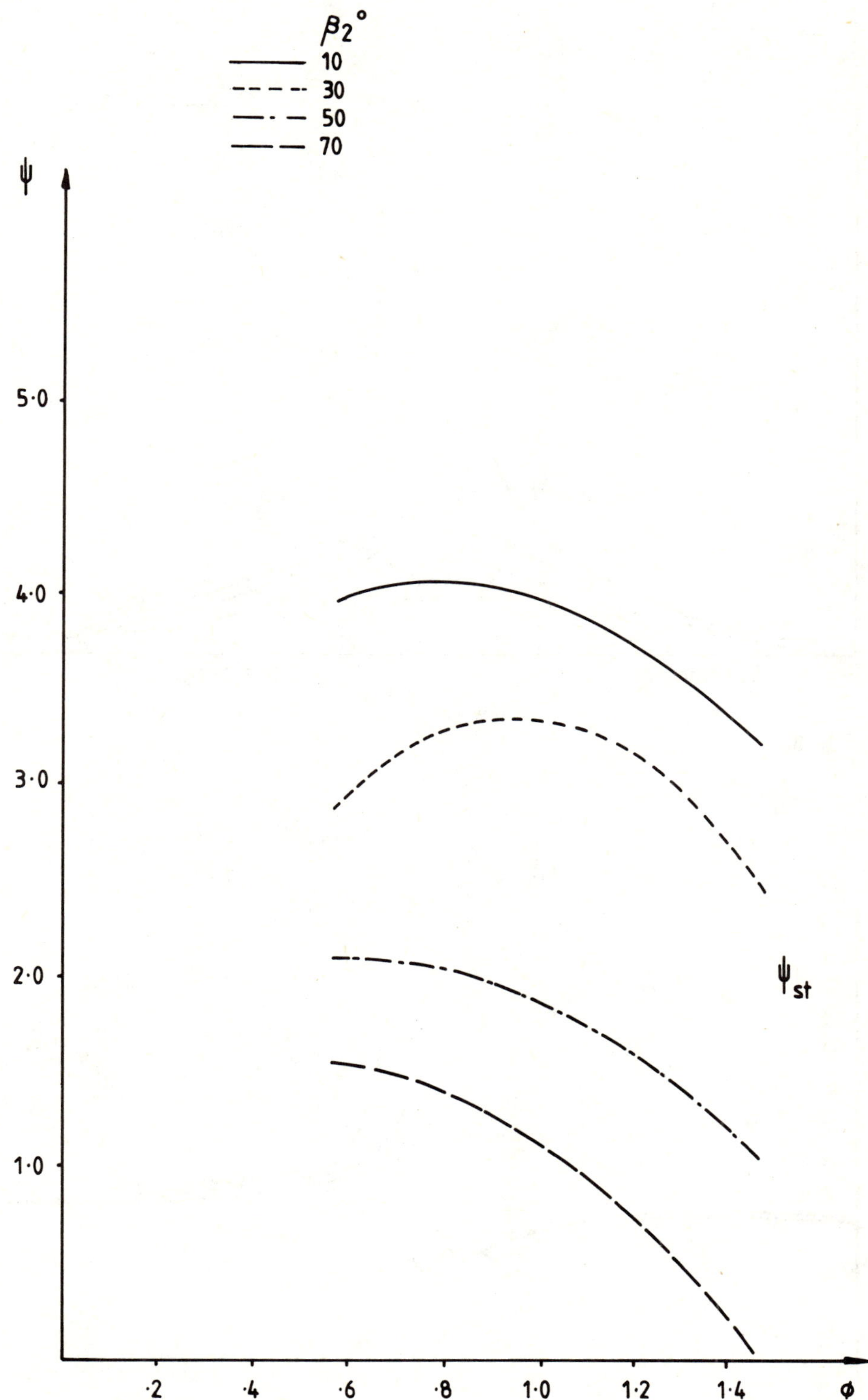

Fig. 7.

B/D %
0
10
20
30
VR
(m/s)
30
20
10
Suction Arc
Discharge Arc

Fig. 8.

VT
(m/s)
60
40
20
0
- 20
- 40
- 60
Suction Arc
Discharge Arc

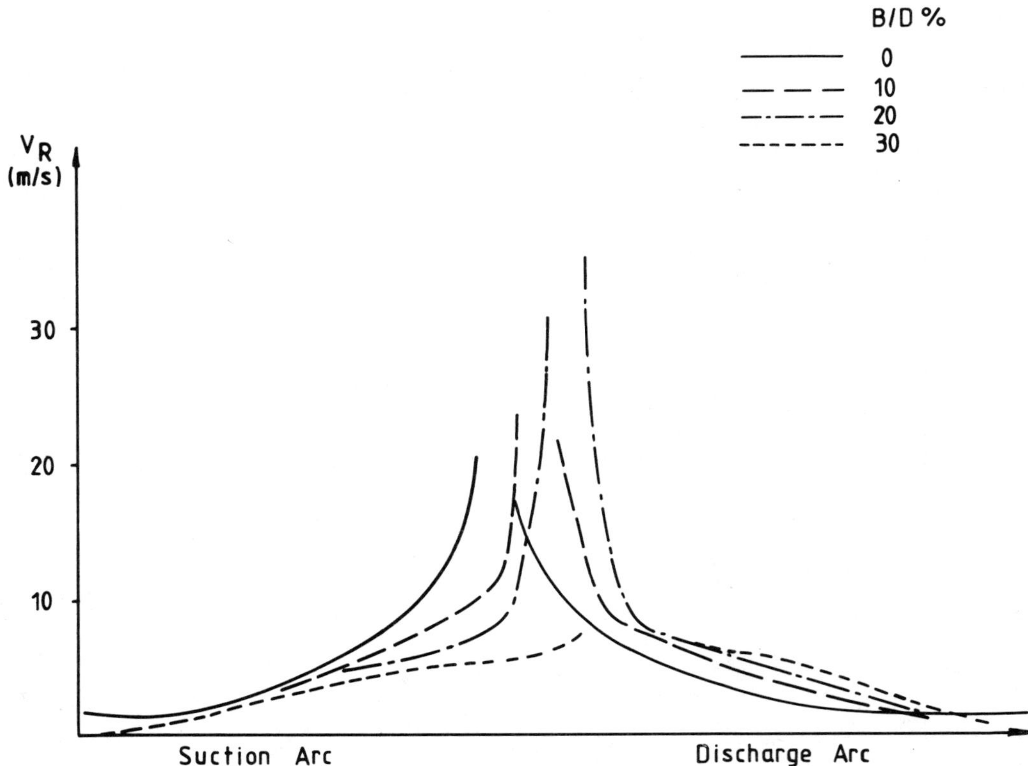

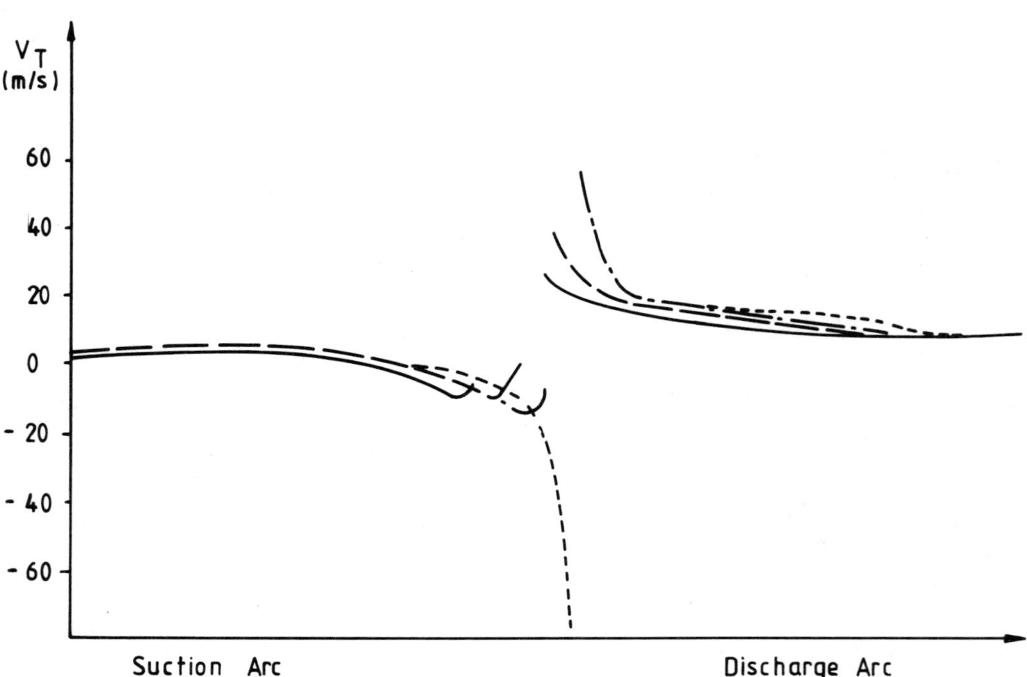

Fig. 9.

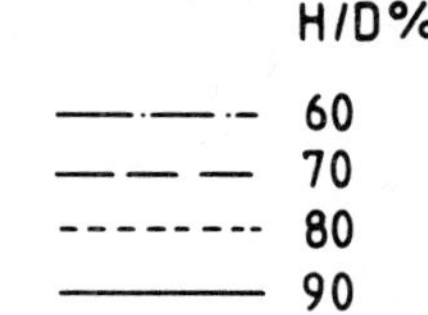

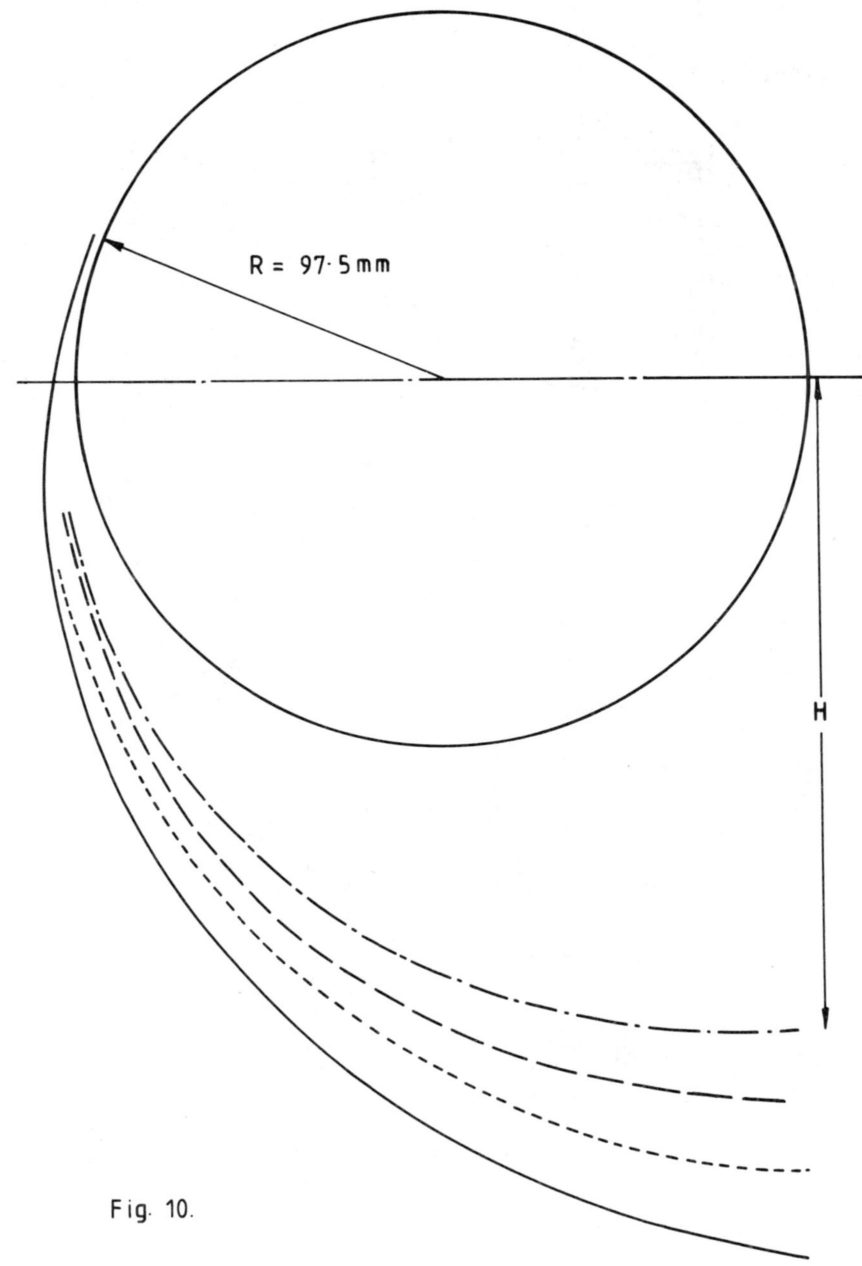

Fig. 10.

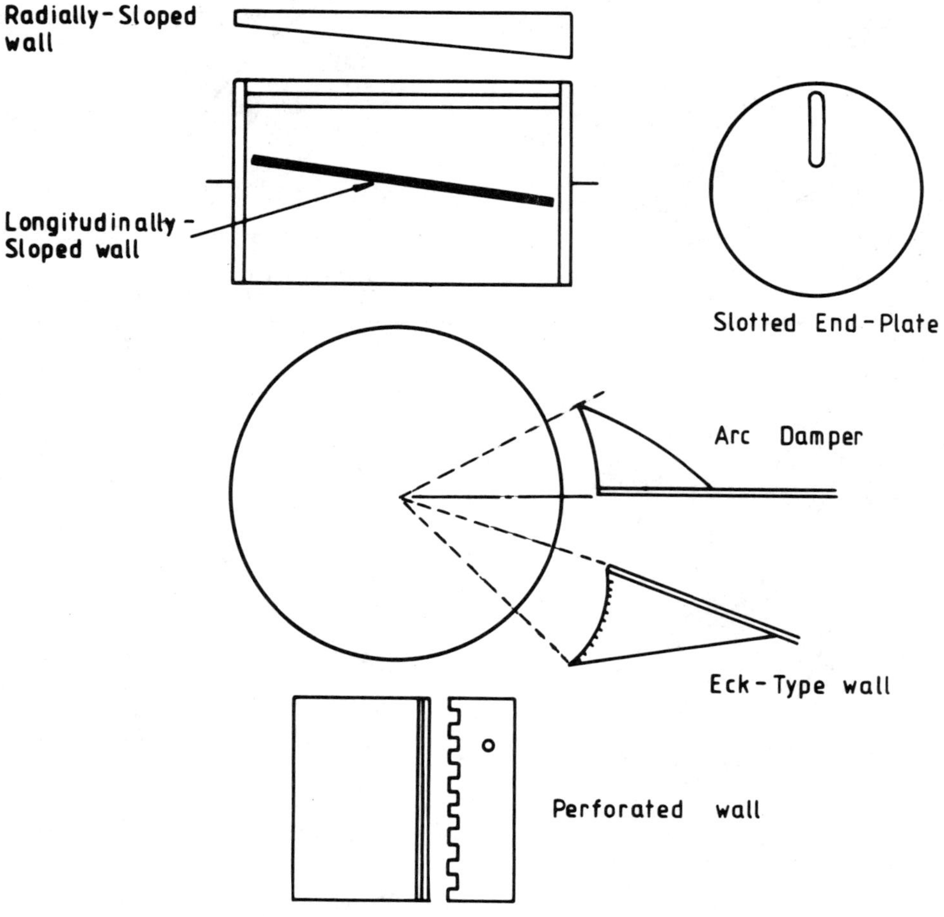

Fig. 11. NOVEL DESIGNS.

η_T %
50
40
30
20
ψ_{st}
3·0
2·0
1·0
0·5
1·0
ϕ
Conventional rear wall design.
Computor - aided design.

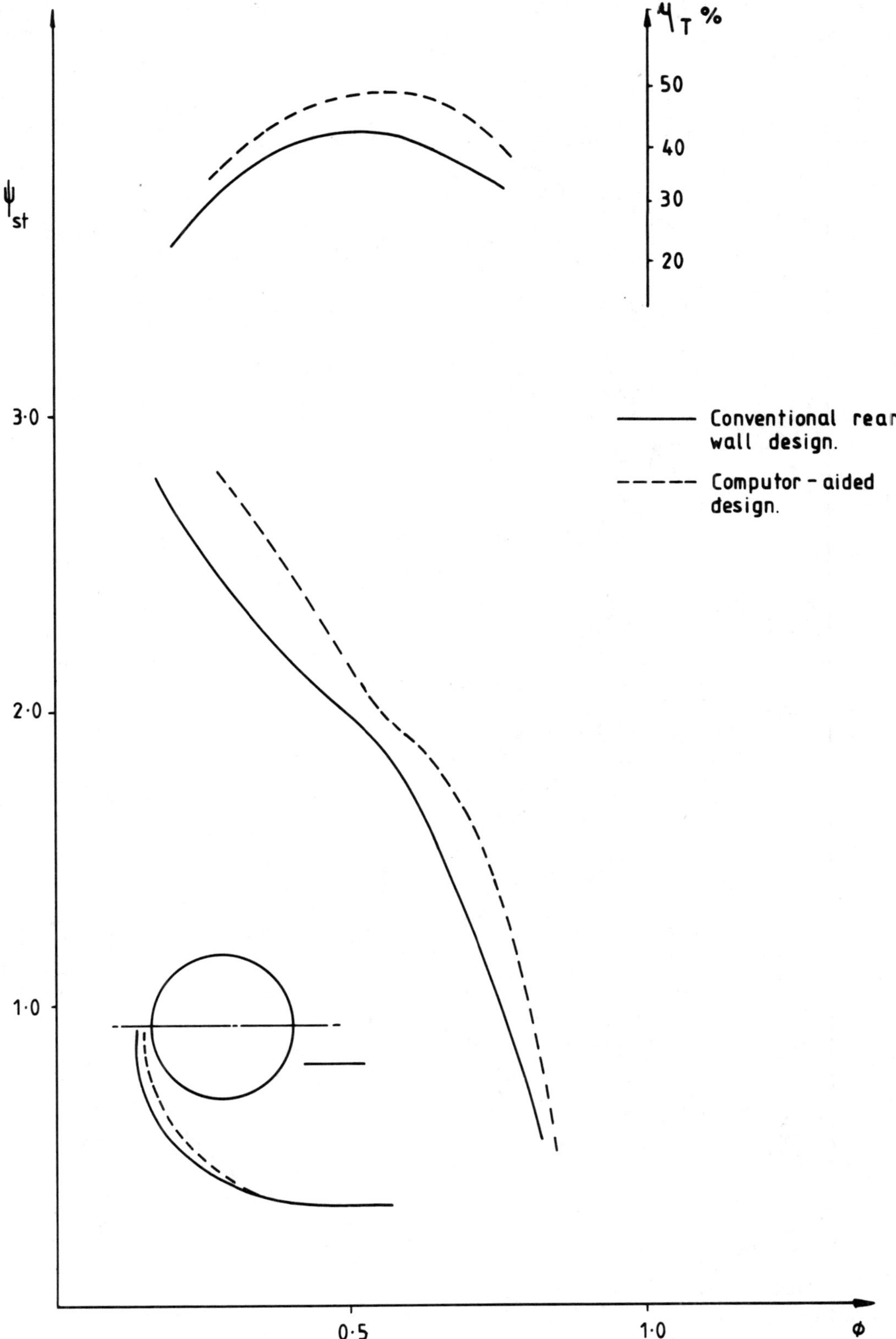

Fig. 12.

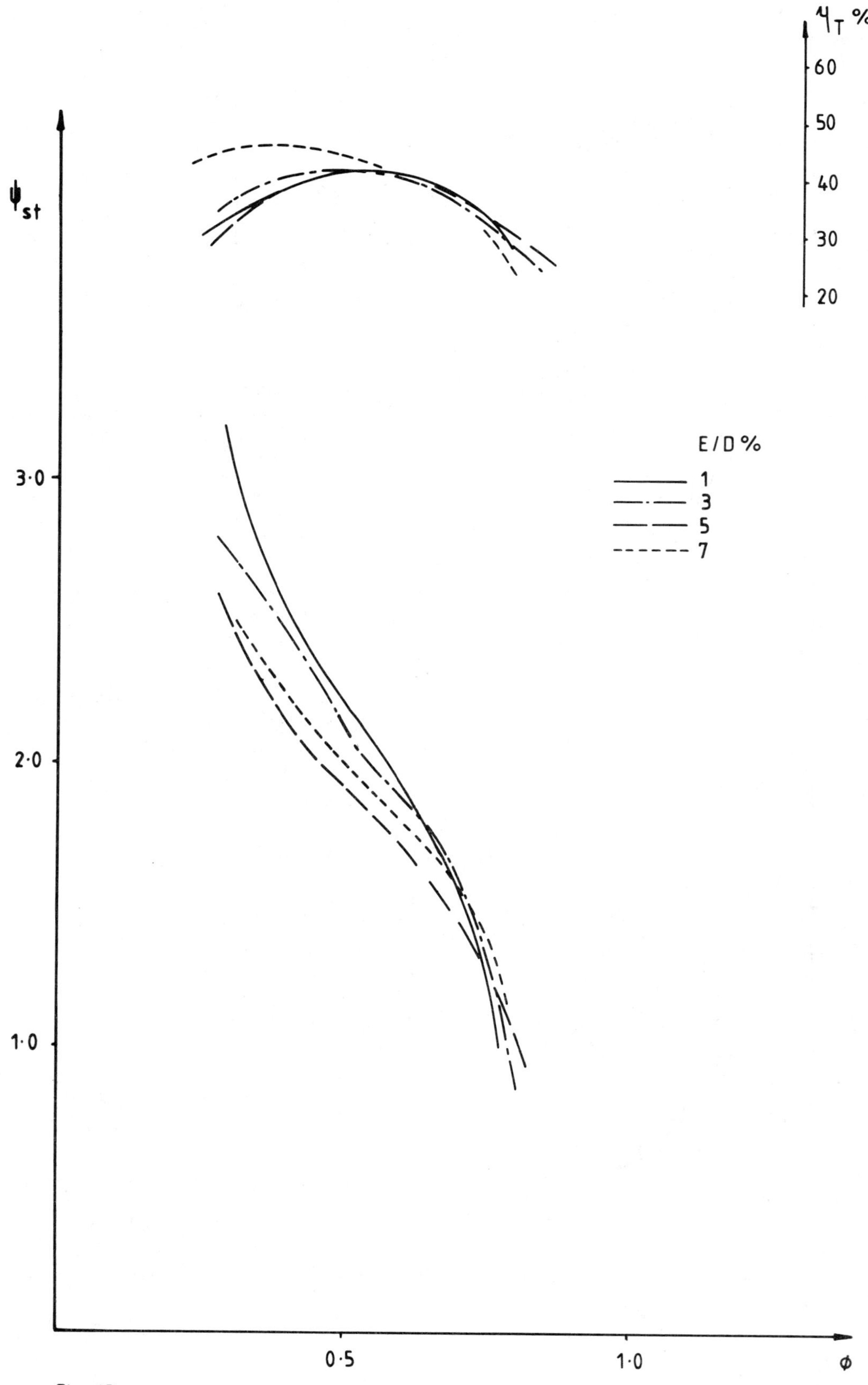

Fig. 13

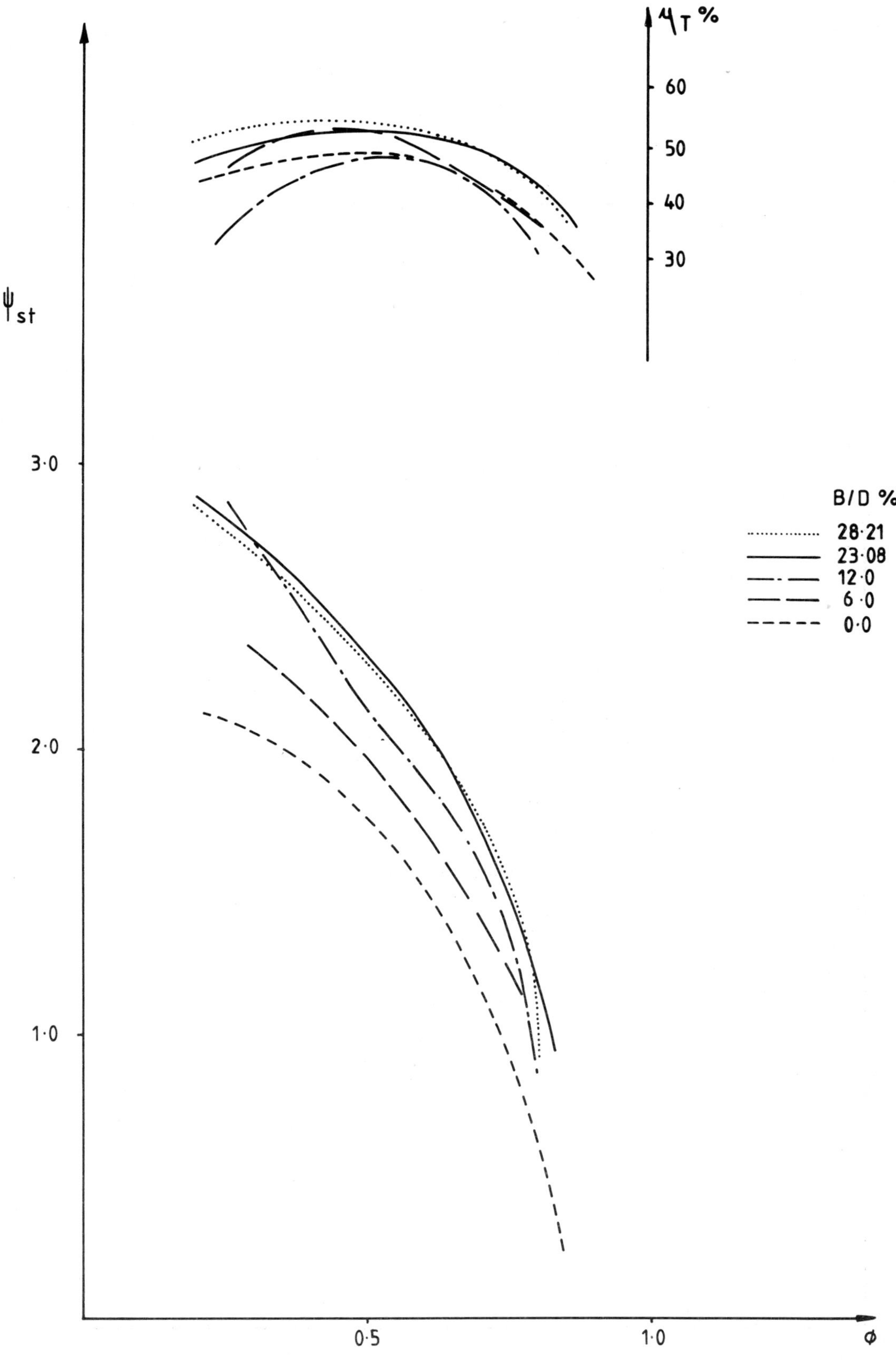

Fig. 14

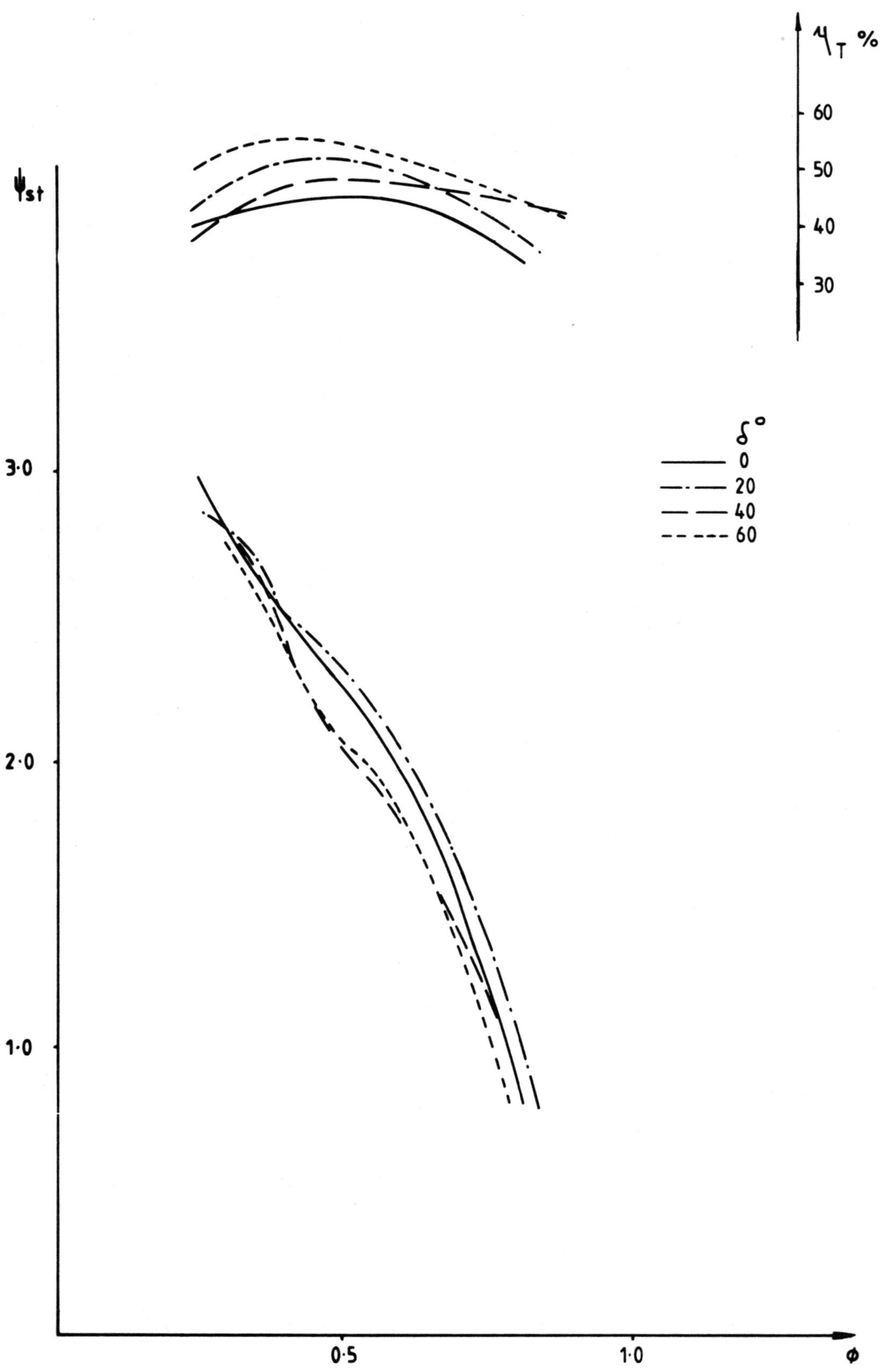

Fig. 15

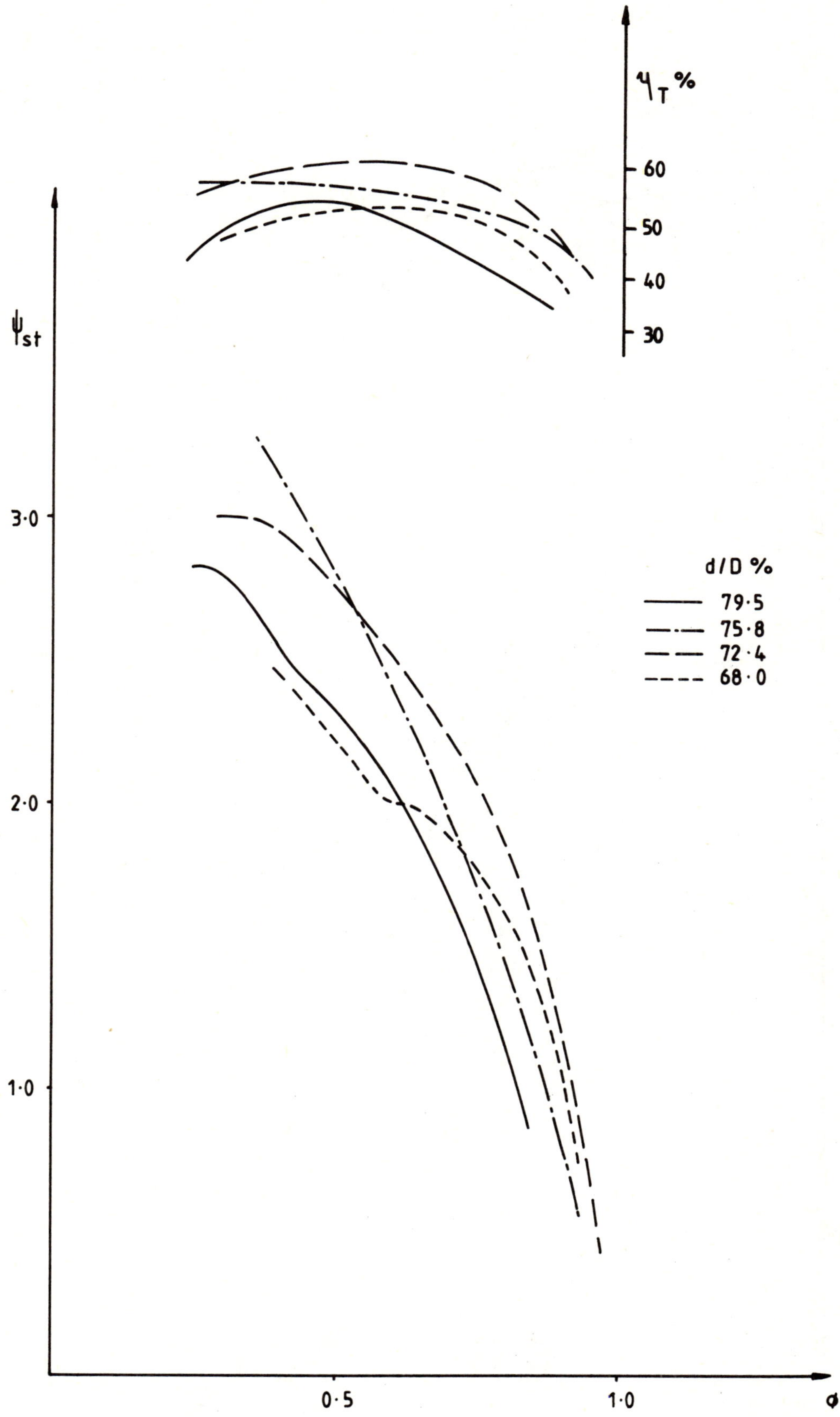

Fig. 16

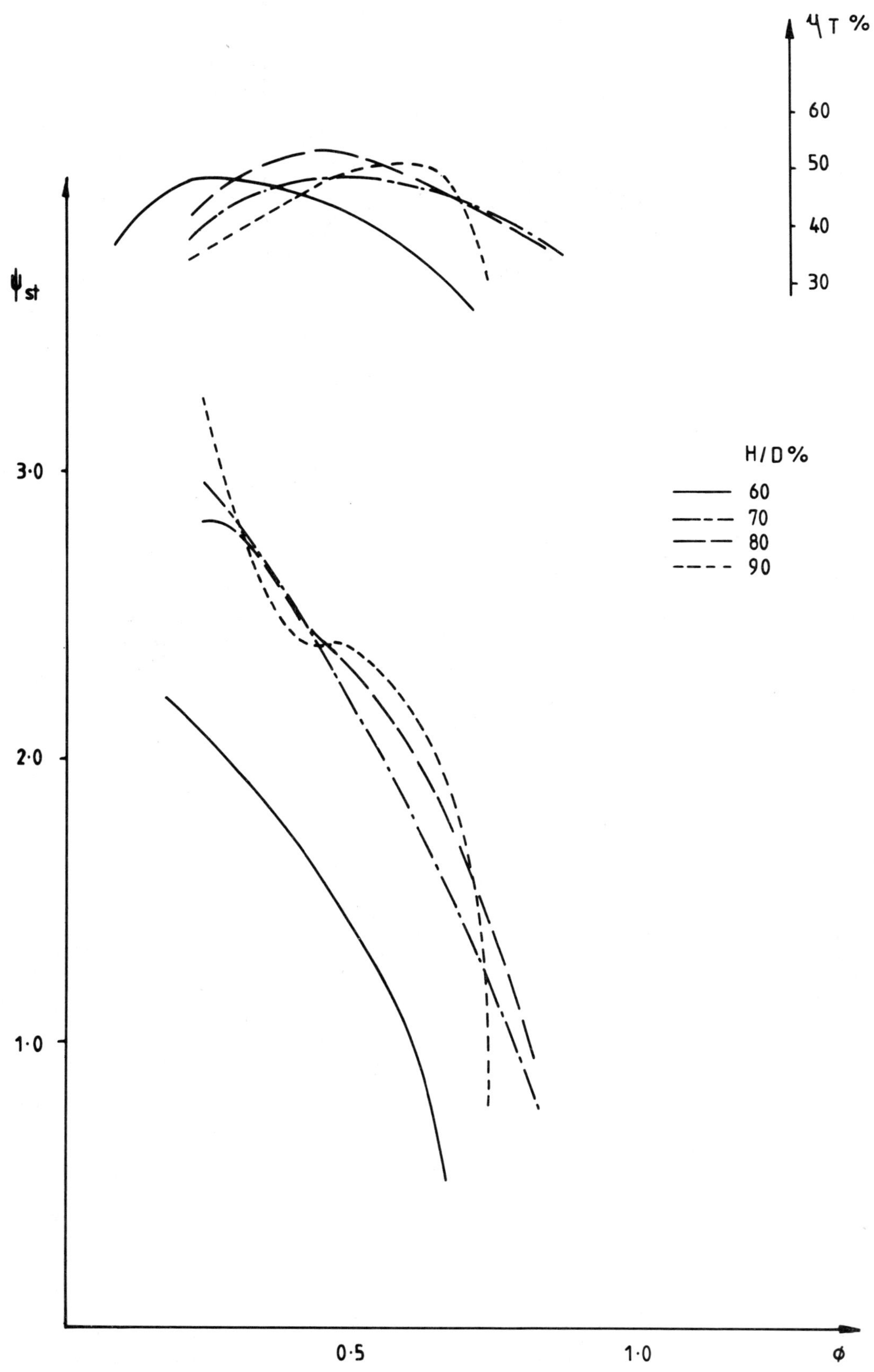

Fig. 17.

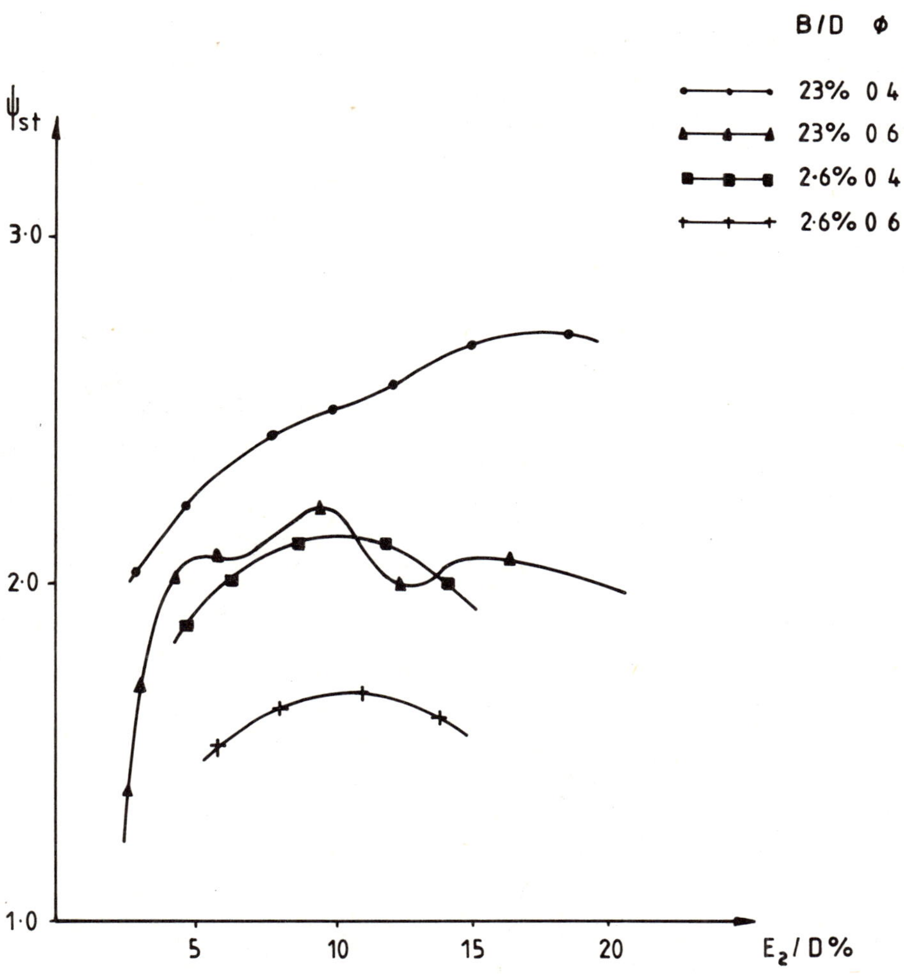

Fig. 18.

η_T %
60
50
40
30
ψ_{st}
3·0
2·0
1·0
0·5
1·0
ϕ
Plain Wall
40° Suction-Arc Damper
ECK-Type Wall

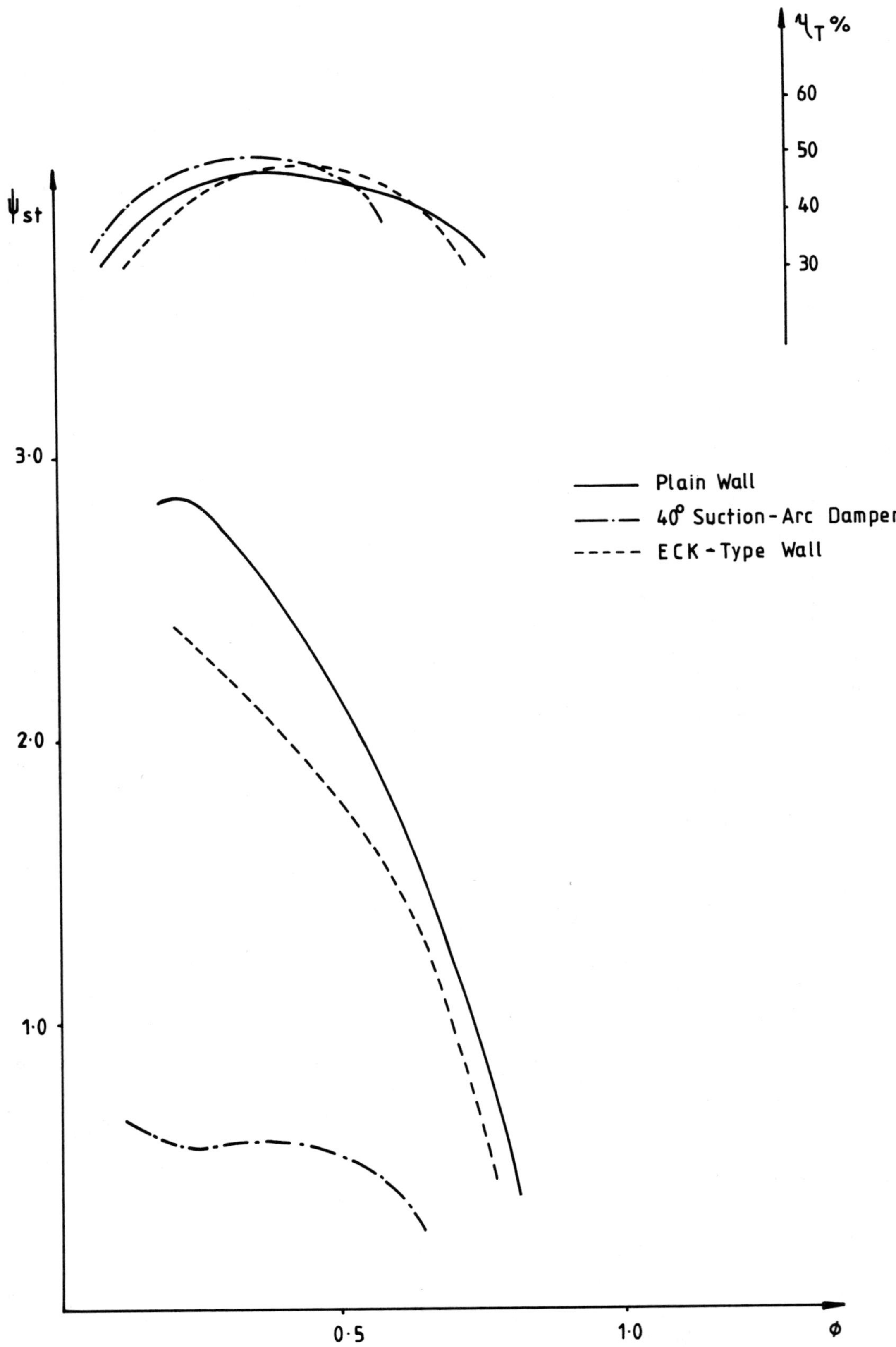

Fig. 19.

International Conference on

Fan Design & Applications

Guildford, England: September 7-9, 1982

PAPER J2

THREE-DIMENSIONAL EFFECTS DUE TO THE THROUGH-FLOW EDDY IN A MIXED-FLOW FAN ROTOR

G.W. Fairbairn

GEC Gas Turbines Ltd., U.K.

R.I. Lewis

University of Newcastle Upon Tyne, U.K.

Summary

It is well known that the relative eddy in a mixed-flow turbomachine plays a significant role in the energy transfer processes due to both coriolis forces and slip. However its effect upon the through-flow is frequently ignored. As shown in this paper significant twist of the meridional surfaces can occur due to the through-flow eddy which could invalidate the frequently made design assumptions of a basically axisymmetric flow. A calculation procedure is proposed which compares well with experimental investigations undertaken upon a large mixed-flow research fan. The effects are most pronounced for highly loaded machines without pre-whirl.

Organised and sponsored by
BHRA Fluid Engineering, Cranfield, Bedford MK43 0AJ, England.

0263 - 421X/82/01 00 - 0001 $5.00
The entire volume can be purchased from
BHRA Fluid Engineering for $82.00

NOMENCLATURE

a,b,c,d,e,f	= Points on sector perimeter
AA,BB etc	= Conical sections normal to mean meridional surface of revolution
c_m	= Mean meridional velocity
K_1, K_2	= Constants
r, θ	= Polar coordinates on transformed sector plane
R, Θ, x	= Polar coordinates
R_2	= Maximum radius of rotor
s_1, s_2	= Stream surface displacements
S_1, S_2	= Wu surfaces
T	= Time
ΔT	= Time increment
u,v	= Velocity components in r, θ directions
U_2	= Blade velocity at R_2
ΔZ	= Gap between planes AA,BB etc
α	= Cone angle of meridional flow
γ_{net}	= Net passage vorticity
τ	= Stream surface twist
ϕ	= Flow coefficient c_m/U_2
ψ	= Pressure rise coefficient, stream function
ψ_o, ψ_1, ψ_2 etc	= Meridional stream surfaces
Ω	= Rotor angular speed
Ω_b	= Apparent sector angular speed

Suffixes

A,B	= Planes AA and BB
m	= Mean meridional streamline
i	= i^{th} sector plane
1,2	= Inner and outer sector radii in r, θ plane

1. INTRODUCTION

If the absolute flow at entry to a turbomachine is irrotational, then the flow field viewed relative to the rotor, Fig.1, will have a vorticity distribution or "relative eddy" of strength -2Ω throughout. If we adopt the familiar quasi-three-dimensional design model, whereby elementary two-dimensional mixed-flow cascades are superimposed upon an assumed axisymmetric meridional flow, the relative eddy may be resolved into two components parallel and normal to each meridional surface. The normal component $-2\Omega\sin\alpha$ interacts with the blade to blade flow generating coriolis forces and also gives rise to "slip". This component is now normally included in advanced mixed-flow cascade analyses (Refs. 1 to 4). The parallel or "through-flow eddy" $-2\Omega\cos\alpha$, on the other hand, is usually ignored for design purposes in view of the extreme complications involved in calculating its effect.

The full three dimensional flow through an arbitrary turbomachine has been expressed in the famous classical paper by Wu (Ref.5) in terms of superimposed two-dimensional flows on S_1 and S_2 surfaces, Fig.2. Although there are conceptual differences between Ref.5 and the present work there is a close relationship between S_2 surface flows and the cascade flow and between the S_1 and S_2 combined flows and the through-flow eddy regime.

It is the main objective of this paper to investigate the second of these, namely the rotational effects induced by the through-flow eddy, with reference to mixed-flow fans in particular. An analytical procedure is developed in section 2 and applied first in section 3 to axial fans which present simpler geometry. Finally we proceed to mixed-flow fans in section 4. The procedure makes use of a basic theory outlined in detail in earlier publications, (Refs. 6 and 7) but which is too complex to repeat here except in brief outline, section 2.3. Finally the analysis is applied to the rotor of a large mixed-flow research fan at Newcastle University described in Ref.8. A technique for defining the average stream surface twist τ at each section plane AA, BB, CC etc., Fig.1, is defined in section 5 and compared with experimental measurements in section 6 undertaken by means of a thermal tracing technique described in Ref.9.

2. BRIEF OUTLINE OF THE PROBLEM AND METHOD OF SOLUTION

2.1 The problem

Fig.1 illustrates both the influence of the through-flow eddy and our method of analysis. A series of conical sections AA, BB etc are taken roughly normal to the meridional flow surfaces. A typical section such as CC will thus be in the form of a sector of the annulus abcd defined by hub, casing, pressure surface and suction surface respectively. Each section is filled with vorticity $-2\Omega\cos\alpha$ normal to abcd which thus generates a general rotation of the flow as viewed along the rotor passage. At entry, section AA, a surface of revolution ψ_0 from upstream will thus form a circular arc ef. Proceeding to section BB under the influence of the through-flow eddy the stream surface ψ_0 will then be twisted anticlockwise. Proceeding progressively through the blade passage we see that there is a cumulative twist which for heavily loaded machines could be considerable and which might seriously threaten the original assumption of axisymmetric meridional flow. In addition to this it would appear that particles at e and f will be shifted through decreased and increased radii respectively with consequent differences in work input. For severe twists the use of two-dimensional mixed-flow cascade modelling might thus begin to lose credibility.

2.2 The proposed analysis

The proposal is that we upgrade the quasi-three-dimensional design model previously mentioned by introducing an additional series of two-dimensional flows in the AA, BB etc planes which account for the influence of the relative eddy and which link the assumed two-dimensional cascade flows. The full three-dimensional flow would thus be modelled by superposition of a large number of coupled two-dimensional flows which could be solved iteratively.

In practice the coupling terms would require such major modifications to the cascade and through-flow two-dimensional theories that it would be less trouble to aim directly at a fully three-dimensional analysis. For the present purpose therefore we have ignored the coupling process and attempted only a first order estimate of the through flow eddy, ignoring blade to blade variations. As shown in section 6 little seems to have been lost in neglecting the coupling effects.

The main assumption then is that the flow in each sector may be treated as a two-dimensional flow under the influence of the through-flow eddy $-2\Omega\cos\alpha$ applicable to that particular plane. The procedure is then as follows.

(i) Compute u,v velocities for a grid of points P located on plane AA. For example line ef could lie on this grid, Fig.1.

(ii) Calculate the new position of the streamlines passing through these grid points as the fluid proceeds to plane BB. Thus the coordinate change of a given fluid particle would be

$$r_B = r_A + u\Delta T \qquad \theta_B = \theta_A + v\Delta T \tag{1}$$

where the time step ΔT_i is related to the meridional velocity c_m (here assumed constant over each section AA, BB etc) and the gap ΔZ between the planes, through the equation

$$\Delta T_i = \frac{\Delta Z}{c_m} \tag{2}$$

(iii) Compute u,v velocities on the next plane BB for each streamline location.

(iv) Proceed to plane CC.

etc

2.3 Rotational flow in a sector

A full analysis for computation of u,v velocities in the sectors is given in Ref.7. In the case of an axial fan, Fig.3, the flow lies in a sector plane normal to the axis of rotation. In plane polar coordinates, the flow may then be expressed in terms of a stream function ψ which satisfies the governing equation

$$\frac{\partial^2\psi}{\partial r^2} + \frac{1}{r}\frac{\partial\psi}{\partial r} + \frac{1}{r^2}\frac{\partial^2\psi}{\partial\theta^2} = -2\Omega \tag{3}$$

$$\text{where } u = \frac{1}{r}\frac{\partial\psi}{\partial\theta} \qquad v = -\frac{\partial\psi}{\partial r} \tag{4}$$

The solution of these equations is the main subject matter of Ref.7 and no further detail will be given at this point. The most elementary application of the solution is to the straight walled rotating passage depicted in Fig.4. In this special case the blades are parallel to the axis resulting in a truly two-dimensional flow for which the solution is exactly applicable. At any sector the streamline pattern conforms to that calculated and plotted also in Fig.4. For the duct geometry summarised in Fig.5 the predicted streamline shifts are thus very considerable as one might expect for this case. It is of interest to point out here that for such a rotor preceded by guide vanes, designed to produce solid body pre-rotation of angular velocity Ω, there would in fact be no vorticity seen by the relative flow and thus no through-flow eddy or stream surface twist. The presence and design of inlet guide vanes thus plays an important role in determination of through-flow eddy effects. In the present paper however calculations are undertaken only for fans without inlet guide vanes.

In more realistic machines, on the other hand, blades are staggered and cambered resulting in a helical passage form which gives rise to certain difficulties which we will consider next.

3. APPARENT ROTATION OF SECTOR WALLS - AXIAL FAN

If we consider sectors AA, BB etc of a practical axial fan, Fig.3, it is immediately obvious that the duct itself is in apparent anticlockwise rotation as one proceeds with the flow. The consequent angular velocity of the sector boundary relative to sector i, proceeding from i to i + 1, is given by

$$\Omega_b = \frac{\theta_{m,i} - \theta_{m,i+1}}{\Delta T_i} \tag{5}$$

where $\theta_{m,i}$ and $\theta_{m,i+1}$ are the angular coordinates of the blade camber line at the ith and $(i+1)^{th}$ sections respectively. Unpleasant though this new found problem is, it cannot be ignored since Ω_b may be of similar magnitude to Ω. The solution to the difficulty is however fortunately simple since, if we rotate our coordinate system with the duct as we progress from sector to sector, all that is needful is to modify the passage vorticity as follows

$$\gamma_{net} = -2\{\Omega - \Omega_b\} \tag{6a}$$

for the axial fan, and for the mixed-flow fan

$$\gamma_{net} = -2\cos\alpha\{\Omega - \Omega_b\} \tag{6b}$$

At this point it is of interest to note that in the case of an Archimedean screw, $\Omega = \Omega_b$, resulting in no net through-flow passage vorticity γ_{net}. Indeed in this case, which is equivalent to a flat plate closely pitched cascade with zero angle of attack, there is zero blade loading or work done. As will be shown later this agrees with the general result that through-flow stream surface twisting increases with blade loading and diminishes to zero for unloaded blades, which was a most significant finding of this research.

4. APPLICATION OF THE PROCEDURE TO MIXED-FLOW FANS

The stages in the calculation of the through-flow eddy in a mixed-flow fan are in principle as already described but with certain modifications as follows.

(i) The sectors AA, BB etc lie on conical rather than plane surfaces. However the matter is simply resolved by applying a conformal transformation which unwraps the cone onto a plane. The cylindrical polar coordinates (r,θ) then transform onto the flat (r,Θ) plane according to the relationships, Fig.6

$$R = r\cos\alpha \qquad \Theta = \theta/\cos\alpha \tag{7}$$

This transformation is applied first to each of the chosen sectors and solutions for each sector form the basis of the step by step procedure.

(ii) Two geometrical problems arise as illustrated by Fig.7. The first problem is associated with blade lean angle λ. Unfortunately the theoretical solutions (Ref.7) apply only to sectors defined by radial lines. This problem is handled by defining initially equivalent idealised sectors with zero lean angle and making the assumption that meridional streamline twist will not be substantially affected.
The second geometrical problem arises due to annulus taper. Unlike the cylindrical annulus considered in section 3, the ratio of transformed inner and outer radii, r_1/r_2, changes from section to section. There is thus a second distortion of the sector which presents difficulties in transferring flows from say sector i to sector i + 1, Fig.7. To deal with this, it is first assumed that there is no change in sector shape proceeding to sector i + 1, Figs.7(c) and 7(d) so that the new position of each stream sheet ψ_Θ may be calculated according to equations (1) and Ref.7. A matching operation is then undertaken to relocate each fluid particle onto an equivalent location in the actual (zero lean) sector at i + 1. This process involves a linear squeezing of the flow pattern into sector i + 1 in moving from Fig.7(d) to 7(e).

Although some accuracy may be lost in these procedures, they are easy to implement and result in considerable simplification. Although the problem of lean could be removed by a more advanced analysis, it is not easy to see how to deal in a more accurate manner with the second problem of change in annulus height.

(iii) Some attempt was made to account for blade to blade variations obtained previously from cascade calculations. Meridional flow, thus obtained, varied in the θ direction, resulting in time steps ΔT which were also functions of θ. Thus

$$\Delta T(\theta) = \frac{\Delta Z}{c_m(\theta)} \tag{8}$$

This influenced the shifting of the streamlines over each ΔZ step but for the purpose of calculation of u,v velocities due to the net through-flow vorticity, the ΔT value used in equation (6)b was the average value for each sector.

5. DEFINITION OF A TWIST PARAMETER

Mid-way through this project it was discovered that there is a close relationship between the blade loading and the amount of stream surface twist, as will be illustrated later. A twist parameter τ was therefore defined as follows and as shown in Fig.8 to express the general degree of rotation of the mean stream surface ψ_0.

$$\tau = \frac{s_1 + s_2}{r_2 - r_1} \tag{9}$$

5.1 Influence of flow coefficient upon twist

A cross-section of the Newcastle rig is shown in Fig.9 and its characteristic curve is shown in Fig.10. Calculations of the through-flow eddy of this rig have been made for a range of flow coefficients and the more detailed results of these tests will be compared with experiment in the following section. Before presenting them it will be of interest to examine the relationship between twist and flow coefficient.

To achieve this numerical solutions were first derived for a good number of flow coefficients and τ was calculated according to equation (9). From the results, Fig.11, it is significant to note a general trend that τ decreases with ϕ. For a value of $\phi = 0.5$ the twist is zero and actually becomes negative for higher values. For $\phi<0.5$ on the other hand τ increases fairly rapidly.

It is particularly important now to compare Fig.11 with the characteristic, Fig.10. We observe a direct correspondence between the flow/head (ϕ,ψ) curve and the flow/twist (ϕ,τ) curve. Thus for $\phi<0.5$, when the machine is acting as a fan, positive twist occurs. For $\phi>0.5$, when the machine is actually operating as a turbine, the eddy twist is reversed. At the zero head condition for $\phi = 0.5$ there is no twist at all. At this condition the machine is behaving as an Archimedean screw and is neither absorbing or generating "Euler" work. Thus we deduce also that more heavily loaded fans will experience stronger through-flow eddies at the design point than will lightly loaded fans. Maximum design errors from this source will thus originate from heavily loaded fan designs, a rather unfortunate conclusion.

This relationship can be demonstrated in a simple way by the following arguments. The relative eddy strength $2\Omega\cos\alpha$ is dependent upon the rotational speed of the machine Ω and not upon the flow rate. The magnitude of stream surface twist will depend upon the eddy strength but also upon the time ΔT taken by a fluid particle to move through the blade passage. This implies that stream surface twist will move with reduced flow coefficient in agreement with Fig.11. On the other hand the passage twist due to stagger and camber is related to circumferential shift $\Delta\theta$ along the camber line, by, (c.f. equation (5))

$$\text{passage twist} \propto \frac{\Delta\theta}{\Delta T}\,\Delta T \qquad (10)$$

and is thus obviously independent of flow rate.

The net twist is thus not directly proportional to $1/\phi$ but certainly decreases with ϕ. Let us extend this simple analysis further under the following assumptions:

(a) The meridional velocity is assumed constant throughout the blade passage.
(b) The term $(\theta_{m,i+1} - \theta_{m,i})/\Delta T_i$, equation (5), may be replaced by $\Delta\theta/\Delta T$, the average rate of change of angular position along the camber line.

Following these assumptions the passage vorticity, equation (6)b, becomes

$$\gamma_{net} = -2\Omega\cos\alpha(\Omega - \Delta\theta/\Delta T) \qquad (11)$$

If we now introduce the flow coefficient, defined

$$\phi = \frac{c_{m2}}{U_2} \qquad (12)$$

ΔT may be eliminated from equation (11), whereupon

$$\gamma_{net} = -2\Omega\cos\alpha\left(1 - \frac{\Delta\theta R_{2m}\phi}{s_m}\right) \qquad (13)$$

The case of zero twist corresponds to

$$\gamma_{net} = 0$$

$$\phi = \frac{s_m}{\Delta\theta R_{2m}} \qquad (14)$$

It can be demonstrated that this case corresponds to blading with a camber line coincident with the path traced out relative to the rotor by a fluid particle with zero absolute circumferential motion, the Archimedean screw case. This would logically represent the condition of zero loading. Thus for zero loading there would be zero twist.

For all other conditions

$$\tau \propto \gamma_{net}\Delta T$$

which after substitution from equation (13) becomes

$$\tau = \frac{K_1}{\phi} - K_2 \qquad (15)$$

where K_1 and K_2 are constants. This is the equation of a rectangular hyperbola and is shown on Fig.11 as a broken line. Although some discrepancy exists between this hyperbola and the value of twist calculated by the more advanced theory, the similarity of general form is apparent confirming the above arguments.

6. MEASUREMENTS IN THE NEWCASTLE MIXED-FLOW RESEARCH FAN

Reference has already been made to the Newcastle University mixed-flow research fan rig which is fully described in Refs.6 and 8. The design was accomplished with the help of a mixed-flow version of Martensen's cascade analysis developed by Fisher, Ref.4. The assumed axisymmetric and conical meridional mean stream surfaces were transformed conformally onto flat planes. The appropriate infinite straight cascade on the mean plane was an uncambered NACA 65-0006 profile staggered at 60°, which was then transformed back to the conical surface. Conditions just described were applicable to the mean surface and sufficient twist was introduced to maintain constant loading on all stream surfaces and consequently free vortex flow. The resulting design duty point was located at $\phi = 0.4$, $\psi = 0.184$. As demonstrated by Fig.10 a fairly reasonable agreement was obtained between the test (ϕ,ψ) characteristic and that predicted by means of the mixed-flow cascade theory combined with an assumed axisymmetric flow, with maximum disagreement at the smallest flow rates. Agreement was close for flow coefficients in excess of $\phi = 0.3$ for which there was minimal twist, Fig.11. For $\phi<0.3$, as twist increased rapidly, theory and experiment began to exhibit more marked disagreement although it is not clear below which ϕ value blade stall would begin to produce the major influence of head loss. Over the range $0.2<\phi<0.3$ for which stall would progress, predicted twist more than doubled and it is likely that this resulted in a reduction of work input. However other real factors came into play which disturbed this situation as discussed in section 6.

The predicted twist of the surfaces of revolution at the design point $\phi = 0.4$ is shown in Fig.12. It is clear from this that considerable departure from axisymmetric flow might be expected at further reduced flow rates.

6.1 Thermal Tracing of Meridional Stream Surfaces

To verify the theoretical predictions a thermal tracing technique was developed which is fully described in Refs. 6 and 9. A system was required which could be mounted within the rotor with minimal interference to the flow, which could withstand centrifugal effects with geometrically acceptable strains and which would give a resolution of streamline location to within ± 1 mm in an airstream of about 10 m/s at a distance of 600 mm downstream of the tracer source. In addition the method was required to be non-toxic. None of the known flow visualisation techniques was able to meet these stringent requirements and so a remarkably simple method of thermal tracing was developed which not only met the specification but which could be set up with elementary 'in-house' equipment at negligible cost. This method is highly recommended to other experimenters.

The method involved the location of an electrically heated wire perpendicular to the main flow direction at the impellor inlet, Fig.13. The centre of the thermal wake from the wire could then be located further downstream by thermocouple traverses which proved to be not excessively difficult. This instrumentation was first developed in a wind tunnel where it was discovered that the thermal wake remains extremely well defined up to a metre downstream and from the symmetry of the thermal wake it is possible to induce the streamline location to within ± 1 mm. The main difficulty experienced was that of accurately prescribed positioning of the heater wires which could not for example be located on circular arcs in accord with the ψ_0 surfaces of revolution. In practice the wires were straight.

The stream surface twist flow coefficients of $\phi = 0.4$ and $\phi = 0.29$ are shown in Figs.14 and 15 respectively, looking back into the exit plane. Note that the stream surfaces investigated here are not the axisymmetric design surfaces but are the 'pseudo-surfaces' close to them but defined by the straight heater wires. The non-twisted surfaces which would have occurred in the absence of through-flow rotation are also shown for comparison to reveal the degree of rotation actually experienced.

6.2 Discussion of Measured Results

At $\phi = 0.4$, Fig.14, the general rotation levels agreed well with predictions and the two inner surfaces were in quite close agreement. For the outer wire however the actual twist was in excess of that predicted and the traces showed an inward displacement of the meridional flow from that predicted by inviscid theory. This displacement was almost certainly related to a measured drop in meridional velocity within the outer 10% of the passage height, i.e., the developing annulus boundary layer. The discrepancy in twist of the outer surface may be due partly to this but also to tip clearance effects which help to develop the annulus boundary thickness and losses and which introduce additional shed passage vorticity. Fairly similar levels of agreement between theory and experiment were obtained at the reduced flow coefficient of 0.29, Fig.15.

However at reduced flow rates the prediction of stream surface twist began to break down quite quickly as illustrated by Fig.16, where twist is plotted versus flow coefficient over the range $0.25<\phi<0.4$ at which experimental traverses were completed. As can be observed for the flow range 0.3 to 0.4 measured and predicted twists were in very close agreement which was most encouraging, but as ϕ

was reduced below 0.3 τ began to reduce to values well below the predicted curve. Ultimately at ϕ = 0.2 the experimental twist reduced to approximately zero. There are two major factors which might cause this phenomenon. Firstly, according to design predictions separation was expected to begin to occur as ϕ reduced below 0.3, which certainly seems to agree with the above observations. It is probable that once major separations occur the potential flow calculations would cease to be representative and the jet-wake model approach of Whitfield, Ref.10, would offer a better explanation of the flow analogous to that which occurs in centrifugal machines. Secondly the extreme twists even at ϕ = 0.3 have moved the situation from an assumed distribution of reasonably independent two-dimensional flow into a strong three-dimensional flow.

An attempt was made to detect the onset of stall below ϕ = 0.3 by means of smoke traces viewed stroboscopically. These experiments revealed that there was very little evidence of separation at ϕ = 0.36 whereas at ϕ = 0.22 the separation was extensive. However it was not possible to demonstrate a sudden onset of major separation at ϕ = 0.3.

Another important factor revealed by the traverses was an indication at ϕ = 0.22 that the flow approached that of a forced vortex. This vortex was of opposite sense to the relative eddy and would thus result in cancellation of the stream surface twist. The vorticity giving rise to the forced vortex would originate from two sources, namely vorticity shed from the blades and so called "smoke ring" vorticity, Ref.11, due to gradients of stagnation pressure caused by radial variations of work input. The first of these, which would be strongly influenced by local blade stall, is likely to have been the predominating influence.

7. CONCLUSIONS

The above analysis has demonstrated a viable method for predicting meridional stream surface rotation due to relative eddy in the rotor passages of a mixed-flow turbomachine. The advantage of the method is that it provides an extension of the traditional quasi-three-dimensional approach which avoids the necessity for a full three-dimensional analysis. A thermal tracing technique applied to a large mixed-flow fan rig, has confirmed that stream surface rotations do occur and that these are well represented by the theory over the stable operating range of the fan. Although other three-dimensional effects are present in fans such as tip clearance and annulus boundary layer vortex effects, for axial and mixed-flow fans with low blade numbers and long blade chords, the relative eddy would normally be the principal cause of stream surface rotation within the rotor passages. These latter effects are thought to have been contributory to the collapse of the through-flow eddy rotation at reduced flows for regions of the fan characteristic where blade stall was occurring.

The stream surface twist discussed in this paper would have a number of effects upon the design of a mixed-flow fan.

(i) Although stream surface twist is modest at the design point it increases rapidly with decreasing flow coefficient up to the point of blade flow separation. Thus the maximum twist occurs when the blade row is about to stall. No doubt the point of separation could be more closely predicted if the rotational effects were taken into account.

(ii) For rotors designed to achieve higher aerodynamic loading than those investigated here, larger twists would occur at the design point. The traditional quasi-three-dimensional design method might then break down.

(iii) The effect of the through-flow eddy upon the circumferential velocities has not been considered here but should be accounted for in the quasi-three-dimensional design process. This will result in reduced loading at the tip and increased loading at the hub unless corrected for. Worster, Ref.12, has noted this point.

(iv) The velocities normal to the supposed surfaces of revolution will affect the mixed-flow cascade solutions in those surfaces although this is expected to be of second order.

(v) At zero head coefficient, i.e., zero work input, there is zero through-flow eddy. This condition is analogous to the Archimedean screw.

(vi) The presence of non-free vortex inlet guide vanes could substantially change the effects of the through-flow eddy.

ACKNOWLEDGMENTS

The authors are extremely grateful to Miss Roberta Stocks for the preparation of this manuscript and to Mr. P. Elliott and Mr. A. McGilligan for help with the diagrams.

REFERENCES

1. Pollard, D. "The extension of Schlichting's analysis to mixed-flow cascades." Inst.Mech.Engs., Proc.180, 35, 1965.

2. Railly, J.W. "The design of mixed flow cascades by Ackeret's method." Engineer, 224 (5827), 405-416, 1967.

3. Lewis,R.I., Fisher,E.H.,Saviolakis,A. "Analysis of mixed-flow rotor cascades." ARC, R.& M.No.3703, 1972.

4. Fisher, E.H. "Performance of mixed-flow pump and fan cascades." Ph.D.Thesis, University of Newcastle Upon Tyne, 1975.

5. Wu, C.H. "A general theory of three-dimensional flow in subsonic and supersonic turbomachines of axial, radial and mixed-flow types." NACA TN 2604, 1952.

6. Fairbairn, G.W. "Three-dimensional flows through the rotor passages of mixed-flow fans." Ph.D.Thesis, University of Newcastle Upon Tyne, 1976.

7. Lewis, R.I., Fairbairn,G.W. "Analysis of the through-flow relative eddy of mixed-flow turbomachines." Int.J.Mech.Sci., Vol.22, pp 535-549, 1980.

8. Williams,J.E., Fisher,E.H., Fairbairn,G.W. "A test facility for the investigation of flows in mixed-flow and other rotors." Int.Conf. on Pumps and Turbine Design and Development, N.E.L., East Kilbride, 1976.

9. Fairbairn, G.W., Fisher, E.H. "The application of a thermal tracing method to stationary and rotating fan cascades." Fluid Mechanics Silver Jubilee Conference, N.E.L., East Kilbride, 1979.

10. Whitfield,A. "Slip factor of a centrifugal impeller and its variation with flow rate." Proc.I.Mech.E., Vol.188, 32/74, 1974.

11. Lewis, R.I. "A theoretical investigation of the rotational flow of incompressible fluids through axial turbo-machines with tapered annulus walls." Int.J.Mech.Sci., Vol.6, pp 51-75, 1964.

12. Worster,D.M. "The calculation of fully three-dimensional flows in impellers using a finite element method." Internal Report, Dept. of Mech. Eng., Heriot Watt University, 1973.

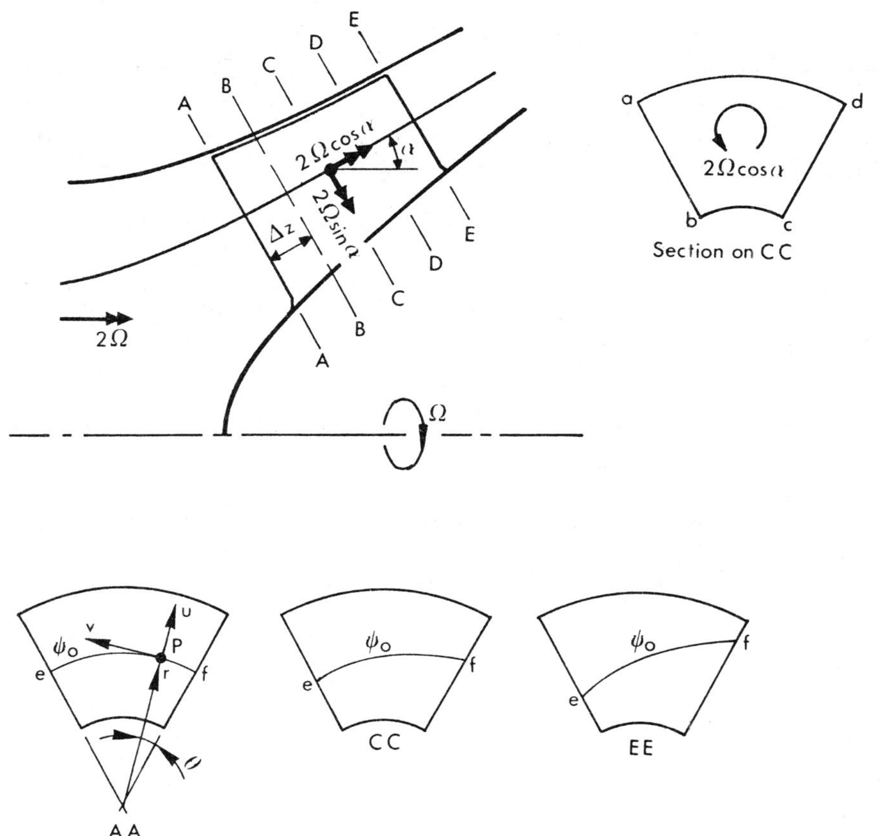

Figure 1. The through-flow relative eddy of a mixed-flow fan

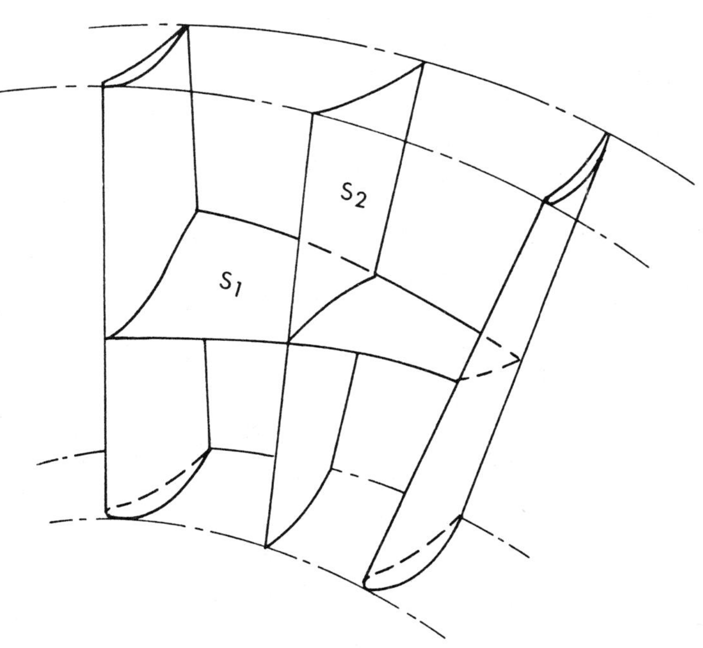

Figure 2. Wu S_1 and S_2 surfaces.

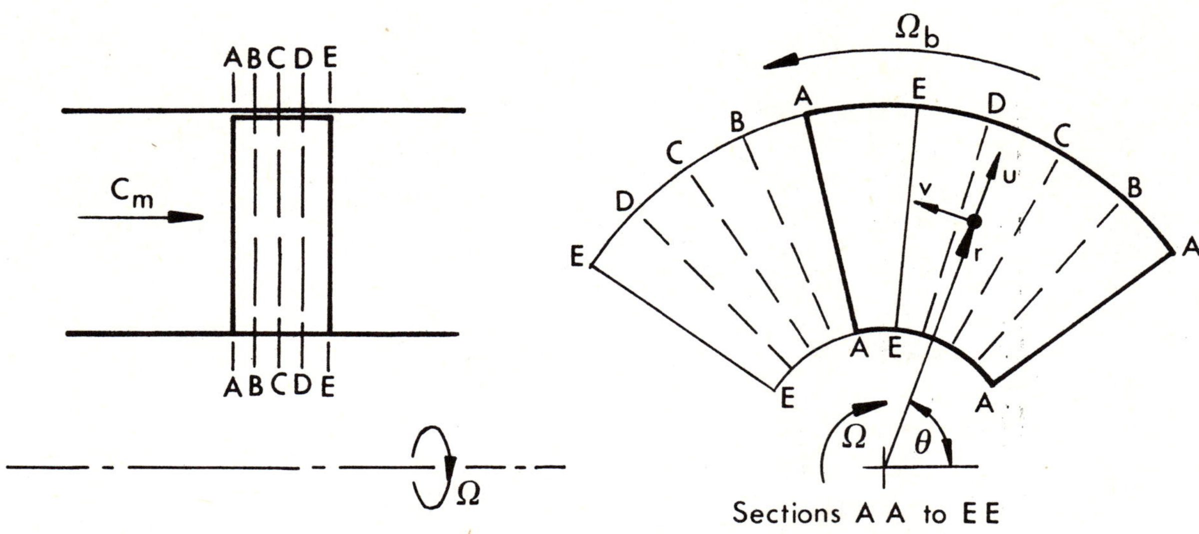

Figure 3. **Anticlockwise rotation of passage sectors proceeding from inlet AA to exit EE of an axial fan.**

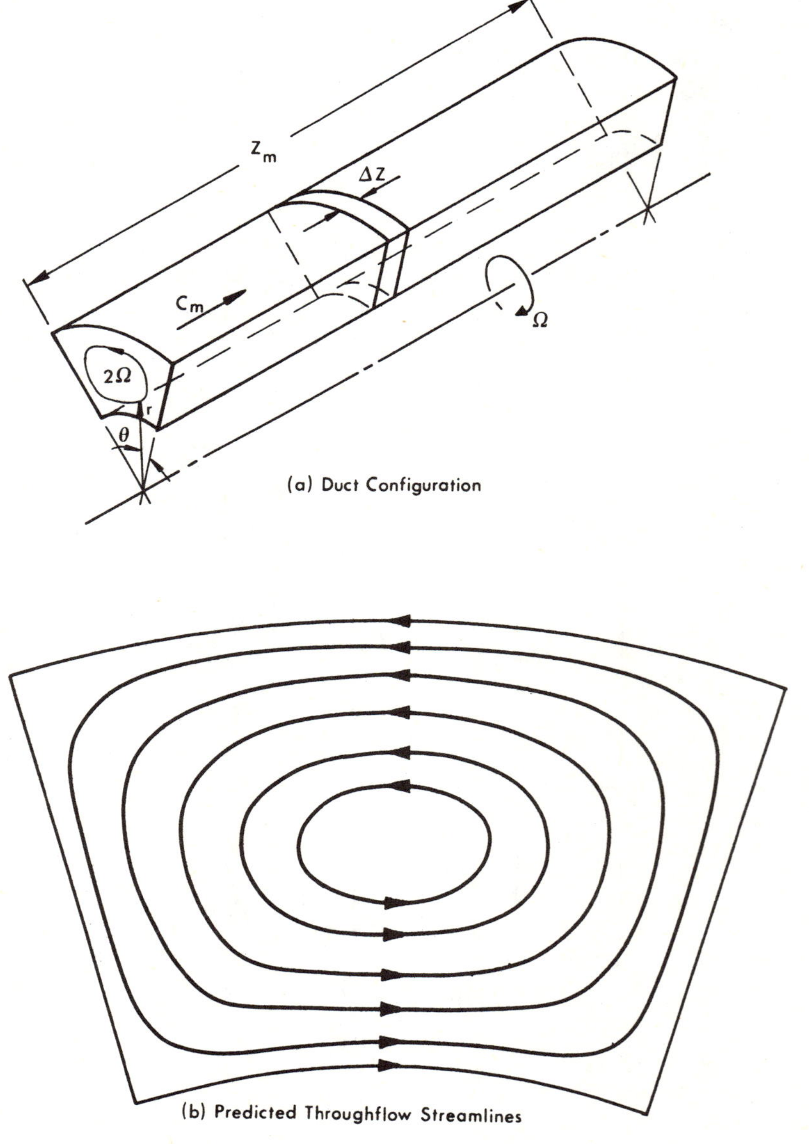

(a) Duct Configuration

(b) Predicted Throughflow Streamlines

Figure 4. **Flow through constant section rotating sectoral duct.**

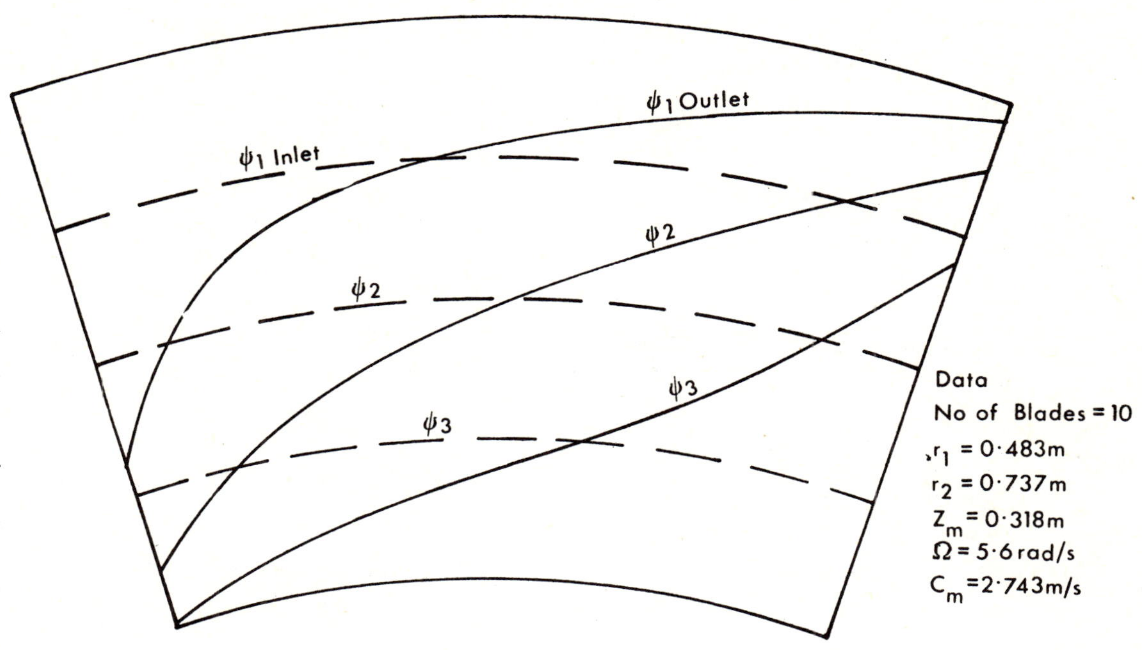

Figure 5. Stream surface shift in rotating sectoral duct.

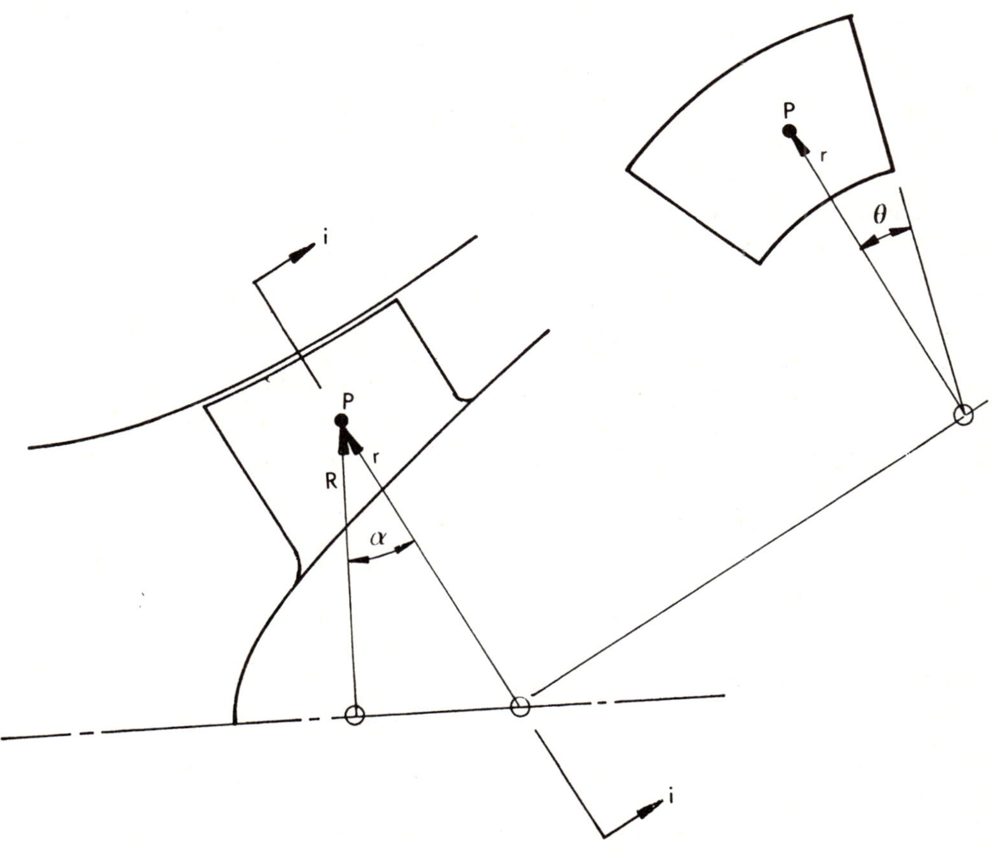

Figure 6. Transformation of conical section i onto (r,θ) plane.

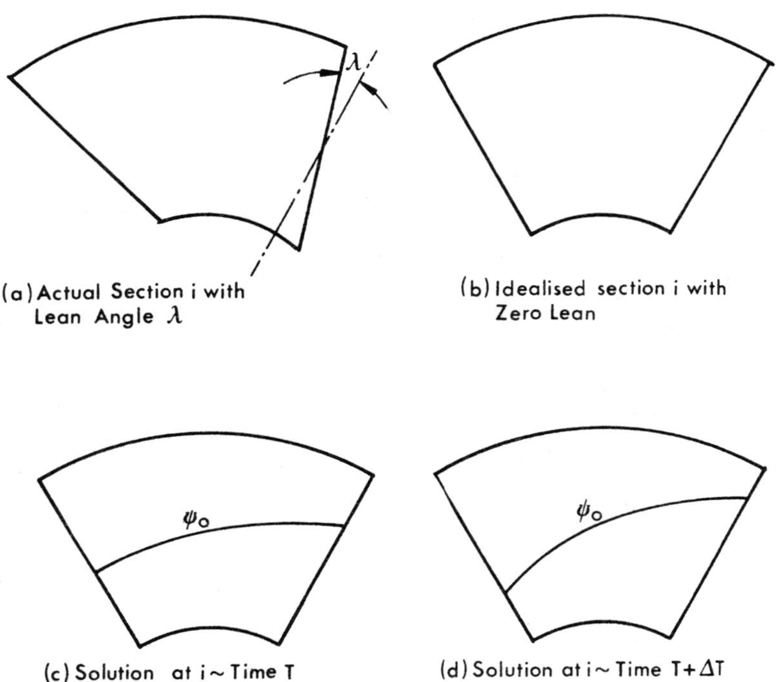

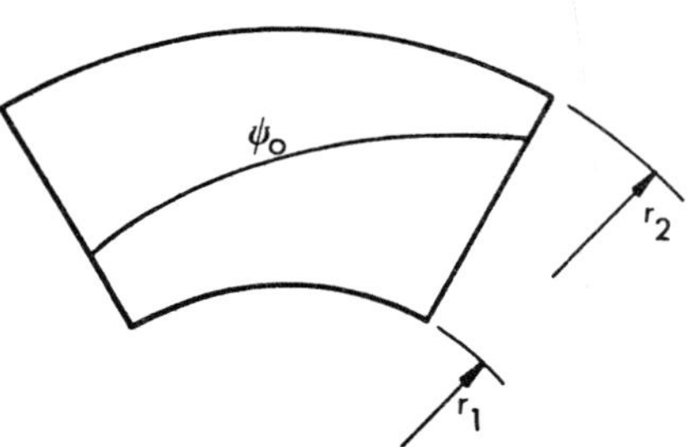

Figure 7. Geometrical procedure for eddy development between sectors i and i + 1 during time step ΔT.

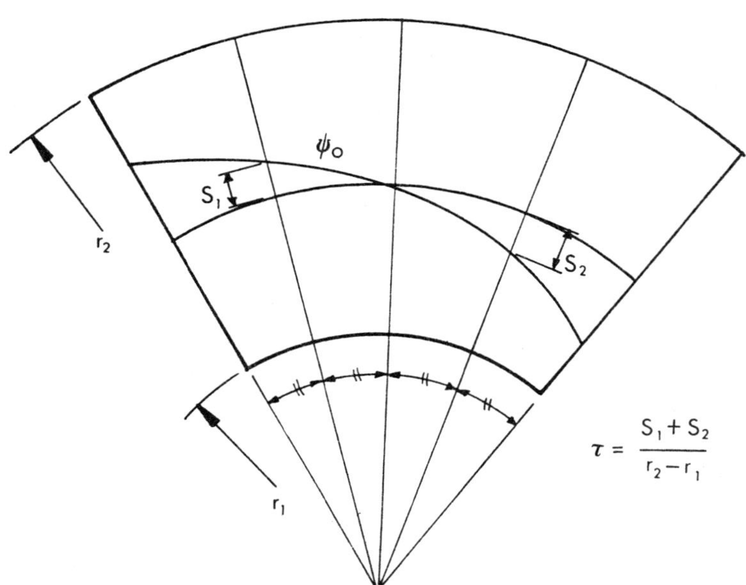

Figure 8. Definition of stream surface average twist τ.

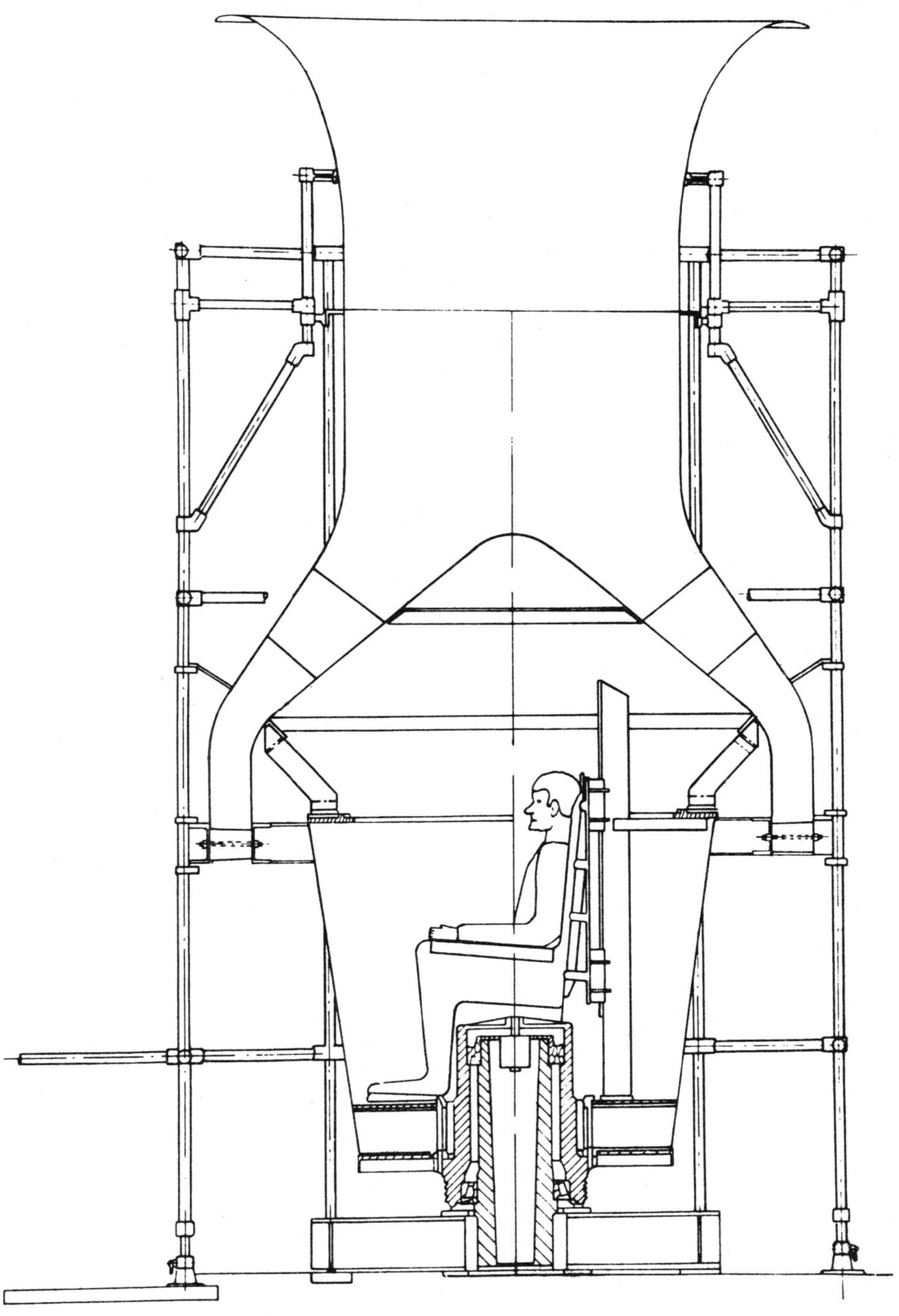

Figure 9. The Newcastle mixed-flow research fan.

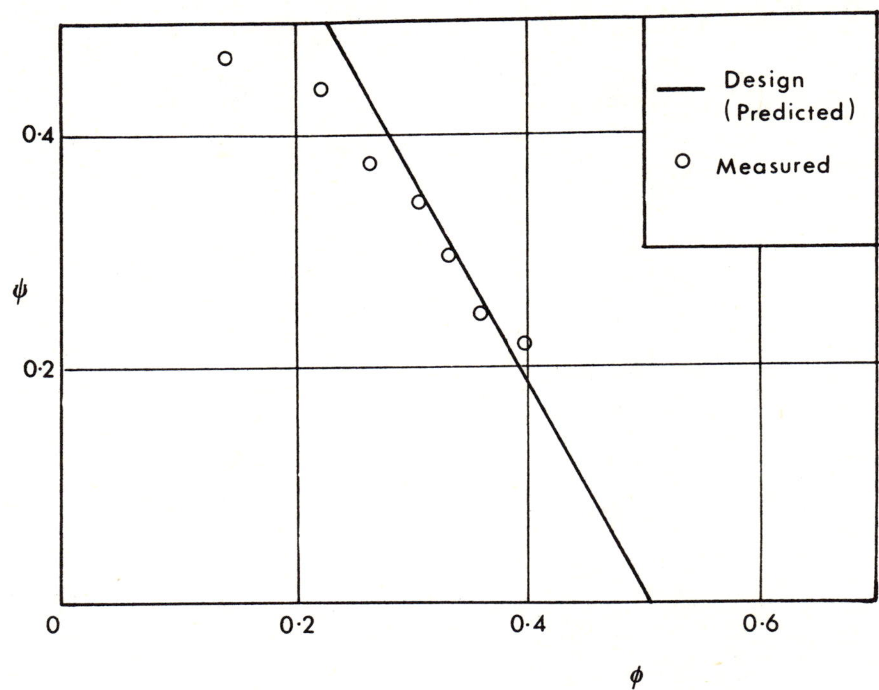

Figure 10. **Newcastle mixed-flow research fan characteristic.**

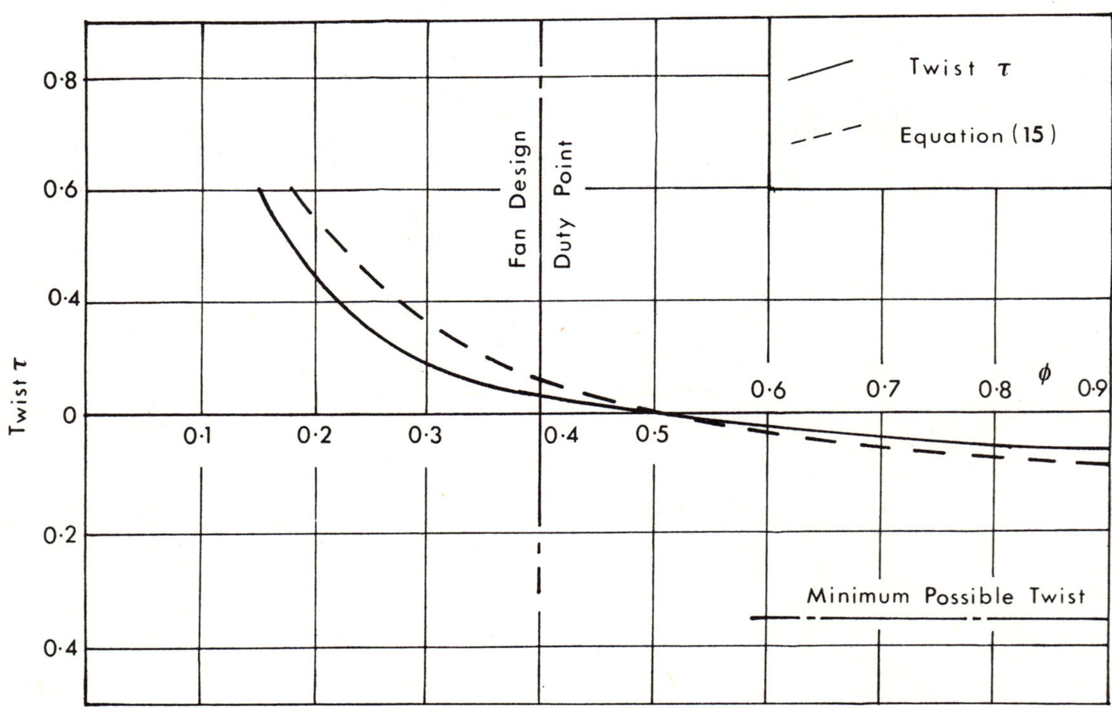

Figure 11. **Effect of flow coefficient on stream surface twist.**

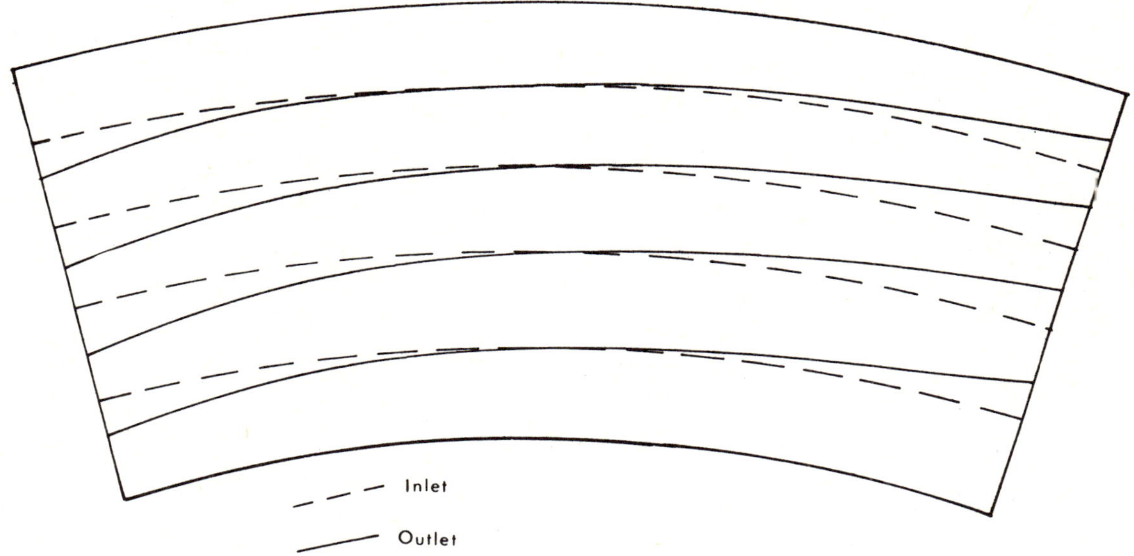

Figure 12. Predicted stream surface twist at design flow. ϕ = 0.5, Newcastle rig.

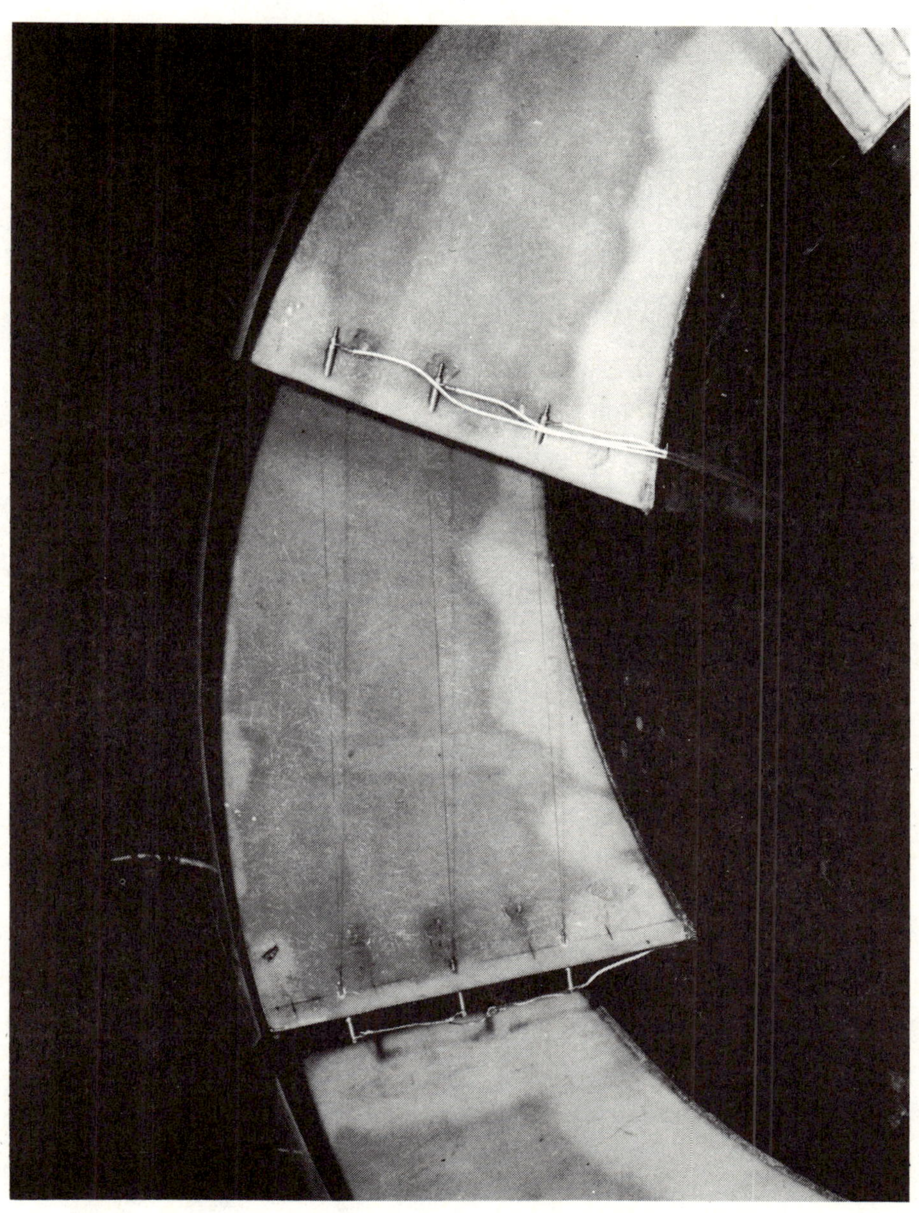

Figure 13. Thermal tracing heater wires at rotor inlet.

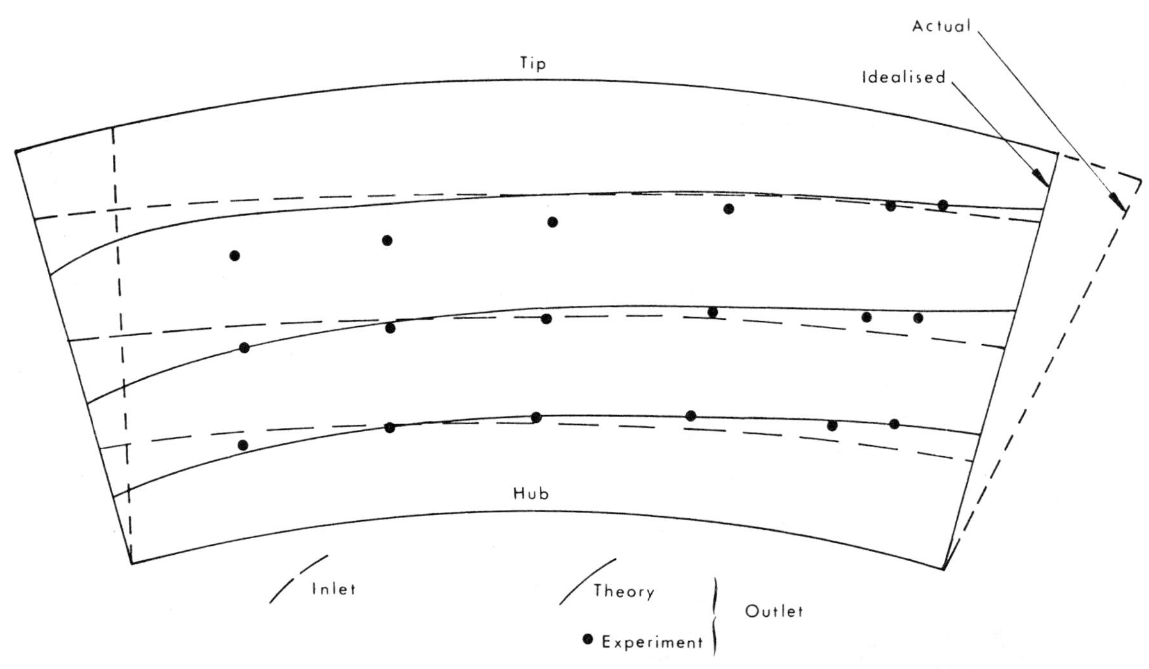

Figure 14. Stream surface twist, $\phi = 0.4$, Newcastle rig.

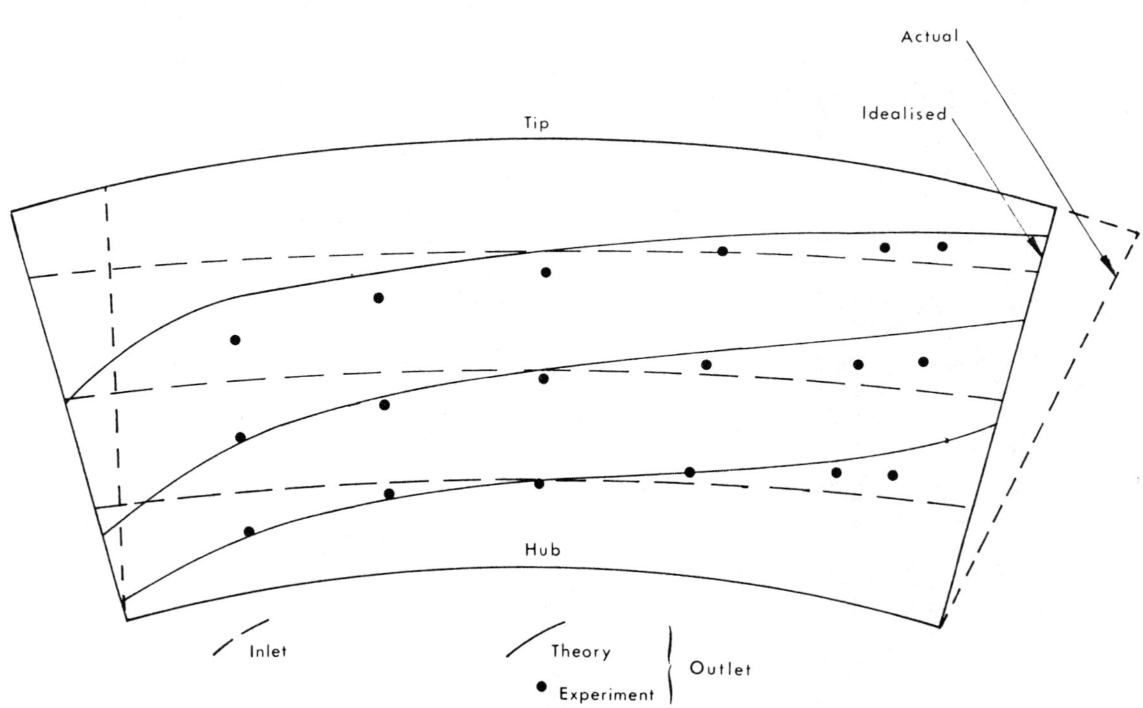

Figure 15. Stream surface twist, $\phi = 0.29$, Newcastle rig.

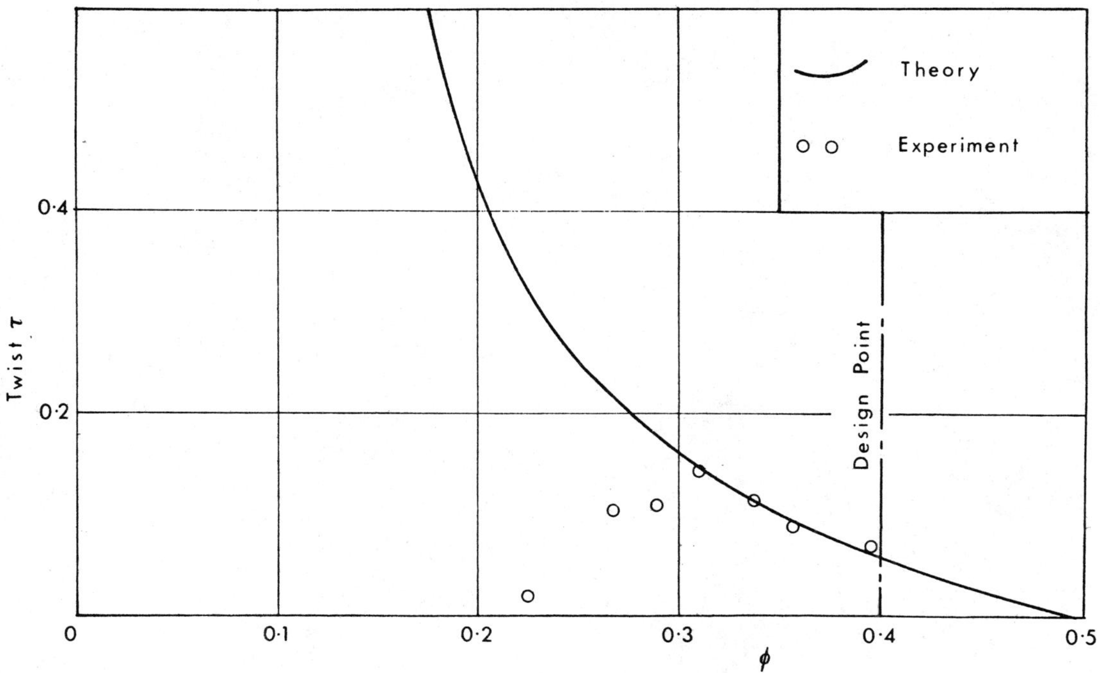

Figure 16. Effect of flow coefficient on stream surface twist.

International Conference on

Fan Design & Applications

Guildford, England: September 7-9, 1982

PAPER J3

PERFORMANCE AND AERODYNAMICS OF A CROSS FLOW FAN

P.R. Tuckey

University College London, U.K.

M.J. Holgate

University of Durham, U.K.

B.R. Clayton

University College London, U.K.

Summary

Experimental results are presented which show the internal flow régime of a large cross flow fan of rotor length 1.0 m and outside diameter 0.625 m. Details of the flow field are examined for a range of flow coefficient Φ between 0.4 and 0.8. Measurements are transformed into streamlines for the purposes of illustration and the corresponding variations of the total pressure coefficient are also given. Flow visualization techniques have been applied to a 1:6.25 dynamically similar scale model operating in water and photographs representing the flow are discussed. A Rankine-type vortex near the periphery is confirmed but it is found to remain substantially stationary over the flow range investigated. The total pressure distribution indicates the growth and weakened circulation of the forced vortex core as flow rate reduces. These data are used to improve a previous theoretical model in which the location of the vortex is used to define the operating point on the fan characteristic.

Organised and sponsored by
BHRA Fluid Engineering, Cranfield, Bedford MK43 0AJ, England.

0263 - 421X/82/01 00 - 0001 $5.00
The entire volume can be purchased from
BHRA Fluid Engineering for $82.00

NOMENCLATURE

a = factor relating radial vortex location

B = blockage factor

c = chord length

D = diameter

h = distance

k' = suction region flow function

K = constant value

L = fan length

m = strength of source (and sink)

N = blade number

p = pressure

q = local flow velocity

Q = volume flow rate

r = radius

$\tilde{t}$ = total pressure coefficient

u = blade velocity

V = fluid velocity

α = casing location

β = blade angle

$\hat{\beta}$ = deviation from blade angle

Γ = vortex strength

ε = casing clearance

θ = angle

Θ = peripheral vortex location

μ = viscosity of air

ρ = density of air

Φ = $\frac{Q}{LD_2u_2}$ flow coefficient

ψ = stream function

Ψ = $\frac{p}{\frac{1}{2}\rho u_2^2}$ pressure coefficient

ω = rotational speed

Subscripts

1 = fan inner periphery

2 = fan outer periphery

r = radial

s = static

t = total

to = total on fan centre-line

θa = tangential on free/forced vortex boundary

θ = tangential

1. INTRODUCTION

Although cross flow fans have been designed and built for nearly a century there still remain ambiguities and a lack of precision in the description of the aerodynamic behaviour of these devices. Many of the difficulties arise from the presence of numerous geometric variables which are deemed to have an influence on performance and even now optimisation criteria are not universally accepted. Nevertheless, the dominant feature of a cross flow fan (and to a large extent its vertical axis wind turbine counterpart) lies in the inherent unsteady, non-uniform aerodynamics of the rotor. The passage of blades through vortex cores, the reversal of flow direction relative to the blade during one rotation and the presence of highly turbulent flow zones present problems of extreme difficulty for the analyst. An appraisal of these problems is given by Clayton (Ref. 1) who additionally brought together a wide range of data in order to provide design guide lines. Since that work was completed further information on the scaling of cross flow fans has been given by Holgate and Haines (Ref. 2). Japanese research workers have also completed programmes on small rotors e.g. Refs 3-6.

Yamafuji (Refs 3, 4) conducted experiments with a small (160mm o.d.) isolated rotor in an effectively unbounded flow contained by a water tank. He was able to show that a stable eccentric vortex could be formed, located close to the periphery of the rotor provided that the Reynolds number based on outer peripheral velocity and blade chord length exceeded about 250. Under these circumstances a through flow across the rotor was established. Using the implications of this and other experimental work an actuator disc theory was developed in Ref. 4. Good agreement with experiment was found including some examples in which simple casing geometries were specified.

Murata and Nishihara (Ref. 5) used a somewhat larger rotor of 240mm o.d. and examined the influence of a variety of casing parameters on performance. These generally tended to confirm the optimised values used for the present design. In Ref. 6 these workers investigated the shape of the fan performance curves in relation to controlled adjustments of the casing. In this and the previous work the vortex centre formed near the periphery of the rotor and was observed to move away from the vortex wall (or tongue) and towards the rotor centre line as the flow rate was reduced from its maximum value. Further systematic studies were conducted by Murata et al (Ref. 7) into the effects of guide vanes within the rotor which encircled the vortex region.

Previous workers have tended to use quite small rotors (i.e. less than 300mm o.d.) but there has been some evidence to show that large rotors may behave differently. In order to clarify some of the areas of doubt concerning the aerodynamics of cross flow fans an experimental and theoretical research programme has been undertaken. Quantitative data were obtained from a large rotor installed on a purpose-built wind tunnel and also from a smaller scale rotor placed in a water tank. This latter facility also allowed the adoption of reliable flow visualization techniques. The theoretical studies were aimed principally at improving the approach adopted by Ikegami and Murata (Ref. 8).

2. AERODYNAMIC FACILITY

The need for a simple, straight-through, open-circuit wind tunnel with a conventional test section led to the specification of a cross flow fan as the primary source of air supply. It was also possible to investigate the aerodynamic performance of the fan itself. A side elevation of the principal features of the complete rig is shown in Fig. 1. The requirements for the wind tunnel test section indicated that the fan needed to supply air at $10m^3s^{-1}$ and develop a static pressure rise of 1kPa. An examination of previous data showed that a value of the flow coefficient $\Phi = 0.6$ would correspond to optimum efficiency and approximately maximum static pressure coefficient $\Psi_s = 2.2$ and also result in a delivery flow reasonably free from pulsations i.e. a stable flow. These considerations resulted in the following rotor parameters; length L = 1.0m, outside diameter D_2=0.625m, rotational speed of 6.67 rev s^{-1}. A variable speed drive was incorporated between the motor and the fan so that fan speeds between 5 and 16 rev s^{-1} could be set. Experiments on this large fan which confirmed the expected performance have been previously reported by Haines and Holgate (Ref. 2).

2.1 Geometry of rotor

The main geometric parameters of a cross flow fan rotor are illustrated in Fig. 2. Numerical values of these parameters for the rotor tested are summarized in Table 1 and were the same as those used in the previous research project (Ref. 2). The 24 identical blades are forward curved and symmetrically arranged round the circumference of the rotor. Each end of the rotor is blanked-off by plates which carry the bearings and housings for fixing to the main support frame.

The appraisal in Ref. 1 led to the formulation of some design 'rules'. For example, a change in the details of the impeller geometry was found to have a far smaller effect on performance than a change in the geometry of the surrounding casing. The present fan blades are therefore of constant cross-section and rounded at the leading and trailing edges. Since each blade experiences a flow reversal on passing from the suction to the discharge region a circular arc camber has been used. It was felt that any attempt to add a variable thickness profile to the blades was unlikely to improve performance or reveal any otherwise unseen flow behaviour; constructional complexity would also be increased. Using the continuity and momentum equations it was shown by Eck (Ref. 9) that $\beta_1 = 90°$ although little effect of changes in this angle over the range 60-100° is usually observed. Values of β_2, on the other hand, have been optimised on the basis of measurements with the result that $\beta_2 > 22°$ for stability with, say, 26° representing an optimum. The number of blades appears to be of secondary importance with the final choice resulting from a compromise between torsional and bending rigidity, friction losses and the cascade effect on flow deflection. A central shaft should be avoided and has been so in the present case. The diameter ratio D_1/D_2 also appears not to be a critical parameter within the range 0.70 - 0.85 and an average value was accepted.

It is possible to increase the flow rate linearly with increase of rotor length at a given speed of rotation. There are limitations to this, however, owing to the tendency for the line vortex to break down into shorter line vortices with a consequent deterioration in overall performance. Alternatively, if the rotor is too short significant secondary flows develop between the ends and these grossly distort the flow through the rotor core and at the discharge periphery.

2.2 Geometry of the vortex and rear walls.

Following the earlier controversy surrounding the function and geometry of the vortex wall there now seems little doubt that a simple straight wall with a rounded leading edge produces good fan performance. A compromise between noise and efficiency led to a clearance (see Fig. 2) of $\varepsilon_2/D_2 = 0.04$, a value easily attainable with straightforward manufacturing techniques. The peripheral location of the leading edge of the vortex wall was based on the results of Murata and Nishihara (Ref. 5), Ikegami and Murata (Ref. 10), Porter and Markland (Ref. 11) and Preszler and Lajos (Ref. 12) and a value of $\alpha_2 = 36°$ was chosen.

It has only been recently that the importance of the rear wall in helping to stabilize the vortex core has been appreciated. A logarithmic spiral shaped rear wall was advocated in Ref. 11 and has been used herein. A schematic layout of the present fan and surrounding ductwork is shown in Fig. 3 and it can be seen that no straight wall diffuser is attached to the discharge casing beyond the point where the tangent to the rear wall becomes parallel to the vortex wall. An examination of previous work indicated no clear recommendation for the incorporation of a straight walled diffuser just downstream from the fan. The effectiveness of a diffuser is greatly reduced by the presence of a strongly non-uniform velocity at its entrance and this is precisely the condition at exit from the cross flow fan. Diffusion of the flow takes place mainly in the region enclosed by the rear wall which acts as the volute plus the natural diffusion due to the vortex location. It was found in Ref. 2 that diffusion became too great, leading to flow separation and reversal, when a logarithmic rear wall was used with its leading edge located diametrically opposite the leading edge of the vortex wall. To combat this problem a logarithmic spiral was developed from the leading edge at $\alpha_1 = 0$ (see Fig. 2) but with a circumferential extension attached to the leading edge extending forwards to $\alpha_1 = 20°$ at a clearance $\varepsilon_1/D_2 = 0.05$ as shown in Fig. 3. This geometry led to an optimum rotor, of large size, which combined the attributes of high efficiency and low noise levels.

Four symmetrically placed pressure tapping holes were located in the casing in a plane 3 rotor diameters downstream from the leading edge of the vortex wall. The mean response from these tappings was used to calculate the static pressure rise across the fan and hence the static pressure coefficient Ψ_s. The test section of the wind tunnel had been previously calibrated in terms of flow rate related to the centre line velocity and thus the flow coefficient Φ and the total pressure coefficient Ψ_t were easily determined. The performance curves for the present fan (at a rotational speed of 6.67 rev s^{-1}) in terms of these parameters are shown in Fig. 4. It can be seen that the design requirement is met and subsequent data will be related to these curves.

3. HYDRODYNAMIC FACILITY

Flow visualization techniques are often used to indicate the general nature of the flow behaviour but with care some may be developed to yield quantitative data particularly for water flows (Ref. 13). Hydrolysis of water generates hydrogen bubbles from a cathode when d.c. is supplied between electrodes. If the cathode takes the form of a thin wire a very fine cloud of bubbles is formed which appears as a line when viewed edgewise. It is also possible to pulse the supply to obtain time-streak markers. However, the highly turbulent flow in and around a cross-flow rotor was found too great to allow any reasonably sustained stream to be observed especially on passing through blade passages where the bubbles were totally dispersed, although general flow patterns could be discerned. It was therefore decided to opt for a method by which neutral density polystyrene beads could be introduced into the flow.

The test facility consisted of a rectangular tank with sides made from heavy gauge perspex sheet. Deflections of the assembly when full of water were minimised by supporting the sides and bottom with a steel bracing framework. The model rotor, built to a geometric scale of 1:6.25, was installed towards one corner of the tank with its axis of rotation vertical. The drive shaft attached to the lower end of the rotor passed through a seal in the base of the tank and was attached by pulleys and a timing belt to an a.c. motor. Dynamically similar operating conditions were achieved by driving the rotor at a speed appropriate to equality of Reynolds number at a given operating condition. The blades of the rotor were constructed from stainless steel sheet but the end walls and the surrounding casing were constructed from perspex sheet to allow maximum light transmission for the purposes of observation and photography. Figure 5 shows the general layout of the model system. The vortex wall clearance and the circumferential extension of the rear wall could each be adjusted. It was on the basis of a previous flow visualization study with this easily adaptable arrangement along with the model aerodynamic studies reported in Ref. 2 that the large wind tunnel fan was designed.

Polystyrene beads of 0.5mm diameter were used as tracers after first soaking them in water to ensure a thoroughly wetted surface. Illumination was provided by three high-power, halogen strip lamps forming a 6mm slice of light transversely across the tank at the mid-span position of the rotor. A camera was mounted above the tank and was positioned directly above the rotor. By choosing a suitable shutter speed it was possible to record photographic streaks which represented the displacements of the beads and the corresponding directions of motion. Local velocities could then be estimated from the known camera shutter speed. Care was taken to ensure that the start and end of a streak was due solely to the chosen exposure time and not because the bead had a large axial velocity component which caused it to pass across the light beam. In this way the motion could be regarded as substantially two dimensional.

4. FLOW VISUALIZATION STUDY

The vortex wall and rear wall clearances of the model described in Section 3 were both set to 0.05 D_2 as it was unclear how these parameters would affect the conditions for dynamic similarity with the large rotor. Dynamic similarity with respect to other variables was maintained throughout all the tests.

For each resulting photograph, velocities in the various regions were determined from a knowledge of the camera shutter speed. The information extracted from each of the photographs took the form of:

(i) the effective inflow arc,
(ii) the general behaviour of the flow in the suction region,
(iii) the flow velocities in the suction region,
(iv) the size and velocity of the discharge jet over the rear wall,
(v) the general behaviour of the flow in the interior region,
(vi) the flow velocities in the interior region,
(vii) the position of the vortex core,
(viii) the effective outflow arc,
(ix) the general behaviour of the flow in the discharge region, and
(x) the flow velocities in the discharge region.

From the resulting data it was possible to identify discrete zones of flow within the suction, interior and discharge regions as shown in Fig. 6.

The zones identified within the suction region were:
(a) an inflow zone covering an arc beginning at the vortex wall,
(b) a discharge jet from the rotor interior covering an arc beginning at the leading edge of the rear wall, and
(c) a zone of entrained flow resulting from (b).

The zones identified in the interior region were:
(d) a forced vortex core,
(e) a recirculating flow return path from the discharge region,
(f) a return flow path from the discharge region below the vortex wall,
(g) a throughflow path,
(h) a throughflow path from the suction region to the discharge jet over the rear wall, and
(i) a low energy random zone.

The zones identified in the discharge region were:
(j) a turbulent recirculation zone from the fan interior,
(k) a diffusing throughflow zone, and
(l) a flow path under the vortex wall returning essentially laminar flow to the fan.

The size of these zones and the associated velocities were found to be entirely dependent on the flow rate passing through the fan. Several casing configurations were examined but here we shall deal primarily with the geometry relating to the full scale wind tunnel fan having the 20°, constant radius extension above the rear wall.

For the fan operating under medium throttle conditions the flow in the suction region exhibited the three regions mentioned previously. Figure 7 shows the high velocity region close to the vortex wall where the majority of the flow enters the fan (region a), with velocities in this region estimated at $0.5u_2$. The geometric suction arc consists of 80% inflow with the remainder comprising the jet region b and its associated entrainment c. The velocity of the flow within the jet region is the order of the blade speed, showing clearly that the origin of the flow lies within the fan interior. The interior flow clearly shows the vortex core (region d) and the turbulent low energy region diametrically opposite (region i).In Fig.7 the outflow arc covers 65% of the geometric arc before meeting the recirculation region (region j) where vortices are seen to be shed from the main vortex core. The flow under the vortex wall is returning towards the fan (region ℓ). The velocities in the discharge region are about $3.0u_2$ (region k). As the fan outflow was throttled, the vortex moved by about 15° peripherally towards the rear wall but did not move significantly in a radial direction. Figure 8 is a high contrast, slow shutter speed photograph which shows the vortex core and a surrounding fainter region which corresponds to the flow returning from the discharge region below the vortex wall (region e).

Figure 9 illustrates the 'zero' flow criterion and shows the large recirculating region in the fan discharge (region j). It should be understood that zero flow relative to the fan cannot exist by throttling the outlet duct, as flow is always maintained through the fan owing to recirculation within the duct. A true zero flow condition can be imposed with a damper installed adjacent to the fan outlet. The vortex then moves to the centre of the rotor and becomes a forced vortex. The suction region velocities for a throttled fan are little different from those of the full flow condition. However, the inflow arc then diminishes to 50% of the geometric arc with the rear wall jet region becoming increasingly dominant as it contains flow velocities of around $3.0u_2$. This jet region appears to be a major source of losses and hence poor efficiency of cross flow fans. On removing the 20° extension to the rear wall the jet region became more extensive and contained flow velocities lower than those previously recorded. Furthermore, the total inflow arc was somewhat smaller than that observed with the 20° extension. When the rear wall extension was set at 50° the jet became smaller but flow velocities higher. The total inflow arc was also smaller than the 20° extension case. Figures 10 and 11 show these two cases at 'zero' flow where they were found to be most dominant.

Table 2 summarizes the photographic records taken for the 0°, 20° and 50° extensions to the rear wall, with specific parameters labelled on Fig. 6. Figure 12 shows a more general view of the total flow fields mentioned previously and clearly indicates that the cause of pulsations found during the use of many of these units stems from the shed vortices and recirculation in the outlet duct. The flow visualization reveals that the interior flow field is rather different from those proposed in mathematical models and this divergence will be discussed later. The disturbed region within the impeller was always most predominant under low throttle conditions and the pressure distributions in the full scale wind tunnel fan showed similar trends.

5. AERODYNAMIC MEASUREMENTS

The results of traverses inside the rotor of the wind tunnel fan were related to a polar coordinate system located on the rotor centre-line, as shown in Fig. 13. Readings were taken at a number of r, θ grid coordinates using an improved 3-hole yaw probe. The probe was fixed to the end of a variable geometry 'Z' drive, with one arm concentric with the shaft of the fan, so that any radial location at mid-span could be adopted. Yawing was effected through a series of bevel gears attached to the hollow shafts which carried the pressure tubes. Measures were obtained of:

(i) the angle of the flow relative to a radial line between the fan centre-line and the probe head,
(ii) the velocity of the flow, and
(iii) the total pressure.

For a comparison with previous work the streamline pattern in the rotor was required. Figure 14 shows two adjacent streamlines associated with stream functions ψ_1 and ψ_2, with local velocities of q_1 and q_2 and making angles of θ_1 and θ_2 with respect to a radial line in the coordinate system. The streamlines are separated by the radial distance Δr. By definition,

$$\psi_2 = \psi_1 + \bar{u} d_n$$

where, d_n is the normal distance between the streamlines, and

$$\bar{u} = \tfrac{1}{2}\{q_1 \sin(\theta + \theta_1) + q_2 \sin(\theta + \theta_2)\}$$

Therefore,

$$\psi_2 = \psi_1 + \tfrac{1}{2}\{q_1 \sin(\theta + \theta_1) + q_2 \sin(\theta + \theta_2)\}\Delta r$$

and the stream function for the streamline crossing the rotor axis was arbitrarily chosen to be zero. The radial and tangential components of the velocities may be related to the stream function by the expressions,

$$\frac{\partial \psi}{\partial \theta} = - r V_r \qquad (1)$$

and

$$\frac{\partial \psi}{\partial r} = V_\theta \qquad (2)$$

From the experimental results, all the velocity and direction data were transformed into radial and tangential components, so that the above expressions could be used, in the form

$$\Delta\psi_\theta = -r V_r \Delta\theta \qquad (3)$$

and

$$\Delta\psi_r = V_\theta \Delta r \qquad (4)$$

The calculation procedure is as follows:

(i) Plot V_r against θ, at all radii traversed (see Fig. 15 for $\Phi = 0.8$).

(ii) Plot V_θ against r, at all peripheral positions traversed (see Fig. 16).

(iii) Begin the calculation procedure on the fan centre line at $\psi = 0$, designated coordinate r(0), θ(0).

(iv) Using eqn (4) find the change in stream function between r(0), θ(0) and r(1), θ(270°), i.e. $\Delta\psi_1$.

(v) Again using eqn (4) find the change in stream function between r(0), θ(0) and r(1), θ(275°), i.e. $\Delta\psi_2$.

(vi) Using eqn (3) find the change in stream function between r(1), θ(270°) and r(1), θ(275°), i.e. $\Delta\psi_3$.

(vii) Two values of stream function are then found at coordinate r(1), θ(275°), and these are then averaged, e.g. at r(1), θ(275°)

$$\psi = \tfrac{1}{2}(\Delta\psi_1 + \Delta\psi_2 + \Delta\psi_3).$$

(viii) This method is repeated as the first radius r(1) is scanned ending finally at θ(270°) which is averaged with $\Delta\psi_1$. The second radius r(2) is treated similarly except that the radial changes in stream function derived from eqn (4) are added to those present on the first radius r(1) and so on.

(ix) Curves are then drawn of ψ against r for constant θ and ψ against θ for constant r as shown in Figs 17 and 18, for example, from which coordinates are extracted for the final flow diagrams, as shown in Figs 19(a)-(e) for $\Phi = 0.4$ to 0.8.

Total pressures were measured inside the rotor and then used to form the dimensionless total pressure coefficient

$$\tilde{t} = (p_t - p_{to})/\tfrac{1}{2}\rho u_2^2 \tag{5}$$

Results are presented in Figs 20 (a)-(c) which indicate the locations and ranges of $\tilde{t}$ for three values of Φ. The data show that first, the total pressure along a particular streamline remains constant as it should, of course, and second, an indication of the type of vortex flow in existence may be established as follows. The pressure variation perpendicular to a streamline is given by

$$\frac{\partial p_s}{\partial r} = \rho \frac{V_\theta^2}{r} \tag{6}$$

For a free vortex (irrotational flow);

$$rV_\theta = \text{constant, } K \tag{7}$$

so that eqn (6) becomes

$$\frac{\partial p_s}{\partial r} = \rho \frac{K^2}{r^3} \tag{8}$$

Integration with respect to r gives

$$p_s = -\frac{\rho K^2}{2r^2} + \text{constant} \tag{9}$$

Now

$$p_t = p_s + \tfrac{1}{2}\rho V_\theta^2$$

$$= \frac{-\rho K^2}{2r^2} + \tfrac{1}{2}\rho\left(\frac{K}{r}\right)^2 + \text{constant} \tag{10}$$

and so in a free vortex p_t is constant. For a forced vortex

$$V_\theta = r\omega \tag{11}$$

and it is easy to show that

$$p_t = \rho V_\theta^2 + \text{constant} \tag{12}$$

Figure 21 shows typical velocity and pressure distributions in the rotor for the expected Rankine type vortex along with some measured data for velocity.

The conclusions from Figs 19-21 may now be summarized.

(a) For all flow rates, a Rankine type vortex was present within the rotor.

(b) The dimensionless total pressure was severely depressed in the vicinity of the vortex core and became further depressed as the flow rate increased and the vortex strengthened.

(c) A region of depressed total head (low energy) was present almost diametrically opposite that of the vortex core, which became more depressed and larger in area as the flow rate increased (see Section 4 for confirmation).

(d) The streamlines do not remain concentric with the vortex core as the fan is traversed radially. They are seen to flatten and indeed exhibit reversed curvature towards the rear wall.

(e) The streamlines indicate a general acceleration of the flow across the fan interior.

(f) The spacings between the streamlines yield a velocity profile across a diameter, including the vortex core, which has a peak at the boundary between the free and forced vortex regions and progressively falls towards the rear wall side.

(g) For $\Phi = 0.8$ the velocities in the quadrant containing the vortex core are higher than those for $\Phi = 0.4$, whereas the velocities in the opposite quadrant are similar for both flow rates.

(h) The position of the vortex core changed little in the range $\Phi = 0.5$ to 0.8 but began to move peripherally and radially in the range $\Phi = 0.4$ to 0.5.

6. THEORY

Only a brief summary of the theory of cross flow fans is given here. A more comprehensive summary will be available shortly (Ref. 14). Notable work in the past has been undertaken by, for example, Coester (Ref. 15) who devised the first analysis, Ilberg and Sadeh (Ref. 16) and Moore (Ref. 17). These theories were formulated for a 'one-flow' situation and therefore could not be extended to overall performance prediction.

The work of Ikegami and Murata (Ref. 8) however, provided an inviscid flow analysis based on rotor geometry and a flow field which allowed mobility of the vortex within the impeller. Characteristics were deduced for a cross flow fan which consisted of a simple linear casing acting as a dividing streamline to separate the suction and discharge regions, and an impeller with an infinite number of blades. Laplace's equation was applied to the interior flow which was modelled by including a second vortex of equal strength Γ and direction as the internal vortex but located outside the impeller as shown in Fig. 22. The stream function for this combination can be expressed in the form

$$\psi = \frac{-\Gamma}{4\pi} \ln \left\{ (r^2 + a^2 r_1^2 + 2arr_1\cos\theta)(r + \frac{r_1^2}{a^2} + \frac{2rr_1}{a} \cos\theta) \right\} \qquad (13)$$

The vortex strength

$$\Gamma = 2\pi r_1 u_1 \qquad (14)$$

since the flow velocity at the rotor periphery equals the rotor velocity. The flows on the suction side and within the impeller are related by virtue of the expression

$$\psi_{r = r_1} = \psi_{r = r_2} + \text{constant} \qquad (15)$$

The flow rate per unit length can be expressed in the form

$$Q = \psi_{\theta = 0} - \psi_{\theta = \pi} \qquad (16)$$

so that in dimensionless terms

$$\Phi = \frac{Q}{2r_2 u_2} = \left(\frac{r_1}{r_2}\right)^2 \ln\left\{\frac{1 + a}{1 - a}\right\} \qquad (17)$$

As the impeller does not impart uniform energy transfer to each streamline, the pressure rise was evaluated as a mean value of energy transfer to each streamline passing through the rotor. The dimensionless mean total pressure coefficient was found to be

$$\Psi_t = \frac{\bar{p}_t}{\frac{1}{2}\rho u_2^2} = \frac{2a}{\ln\left\{\frac{1+a}{1-a}\right\}} \int_0^{\pi} \left\{ 1 + \left(\frac{r_1}{r_2}\right)^2 \left(\frac{2a \sin\theta}{1+a^2-2a\cos\theta} \cot\beta_2 - 2\sum_{n=1}^{15} nk'_n \sin n\theta \right) \left(\frac{\sin\theta}{1+a^2-2a\cos\theta} \right) \right\} d\theta \qquad (18)$$

and values of Φ and Ψ_t were plotted to show the effect of changes in the rotor geometry and the radial vortex location.

Although the mathematical analysis presented in Ref. 8 for the radial motion of the vortex was correct, a number of serious computation errors occurred especially for large values of a ; matters became worse when peripheral movement of the vortex was permitted. The authors have corrected these errors by including higher harmonics in eqn (18) to produce the data in Fig. 23 . The theory of Ikegami and Murata (Ref. 8) has been refined by including the effects of blade number and blade thickness-to-chord ratio. Equation (15) is replaced by

$$B\psi_{r=r_1} = \psi_{r=r_2} + \text{constant} \qquad (19)$$

where B is a 'blockage' factor. Figure 24 shows that blades of slender profile are associated with a high pressure rise; the number of blades has a small effect, as shown in Fig. 25.

The flow visualization and aerodynamics studies showed that the curvature of the streamlines within the rotor tended to increase more rapidly than expected across a radial line through the vortex core; in some cases a reversed curvature occurred towards the rear wall leading edge. To model this reversed curvature a source and a sink were positioned opposite the vortex core as shown in Fig. 26. The new expression for the stream function in the internal region was the summation of $\psi_{ss} = m(\theta_1 - \theta_2)/2\pi$ with ψ from eqn (13). To satisfy the boundary condition on the rotor periphery the vortex strength Γ varied with position as shown in Fig. 27. The resulting performance mesh was not significantly different to that in Fig. 23. The vortex strength varied considerably for peripheral movement although not for radial movement apart from the region close to the inner periphery of the blade row. An example of the modified flow diagram is given in Fig. 28.

Flow losses associated with cross flow fans fall into two categories;
(i) shock losses at blade leading edge owing to the disparity between the inlet relative flow angle and the blade angle β_2, and
(ii) losses within the vortex itself.
Shock losses were examined by using Laplace's equation for the velocities in the suction region. Figure 29 shows the variation of the tangential and radial velocities around the outer periphery of the rotor, together with the deviation of the relative velocity from the blade angle. It is seen that the majority of losses occur near the vortex wall, where the radial velocities are highest. Vortex losses were estimated from the energy losses of an element within the flow. Neglecting radial flow the total energy loss was found to be $E = 4\pi\mu V_{\theta a}$ where $V_{\theta a}$ is the tangential velocity component at the interface between the free and forced vortex zones, but the estimate appeared unrealistically low. When radial velocities were included, terms containing fluid density entered the original equations as a result of tangential pressure gradients. The size of this term suggests that momentum changes could account for a large proportion of the energy dissipated within a cross flow fan.

7. CONCLUSIONS

The naturally formed vortex changes intensity and size according to its location. This location could be changed by sending a wake, generated outside the rotor, through the machine. The system resistance (which was unaltered) was not therefore the only agent responsible for vortex position. The internal streamlines are not circular. A Rankine combined vortex has been confirmed from the total pressure distribution in the rotor. For the present fan the vortex core is always close to the inner periphery. Important modifications to an existing theory allow for some losses, variable vortex strength and reflex curvature of streamlines in the suction region of the rotor.

ACKNOWLEDGEMENTS

The work reported here was carried out in the Department of Engineering, University of Durham, under a grant from the Science (and Engineering) Research Council. The grantholder, M.J. Holgate and P.R. Tuckey, who was supported by the grant, gratefully acknowledge that assistance.

REFERENCES

1. Clayton, B.R.: "A review and appraisal of cross flow fans". Building Services Engineer (I.H.V.E.), 42, Jan. 1975, pp 230-247.

2. Holgate, M.J. and Haines, P.: "Scaling of cross flow fans - an experimental comparison". Proc. I. Mech. E. Conference on Scaling for Performance Prediction in Rotodynamic Machines, Stirling, 1977.

3. Yamafuji, K.: "Studies on the flow of cross-flow impellers (1st Report, Experimental study)". Bull. Jap. Soc. Mech. Engrs, 18, Sept. 1975, pp 1018-1025.

4. Yamafuji, K.: "Studies on the flow of cross-flow impellers (2nd Report, Analytical study)". Bull. Jap. Soc. Mech. Engrs, 18, Dec. 1975, pp 1425-1431.

5. Murata, S. and Nishihara, K.: "An experimental study of cross flow fan (1st Report, Effects of housing geometry on the fan performance)". Bull. Jap. Soc. Mech. Engrs, 19, Mar. 1976, pp 314-321.

6. Murata, S. and Nishihara, K. "An experimental study of cross flow fan (2nd Report, Movements of eccentric vortex inside impeller)". Bull. Jap. Soc. Mech. Engrs, 19, Mar. 1976, pp 322-329.

7. Murata, S., Ogawa, T., Shimizu, I., Nishihara, K. and Kinoshita, K.: "A study of cross flow fan with inner guide apparatus". Bull. Jap. Soc. Mech. Engrs, 21, Apr. 1978, pp 681-688.

8. Ikegami, H. and Murata, S.: "A study of the cross flow fan: Part 1, A theoretical analysis". Technology Rep. Osaka Univ., 16, 1966, No. 731.

9. Eck, B.: " Fans". Oxford, Pergamon Press, 1973, pp 156-182.

10. Ikegami, H. and Murata, S.: "Experimental study of the cross flow fan". Science of Machine, 18, 3, 1966, pp 557-566. (In Japanese)

11. Porter, A.M. and Markland, E.: "A study of the cross flow fan". Jour. Mech. Eng. Sci., 12, 6, 1970, pp 421-431.

12. Preszler, L. and Lajos, T.: "Experiments for the development of the tangential flow fan". Proc. 4th. Conf. on Fluid Machinery (Budapest, 1972), Akadémiai Kiadó, pp 1071-1082.

13. Clayton, B.R. and Massey, B.S.: "Flow visualization in water: a review of techniques". J. Sci. Instrum., 44, 1967, pp 2-11.

14. Tuckey, P.R. and Clayton, B.R.: Ph.D Thesis, University of Durham, in preparation.

15. Coester, R.: "Theoretische und experimentelle untersuchungen an querstromgeblasen". Mitt. aus dem Institut für Aerodyn., E.T.H., Zurich, No. 28, 1959. (In German)

16. Ilberg, H. and Sadeh, W.Z.: "Flow theory and performance of tangential fans". Proc. Inst. Mech. Engrs, London, 180, 1965-66, pp 481-496.

17. Moore, A.: "The tangential fan - analysis and design". Conference on Fan Technology and Practice, Institution of Mechanical Engineers, London, April 1972.

Table 1 Rotor geometry

Blade angles; β_1 = 90°, β_2 = 26°

Blade profile; circular arc

Number of blades; N = 24

Diameter ratio; D_1/D_2 = 0.78

Length to diameter ratio; L/D_2 = 1.6

Table 2 Summary of photographic records

α (deg)	Throttle (increases with more*)	Arc_{in} (deg)	V_{in}	V_{jet}	h_{jet}	θ_{vortex} (deg)	vortex size	θ_d (deg)	V_{int}	Arc_{out} (deg)	V_{out}	outflow quality (the more* the better)
0	*	154	$0.75u_2$	$1.0u_2$	$0.2r_2$	275	$0.17r_2$	81	-	96	$3.0u_2$	****
0	**	131	$0.50u_2$	$1.5u_2$	$0.4r_2$	295	$0.30r_2$	62	$1.7u_2$	84	$2.0u_2$	*****
0	***	116	$0.63u_2$	$1.0u_2$	$0.51r_2$	289	$0.50r_2$	61	$3.0u_2$	71	$1.7u_2$	**
20	*	158	$0.48u_2$	$1.0u_2$	$0.19r_2$	279	$0.37r_2$	95	$3.0u_2$	91	$3.0u_2$	***
20	**	128	$0.58u_2$	-	$0.25r_2$	290	$0.50r_2$	65	$1.5u_2$	80	$2.5u_2$	**
20	***	101	$0.57u_2$	$2.0u_2$	$0.31r_2$	292	-	75	$1.0u_2$	72	$3.0u_2$	***
50	*	153	$0.70u_2$	$0.35u_2$	v. small	281	-	75	-	91	$3.6u_2$	**
50	**	121	$0.63u_2$	$0.60u_2$	$0.21r_2$	290	-	57	-	84	$3.0u_2$	***
50	***	99	$0.70u_2$	$2.5u_2$	$0.38r_2$	292	-	29	-	71	$3.0u_2$	*

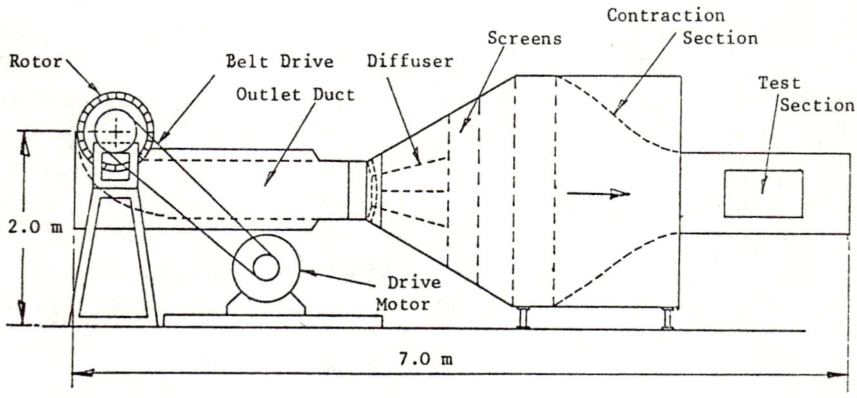

Fig. 1 Layout of aerodynamic test rig

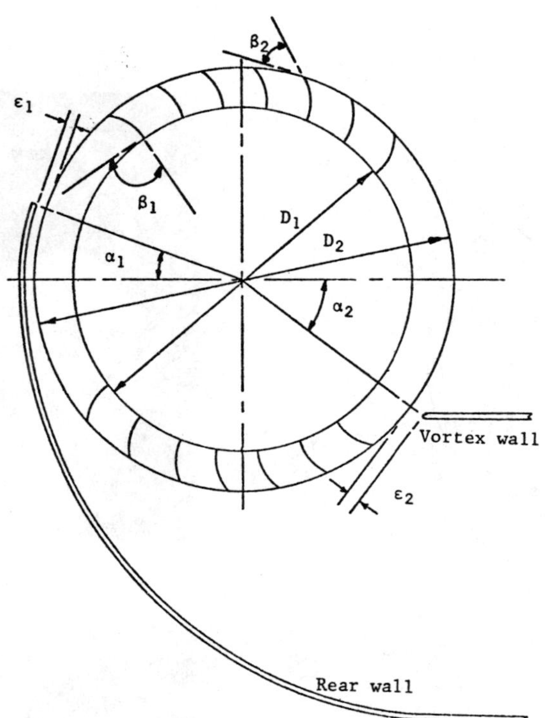

Fig. 2 Geometry of the cross flow fan

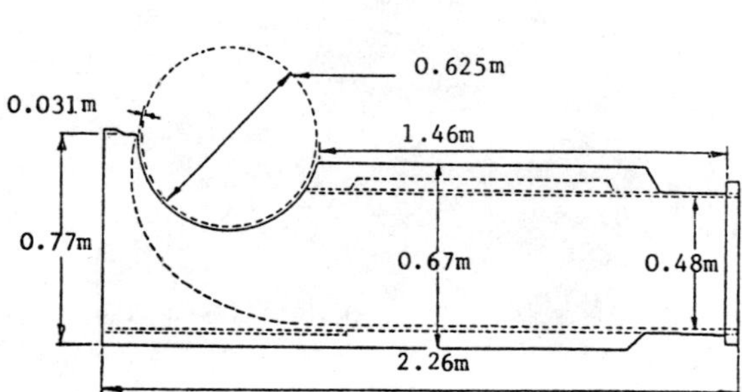

Fig. 3 Geometry of fan under test

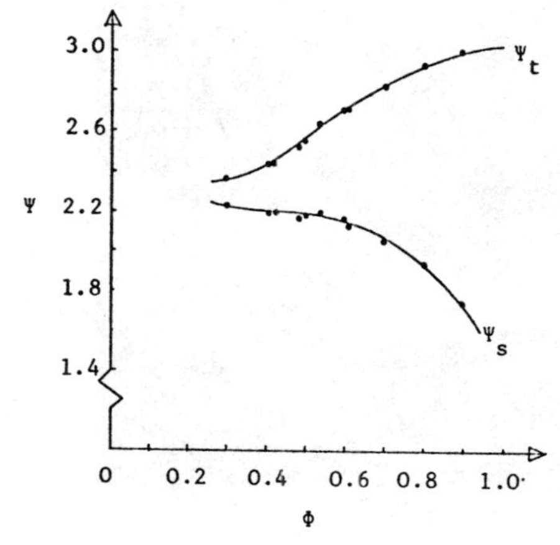

Fig. 4 Fan performance curves

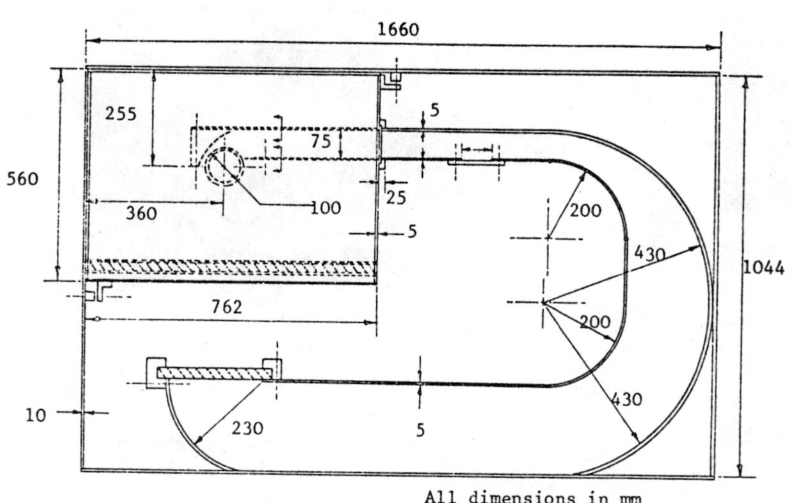

Fig. 5 Layout of hydro-dynamic rig

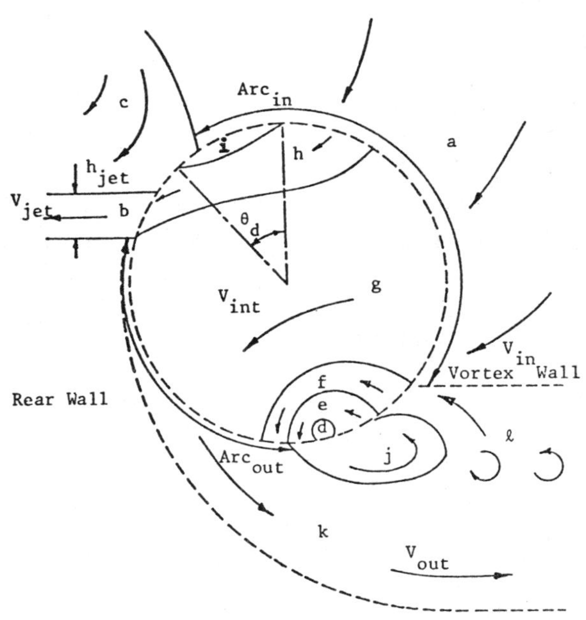

Fig. 6 Regions of flow

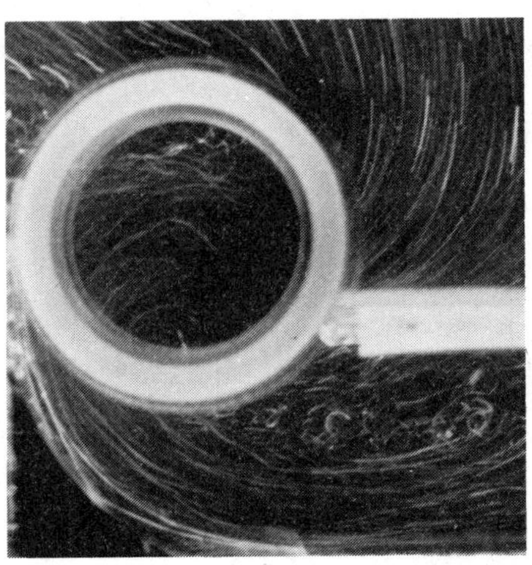

Fig. 7

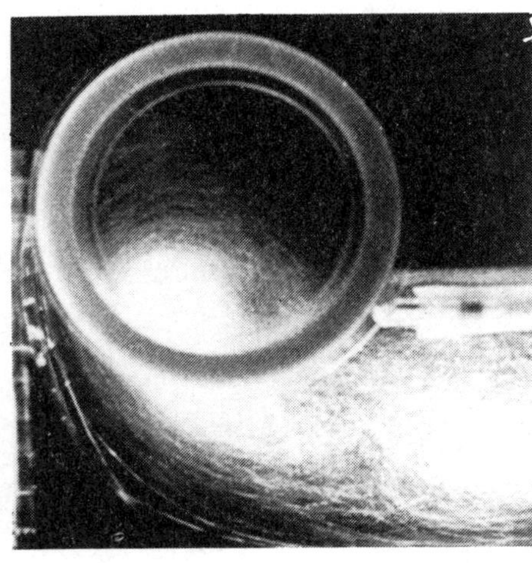

Fig. 8

Fig. 9

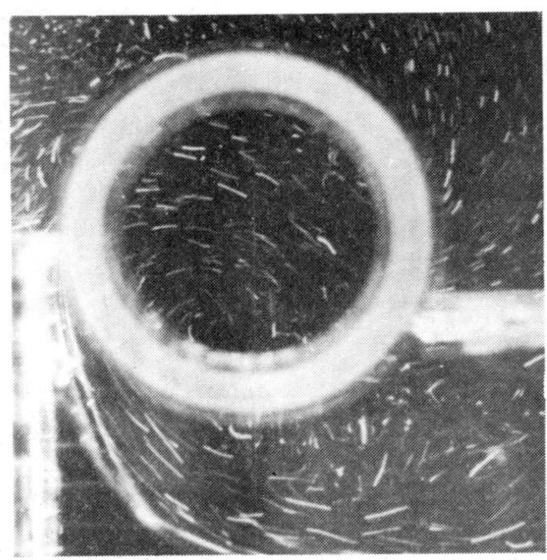

Fig. 10

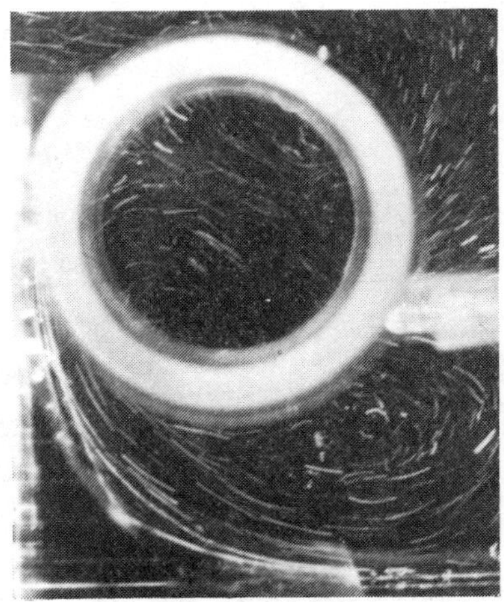

Fig. 11

Figs. 7-11 Show flow behaviour under different throttle conditions as described in text

Fig. 12 General flow condition

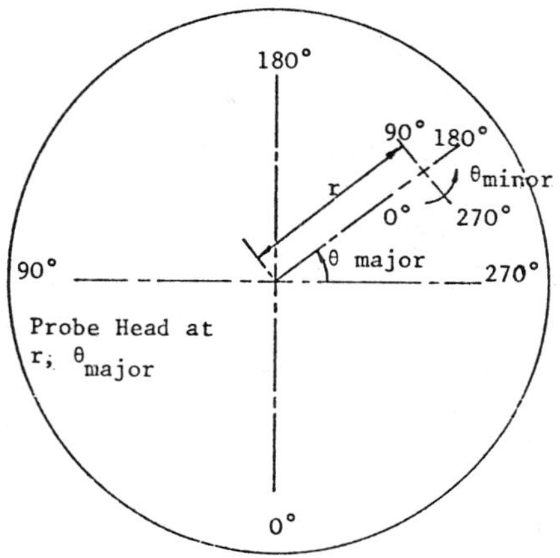

Fig. 13 Coordinate system

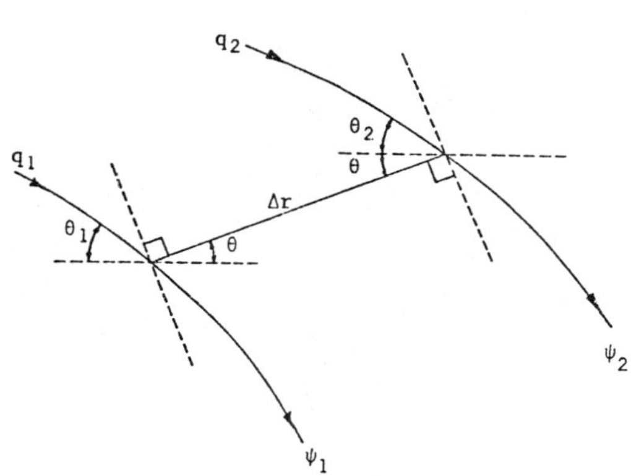

Fig. 14 Streamline geometry

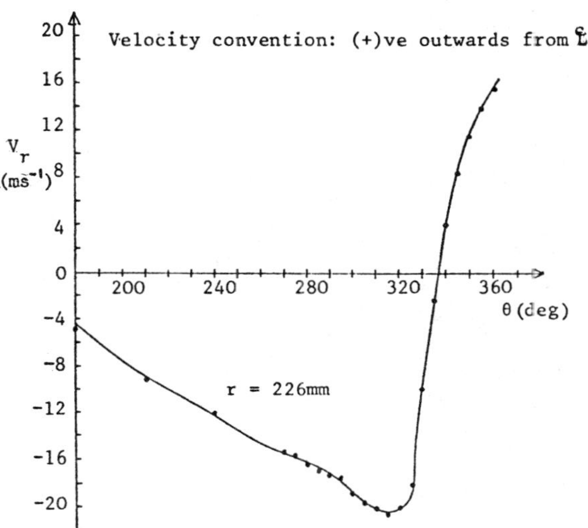

Fig. 15 Variation of radial velocity for Φ = 0.8

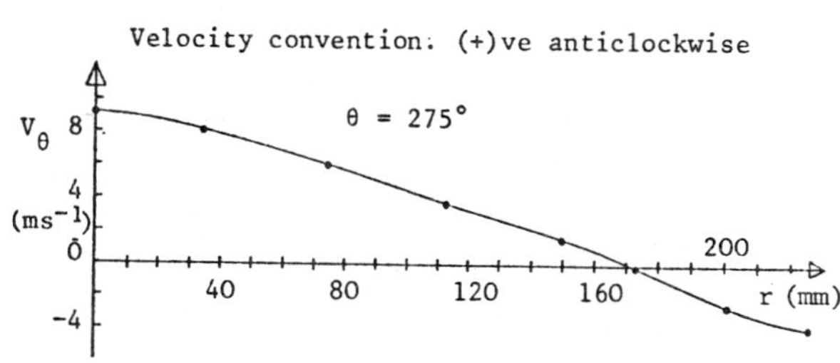

Fig. 16 Variation of tangential velocity for Φ = 0.8

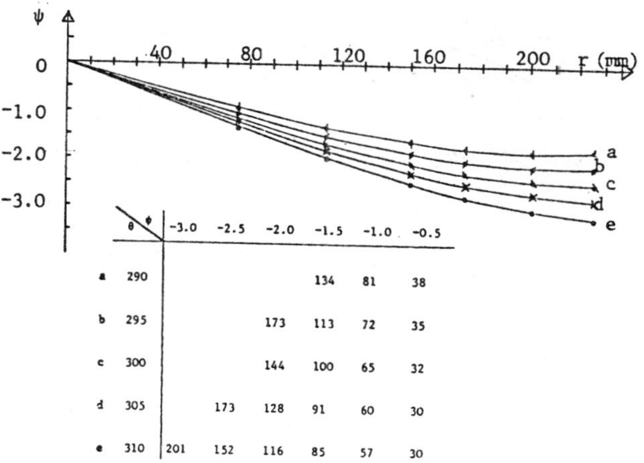

	θ \ φ	-3.0	-2.5	-2.0	-1.5	-1.0	-0.5
a	290				134	81	38
b	295			173	113	72	35
c	300			144	100	65	32
d	305		173	128	91	60	30
e	310	201	152	116	85	57	30

Fig. 17 Variation of ψ with r for given angle θ; Φ = 0.8

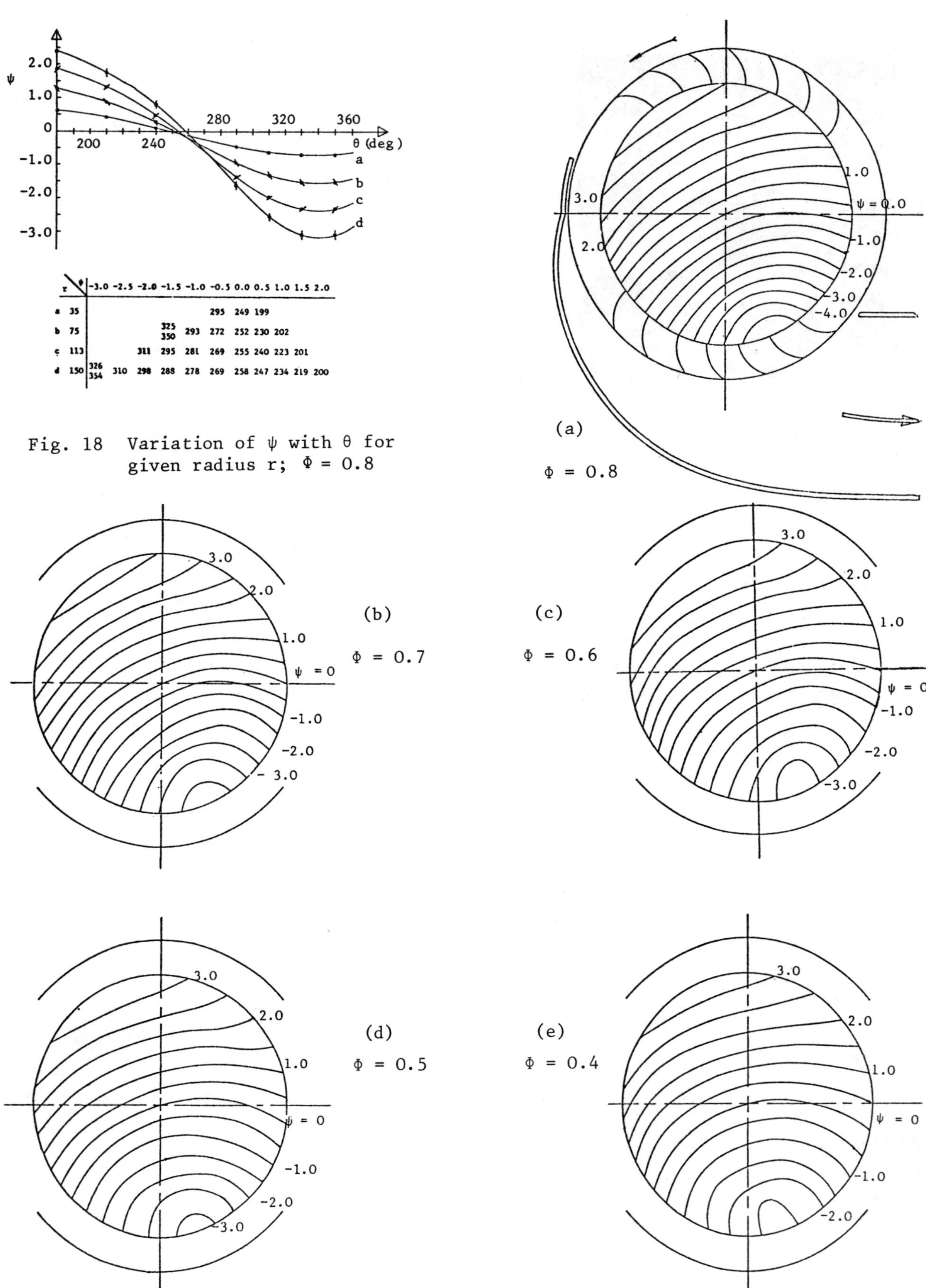

r \ ψ		-3.0	-2.5	-2.0	-1.5	-1.0	-0.5	0.0	0.5	1.0	1.5	2.0
a	35						295	249	199			
b	75				325 350	293	272	252	230	202		
c	113			311	295	281	269	255	240	223	201	
d	150	326 354	310	298	288	278	269	258	247	234	219	200

Fig. 18 Variation of ψ with θ for given radius r; Φ = 0.8

Fig. 19 (a) - (e) Streamline patterns deduced from pitot traverses

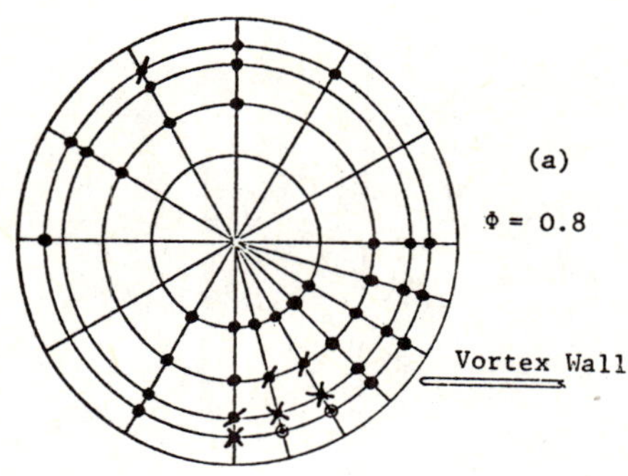

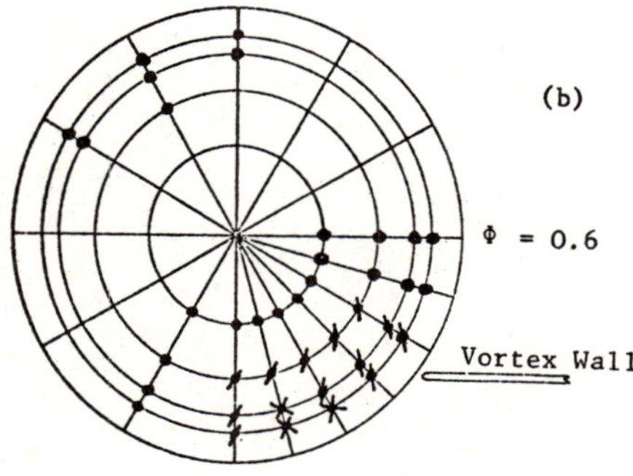

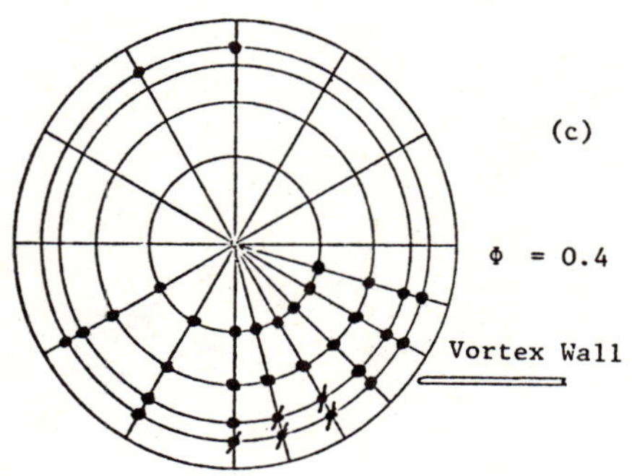

Symbol	$\tilde{t}$ Range
Nothing	0
●	-2 to 0
✗ (single bar)	-4 to -2
✗	-6 to -4
◎	-8 to -6

Fig. 20 Distribution of total pressure coefficient $\tilde{t}$

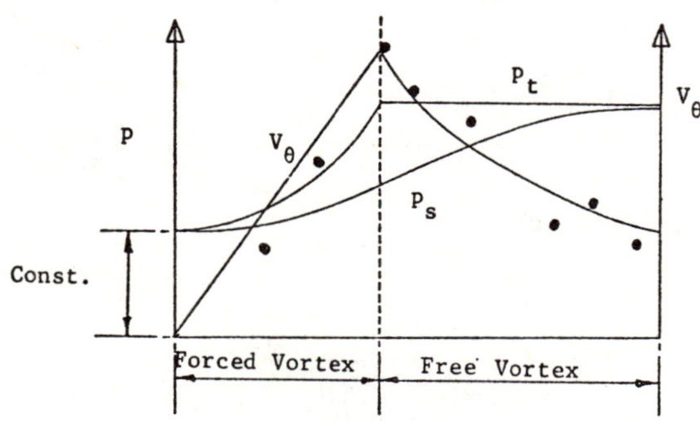

Fig. 21 Theoretical velocity and pressure distributions in a Rankine vortex

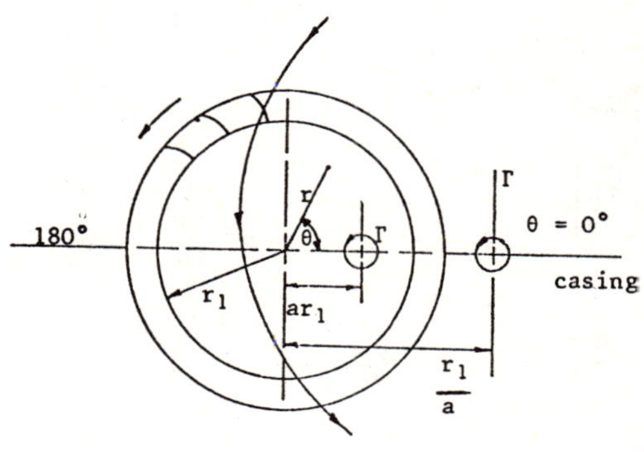

Fig. 22 Flow model used in Ref. 8

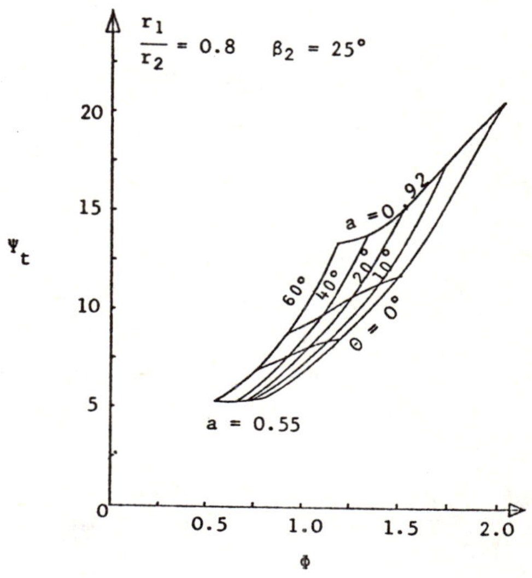

Fig. 23 Theoretical performance

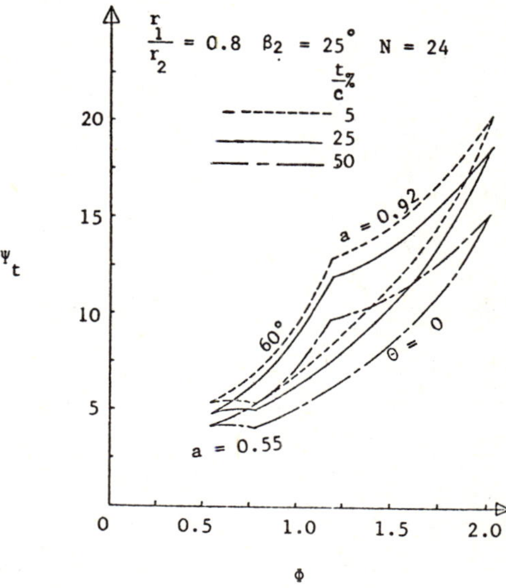

Fig. 24 Effect of blade thickness on theoretical performance

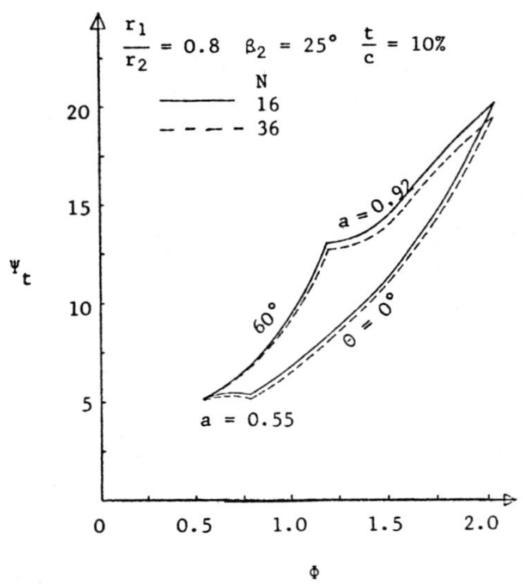

Fig. 25 Effect of blade number on theoretical performance

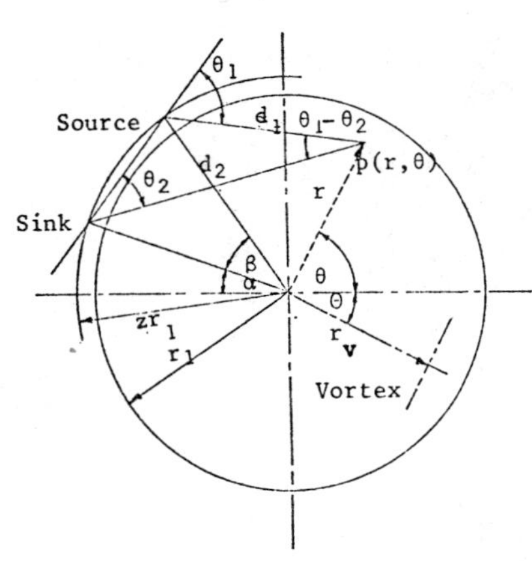

Fig. 26 Added source and sink to modify theoretical internal flow

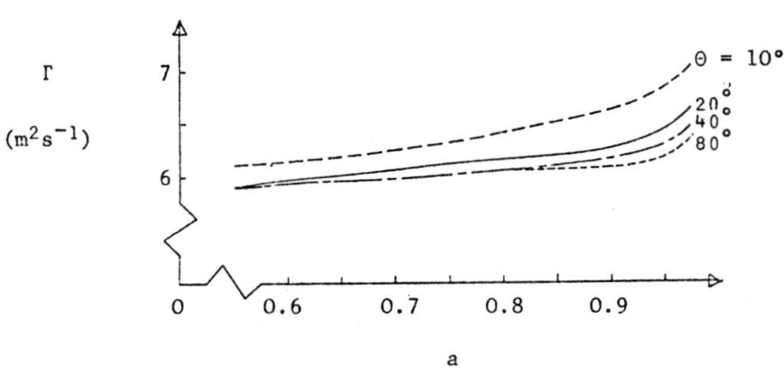

Fig. 27 Variation of vortex strength in rotor

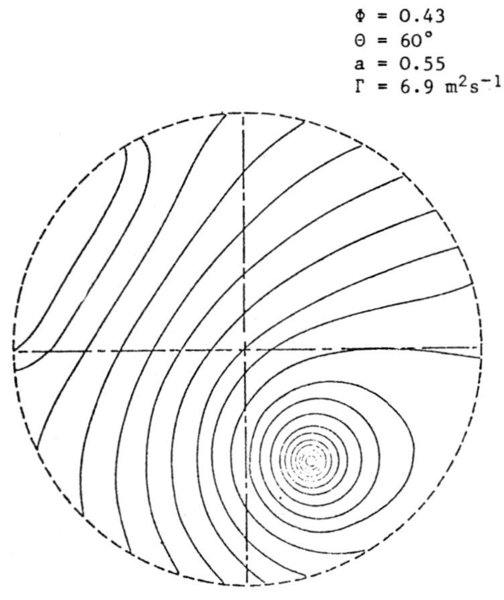

Fig. 28 Flow field with modifications of Fig. 26

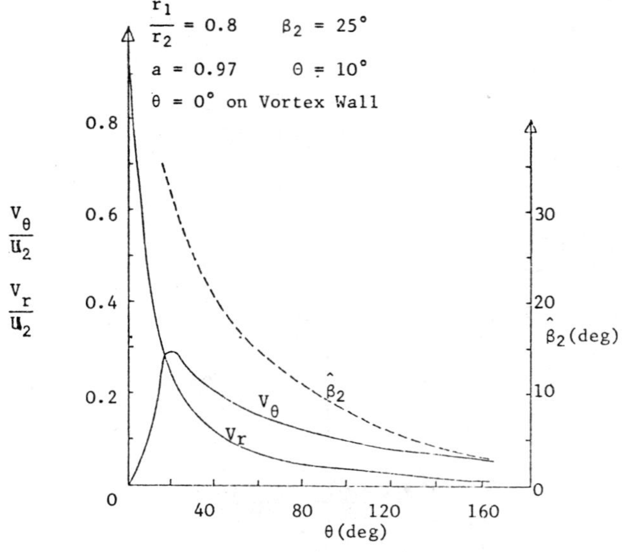

Fig. 29 Variation of velocity and deviation in suction region

International Conference on

Fan Design & Applications

Guildford, England: September 7-9, 1982

PAPER K1

COMPARATIVE ENERGY SAVINGS IN FAN SYSTEMS

W.R. Woods Ballard

Woods of Colchester Limited, U.K.

Summary

A number of methods of achieving significant energy savings are currently used in fan systems. A review of the main methods is given and a comparison of the energy savings and pay back periods for a typical load cycle is presented.

Organised and sponsored by
BHRA Fluid Engineering, Cranfield, Bedford MK43 0AJ, England.

0263 - 421X/82/01 00 - 0001 $5.00
The entire volume can be purchased from
BHRA Fluid Engineering for $82.00

NOMENCLATURE

P_E = Electrical input power

P_M = Mechanical output power from motor

P_R = Mechanical power absorbed by impeller

K = System component loss factor

q_V = Volume flow rate

ρ = Air density

V = Local velocity

η_M = Motor efficiency

η_D = Drive efficiency

η_F = Fan efficiency

1. INTRODUCTION

The most common variable duty fan system is the so-called variable air volume or VAV system. Studies (1) and practical experience have shown this to be the most energy effective method of providing the necessary air power to drive a building air conditioning system.

As it absorbs the greater part of the fan power in operating a VAV system the supply fan has been chosen for detailed study.

A number of different methods of providing control of the fan output are examined, including both variable fan geometry and variable fan speed. In addition centrifugal and axial fan types are included.

2. SYSTEM DESIGN AND FAN INSTALLATION

2.1 The importance of the system velocity and its relation to energy usage cannot be over emphasised since clearly there is little virtue in optimising the fan selection when, with closer attention to system design, a less powerful fan could have been installed in the first place.

For each system component, be it a filter, coil or a length of duct, the power needed to move the required flow of air is:

$$P_E = \frac{K.\, q_V \frac{1}{2}\rho V^2}{\eta_M \eta_D \eta_F}$$

As the input power is proportional to V^2, it can be seen that for a given air flow rate, loss can be reduced by 42% if the duct and component dimensions are increased by 10%. Even if this exercise can only be applied to the duct work alone, a reduction of some 20% of fan static pressure should be possible. Therefore, where space allows, generously sized system components and duct work should be used in the interest of energy consumption, despite their higher first cost.

2.2 Now having examined the energy consuming aspects of the system, corresponding care must be taken in the fan installation. The inlet must be free of obstructions and a proper bellmouth entry fitted where appropriate. If a screen guard is attached to the inlet, it should be placed in a low velocity region and designed not to distort the flow through the bellmouth. On the discharge side, where flow velocities are reducing and mixing is taking place, great care must be taken to provide a smooth and stable transformation of the flow from the fan discharge to the duct system.

3. CONTROL AND DRIVE SYSTEMS

The basic methods of providing variable volume in a fan system are by stepless speed change and by use of variable geometry fans indicated schematically in Fig. 1. The two most commonly used variable geometry arrangements are: inlet vane angle control as applied to the centrifugal fan and variable blade pitch control on the vane axial.

There are many commercially available means of providing variable speed and three variable speed systems are studied in relation to a VAV fan system. These are (i) the variable slip or eddy current coupling which is a commonly used and reliable drive, (ii) a proprietary unit incorporating a belt drive whose speed ratio can be controlled, and thirdly, (iii) an A.C. inverter controlled motor drive, which though less common, seems likely to become more popular in the future, owing to its potential for energy savings. Alternative methods which have been omitted are:

(a) thyristor converter with D.C. motor;

(b) slip ring induction motor;

(c) fluid coupling with induction motor;

(d) Schrage motor.

Control of fan output by damper or by-pass control is not considered as the energy savings are small.

4. SYSTEM CHARACTERISTIC

Fig. 2 illustrates the selected system pressure/flow characteristic. It will be seen that this contains a proportion at constant pressure loss with the major part varying as the $(\text{flow rate})^2$.

The fixed resistance at zero flow (200 Pa) is required to operate the air terminal devices installed at the end of the supply system. Although only representing a small proportion (10%) of the total system loss, this resistance plays a very influential role when selecting fans for variable speed control.

5. FAN SELECTION

5.1 In order to match the maximum system flow rate the fans have been selected to provide $9m^3/s$ against a system pressure loss of 2kPa. For both the axial and the centrifugal fan, the same basic fan has been selected whether the control and drive system relates to variable speed or variable geometry. The fan sizes have been selected to give near peak efficiency at maximum flow rate, whilst the discharge connections are sized for approximately the same discharge velocity.

Both fans are assumed to be in a type B installation (2), that is with a free space surrounding the inlet and with a ducted discharge. The main features of the fans selected are summarised in Table 1 and it will be evident that the duty does not favour the vane axial fan.

TABLE 1: FAN DETAILS

Fan	Centrifugal	Vane Axial
Impeller Type	Backward Curved DI	Axial
Impeller Dia. mm.	800	710
Rev/Min. Max.	1500	2950
Total Efficiency (at max. flow)	.83	.74
Discharge Vel. m/s	11.1	14.2
Discharge Dims. mm.	900 x 900	900 dia.

5.2 The operating line and pressure/flow characteristic for the centrifugal fan is shown on Fig. 3 and it should be noted that at $6.5m^3/s$ the operating line moves to the left of that shown in the manufacturer's catalogue for good selection. This is not too important however, as the characteristic of the chosen fan is stable down to zero flow and efficiency is still high even at 50% of maximum flow. The fan could have been selected at a higher speed and smaller diameter to give a maximum flow operating point closer to peak efficiency, but this would have involved a fan of heavy duty construction.

The corresponding data for the axial flow fan is shown on Fig. 4 and in this case the fan operating line crosses that for peak pressure at $4m^3/s$. As the fan cannot develop sufficient pressure below this flow rate it cannot be controlled to lower flows. In practice it would be unwise to operate below 50% of maximum flow in this particular case. As the axial flow fan selected is of the direct drive type, there is little scope for fine tuning of the selection. A high tip speed is required, together with adequate blade area, to give the pressure,and the blade pitch angle is selected to give the desired flow rate.

6. IMPELLER POWER - P_R

The fan manufacturers' catalogues, together with Fig. 3 and Fig. 4, have been used to calculate the power absorbed by the various impellers with the fan operating on the system characteristic shown on Fig. 2. This is shown on Fig. 5 where it can be seen that variable speed shows up better than variable geometry, and very much better than the inlet vane control. The restriction on the degree of flow modulation on the vane axial with speed control is evident, as is the greater power absorbed by the vane axial at maximum flow.

7. LOSSES

7.1 Motor losses

Motor efficiencies used in calculating the electrical input power are shown on Fig. 6. The efficiency is plotted against the % of motor rating and the appropriate proportion of the motor load has been used at each operating point. For A.C. inverter supplied motors the motor output rating has been assumed to vary with speed and for constant speed motors a rating of 30kW is taken.

Efficiencies for the constant speed motors are based on published data as far as possible, although at low loads additional data has had to be sought. The inverter efficiency is taken to be 93.5% and the relevant motor losses are taken to be 10% higher than for a normal supply.

7.2 Drive losses

The drive system losses associated with the configurations shown in Fig. 1 are presented on Fig. 7. These are given as an actual power loss in kW against flow rate. The intermediate belt drive losses are taken as a constant 3.3% of the maximum rated motor power. Slip, no load and exitation losses for the eddy current coupling are calculated in the same manner as Wright (3). The maximum loss for a fan drive with this type of coupling occurs at 33% slip which, as can be seen from Fig. 7 is a little less than 67% of the maximum flow rate. Loss data for the variable speed drive is taken from a manufacturer's catalogue where full technical supporting data is available. The drive efficiency is given as a function of % load against speed ratio and the drive selected for the study was rated at 30kW.

7.3 Electrical and mechanical losses

Fig. 8 shows the total of all the losses between electrical supply connection and the impeller plotted against flow rate.

8. POWER REQUIREMENTS

8.1 Electrical power input

Adding the electrical and mechanical losses of Fig. 8 to the impeller power characteristics of Fig. 5 gives the electrical input power shown on Fig. 9. At 85% flow there is little to choose between the various fan and drive systems. However, at maximum flow the vane axial absorbs appreciably more power than the centrifugal, on account of the lower fan total efficiency for the particular selection chosen for this comparison duty. Between 50 and 75% of flow, which is likely to account for more than 50% of the total operating time, the centrifugal + inlet vane control shows up worst, followed by the centrifugal + eddy current coupling. The A.C. inverter controlled fan systems show up best of all with the variable pitch axial and variable pulley centrifugal between.

8.2 Fan operating load profile

The fractions of operating time spent at various proportions of maximum flow are shown in Fig. 10, for every 5% step in flow. The profile may not be representative of a specific case but has been chosen as a result of examining published load profile data. It will be noted that no running time is shown below 20% flow and none is shown at maximum flow. This is not intended to show that operation does not take place in these regions, but that it does so for an insignificant proportion of the total operating time.

9. COST ANALYSIS

9.1 Energy costs

Assuming 3000 hrs/annum operation and a cost of electricity at 3.5p per unit, Fig. 11 summarises the energy costs for the various VAV fan systems operating as shown on Fig. 10. The trends shown obviously reflect the respective power inputs of Fig. 9 as indicated above.

9.2 Capital and operating cost

Fig. 12 provides an assessment of the capital and operating cost aspects where energy costs have been calculated as described above. With the exception of the A.C. inverter, all VAV fan system costs are actual prices. The inverter cost has been taken at £150/kW, that is £4,500.

As a guide based on experience where possible, a rating out of 10 for the best, has been given for installation and commissioning, maintenance/servicing and reliability. The well proven variable geometry fans and the eddy coupling drive fan fare equally well. The variable ratio pulley drive is an essentially mechanical system using pneumatic actuation and is fairly complex, and has therefore been marked lower. The inverter systems are not yet well developed by extensive field service, a large number of components are involved and expensive skills are likely to be involved in any servicing.

10. FINANCIAL PAY BACK

For the various VAV fan systems, a simple pay back period has been calculated and is shown on Fig. 13. The basis for calculating the pay back period relates the additional cost of the VAV fan system to the saving in energy cost compared with that for a simple constant volume fan. The cost of a fixed output fan has been subtracted from the VAV fan system cost to obtain the on-cost for the variable output feature. The additional VAV fan system cost does not make allowance for any extra operating costs, for which only an assessment was given on Fig. 12. The result is shown in the first column of Fig. 13.

Although their energy cost savings are greatest, the A.C. inverter controlled fans show the longest pay back periods and on the assumptions made, this is shown as over 2.5 years. However, the variable pitch vane axial fan with a pay back period of 1 year is the optimum choice for energy saving in the VAV system at the duty selected.

The annual energy saving compared with a constant volume fan is shown in column two of Fig. 13. For the A.C. inverter controlled fans this rises to over 60%, and even in the least effective VAV fan system the saving amounts to 35%.

11. CONCLUSIONS

It must be emphasised that some of the conclusions which follow are closely tied in with the assumptions which have been made. Whilst the assumptions made such as the system pressure/flow characteristic, the number of running hours per year, the operating load profile and the unit cost of electricity affect quantitatively the calculations ,qualitatively it is likely that they are generally applicable.

(i) Care must be exercised in selecting fans for variable speed operation when there is a proportion of the required system pressure independent of flow.

(ii) Both variable geometry and variable speed fans show significant energy cost savings in a VAV system. These are between 35% and 60% when compared with a constant volume system.

(iii) In terms of energy consumption the important region of fan operation is between 50 and 75% of maximum flow.

(iv) The A.C. inverter controlled fan systems show the greatest energy cost savings.

(v) The simple pay back period for the A.C. inverter controlled fans is more than 2.5 times that for the variable pitch vane axial.

(vi) In terms of a simple pay back period the variable pitch vane axial is significantly better than the rest.

REFERENCES

1. Jones, W.P.: "Energy Consumption by Variable Air Volume Systems". Haden Young Limited.

2. BS 848: "Methods of Testing Fans for General Purposes: Part 1: Air Performance: 1980.

3. Wright, M.T.: "The Economics of Eddy Current Couplings applied to Pump Drives". Power, March and April 1976.

VAV FANS AND CONTROL SYSTEMS

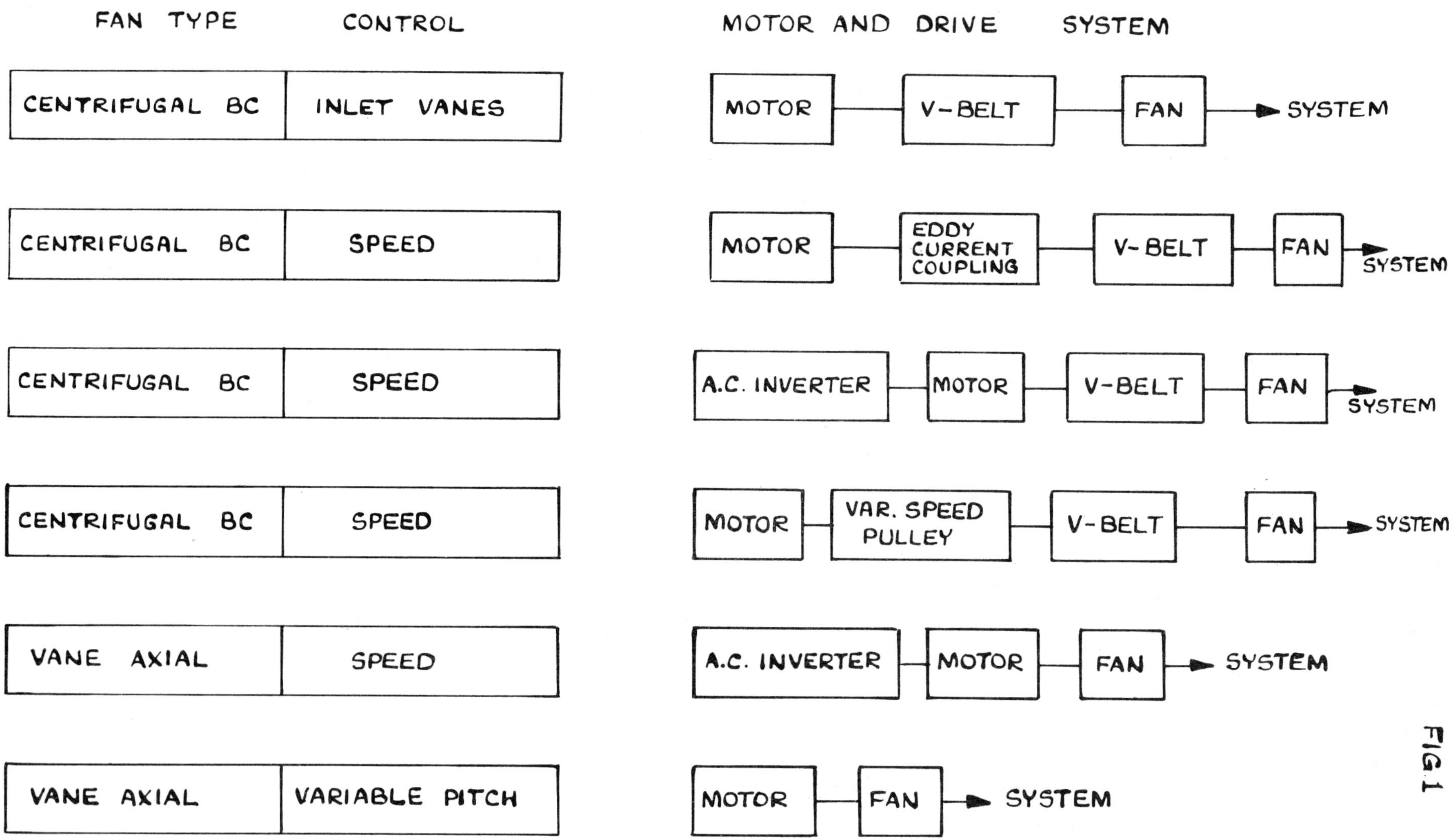

FAN TYPE	CONTROL	MOTOR AND DRIVE	SYSTEM
CENTRIFUGAL BC	INLET VANES	MOTOR — V-BELT — FAN	→ SYSTEM
CENTRIFUGAL BC	SPEED	MOTOR — EDDY CURRENT COUPLING — V-BELT — FAN	→ SYSTEM
CENTRIFUGAL BC	SPEED	A.C. INVERTER — MOTOR — V-BELT — FAN	→ SYSTEM
CENTRIFUGAL BC	SPEED	MOTOR — VAR. SPEED PULLEY — V-BELT — FAN	→ SYSTEM
VANE AXIAL	SPEED	A.C. INVERTER — MOTOR — FAN	→ SYSTEM
VANE AXIAL	VARIABLE PITCH	MOTOR — FAN	→ SYSTEM

FIG.1

FIG. 2.

SYSTEM CHARACTERISTIC.

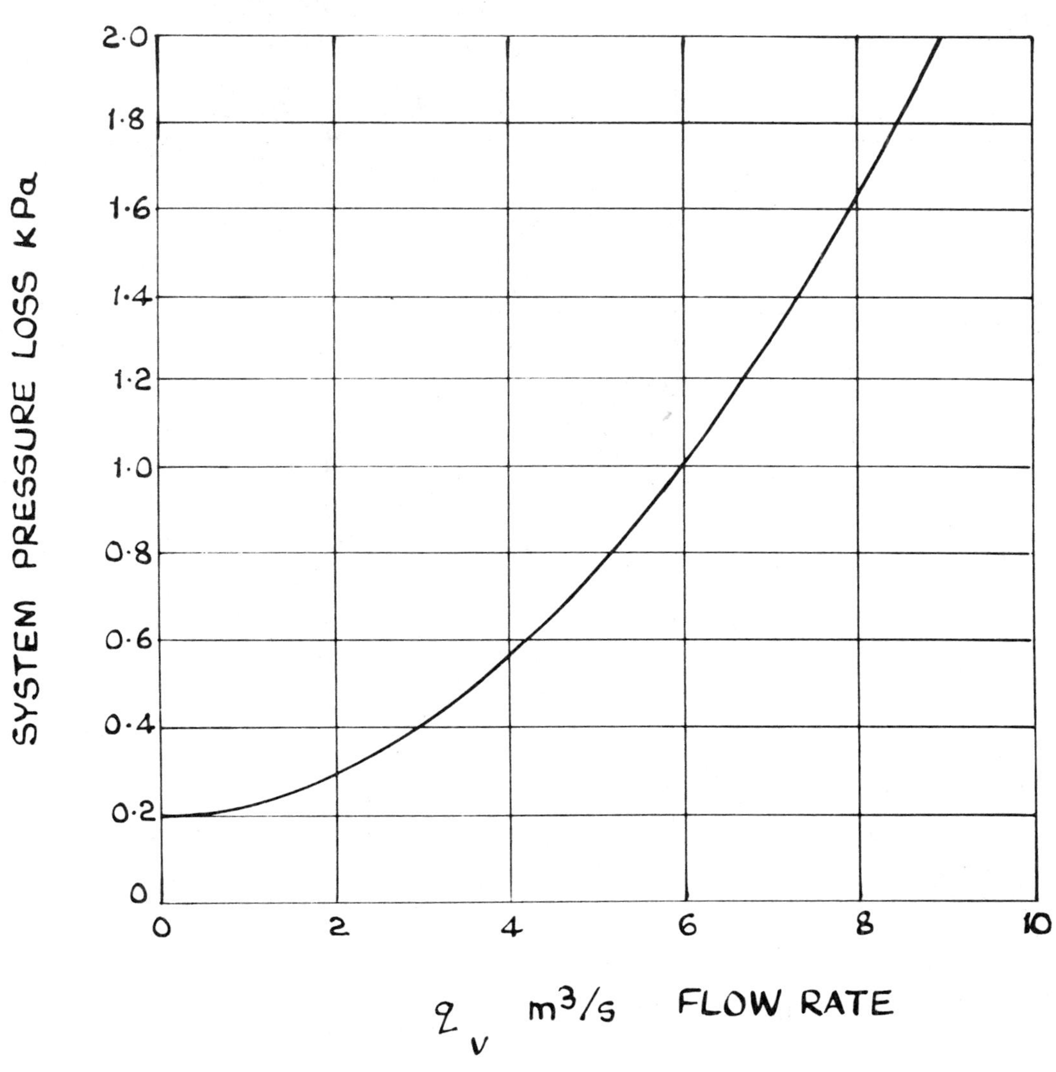

FIG. 3.

VARIABLE SPEED CENTRIFUGAL.
DOUBLE INLET. BACKWARD CURVED.

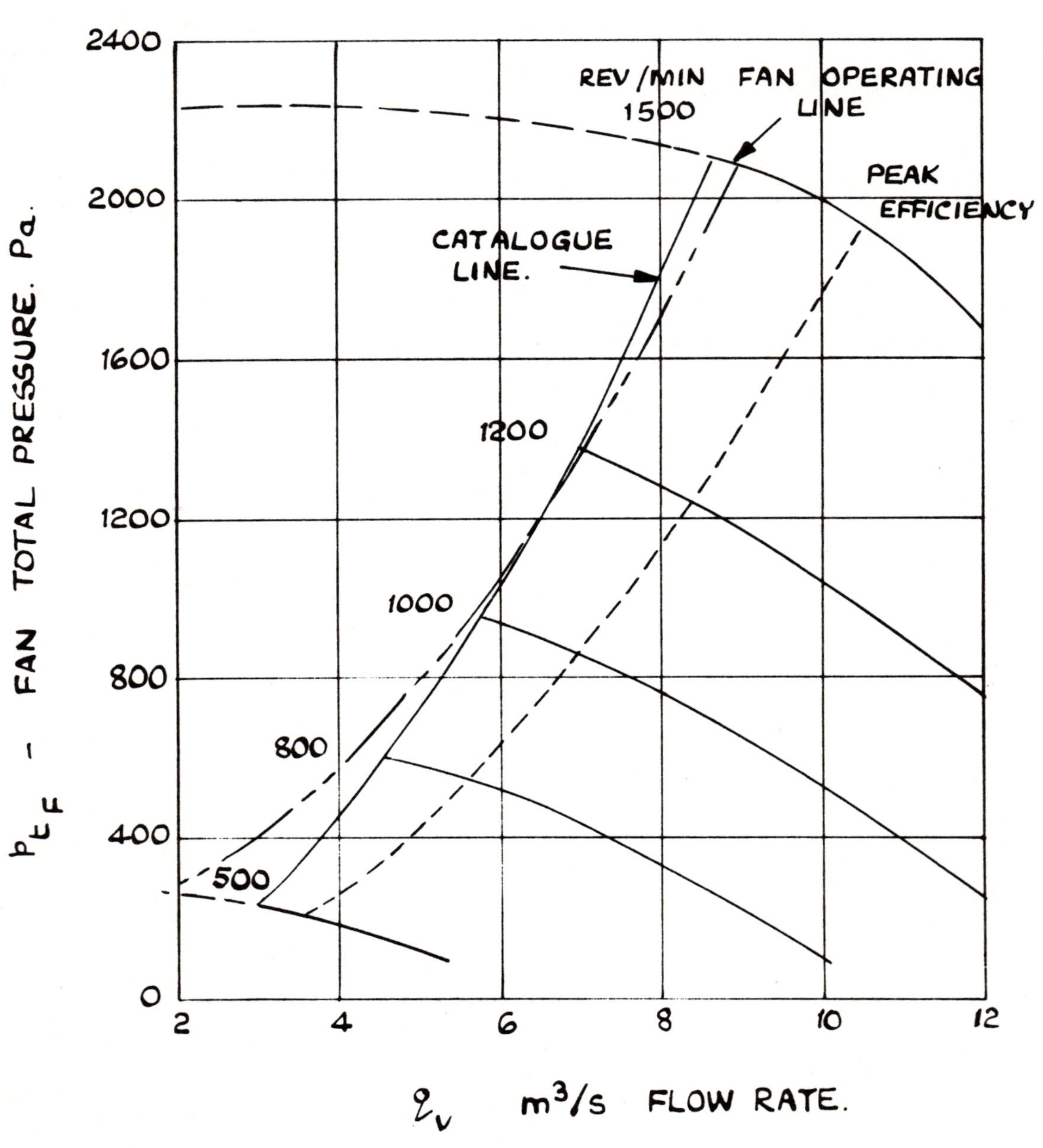

FIG.4.

VARIABLE SPEED VANE AXIAL.

28 DEGREE BLADE ANGLE.

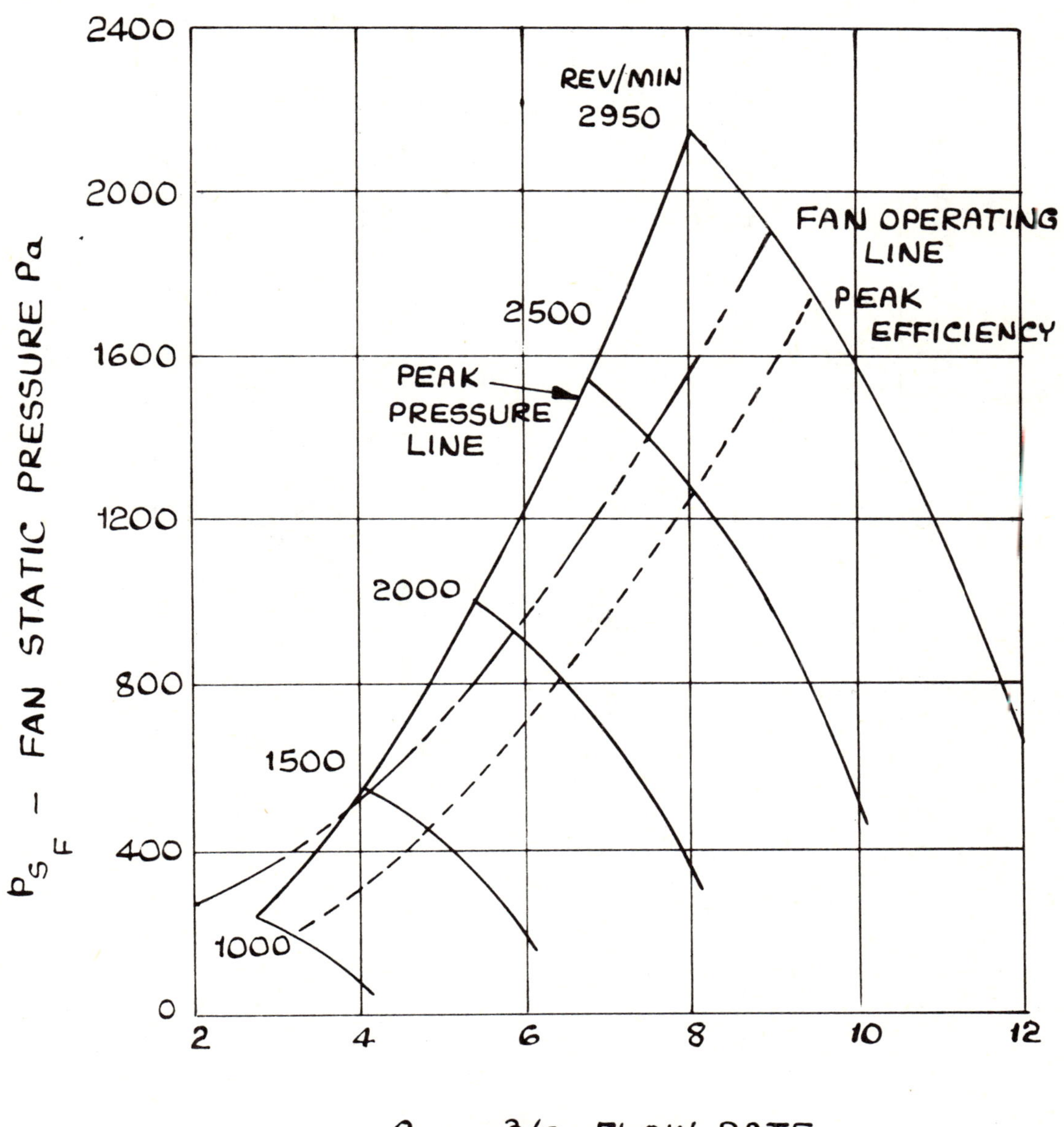

q_v m³/s FLOW RATE.

FIG. 5.

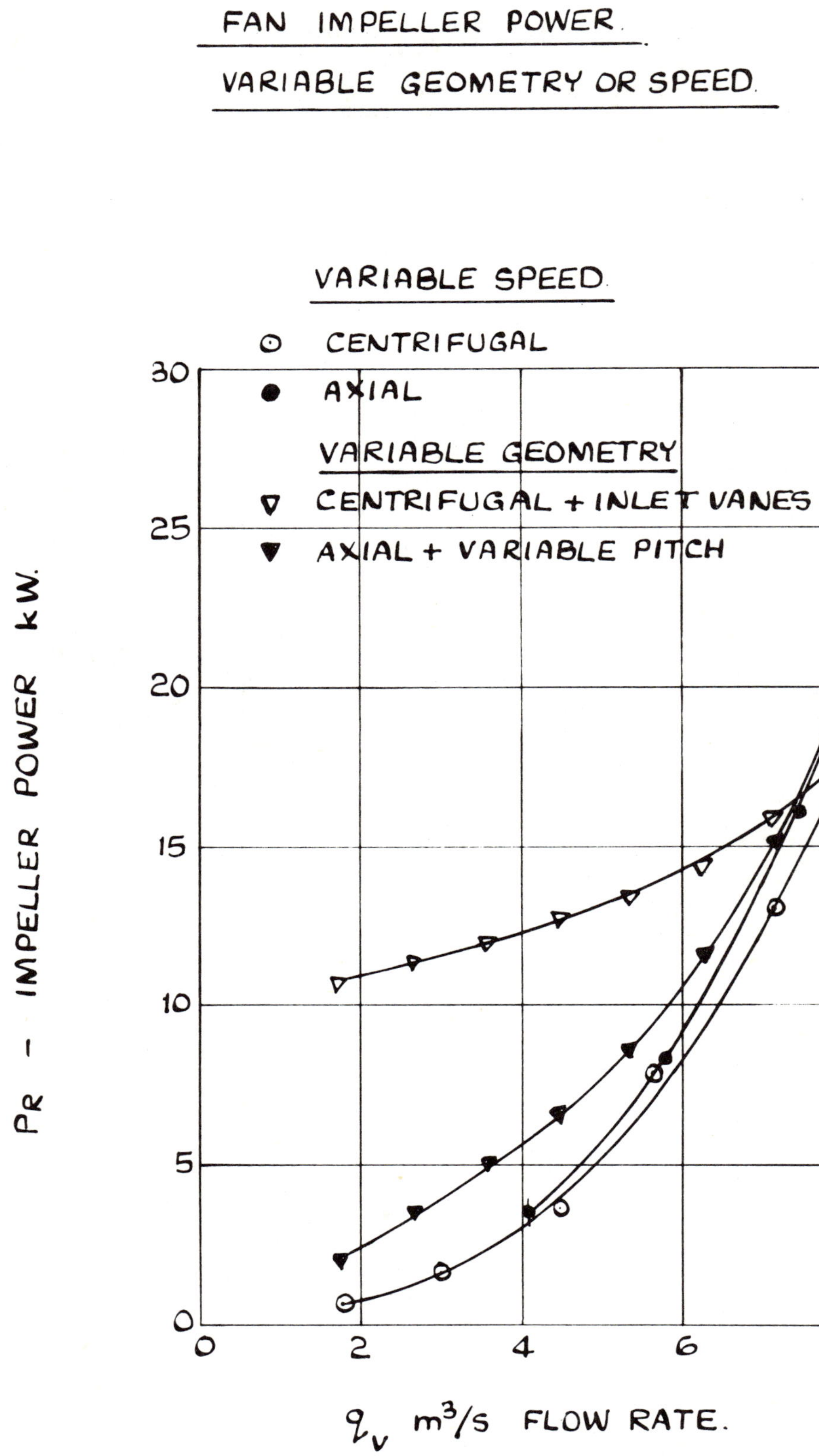

FAN IMPELLER POWER

VARIABLE GEOMETRY OR SPEED

VARIABLE SPEED

○ CENTRIFUGAL

● AXIAL

VARIABLE GEOMETRY

▽ CENTRIFUGAL + INLET VANES

▼ AXIAL + VARIABLE PITCH

P_R - IMPELLER POWER kW.

30
25
20
15
10
5
0

0 2 4 6 8

q_v m^3/s FLOW RATE.

FIG. 6

MOTOR EFFICIENCY

CONSTANT AND VARIABLE FREQUENCY SUPPLY.

– – – – – 1470 REV/MIN. INDUSTRIAL MOTOR.

———— 2950 REV/MIN. AIR-OVER FAN MOTOR.

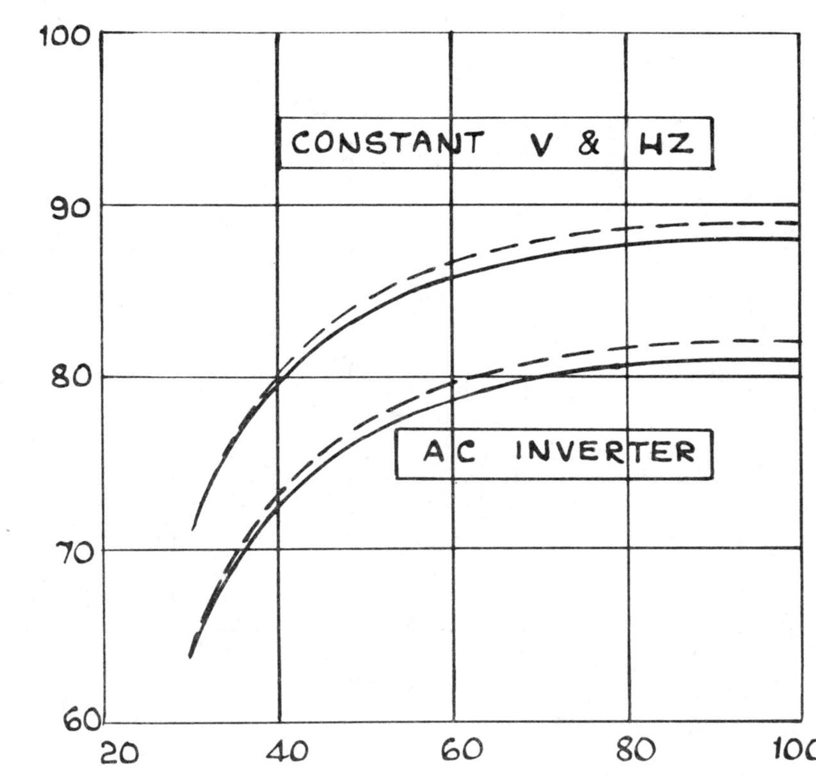

% MOTOR EFFICIENCY.

% MOTOR RATING.

NB.

FOR THE INVERTER SUPPLIED MOTORS THE RATED OUTPUT IS ASSUMED PROPORTIONAL TO THE FREQUENCY SUPPLIED.

FIG. 7.

DRIVE LOSSES.

ELECTRICAL AND MECHANICAL.

- - - - - - - - INTERMEDIATE BELT DRIVE.

——— — — ——— EDDY CURRENT COUPLING.

——————— VARIABLE SPEED PULLEY.

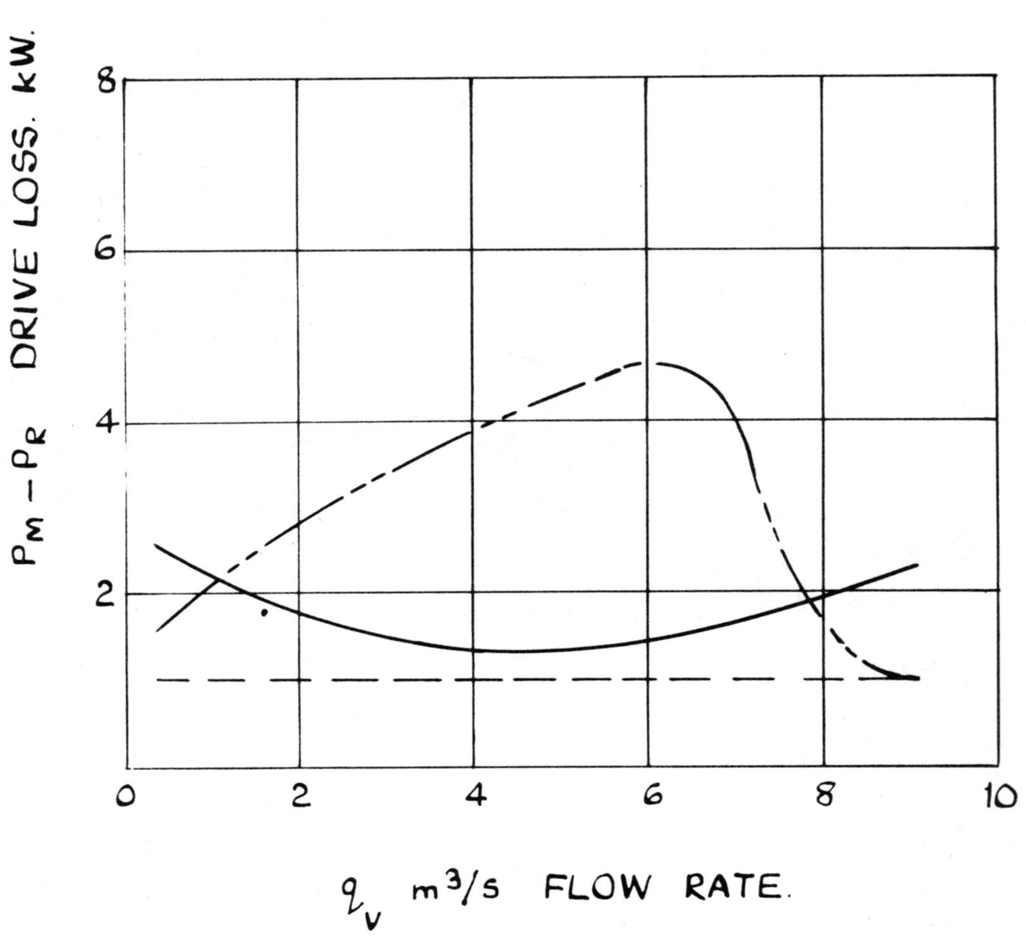

FIG. 8.

ELECTRICAL + MECHANICAL LOSSES.

VAV FAN SYSTEMS

VARIABLE SPEED.

- ⊡ CENTRIF. + INVERTER
- ● AXIAL + INVERTER
- ⊙ CENTRIF. + VARIABLE PULLEY
- △ CENTRIF. + EDDY COUPLING.

VARIABLE GEOMETRY.

- ▽ CENTRIF. + INLET VANES.
- ▼ AXIAL + VARIABLE PITCH.

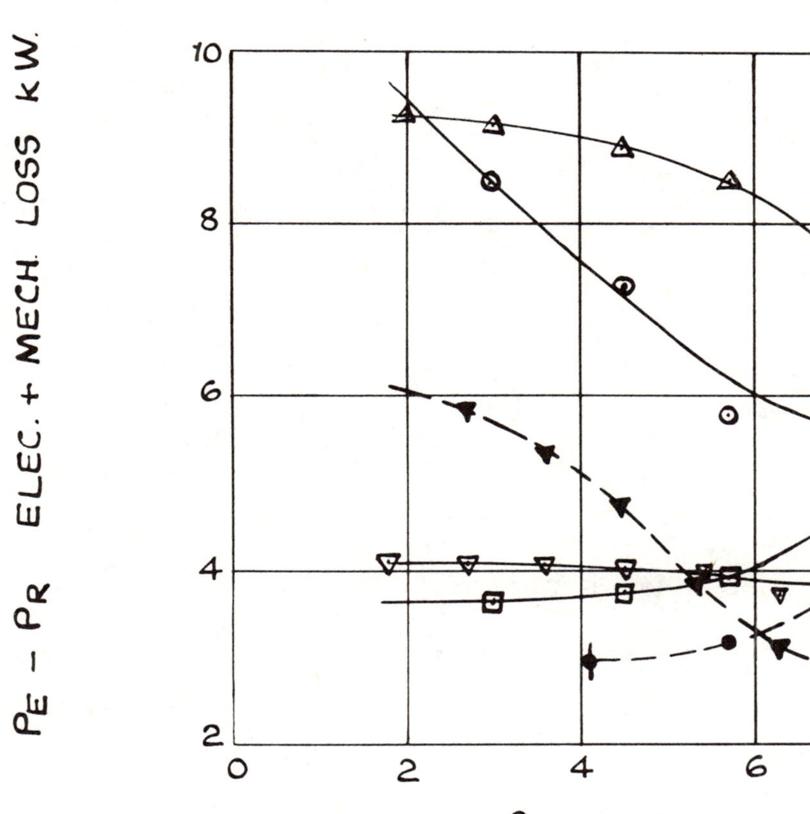

FIG. 9.

MOTOR INPUT POWER

VAV FAN SYSTEMS.

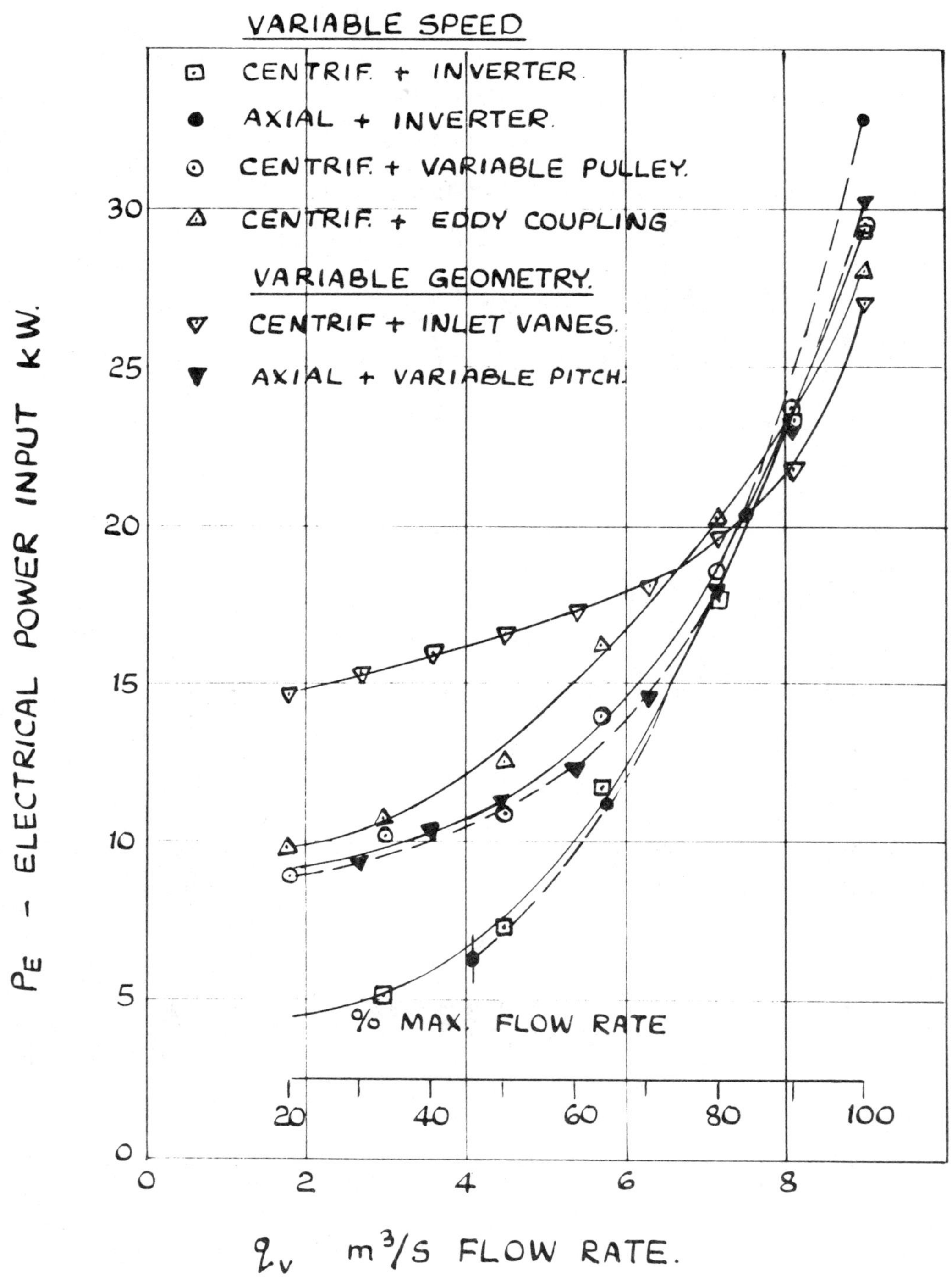

FIG. 10. FAN OPERATION – LOAD PROFILE.

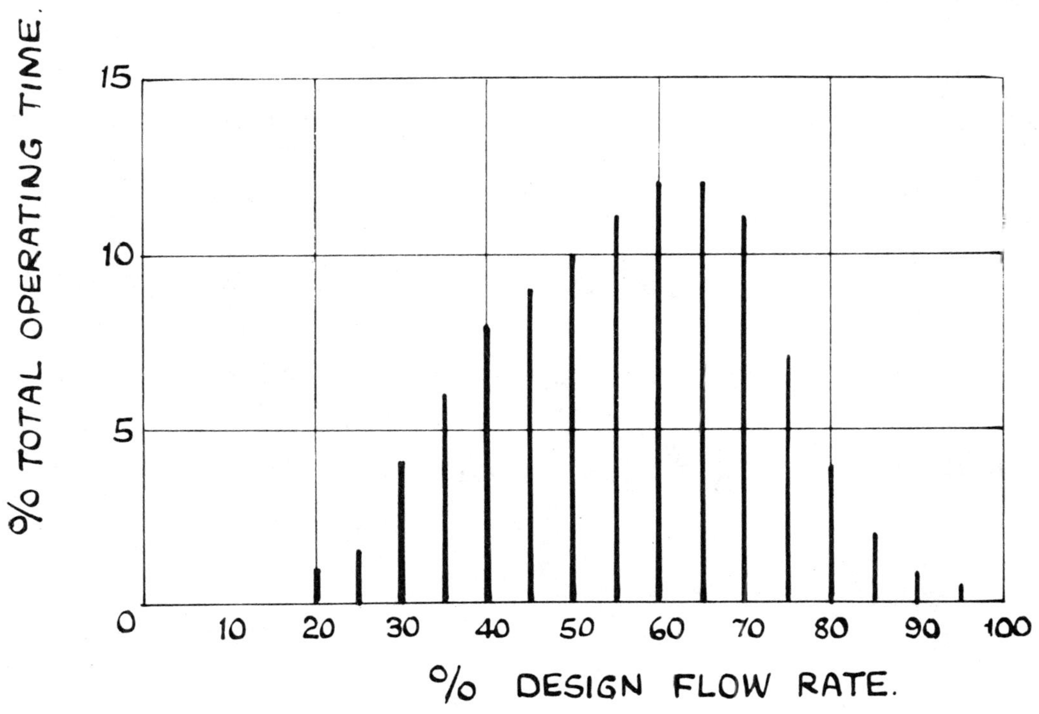

FIG. 11.

VAV FAN SYSTEMS.

RUNNING COSTS £.

VARIABLE SPEED.			1 YEAR	3 YEARS
CENTRIF.	+	INVERTER	1075	3225
AXIAL	+	INVERTER	1085	3255
CENTRIF.	+	VARIABLE PULLEY.	1350	4050
CENTRIF.	+	EDDY COUPLING.	1545	4635
VARIABLE GEOMETRY.				
CENTRIF.	+	INLET VANES	1805	5415
AXIAL	+	VARIABLE PITCH	1325	3975

ASSUMPTIONS:

3000 HRS. PER ANNUM RUNNING.
COST OF ELECTRICITY 3·5p PER UNIT.

VAV FAN SYSTEMS - COST COMPARISONS.

VARIABLE SPEED.	FAN SYSTEM COST £	ANNUAL ENERGY COST £	INST. & COMMISSIONING.	MAINT. & SERVICING.	RELIABILITY
CENTRIF. + INVERTER	5700	1075	5	6	5
AXIAL + INVERTER	6000	1085	6	7	6
CENTRIF. + VARIABLE PULLEY	3655	1350	7	6	8
CENTRIF. + EDDY COUPLING	2860	1545	8	9	9
VARIABLE GEOMETRY.					
CENTRIF. + INLET VANES	2950	1805	8	9	8
AXIAL + VARIABLE PITCH	2650	1325	9	9	9

NOTE:

OPERATIONAL COST FACTORS BASED ON A BEST FIGURE OF 10 WHICH RELATES TO A FIXED BLADE PITCH VANE AXIAL.

FIG. 12.

VAV FAN SYSTEMS – SIMPLE PAY BACK.

VARIABLE SPEED.	COST OF VARIABLE OUTPUT FEATURE £	ANNUAL ENERGY COST SAVING £	SIMPLE PAY BACK, YEARS.
CENTRIF. + INVERTER	4500	1760	2·6
AXIAL + INVERTER	4800	1750	2·7
CENTRIF. + VARIABLE PULLEY	2455	1485	1·7
CENTRIF. + EDDY COUPLING.	1660	1290	1·3
VARIABLE GEOMETRY.			
CENTRIF. + INLET VANES	1750	1030	1·7
AXIAL + VARIABLE PITCH	1450	1510	1·0

NOTES:

CHEAPEST FIXED OUTPUT FAN – CENTRIFUGAL – COST £ 1200

ANNUAL ENERGY COST AT CONSTANT 9 m^3/s £ 2835

(3000 (hv) × 27 (kW) × ·035 (£ 1 UNIT)

FIG. 13.

International Conference on

Fan Design & Applications

Guildford, England: September 7-9, 1982

PAPER K2

A DISCUSSION OF FAN AIR QUANTITY CONTROL WITH PARTICULAR RESPECT TO AXIAL FLOW FANS AND VARIABLE PITCH MECHANISMS

J. Marples

Aerex Limited, Sheffield, U.K.

Summary

The efficient control of air quantity makes a very large contribution toward energy savings when particular applications such as process systems, tunnel ventilation, mining operations, etc., require continually changing airflow requirements.

This paper discusses the effects of changing air flow rates, and changing system resistances on fan operation with particular attention being given to axial flow fans.

Organised and sponsored by
BHRA Fluid Engineering, Cranfield, Bedford MK43 0AJ, England.

0263 - 421X/82/01 00 - 0001 $5.00
The entire volume can be purchased from
BHRA Fluid Engineering for $82.00

1. FAN CHARACTERISTICS AND SYSTEM RESISTANCE CHANGES

The axial flow fan, whilst for a given tip speed has limited pressure developing capabilities compared with the centrifugal fan, accommodates a great potential for air quantity control which cannot be achieved by the centrifugal fan. This is because the lift generated by an aerofoil or cambered blade, is highly sensitive to changes in the direction and value of the approaching fluid velocity. The axial flow fan is totally dependent upon the blade lift component for its pressure development whereas the centrifugal fan, even if it has aerofoil blades, is not so dependent. The axial flow fan blade can also have its angle adjusted with relative ease whereas such adjustment of the aerofoil blade on a high efficiency centrifugal fan is not readily and cheaply available. As a result changes in characteristic shape over a wide range of performance above and below an optimum can be effected on an axial flow fan in a more easy and practical manner, such that it is ideally suited for many applications particularly those where a variation in system resistance over a period is expected as well as adjustment from time to time of air quantity.

The illustration in Fig. 1 shows a typical axial flow fan characteristic curve. The position of optimum fan efficiency is shown and the change which can occur in this value if the performance requirement is varied from the optimum for a given fan is illustrated. The unstable part, or stall region, of the characteristic is clearly defined and the presence of this is one disadvantage of an otherwise highly efficient air moving device.

The extent of this disadvantage is more clearly illustrated by comparing the axial flow fan and centrifugal fan characteristics. The next curve in Fig. 2 shows, to the same scale, the comparison in performance of a backward aerofoil bladed high efficiency centrifugal fan and a medium pressure axial flow fan, the latter being the same fan whose characteristic was shown in the first illustration, and each having a tip diameter of 1 metre and running at a tip speed of 100m/s. Also illustrated is the curve for the axial flow fan if the tip speed of this is increased to 136m/s. It is clearly indicated that merely by increasing the speed, all else remaining equal, the axial flow fan cannot match the centrifugal fan in performance. However Fig. 3 shows how this deficiency can be overcome, for compared with the 1 metre diameter centrifugal fan running at a tip speed of 100m/s, we now have a smaller axial flow fan running at a substantially higher speed, the comparison in this case being that of an axial flow fan having a tip diameter of 0.77m, operating at a tip speed of 167m/s. This comparison serves to illustrate how the use of an axial flow fan can result in a reduction in impeller size compared with the diameter of a centrifugal fan to approximately 70/80% and it is clear that such a size reduction can result in considerable savings in installation costs.

The area of unstable operation has not been removed however and clearly if the system resistance should increase by an appreciable amount, then the axial flow fan would experience unstable operating conditions. This is not a very desirable situation at any time and particularly with modern high efficiency aerofoil blades. It will be seen that the centrifugal fan would accommodate such a situation more readily and certainly without the inherent danger to the integrity of the blade structure that the axial flow fan would experience. A change in fan characteristic to produce a reduced performance which would accommodate the higher system resistance whilst maintaining a reasonably high efficiency is therefore highly desirable.

2. METHODS OF ADJUSTING FAN PERFORMANCE

A change in system resistance, particularly an increase, is not the only reason for wishing to have the flexibility of performance adjustment incorporated into a fan system. Clearly in such areas as mining, process work, etc., where installations must cater for future expansion then changes in air quantity as well as system resistance must be taken into consideration. In times when the costs of energy are continually increasing, efficient and even rapid changes in ventilation rates to meet alterations in demand can be an important contribution to savings in operating costs.

The control of air quantity to meet such changing requirements, whether by frequent variation in demand or a gradually changing system can be achieved by a number of methods. These can be categorised as follows:-

(a) Damper control

(b) By-pass control

(c) Inlet vane control

(d) Variable pitch control

(e) Speed control - (i) By variable speed motor
(ii) By fluid coupling
(iii) By multiple speed motor

All the methods listed can be used with axial flow fans, and all except (d) in practice, with centrifugal fans.

The first two, damper control and by-pass control, are merely simple methods of artifically changing the system resistance, the former increasing it and the latter generally decreasing it by relieving the system. For increasing or decreasing the air quantity through the primary system, both methods have many limitations and contribute little, if anything, to savings in energy costs. In any case it has already been established that controlling air quantity by increasing system resistance on an axial flow fan particularly is not necessarily a desirable thing to consider.

Inlet vane control has much more to offer. It has already been stated that the performance of an aerofoil is highly senstive to changes in the direction of, and changes in the value of, the approaching fluid velocity. Inlet control vanes will provide these changing conditions on an axial flow fan blade and the use of such devices is by no means new. The idea was patented nearly 50 years ago when the introduction of the vortex theory based axial flow fan or screw fan as it was then known was still quite new.

Fig. 4 shows how the performance of an axial flow fan will vary when controllable inlet vanes are introduced. The fan is that whose performance was illustrated in Fig. 1, and demonstrates how the provision of angularly adjustable flat, or slightly cambered plates, at the inlet to an axial flow fan can be used to vary the performance. Contours of constant fan efficiency are shown and at the top is illustrated the variation in fan efficiency against the percentage change in air quantity. Depending on the application for which the fan is being installed, the provision of these control devices with their inherent simplicity of operation could offer a useful solution providing air quantity adjustment and giving far greater power saving than the two previously mentioned.

A closer examination of the curve will reveal however certain disadvantages. Firstly an increasing system resistance would still put the fan into an unstable operating condition regardless of the inlet vane control angle. It will be observed that the pressure at which stall commences initially follows an approximate square law or system resistance line. Secondly, whilst reasonable reductions in air quantity, up to say 25%, can be readily accommodated, greater reductions, for example 50%, can again put the blades into a stalled condition. Such reductions are not uncommon in the mining and tunnelling industries and where changing processes are being accommodated. For large air quantity reductions the inlet vanes would have to become very steeply angled, and as a result would act as a resistance and turbulence generator as well as flow deflectors. Fig. 5 illustrates pictorially typical flow conditions and the effect of the disturbed flow profile at increased vane angles is to induce premature stall into the blade characteristic. Flat control vanes have been illustrated for simplicity, and the introduction or curved or profiled vanes could improve conditions marginally. If the fan has been designed with sufficient power margin available however there is the advantage that worthwhile increases in performance can be obtained with positive angular movement of the inlet vanes for the point at which stall occurs still continues to move upwards within limits on a square law line. Certain applications could benefit from such a change in fan characteristic.

Fig. 6 shows the effect of varying the blade angle above and below the nominal design setting previously illustrated in Fig. 1. Contours of constant fan efficiency are again illustrated and the stall line is clearly indicated. The shallower curve formed by this stall line will be instantly noticeable and thus, somewhat higher system resistances can be readily accommodated if the blade angle is reduced. Equally evident will be the capacity to pass greatly reduced air quantities whilst maintaining aerodynamically stable operation and the curve showing efficiency against percentage air quantity will demonstrate how reduced volume flows can be obtained at higher efficiencies than those obtained by the methods of volume control previously discussed. Although at 50% flow rate the air horsepower is a mere 12½% of the normal air horsepower equivalent to 100% flow, the benefits of saving a few percentage points in efficiency can always be worthwhile.

A further advantage of using blade angle variation for volume control can be particularly appreciated when a system requires the operation of fans in parallel. The problems of starting combinations of fans on tight system resistances are already well known. However the ability to reduce blade angles before running up to full speed thus presenting a much better working margin to stall, before final adjustment when full speed has been obtained, provides an ideal solution to the situation.

Speed control as a means of varying performance has been mentioned. This will provide an air quantity change at almost constant fan efficiency, but whilst catering for most changes in performance requirement it does not allow for a higher system resistance. However where considerable reductions in volume flow are required and changes in system resistance expected also over the operational life of a fan, then a combination of speed variation using a two or even three speed motor with variable blade pitch facilities, can provide a most efficient and effective way of controlling air quantity. Mining, tunnelling and power generation are but three industrial applications for which variable pitch axial flow fans can make an effective energy saving contribution for air quantities as low as 20% or even 10% can be required during part of the operational cycle.

3. METHODS OF VARIABLE PITCH CONTROL

A number of methods by which the blades on an axial flow fan rotor can be adjusted simultaneously are available. The use of any one or the other of these methods is largely dependent upon the size, power, speed, and application for which the fan is being used. It has to be considered whether continuous modulation is specifically required, whether changes in performance over a period are required and the anticipated time span between changes, or whether a simple increase in air quantity as demand increases over much longer periods is all that is necessary. Clearly careful thought must be paid to the economics of putting in an expensive and sophisticated modulating system if only infrequent changes in airflow requirements are envisaged.

The complete system of actuation can vary from the totally mechanical, where angular movement of the blades is initiated by manual means external to the fan, through pneumatic, hydraulic to electrically assisted mechanisms. Electrical assistance would normally be provided by an electro mechanical thruster placed outside the fan casing, but both pneumatic or hydraulically assisted systems can have the thruster units either external to the fan, internal within the fairing, or even integral within the fan rotor hub. Whichever system is considered however, all except that which is manually operated or manually controlled external to the fan, will require a control source or sensor. The sensor will measure the parameter being controlled, namely, air quantity, air pressure, temperature, or pollutent concentration. A small feed-back signal from the particular monitoring sensor will be compared with a similar signal from the controller or actuator. Whilst the output from the two comparing sources remains equal, an electrical balance will occur in the circuitry and the rotor blade angle will remain stationary. Immediately the sensor detects a change in the monitored parameter an electrical imbalance will occur. This imbalance will result in a signal being passed to the actuating device causing the blades to move in the desired direction. The feed-back signal from the actuator will correspondingly change and when it finally matches that from the sensor movement will cease. Such a system is ideal for the automatic variation of performance where important process work, environmental control such as is required in vehicular tunnels, or draught regulation etc., is necessary. However it could be considered that such a system would be too sophisticated when infrequent changes in performance were all that was necessary. In such instances push-button control or even mechanically linked manual operating devices would be perfectly adequate.

4. SOME ASPECTS OF VARIABLE PITCH ROTOR AND ACTUATING SYSTEM DESIGN

When rotating at speed each blade on an axial flow fan rotor experiences a number of forces. These are aerodynamic or airloads in the form of lift and drag, and centrifugal in the form of radial loads associated with the rotating mass of the blade. On normally supported aerofoil blades the air loads attempt to rotate the blade to a more positive angle, whilst the centrifugal forces attempt to turn it to a more negative angle. However the latter forces are far greater than the air loads and so any axial flow fan blade left free to rotate about its own axis would move generally to a lower blade angle. In designing a variable pitch rotor therefore not only have direct centrifugal loads to be considered but blade turning moments also. When one considers that the steel blade on a typical 2 metre diameter fan running at a speed of 1450 rpm, can impose a centrifugal force as great as 12 tonnes, then it will be appreciated readily that the turning moments due to the centrifugal force can be quite substantial. Unless facilities to counterbalance this moment are incorporated then any actuating mechanism would have to provide a permanent restraining force and clearly this would be highly undesirable. As a result virtually all variable pitch rotors, except those of lightweight construction or having a low speed operation, have counterbalance weights fitted to the blade stub shaft which are sized to have a turning moment approximately equal in value to, but acting in opposition with, the blade feathering moment. Fig. 7 is a simple illustration indicating how this is achieved. With correctly proportioned balance weights taking due account of the blade feathering moment, and air load, then friction forces are the only substantial ones remaining to be overcome by the actuating mechanism. The addition of the counterbalance weights however increases the weight of the rotating items and adds to the imposed centrifugal loads on the rotor hub. The thrust bearings supporting each blade stub shaft must be amply rated to accommodate these high centrifugal loads and special bearings packed with anti-fretting lubrication compound or with special cages are invariably used to ensure reasonable life.

Plate 1 illustrates a typical variable pitch rotor assembly. Clearly variations upon this exist but those shown are not untypical of the many types which are in service. Axial movement of the bell housing in the centre will be brought about by either a double acting hydraulic cylinder located within the hub and rotating with it or through a thrust bearing mechanism attached to linkage connected to an actuator mounted external to the fan. One advantage of using an integral hydraulic cylinder is that during operation of the mechanism any reaction loads which occur are contained completely within the fan hub and not transmitted to the main fanshaft bearings. However with effective balancing as described earlier these reaction loads will not in any case be particularly great and a mechanically connected thrust bearing arrangement can readily accommodate these.

The supply of oil for operating the hydraulic cylinder will be supplied through a rotary union connected to an external pumping unit. This rotary union can incorporate a positioner valve or sleeve valve which will direct the oil to one side of the cylinder or the other depending upon whether an increase or decrease in blade angle is required. Fig. 8 shows a simple rotary union/sleeve valve to which oil would be supplied at constant pressure. Movement of the hydraulic piston would be initiated by mechanically moving the sleeve valve the short distance to uncover one set of ports or the other depending upon which direction the cylinder was required to move. This mechanical movement would in turn be initiated from a control unit outside the fan and connected mechanically with the union. Alternatively the rotary union can be a simple fluid transferring device only, with the sleeve valve fitted within the pressure piping external to the fan. Such an arrangement would not require the rotary union mechanically connecting to a controller outside the fan as all positioning would be carried out within the external sleeve valve arrangement. This particular system lends itself more readily to external manual control of blade angle movement rather than the former which will be more suited to a fully modulating type of operation.

Both systems suffer however from the disadvantage of having to transfer hydraulic fluid under high pressure through an inaccessible rotary union type of device. Hence there is a favour in many circles, particularly those where maintenance skills and costs are at a premium, to have the actuating device, whether it be a hydraulic cylinder or alternative means of applying the moving force, outside the fan and transferring all movement via mechanical linkage. Final transfer of motion is effected by a thrust bearing arrangement which can be grease lubricated and amply rated for long periods of trouble free operation.

For precise control of airflow, hysteresis free operation is highly desirable. The oil-hydraulic system of variable pitch control lends itself readily to precise positioning, for the sleeve valve arrangement system with non-return valves in the hydraulic circuit can enable this positioning whether increasing or decreasing the blade angle to be achieved. With the provision of external actuators operating through mechanical linkages and thrust bearing arrangements, high manufacturing standards, and tight tolerances, must be adhered to for backlash, or clearances in any part of the system between the blade and the actuating cylinder cannot be tolerated when any position transmitting device within the actuator has to give a signal which truly represents the position of the fan blade.

Whether the methods of actuating system design follow those described or others equally well known it is self evident that variable pitch control of axial flow fans offers a flexible and efficient method of air quantity variation. It will also be apparent that some degree of sophistication is necessary in order to achieve this variation with any reasonable degree of precision. However as the cost of energy continues to escalate then the capital invested in such mechanisms could soon be repaid by reduced operating costs.

5. ACKNOWLEDGMENTS

The author wishes to record his appreciation to his Co-Directors of Aerex Limited for their permission to prepare, present and publish this paper.

m
400
380
360
340
320
300
280
260
240
220
200
180
160
140
120
100
80
60
40
20
0
2
4
6
8
10
12
14
16
18
20
22
24
26
28
30
m3/s
100
90
80
70
60
50
40
30
20
10
0
%
Effy.
Unstable
(stall) Region
Stable
Region
Higher System
Resistance
Optimum
System
Lower System
Resistance
80%
85%
80%
75%
70%
60%
50%

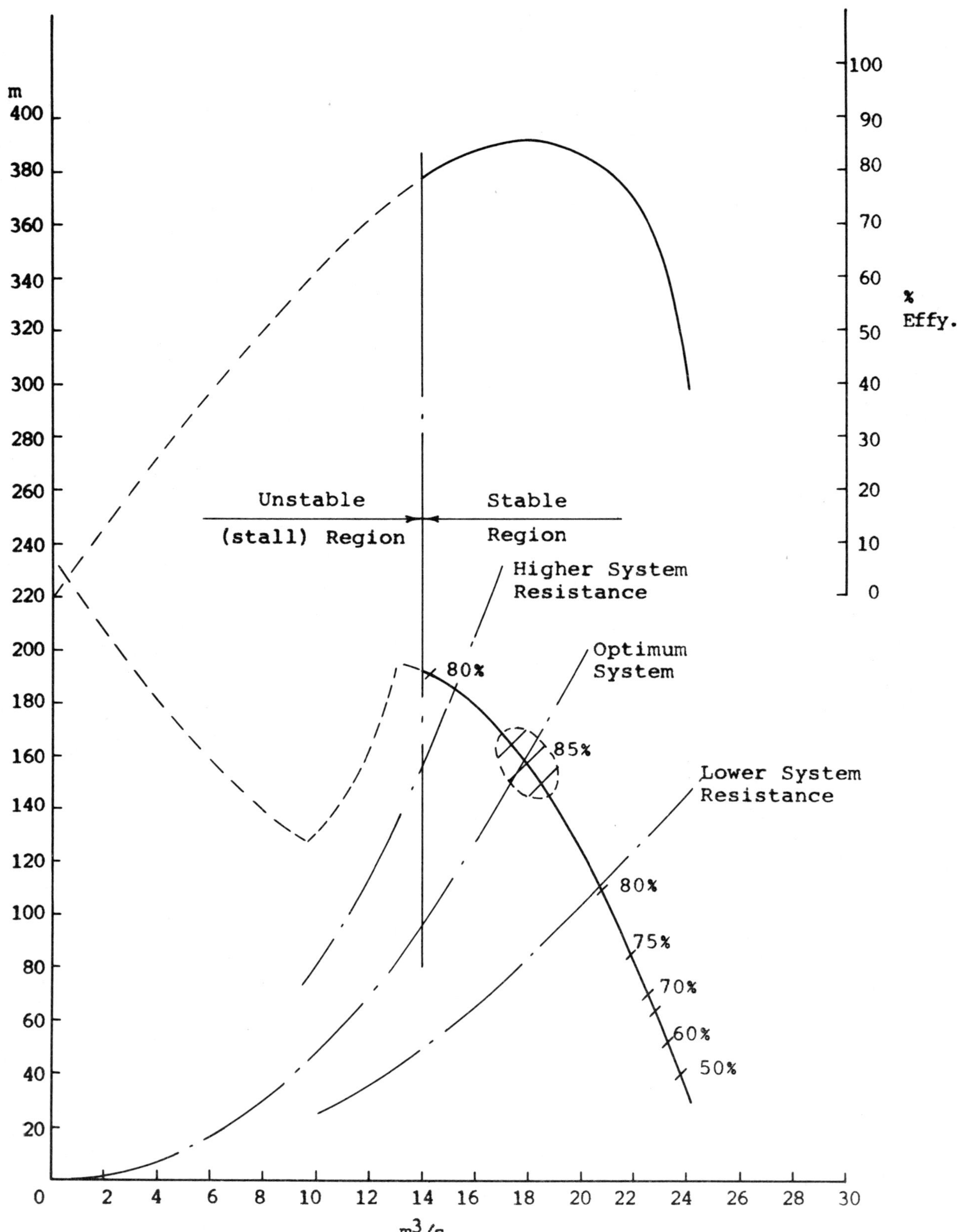

Figure 1.

m
460
440
420
400
380
360
340
320
300
280
260
240
220
200
180
160
140
120
100
80
60
40
20
0
2
4
6
8
10
12
14
16
18
20
22
24
26
28
30
m3/s
87%
Backward Aerofoil bladed
Centrifugal Fan
1m. dia.
100 m/s tip speed
Optimum System
(Centrifugal Fan)
85% (Axial
flow fan)
Aerofoil bladed
Axial Flow Fan
1m. dia.
136 m/s tip speed
Optimum System
(Axial Flow Fan)
85%
Aerofoil bladed
Axial Flow Fan
1m. dia.
100 m/s tip speed

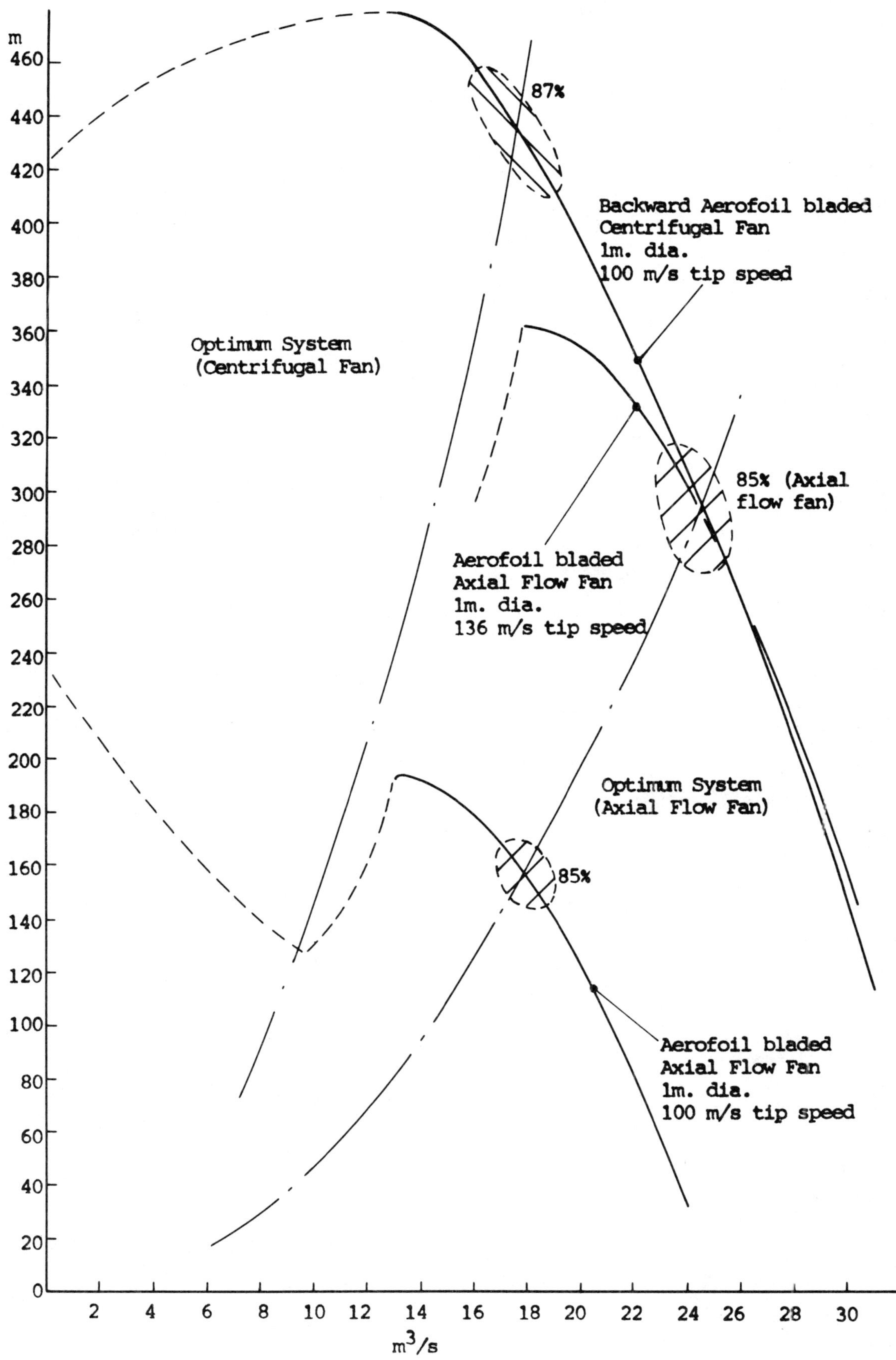

Figure 2.

m
550
500
450
400
350
300
250
200
150
100
50
0
2
4
6
8
10
12
14
16
18
20
22
24
26
28
30
m3/s
Peak Efficiency Zones
Backward Aerofoil bladed
Centrifugal Fan
1m. dia.
100 m/s tip speed
Optimum
System
Aerofoil
bladed Axial
Flow Fan
0.77m. dia.
167 m/s tip speed

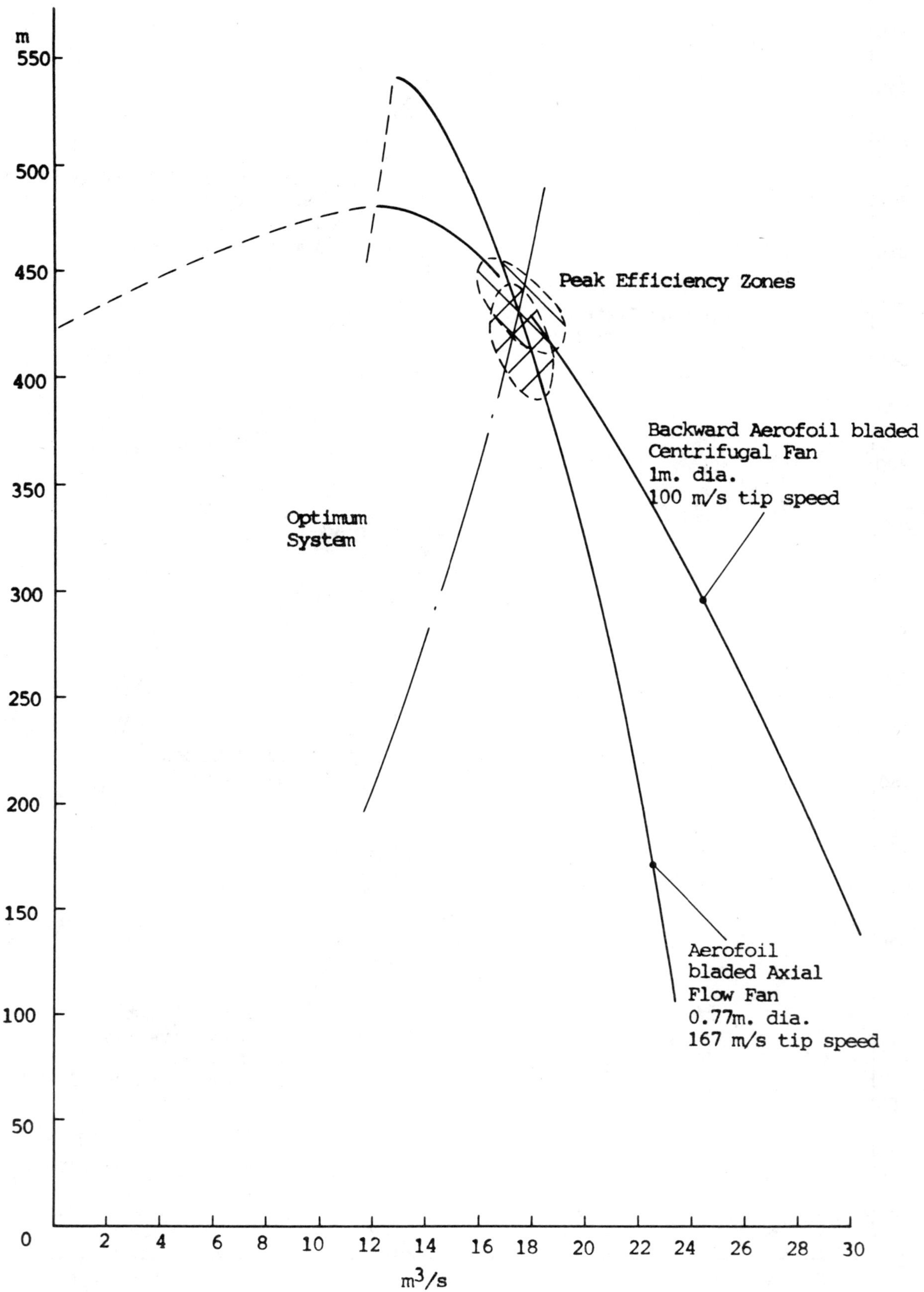

Figure 3.

100
80
60
40
20
%
Effy.
Efficiency/Load
10 20 30 40 50 60 70 80 90 100 110 120
% Design Flow
m
300
280
260
240
220
200
180
160
140
120
100
80
60
40
20
0 2 4 6 8 10 12 14 16 18 20 22 24 26 28
m3/s
High System
Blades 25° Fixed
Various Inlet Vane Angles
Stall Line
Optimum System
85%
80%
70%
60%
40%
20%
-80°
-60°
-40°
-20°
0°
+20°

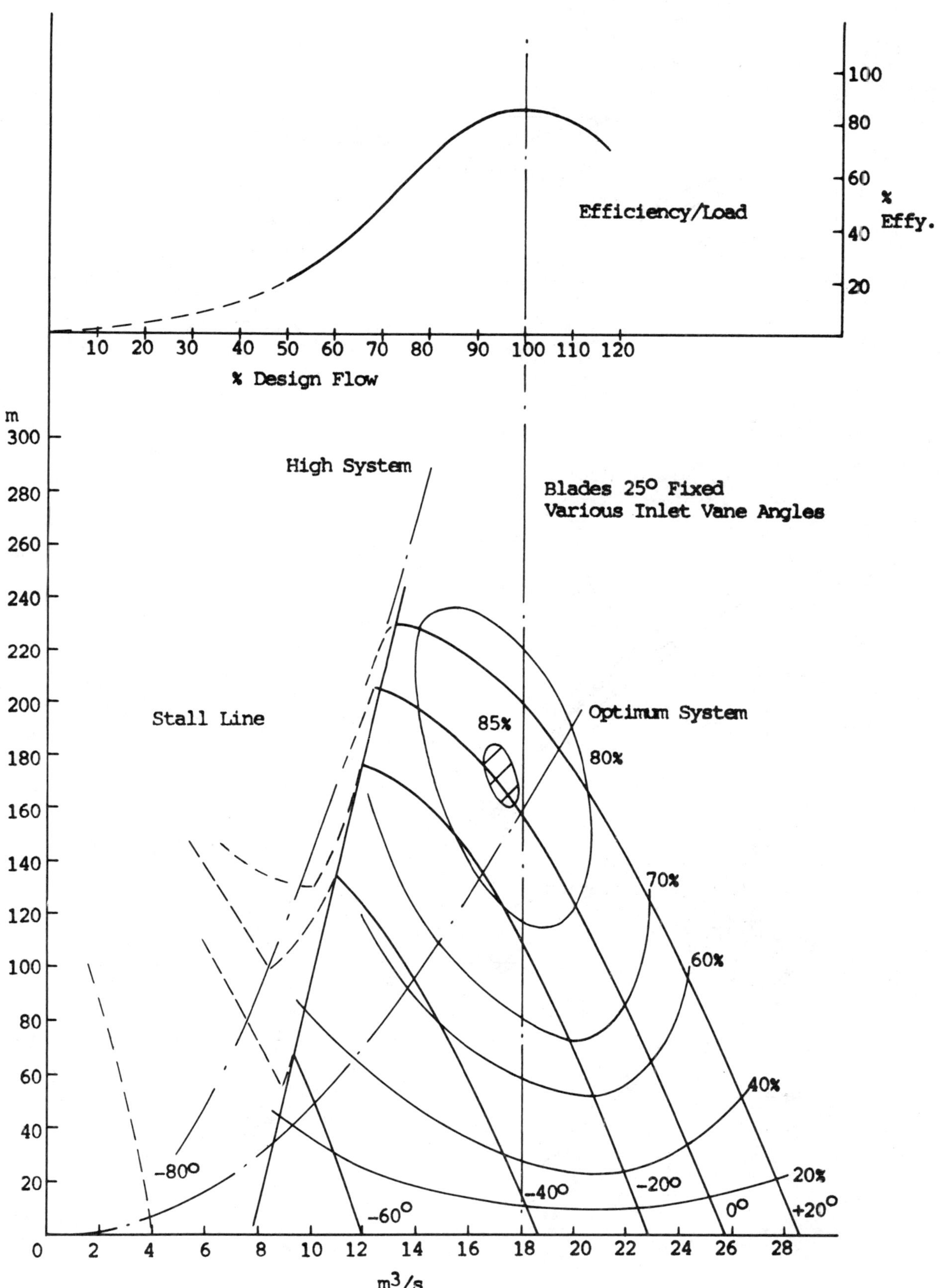

Figure 4.

Airflow
Inlet Vane Element
Velocity Diagram
Rotation
Blade Element
Angle of Approach
Chord Line
Local Separation
Total Separation

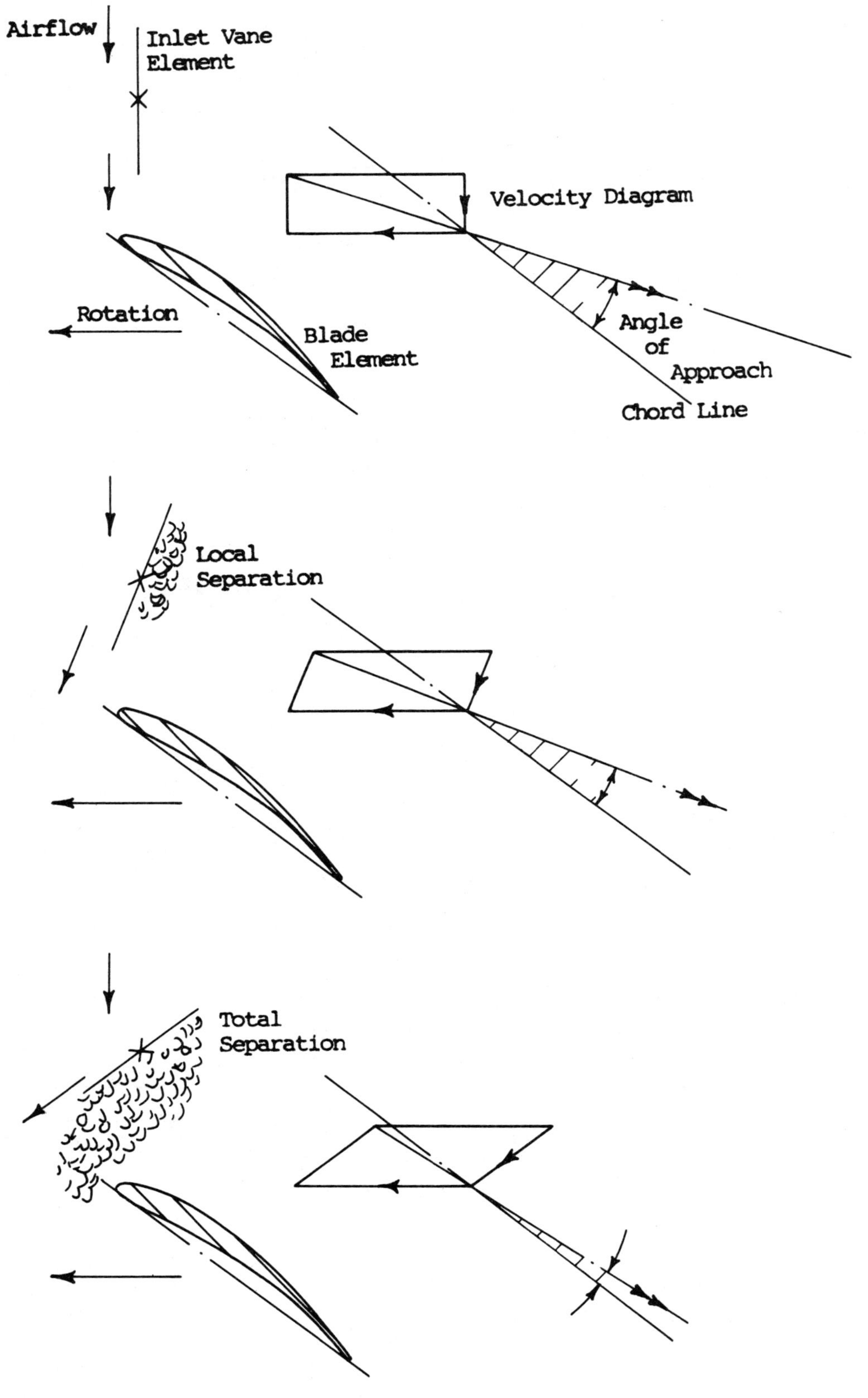

Figure 5.

Efficiency/Load
% Eff'y
100
80
60
40
20
10
10 20 30 40 50 60 70 80 90 100 110 120
% Design Flow
m
300
280
260
240
220
200
180
160
140
120
100
80
60
40
20
0
High System
Optimum System
Stall Line
85%
80%
70%
60%
40%
20%
0°
10°
20°
25°
30°
2 4 6 8 10 12 14 16 18 20 22 24 26 28
m3/s

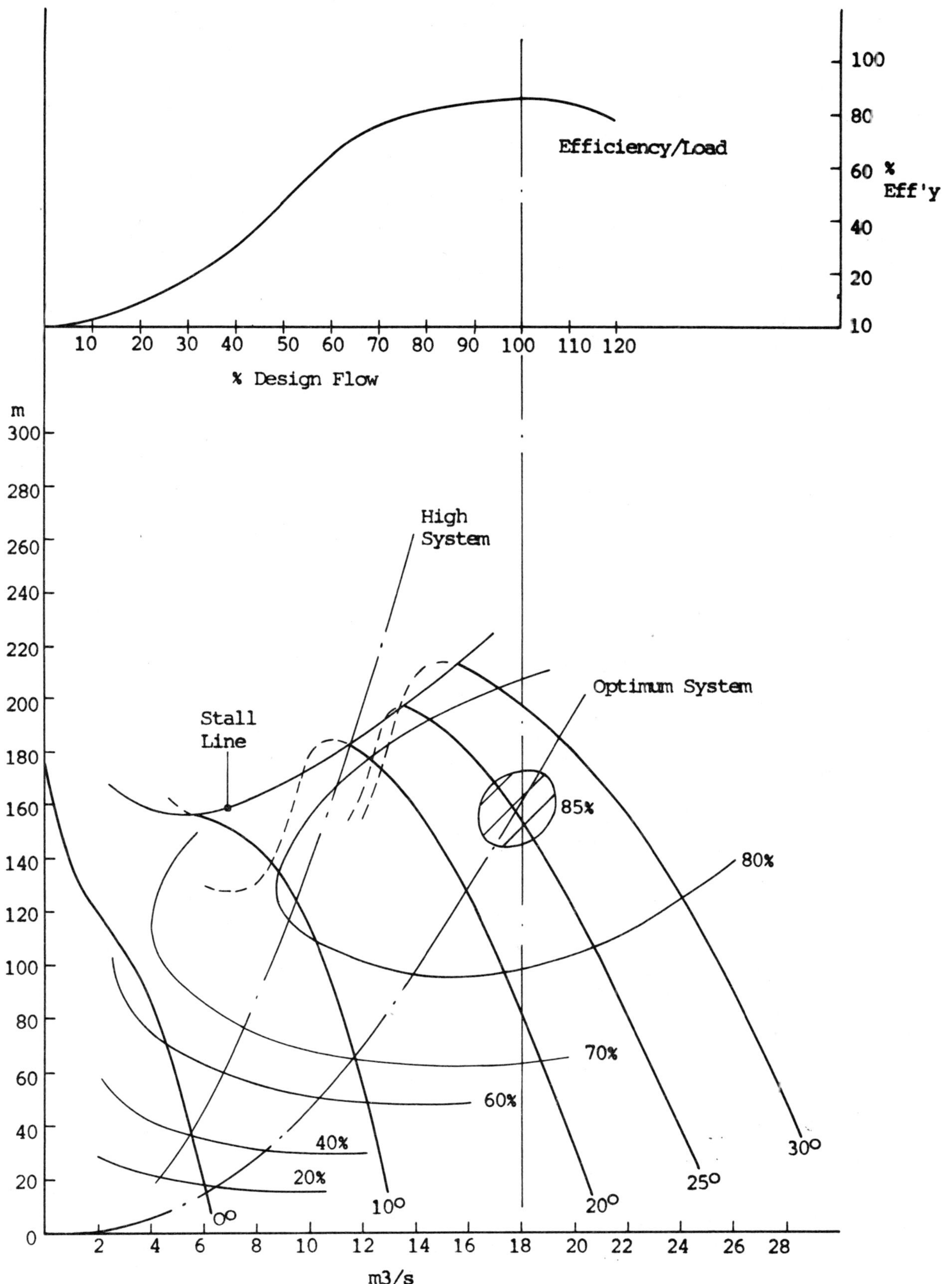

Figure 6.

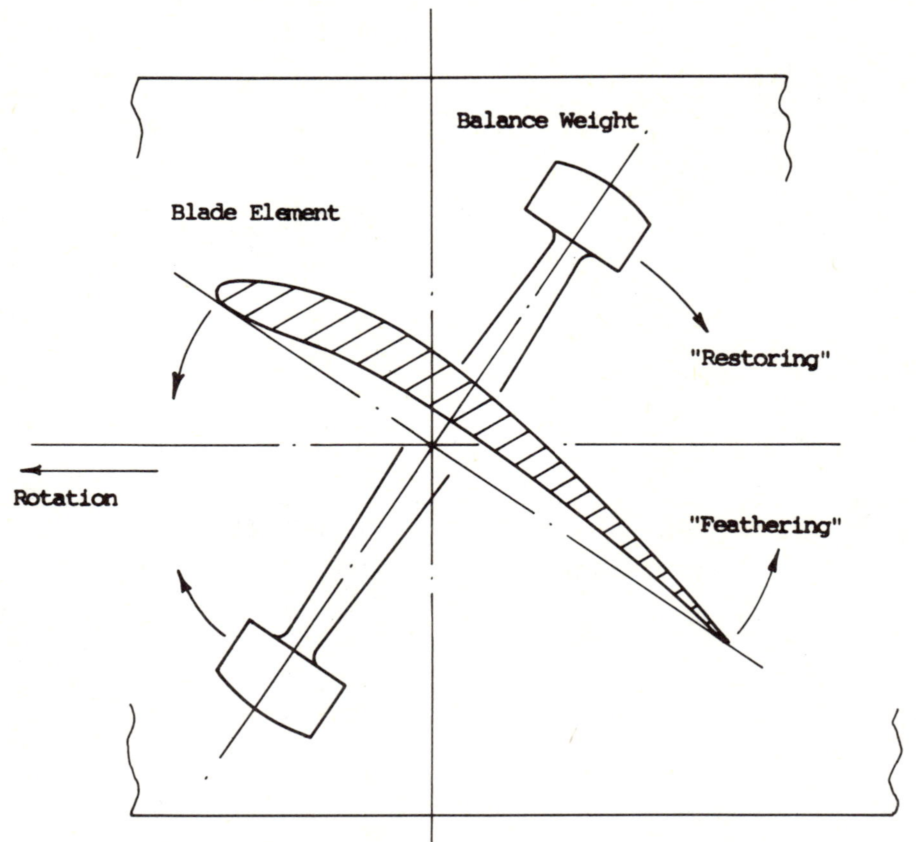

Figure 7.

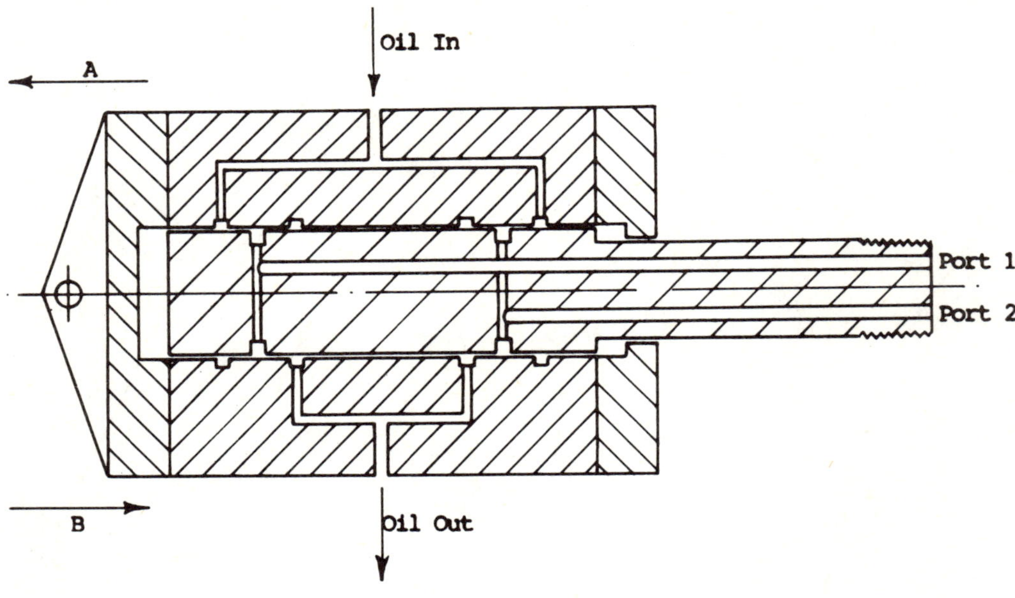

ROTARY SLEEVE VALVE OR POSITIONER

Movement Direction A - Oil in Port 2, Out Port 1.

Movement Direction B - Oil in Port 1, Out Port 2.

Figure 8.

Plate 1

International Conference on

Fan Design & Applications

Guildford, England: September 7-9, 1982

PAPER K3

VARIABLE PITCH AXIAL FLOW FANS

A. C. O'Neill

Davidson and Company Limited, Belfast, U. K.

Summary

The need to design draught systems capable of supplying maximum demand, often of short duration, invariably results in inefficient operation for long periods of plant life. In all major areas of application (mine ventilation, tunnel ventilation, boiler draught systems) it is therefore necessary to provide a volumetric control system. This can range from a simple damper to a more sophisticated system of vane control. Unfortunately this type of control system also reduces fan efficiency and hence more power is required to achieve required output.

The Variable Pitch Axial Flow Fan can significantly reduce power consumption at lower operating conditions without the need for external damper type control.

This paper discusses the operation, design and construction of this type of fan in relation to:-

* Aerodynamic Performance
* Operation
* Mechanical Design

Organised and sponsored by
BHRA Fluid Engineering, Cranfield, Bedford MK43 0AJ, England.

0263 - 421X/82/01 00 - 0001 $5.00
The entire volume can be purchased from
BHRA Fluid Engineering for $82.00

VARIABLE PITCH AXIAL FLOW FANS

INTRODUCTION

In todays heavily industrialised world we face fossel fuel shortages, with accompanying disproportionate increases in the cost of energy. A proven method to conserve energy in mechanical draught plant is to install variable pitch axial fans. Consulting Engineers and Boiler Manufacturers have extended their fan specifications to include this product and provide for its technical and economical evaluation.

In order for these specifications to provide maximum benefit, purchasers and specifiers need to understand the fundamentals of axial fan design, manufacture, quality control, installation, operation and maintenance.

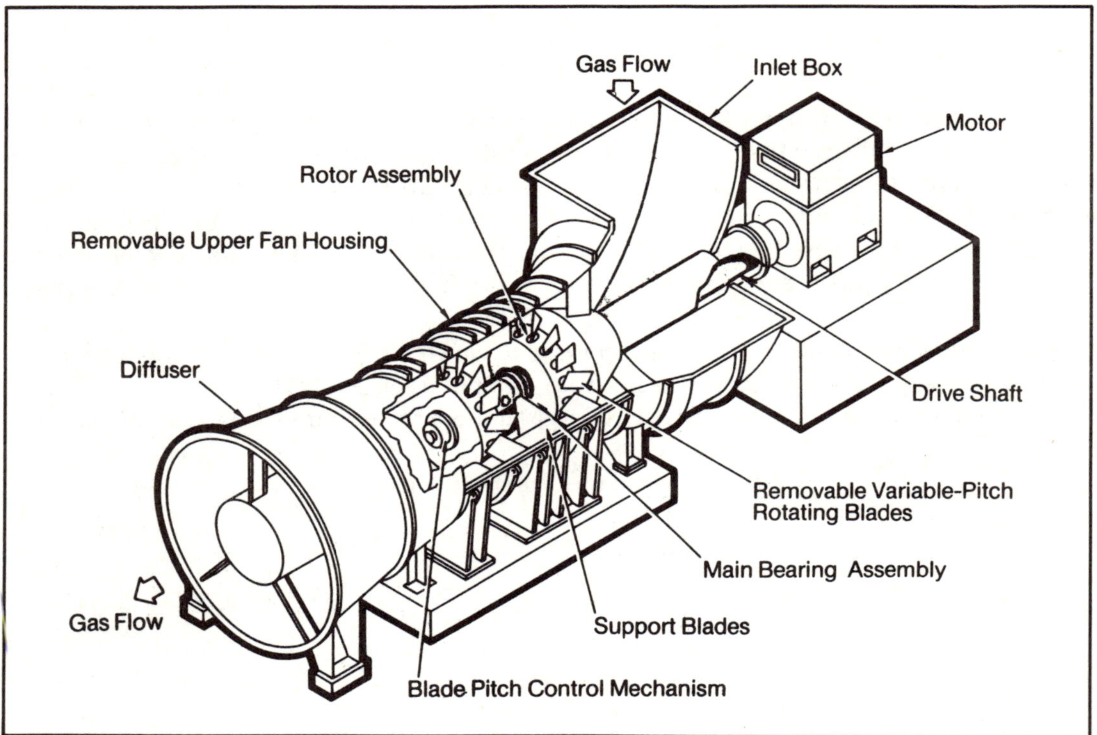

Two Stage Variable Pitch Axial Flow Fan

Fig. 1

The primary physical components of axial fans are shown in Fig.1. This paper will present the essential elements of the design, and operation of variable pitch axial fans. In addition this paper addresses key problem areas which have been encountered in the past concerning the operation of mechanical draught fans: in rotor vibration, aerodynamic performance, system instabilities, erosion and corrosion.

FAN DESIGN.

The following flow chart illustrates the various elements of the design Process used for the major components of an axial fan - beginning with aerodynamic selection, continuing through all rotating and impeller support components, and concluding with the control interface linking the fan unit to the system supervisor/control.

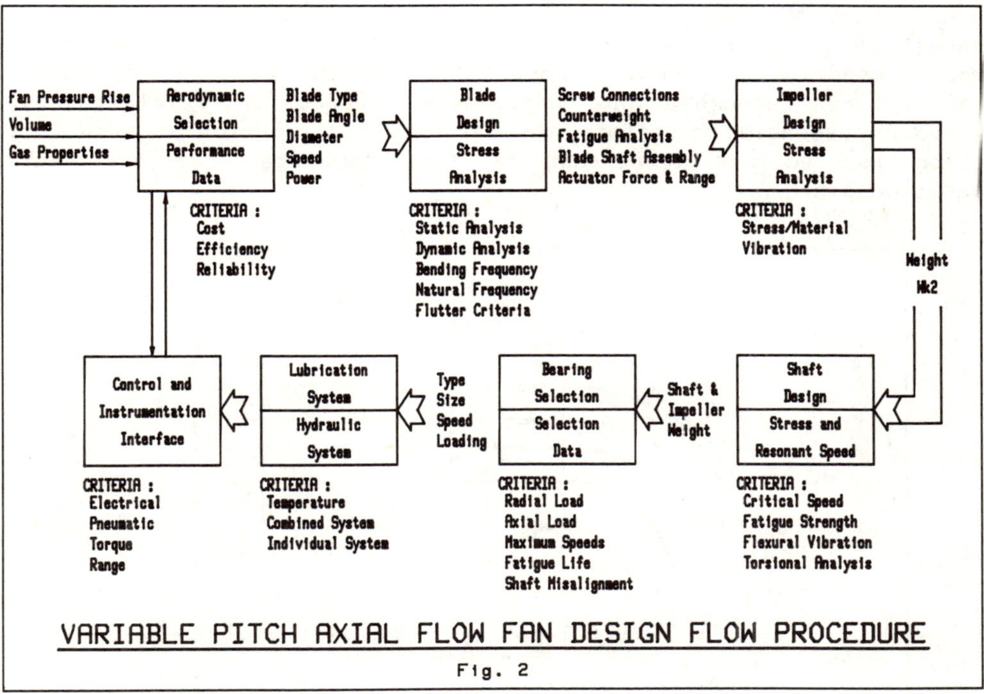

VARIABLE PITCH AXIAL FLOW FAN DESIGN FLOW PROCEDURE

Fig. 2

Aerodynamic Performance

The first step in the design process is the selection or verification of a fan capable of delivering a specified pressure rise and flow, based on operating gas properties.

First it is necessary to understand the derivation of aerodynamic performance data. Aerodynamic characteristics of axial fans lend themselves to mathematical analysis as has been amply demonstrated by extensive research on steam turbines and jet engines. Mathematical models for the flow of air through an axial fan, past the rotating blades, over the guide vanes, and through the diffuser section of the fan are well established. Knowing such basic aerodynamic properties such as hub ratio, chord length, solidity and angle of attack, the performance of an axial fan can be predicted using analytical techniques. So, the purpose of aerodynamic model testing in an axial fan development programme is merely to verify the analytical modeling techniques for the design point and off-design point performance map calculations.

A wide selection of axial fan designs permit fast and accurate computer selection from project conception.

These basic designs have been confirmed by extensive model testing. Full scale performance parameters are derived through proven fan laws such as volume is proportional to speed and pressure to the square of the speed. British Standard BS.848 : 1980, Society of German Engineers, VDI.2044, The Air Movement and Control Asssociation Of America publication AMCA 210-74, define the methods used to determine the pressure volume characteristics of a fan.

Although these standards allow various methods of fan testing, the most commonly used is the discharge duct test. In this test, the exhaust from a fan is blown through a long duct which includes flow straighteners, to achieve an essentially uniform air velocity distribution at a duct cross section located not less than 8½ fan diameters from the fan/duct transition piece.

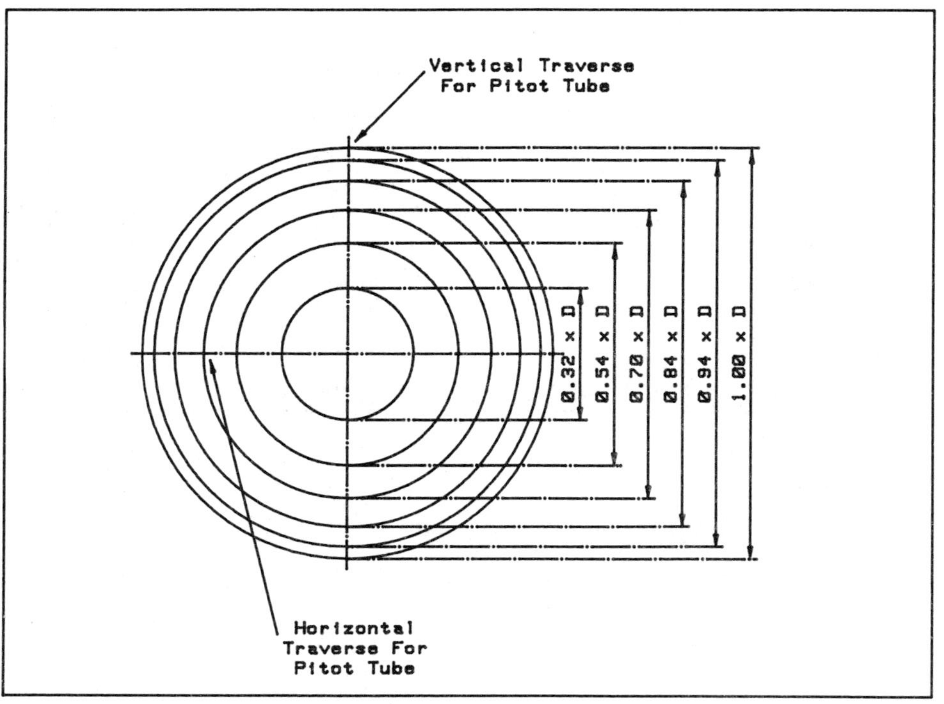

Positions Of Pitot Tubes For Measurements in Std. Airways

Fig. 3a

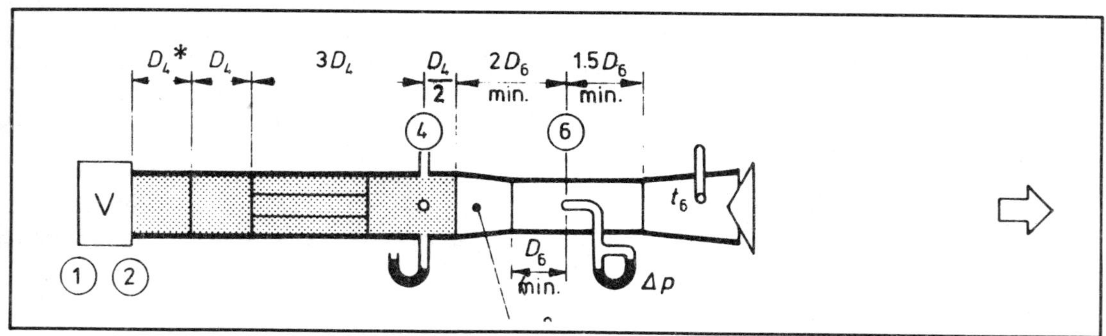

Fan Test Arrangement - Pitot Static Tube Traverse

Fig. 3b

A pitot tube traverse is conducted at that cross section of points shown in Figure 3a. The pitot tube traverse yields static and velocity pressure profiles for the flow in the duct at that point. By integrating this data over the duct area fan performance is established. A throttling device at the discharge end of the duct varies the resistence of the test duct to achieve the fan characteristic curve. This process is repeated for a number of blade angles to give the complete performance map of the fan.

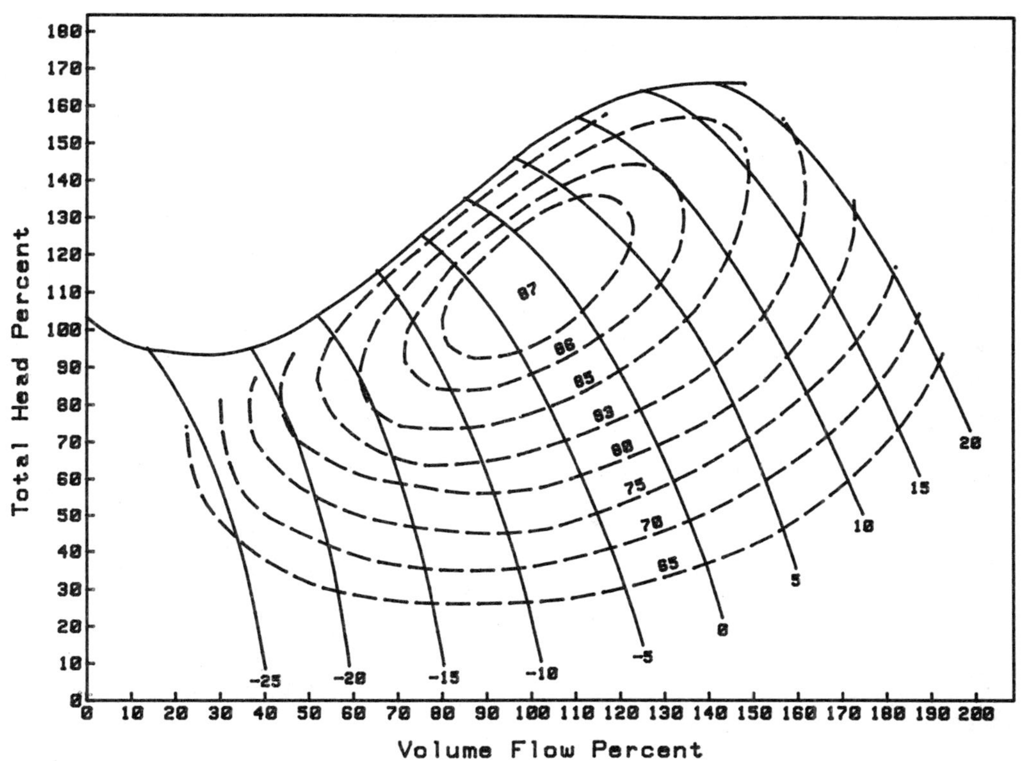

Performance Characteristics - Variable Pitch Axial Fan

Fig. 4

Model test characteristics are now all stored on computer and full scale performance characteristics can be produced in minutes for any specific application by simply supplying the design conditions of volume, pressure and density.

This paper does not seek to be an authoritive guide to fan testing so le those seeking more detailed information should refer to their respective National Standards or British Standard BS.848 : 1980.

Important considerations in selection of a specific fan are the capital cost, life and operating efficiency. A reasonable approximation is that fan cost varies as the square of the impeller diameter. Therefore, the option with the lowest capital cost invariably is a fan with a small diameter operating at high speed.

A critical part of this fan selection is the definition and specification of the Design Duty. As shown on the typical system resistance curve (Fig.5), there is a pressure margin and flow margin between the MCR (Maximum Continuous Rating) point and the Design Point.

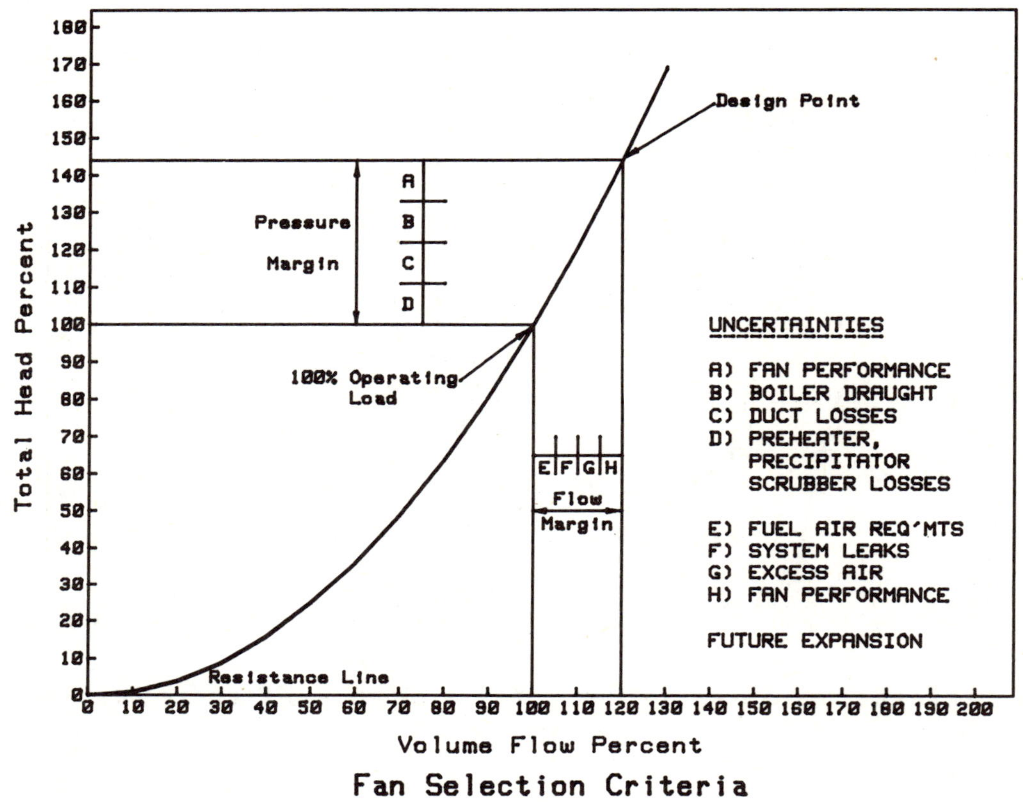

Fan Selection Criteria

Fig. 5

These margins are usually specified by a purchaser to account for many uncertainties at the time the fan is purchased.

For example:-

- Uncertainties about required pressure arise from unknowns in fan performance, boiler draft, pressure losses in the duct, and pressure losses in ancillary equipment.

- Uncertainties about flow arise from unknowns in fuel/air requirements, system leaks, excess air and fan performance.

Many purchasers also include contingencies to allow changes in their system without making changes to components like fans. The total margins between Design Point and MCR can range from 15-45 percent. It is our recommendation that the allowance for performance unknowns be less than 5 percent. We further recommend that the remaining margins be defined with the thorough understanding that excessive margins can prevent the optimum fan selection, resulting in lower operating efficiency throughout the plant life.

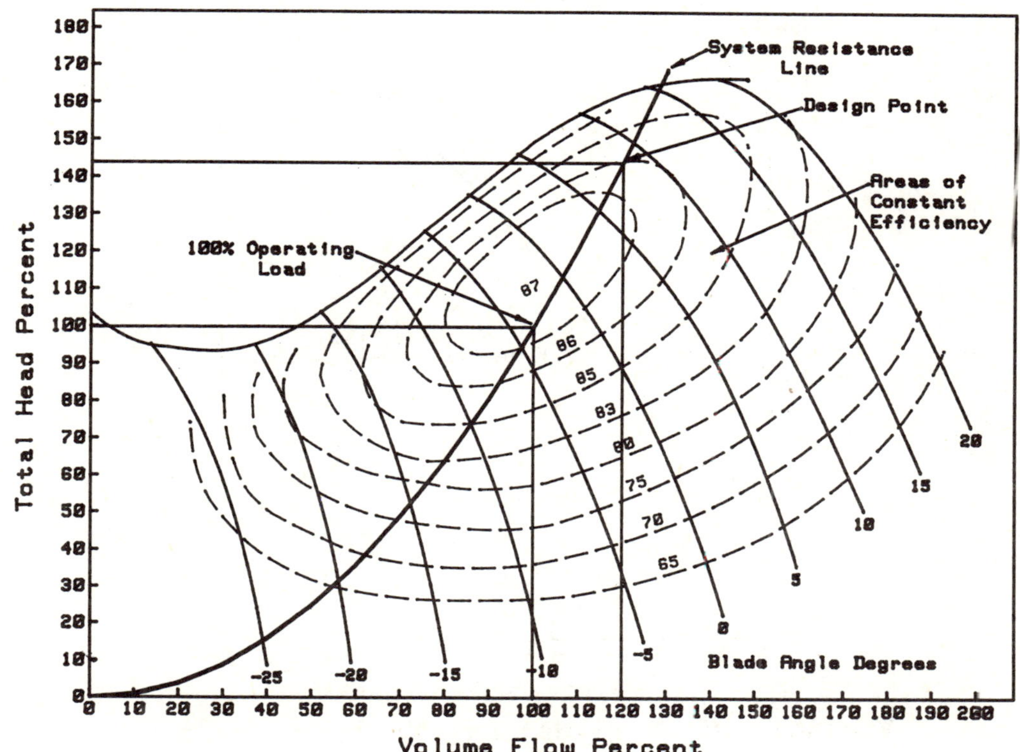

Performance Field For Variable Pitch Axial Flow Fan

Fig. 6

Fig.6 shows the characteristic performance field of a variable pitch axial flow fan.

Two major benefits are:-

- The areas of constant efficiency run parallel to the system resistance line. High efficiency is maintained over a wide operating range.

- There is a large control range above, as well as below, the area of maximum efficiency. This permits the fan to be selected for net boiler conditions, while the Design Point remains within the control range.

The lines of constant blade angle are actually individual characteristic curves for a given blade setting. Because the curves are very steep, a change in resistance produces very little volume change.

As the blade angle is adjusted from minimum to maximum position the flow varies almost linearly as shown in Fig.7.

These two characteristics provide stable fan and boiler control over a wide operating range.

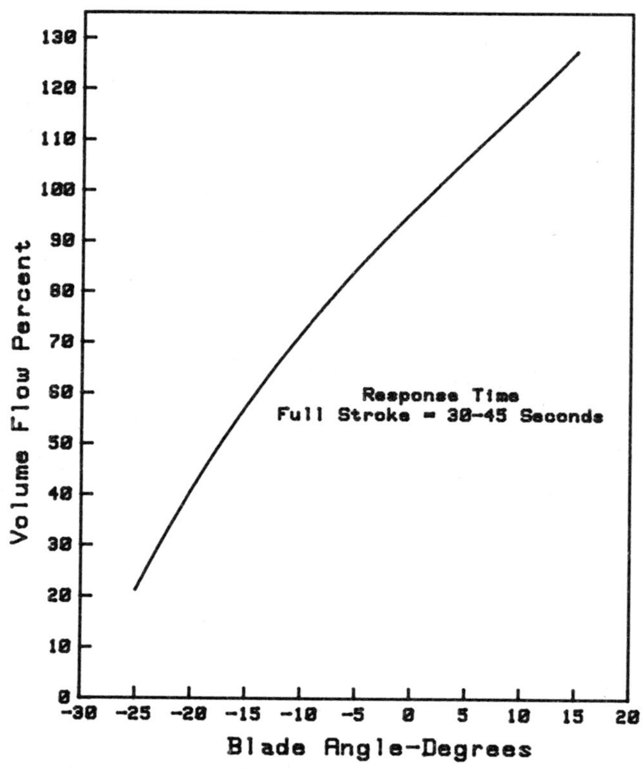

Variable Pitch Blade-Control Characteristics

Fig. 7

Parallel Operation

Variable pitch axial flow fans can be operated in parallel, provided that care is taken to avoid operating either fan in the stall area which will be discussed later in this section. With two fans in operation, the resistance line for one fan is influenced by the other fan, as well as by the boiler conditions. Two fans together will develop the pressure required to overcome the boiler resistance, but their individual volume flows need not be equal.

Figure 8 shows a typical performance field for two variable pitch axial flow fans. In this curve Point 1 shows the total volume flow required by the boiler at net conditions. Point 2 shows the individual fan operating point with proper parallel operation. At net boiler conditions the two fans may operate individually anywhere on the line from Points 1 to 3 provided they do not operate in the range above the stall line ("X-Y"), and the total volume flow produced is equivalent to Point 1. To obtain the most efficient fan operation and to avoid operation in a range close to the stall line, it is best to keep both fans operating at Point 2.

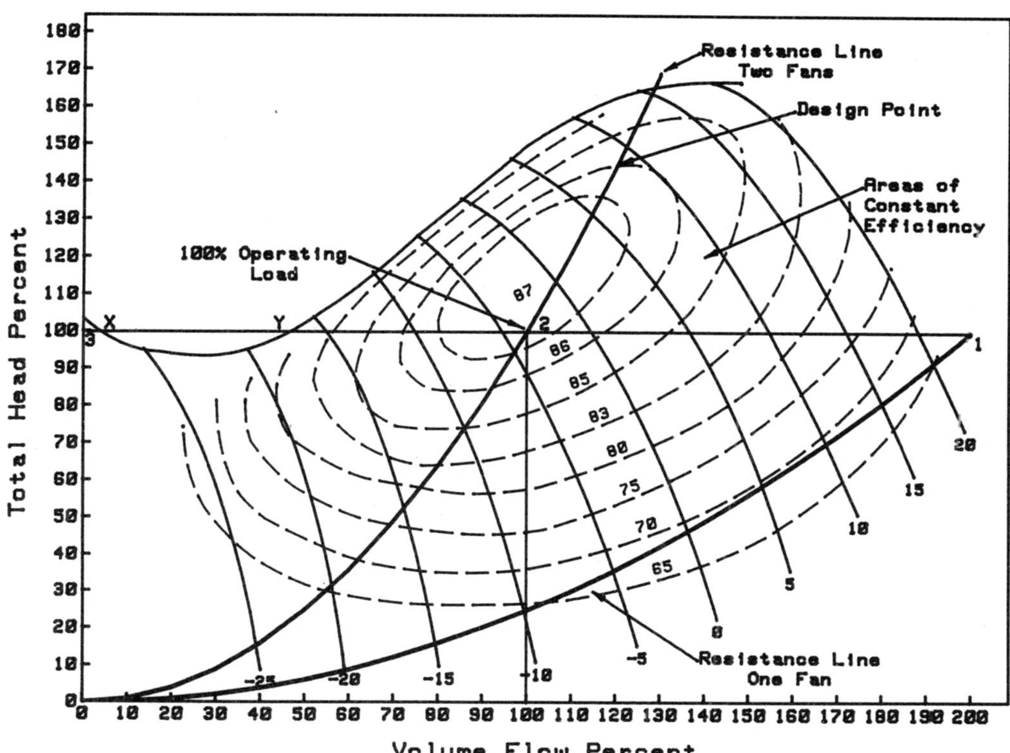

Performance Field For Parallel Operation

Fig. 8

If the fans and motors are sized so that it is possible to operate one fan at Point 1, then it will be necessary to reduce load slightly to put the second fan into operation. The load reduction is to lower the boiler resistance. If the resistance is not lowered, the second fan must start up at Point 3 and then move horizontally to Point 2. When the second fan reaches "X", it will stall.

Axial fans are connected to the boiler control system with an 180° stroke, 100 Nm positioner. The positioner moves a pilot valve in the hydraulic blade adjustment system which positions the blades with minimum hysteresis. This system is explained later.

The suggested control logic for starting, stopping, and supervising the operation of variable pitch axial flow FD and ID fans is very similar to that of the centrifugal fan. Diagrams are included as Appendices A,B,C,D,E, and F.

An additional requirement of the logic is to prevent damage to the fan while always maintaining an open flow path through the boiler to prevent furnace pressure excursions.

Stall Characteristics

Axial flow fans have a unique characteristic called stall. Stall is the aerodynamic phenomenon which occurs when a fan is asked to operate beyond its performance limits. When the fan blades are required to provide more lift than they are designed to produce flow separation occurs around the convex side of the blade. Centrifugal force then throws the air trapped in this separated area in a radial direction, to the outer tip of the blade. At this point pressure builds up until it is relieved throught the blade tip clearance. If this happens the fan becomes unstable and no longer operates on its normal performance characteristic. This process creates a very unstable and oscillating force on the blade, and can cause very severe vibration throughout the entire fan. This characteristic must be understood and considered in boiler operation as it can have an important effect on boiler design.

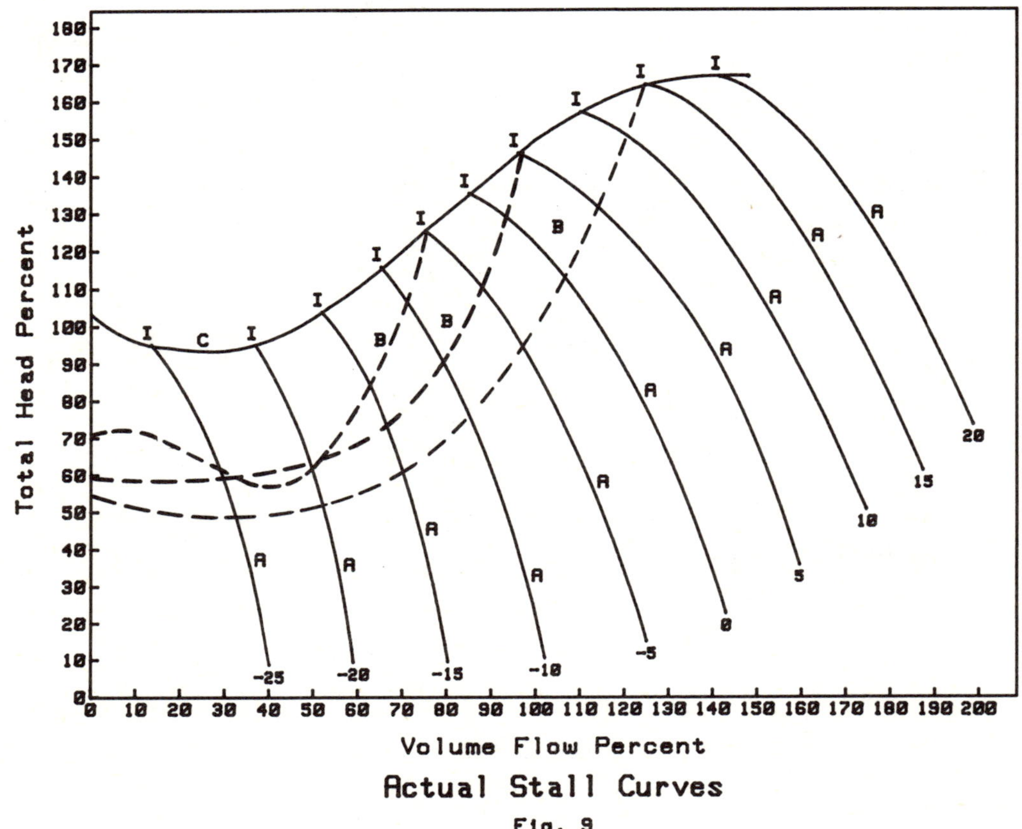

Actual Stall Curves

Fig. 9

The curves in Figure 9, marked "A", are the normal fan performance curves for a constant blade angle.

Each blade angle curve has its individual stall point identified as "I" on the diagram. The curve "C" which connects all the stall points "I" is generally referred to as the "stall line".

The dotted curves, "B", are the characteristic stall curves for three different blade angles. The curves show the path that the fan will follow when operating in a stalled condition.

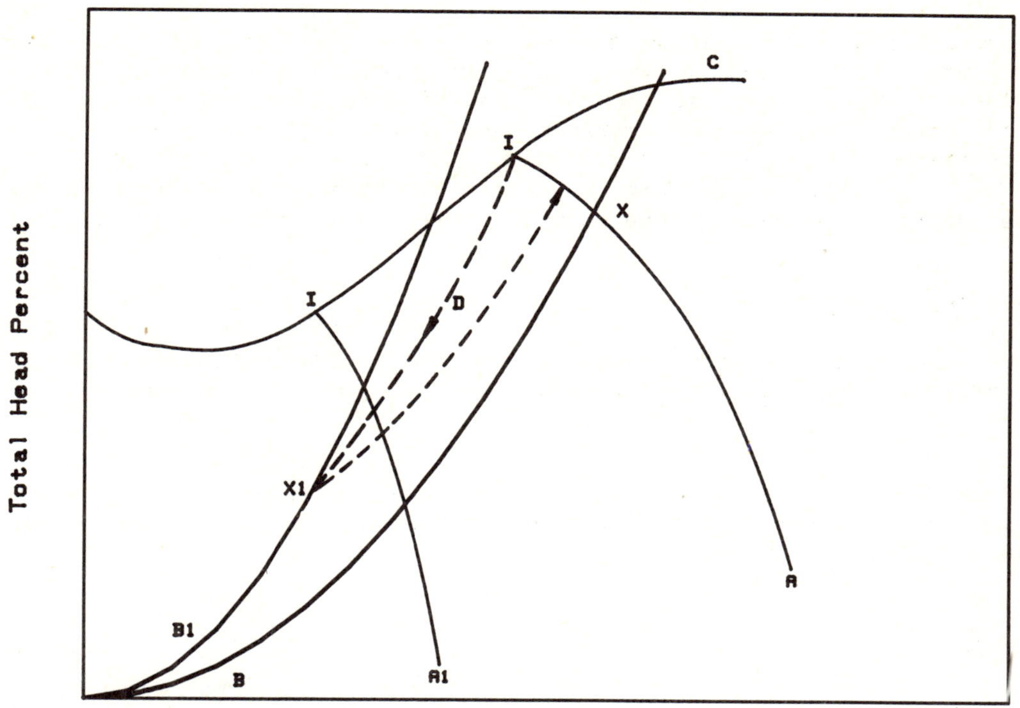

Volume Flow Percent

Stall, as Related to the Boiler

Fig. 10

Figure 10 explains the stall phenomenon in relation to the fan and boiler system.

If the normal boiler system resistance (Curve "B") increases for any reason (for instance, a furnace pressure excursion caused by a main fuel trip) the normal operating point "X" will change to meet a new higher system resistance (Curve "B_1") by travelling along the fan performance Curve "A". If the operating point arrives at point "I", the fan will stall. Because of the relationship between the fan performance Curve "D" in the stall area and the upset system resistance (Curve "B_1"), a new operating point "X_1" will be found where the system resistance (Curve "B_1") and the stall Curve "D" intersect.

If the system resistance remains high (Curve "B_1"), the fan will continue to operate at point "X_1" in an unstable stall condition.

When the system resistance comes down, the fan will recover from the stall condition and return to its normal performance Curve "A".

In the case of an incident as described above the blade angle can be reduced until the fan regains stability. The fan will be stable when the new performance Curve "A_1" provides a stall point "I", which is higher than the upset system resistance (Curve "B_1").

If stablization cannot be attained by blade adjustment, the fan must be shut down. When operating in stall condition flow vibrations occur which are unpredictable in energy and frequency and can cause damage to the rotating blades.

The maximum stable negative furnace pressure generated by an ID fan (pressure differential between the fan stall line and the boiler resistance line) occurs at the condition of zero flow. For a variable pitch axial flow fan, the maximum zero flow static pressure usually approximates the net boiler resistance. This inherent characteristic of the variable pitch axial flow fan allows the structural designer to use lower design pressures for the furnace, flues, ducts and precipitator.

Stall Prevention

When axial fans are sized properly and the system resistance line is parabolic in shape the probability of experiencing stall is low.

The possibility of experiencing a stall increases if a fan is oversized (Figure 11), if the system resistance increases significantly, or if the fans are operated improperly.

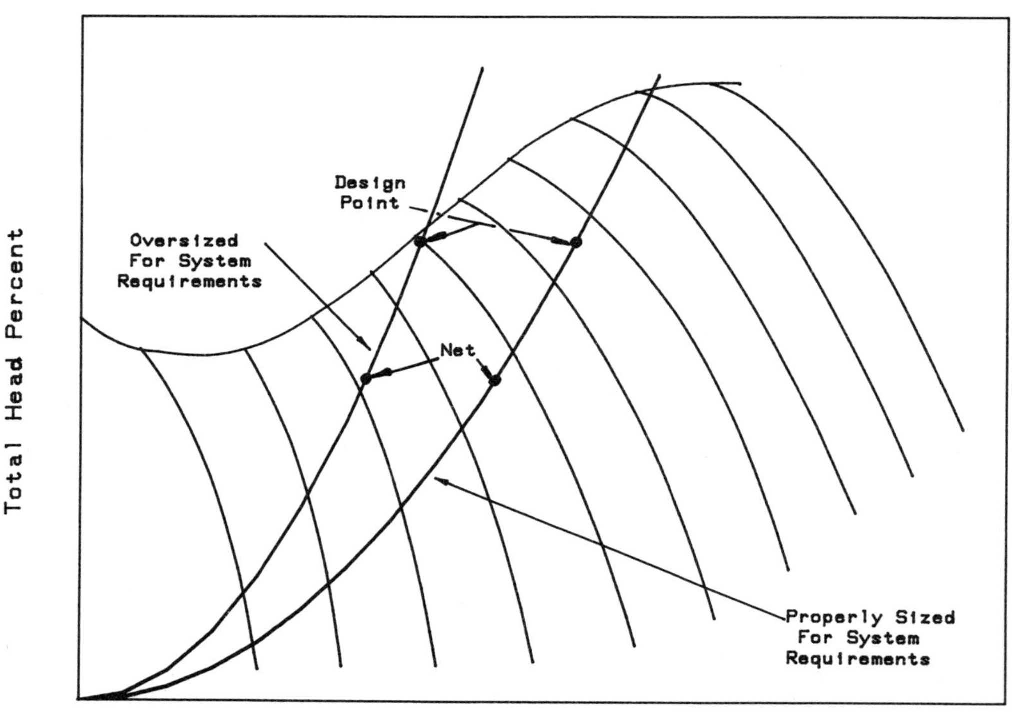

Proper Fan Selection

Fig. 11

The degree of stall protection desired in the control system depends on individual philosophy. A visual indication of pressure and volume (or blade angle) in the control room is a minimum requirement for satisfactory operation.

Fan Design - Mechanical

Axial flow fans designed today are compact and relatively light. By concentrating the load carrying parts of the rotor on a small radius, a low Wk^2 is obtained. The low weight, unbalanced forces, and inertia of the axial flow fan permits either horizontal or vertical arrangement on steel construction. This provides greater arrangement flexibility. Normal installation can be accomplished with reduced concrete requirements.

Typical fan arrangement drawings are shown in Appendix H.

A horizontal fan arrangement is generally the economic preference for forced draught and primary air fans. Induced draught fans can be arranged horizontally, or vertically inside the stack. Vertical arrangements offer additional benefits:

- simplifying the flues (straight horizontal connection between precipitator outlet and fan inlet box is possible)
- reducing draft losses in turning bends
- eliminating the need for sound enclosure (fan, drive, and auxiliary equipment)
- eliminating the need for additional fan space by utilizing existing unused space

Acoustics

Fans produce two distinct types of noise:

- Single tone noise is generated when the concentrated fluid flow channels leaving the rotating blades pass a stationary object (straightener vanes or nose). The distance from rotating blades to the stationary objects affect the sound intensity, with the blade passing frequency and its first harmonic being most predominant. Economic and aerodynamic considerations usually override achieving the optimum acoustic design.
- Broad band noise is produced by high velocity fluid "rushing" through the fan housing, and as the name implies, covers a wide frequency range.

From outside, the apparent source of both types of fan noise is the fan housing. The sound travels out of the inlet box opening through the discharge ductwork, and through the casing of the fan. All three "leaks" must be acoustically analyzed and treated individually to achieve an acceptable installation.

Inlet Noise

FD and PA fans normally have inlets open to either a boiler room or to the plant exterior. The noise emitted to these areas is almost always above desired values.

Installation of an absorption type silencer is the most common way to lower inlet sound pressure level to desired values. This type of silencer is a large box filled with aerodynamically shaped panels (named coulissen) made with perforated plate skin and filled with absorptive mineral wool. The acoustic effectiveness of this type of silencer increases with higher frequency noise up to about 2000 Hz (Figure 12).

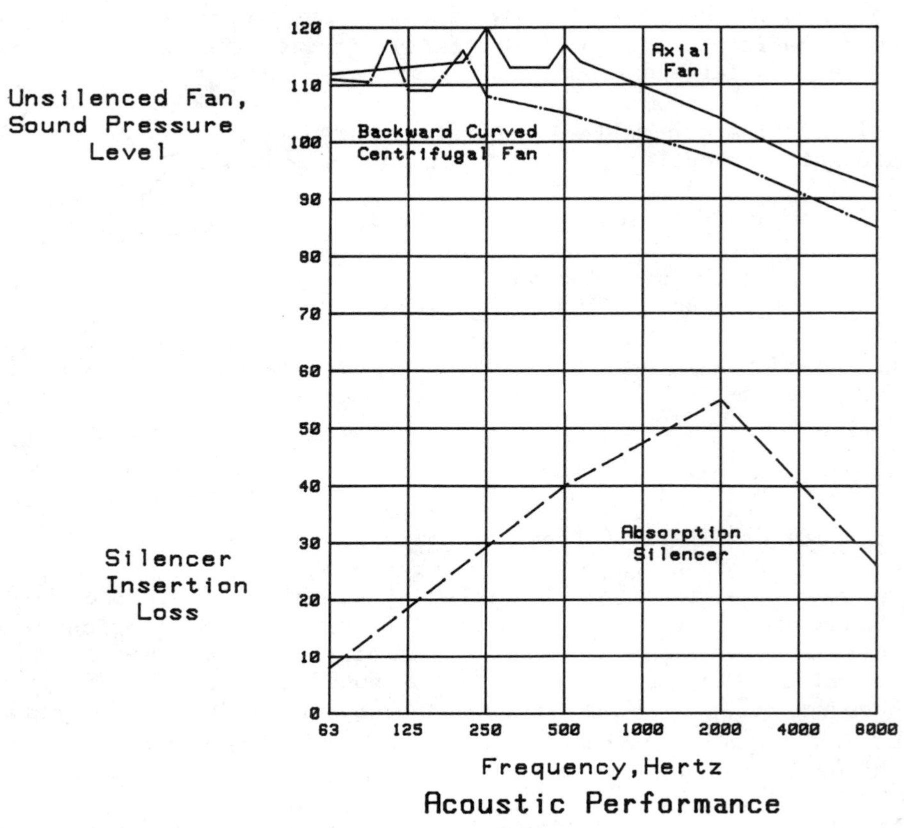

Acoustic Performance

Fig. 12

The single tone base and first harmonic frequencies of the axial fan are usually twice the frequencies of an equivalent centrifugal fan. The silenced sound pressure level of an axial fan with a given length silencer will usually be comparable to the silenced sound pressure level of an equivalent centrifugal fan using the same silencer.

Noise through the Fan Casing

Application of absorptive mineral wool insulation and acoustic lagging on axial fan casing is effective in reducing the sound to acceptable levels. This method avoids the use of expensive sound rooms often used to suppress the lower frequency noise of centrifugal fans. It also permits access to the fan and its auxiliary equipment without exposure to damaging noise.

Discharge Noise

The determintion of the best method to quiet fan discharge noise requires an evaluation of the duct arrangement, fan application, auxiliary power costs, and fan unsilenced sound pressure level.

The most common method of attenuating FD fan discharge noise is to apply acoustic insulation and lagging to the discharge duct from the fan outlet to the air heater inlet. Where a long run of ductwork is used, i.e., tempering air duct in a cold PA system, a discharge silencer may be more economical.

For an axial flow ID fan the thermal insulation and lagging to the flues before and after the fan are usually adequate to control both inlet and discharge noise. On ID fans in residential areas it may be necessary to reduce the stack discharge noise level by using an ID fan discharge silencer. In these cases the objectional noise is the single tone sound at the base frequency and first harmonic. A resonant type silencer is effective for silencing single tone noise. Its performance does not deteriorate with service; absorptive type silencers tend to plug with fly ash and become less effective.

Blades

Axial flow fan blades are the key component in the basic fan design. The selection of blade type, number and material dictate the centrifugal load imposed on the blade shaft assembly, which in turn dictates the impeller hub size and material requirements, and ultimately the required main bearing assembly. It is therefore vital to be able to optimise on the basic blade design, in terms of material selection and operating stress.

Blades can be manufactured in a number of ways, depending primarily on application.

Cast or forged aluminium can be, and is, used for most FD and PA fan applications up to an operating temperature of 90°C. Above this temperature the mechanical properties of aluminium fall away rapidly and require detailed and careful analysis.

For hot gas fans, FD, PA and ID with reasonable dust concentrations, the next alternative is spheroidal graphite (S.G.) iron which like aluminium permits accurate and cost effective manufacture.

Steel blades with wear protection can be used on high temperature applications or when the dust burden is considered to be a problem.

Blade Analysis

The limiting speed at which any axial flow fan blade can operate is a function of operating stress in the blade section, specifically in the transition from the blade actual to the mounting flange. Using pure analytical techniques, considering shape of the blade, material, and operating conditions of the fan, an estimation of stress at discrete sections is performed.

This analysis covers:-

- tensile stress
- bending stress - static and dynamic
- torsional stress
- combined stress
- natural frequency - static and dynamic

In order to verify the results obtained by these calculation procedures, the following controlled tests were carried out.

- measurement of natural frequency - static
- measurement of natural frequency of a series of blades under actual operating conditions at varying loads
- dynamic analysis of blade stress under load
- finite element analysis to verify stress analysis
- finite element analysis to verify natural frequencies

Fig. 13 shows the first five natural frequencies and corresponding mode shapes for the various support conditions illustrated. It can be seen that the first frequency obtained for the blade varies by less than 10 percent. This is because the first mode is the fundamental bending mode which is not strongly influenced by the flexibility of the blade shaft. However, for higher modes, the flexibility of the blade shaft, guide bushing and load bearings has an increasingly significant effect on the blades dynamic behaviour by lowering the frequency of the blade assembly. It is interesting to note that the mode shapes are predominately torsional with the exception of the first mode.

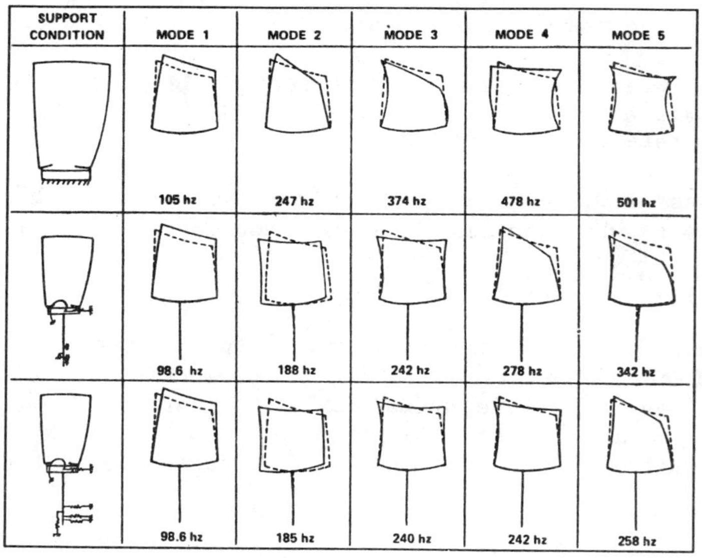

SUPPORT CONDITION	MODE 1	MODE 2	MODE 3	MODE 4	MODE 5
	105 hz	247 hz	374 hz	478 hz	501 hz
	98.6 hz	188 hz	242 hz	278 hz	342 hz
	98.6 hz	185 hz	240 hz	242 hz	258 hz

Blade Natural Frequencies & Mode Shapes

Fig. 13

The foregoing analyses and controlled tests have shown the calculation procedures to be conservative and confirm their suitability for normal design application. For arduous conditions or when requested a full finite element analysis is carried out.

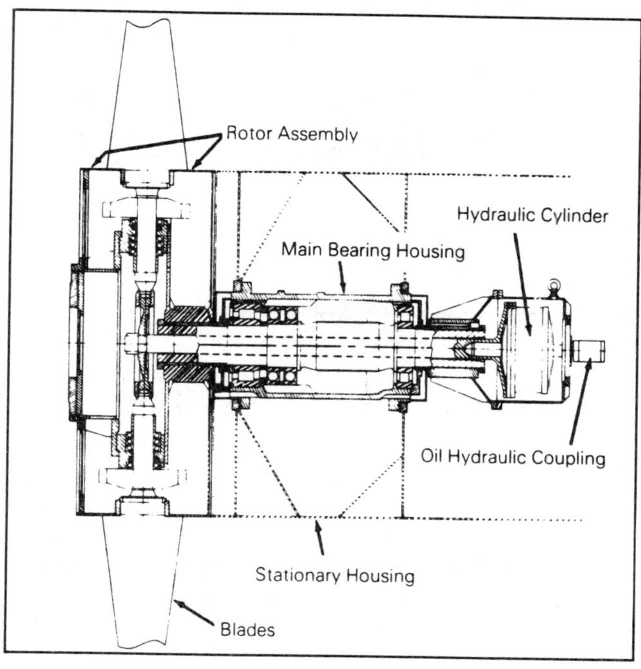

Rotor Assembly

Fig. 14

Main Bearings

The bearings are designed for the radial and thrust loadings of the fan at its design point. Because the rotors are mounted close to the bearings (Figure 14), and the rotating weight is low, conservative catalog anit-friction bearings can be and are selected. (Figure 15).

Sleeve bearings, commonly used in centrifugal fans, can be used in axial flow fans but are uneconomical because the axial thrust loading requires the use of a Kingsbury type thrust bearing.

Lubrication of the main bearings is achieved with an oil splash bath. The bath is continuously circulated through an oil cooling and filtering lube set. In addition, cooling air can be circulated across the main bearing housing by utilizing the pressure differential available between the inlet nozzle and atmosphere (FD and PA fans), or by an auxiliary cooling air fan. Forced outages resulting from a failure of the oil circulating system are minimized since the oil reservoir in the bearing housing is capable of maintaining lubrication for extended periods with the system out of service.

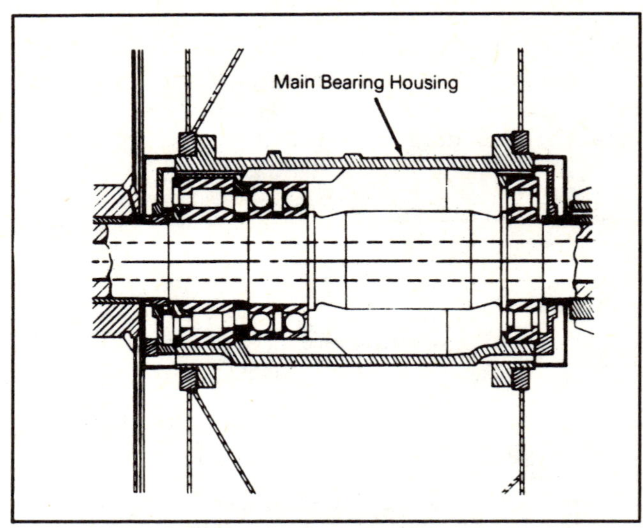

Main Bearing Assembly

Fig. 15

The fan main bearings are designed to have a minimum "B-10" life of 40,000 hours at the design point condition. Depending upon actual operation (power input and temperature) the bearings should be changed on a 3-5 year cycle to prevent a "B-10" type failure.

One main bearing and shaft assembly is relatively inexpensive and should be stocked with each fan type to facilitate maintenance. With this assembly, a main bearing can be changed in 2 shifts with a two-man crew.

Blade Shaft Bearings

Blade shaft bearings are used to minimize the friction in the blade adjustment system. They transfer the centrifugal force due to the blade weight to the fan hub. These bearings are designed for static thrust loading because they operate only through the blade adjustment range, an arc of about 50°.

Stacked angular contact ball thrust bearings (Figure 16) or thrust ball bearings are used for this application. The required number of bearings is a function of the bearing static load coefficient and the thrust. The thrust is a function of angular speed, blade material and blade size.

Prolonged operation with relatively little blade adjustment can cause fretting of the bearing race. Therefore, inspection is recommended after 2 years. Some owners replace the bearings at this time. Depending on the specific design and operating conditions, the bearings may last considerably longer than 2 years.

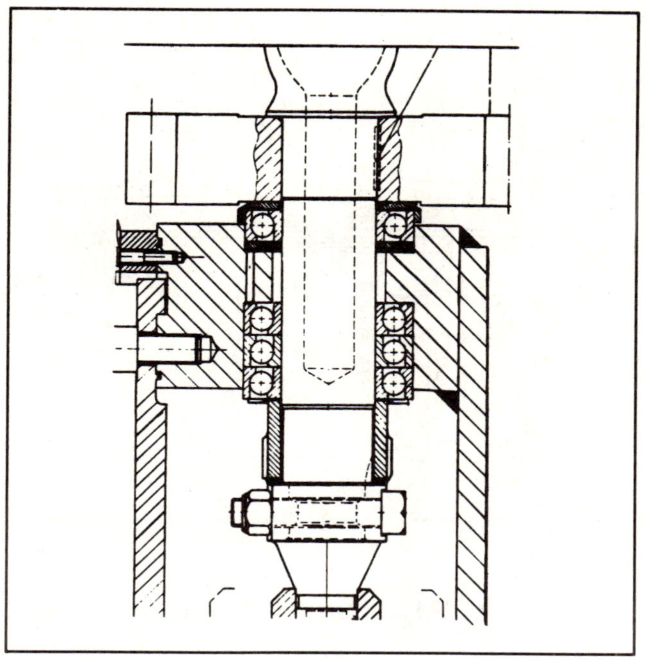

Blade Shaft Bearings

Fig. 16

The bearings are not an item which fails suddenly and an indication of their condition can be monitored by recording the hydraulic pressure required to move the blades.

If fretting occurs, the force required to move the blade increases. When this force is plotted in terms of hydraulic pressure versus time, a fan outage for blade shaft bearing maintenance can be scheduled to coincide with the next unit outage.

The replacement time is approximately 5 shifts with two men.

Hydraulic Adjustment Device

The hydraulic blade adjustment system (Figure 17) consists of the following main elements:

- An adjusting cylinder which moves axially along the fan's centerline and rotates with the fan impeller. This cylinder moves the individual blade adjustment levers.

- A piston inside the cylinder which is also rotating, but fixed axially.

- A positive feedback position rod which mechanically measures the cylinder position.

- A stationary hydraulic servo-mechanism is mounted to the rotating piston rod. Alignment is maintained with anti-friction bearings.

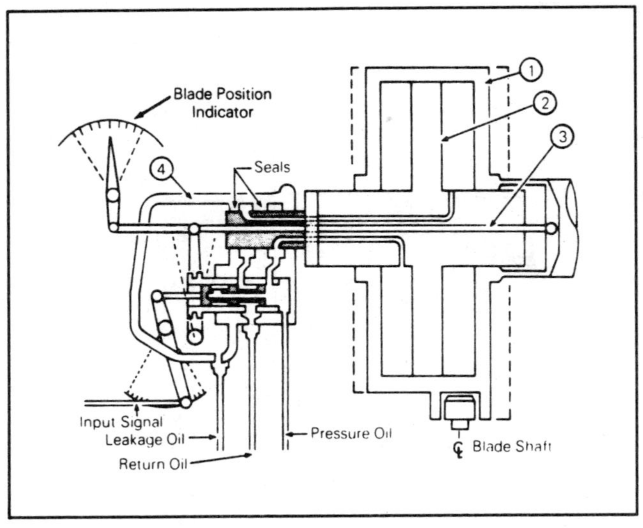

Blade Adjustment Mechanism

Fig. 17

This unit serves the following three functions:

- It serves as the transition piece between the stationary control system and the rotating fan system. This transition is accomplished at the smallest possible diameter to reduce relative speed and seal wear.
- The pilot valve in the unit translates the mechanical input signal from the fan control positioner to a hydraulic signal which drives the adjusting cylinder.
- The unit mechanically reads the feedback position and re-adjusts the hydraulic signal to the cylinder. It also transmits an indication of actual blade position to the exterior of the fan.

Maintenance of the hydraulic blade adjustment system involves replacement of the seals in the stationary hydraulic servo-mechanism. A seal leakage sight glass is provided for visual indication of the rate of change of seal wear so that an outage can be scheduled to change the hydraulic unit.

The cost to overhaul a hydraulic unit is low, and the change-out labour is approximately 1½ shifts of a two-man crew.

Should the hydraulic pressure be lost because of electrical or mechanical failure, the blades are counterweighted to open gradually to their maximum position increasing air flow to its maximum value. The flow through the parallel fan can be adjusted within its performance field so that the unit load is controlled until an outage is scheduled.

Coupling

A hypoid gear coupling is used as the standard design. The couplings are provided with thrust limiting rings to eliminate the need for thrust bearings in the drive motor.

Coupling maintenance recommendations are provided by the coupling manufacturer. His usual recommendation is an annual grease change.

I.D. Fan Blade Wear

Fan blade wear is a function of:

- dust loading
- particle size distribution
- particle hardness
- particle concentration
- the relative velocity at which the particles pass the blade
- the blade material
- angle of attack.

The fan designer has no control over the first 3 parameters. He can only influence the blade life by varying the other parameters.

The blades of a variable pitch axial flow fan are subjected to a relatively uniform velocity across the entire blade surface.

The wear is evenly distributed over the blade of a variable pitch axial fan, since there are no inlet vanes utilized or 90° direction changes which tend to concentrate dust particles.

Angle of attack influences the type of wear experienced (Figure 19). Material is removed in the following two ways:

- Impact wear (high angle of attack)
- Sliding abrasion (low angle of attack).

To resist sliding abrasion, Tungtsen carbide spray is fused to the blade prior to final machining.

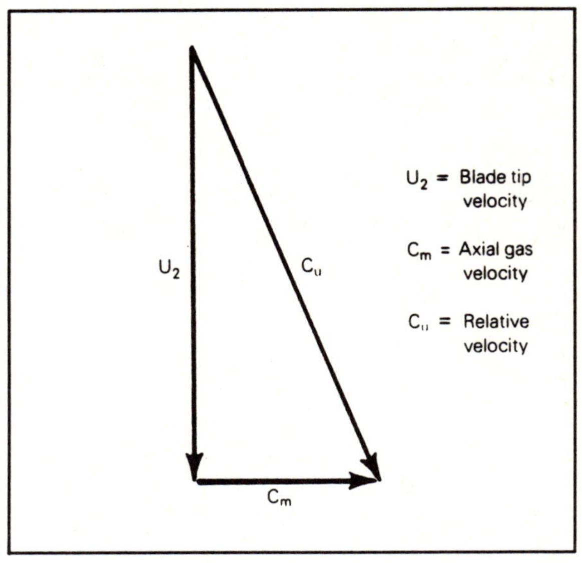

Relative Velocity Diagram

Fig. 18

Figure 18 indicates that relative velocity is a function of blade tip speed and axial gas velocity. By selecting a two-stage fan the blade tip volocity is reduced by a factor of 1.4, and the axial velocity is reduced by a similar factor. Wear is a function of relative velocity to the power of 2.5 - 3.5.

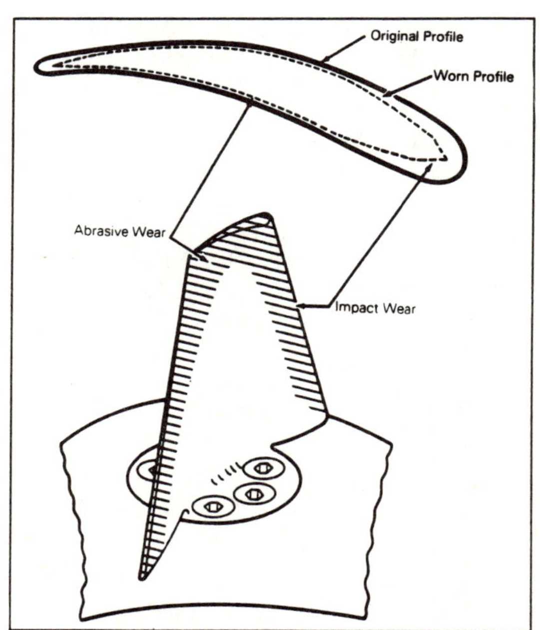

Typical Blade Wear Pattern

Fig. 19

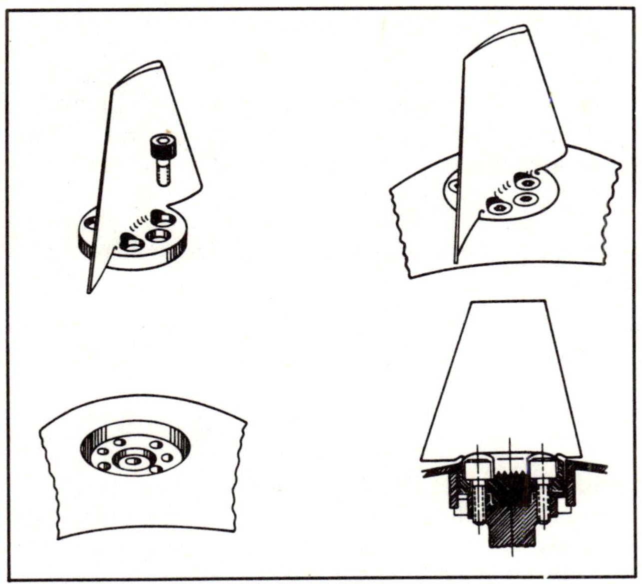

Replaceable Blades

Fig. 20

Even with the above protection and conservative design, wear can occur due to precipitator malfunction. Should blade wear be experienced, the blade set can be unbolted (Figure 20) and a new set installed in 1½ shifts. The worn blades can be taken to the shop and repaired by rewelding the nose and recoating with Tungsten carbide.

To prevent impact wear an impact resistance wear metal is deposited by stick electrode to the leading edge of the steel blade (Figure 21).

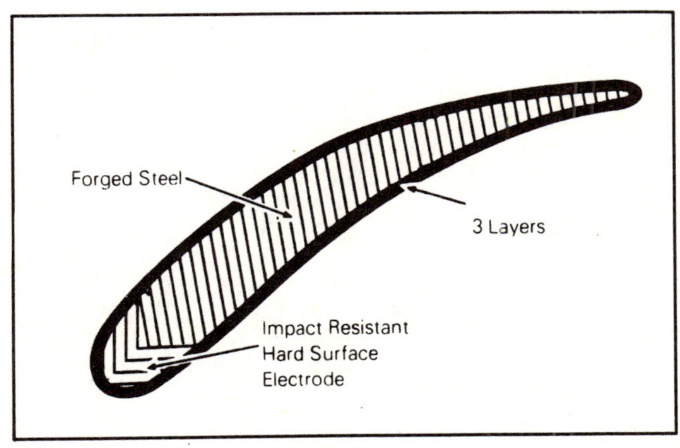

ID Blade Hard Surfacing

Fig. 21

CONCLUSION

The variable pitch Axial Flow Fan has now been developed to a very high specification and is more than adequate to meet the high reliability requirements demanded by modern industrial applications.

The inherent performance characteristics of this type of fan offer substantial savings on operating costs as a result of higher efficiencies obtained over a wide load range.

ACKNOWLEDGEMENTS

The Author wishes to thank the Directors of Davidson and Company for their permission to publish this paper. He would also wish to acknowledge assistance received from colleagues and in particular the help received from Mr. W. R. Reid of the Systems Engineering Group.

F. D. STARTING PROGRAM

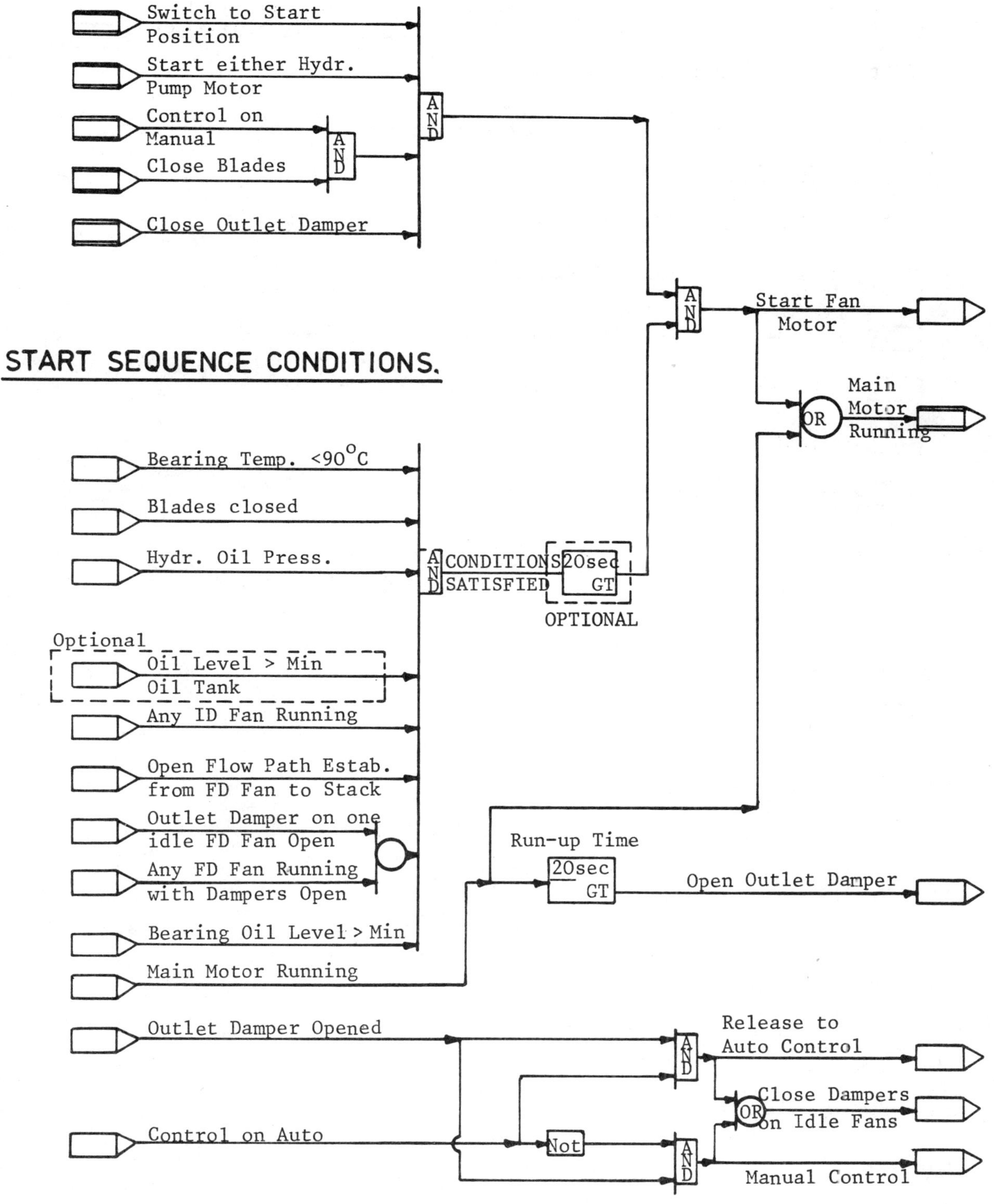

Appendix A.

F. D. SUPERVISION OF OPERATION.

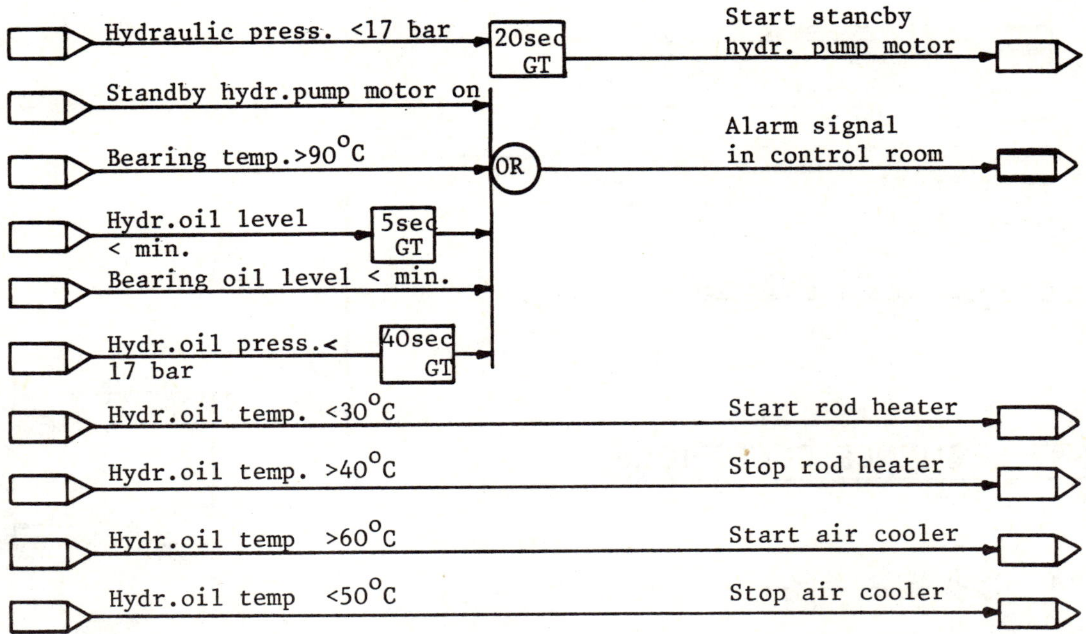

Appendix B.

F.D. SHUTDOWN PROGRAM.

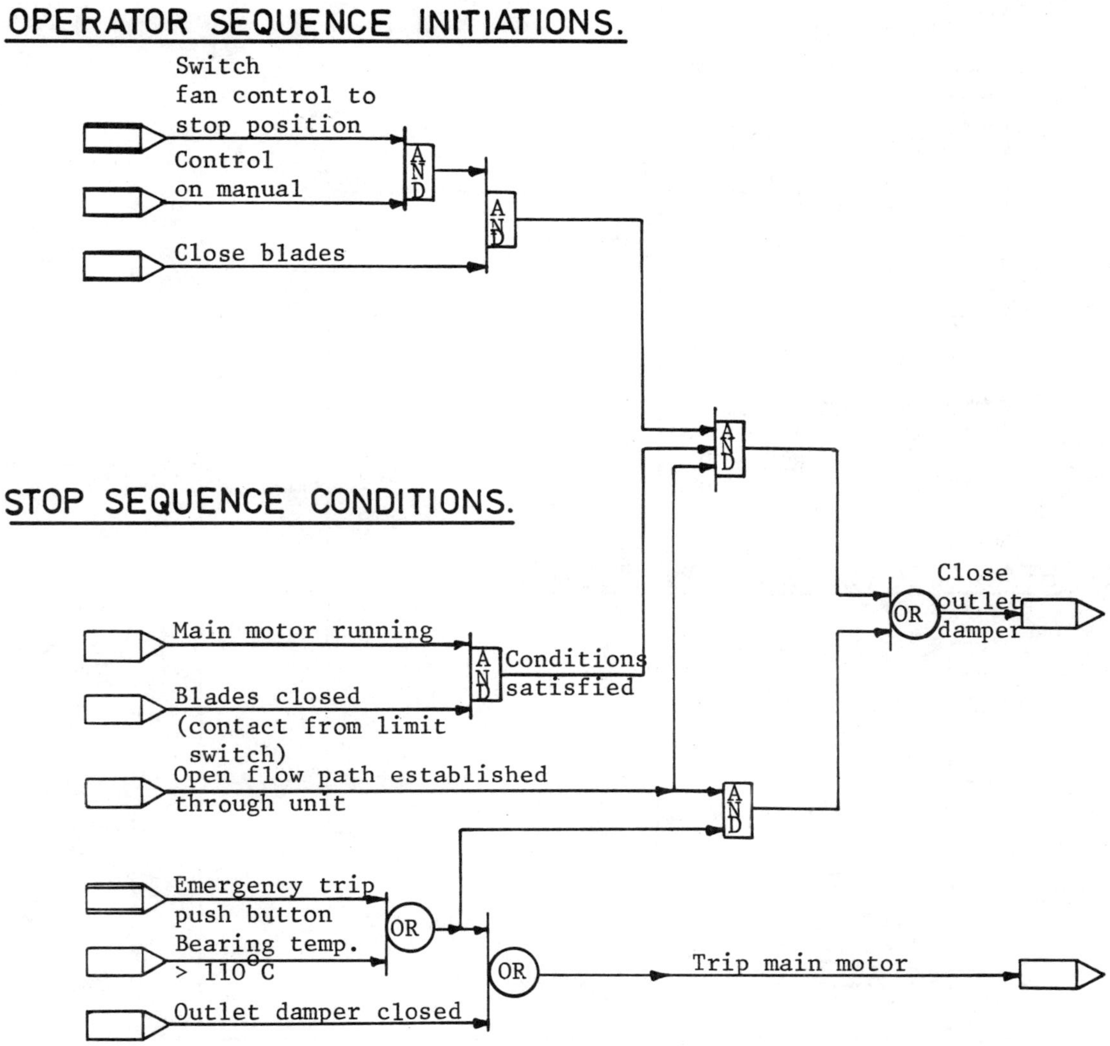

OPERATOR SEQUENCE INITIATIONS.

STOP SEQUENCE CONDITIONS.

Appendix C.

I.D. STARTING PROGRAM.

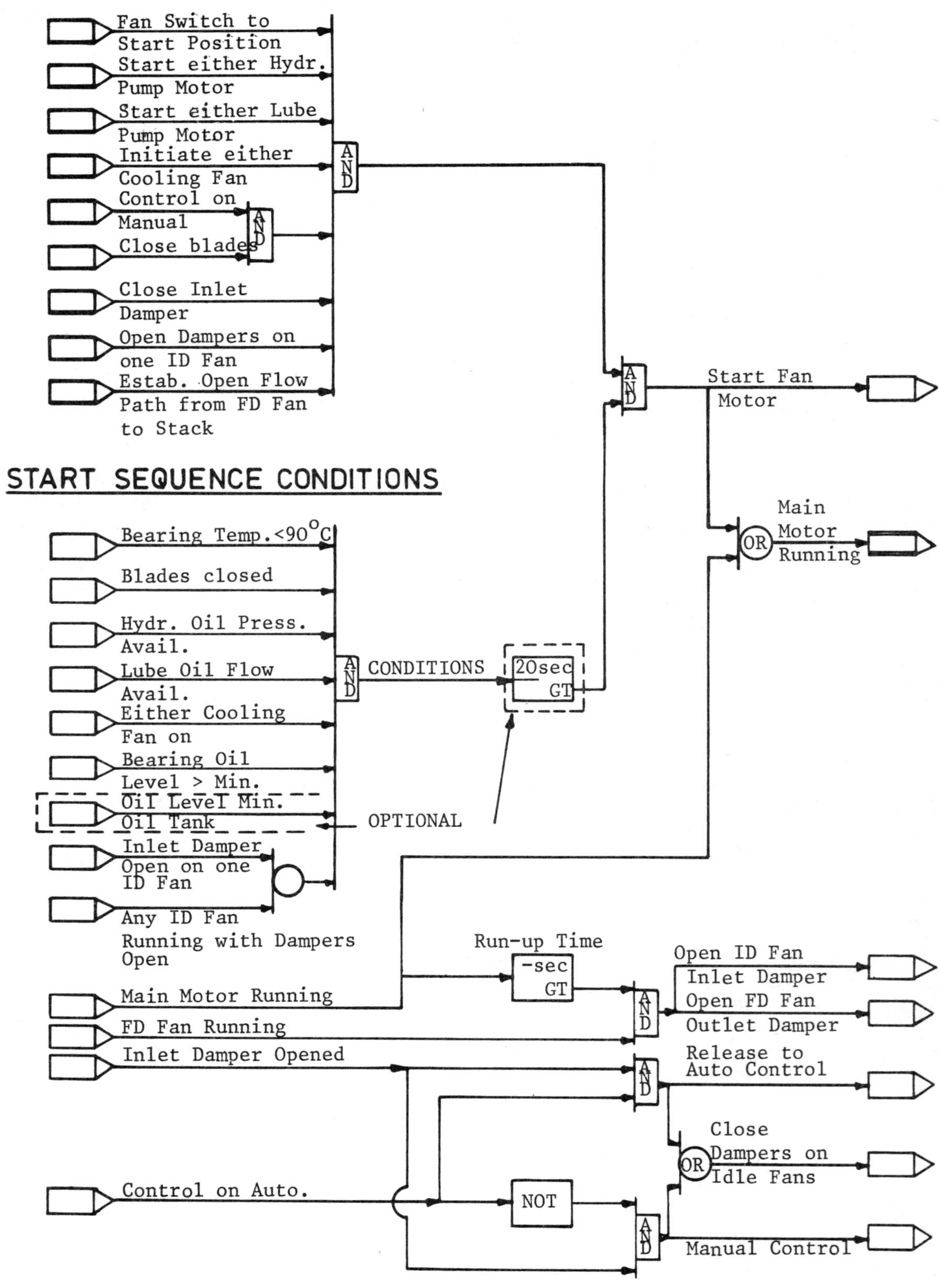

Appendix D.

I.D. SUPERVISION OF OPERATION.

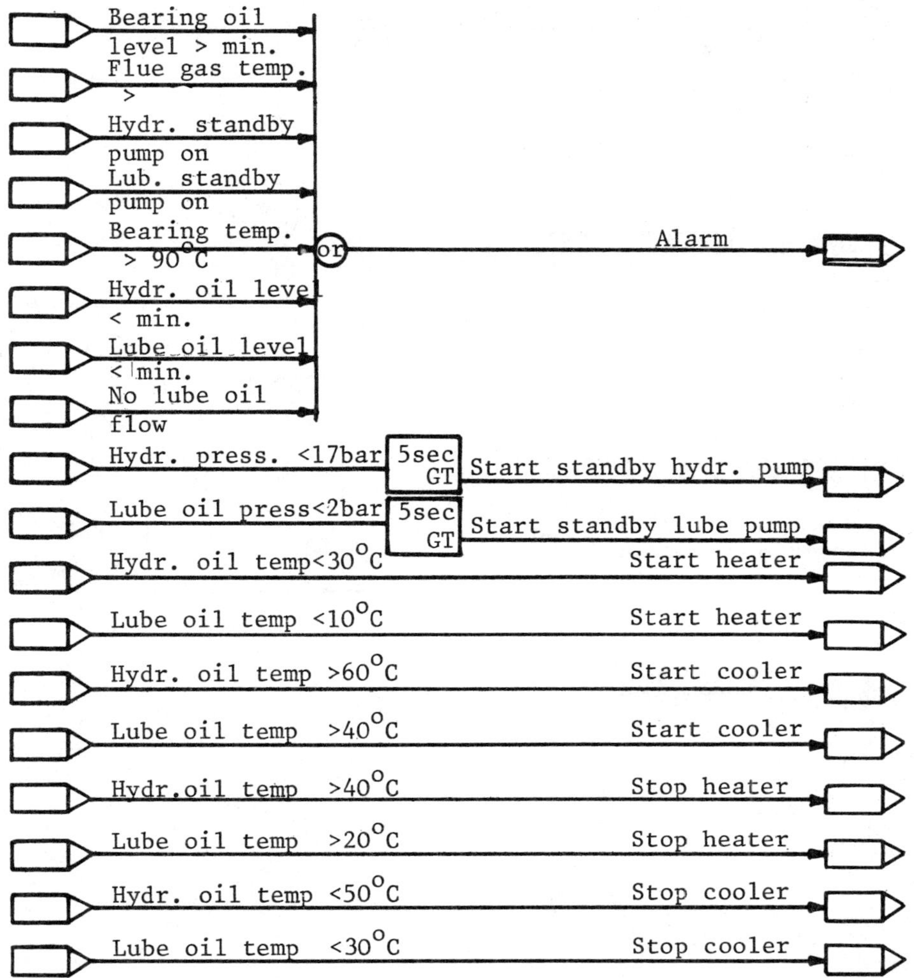

Appendix E.

I.D. SHUTDOWN PROGRAM.

OPERATOR SEQUENCE INITIONS.

Switch fan control to stop position

Control on manual

Close blades

AND

AND

STOP SEQUENCE CONDITIONS.

Main motor running

Blades closed

AND

Conditions satisfied

Open flow path established through unit

Bearing temp >110 °C

Emergency trip push button

or

Inlet damper closed

AND

AND

or

Close inlet damper

or

Trip main motor

Appendix F.

STALL PREVENTION

To prevent the occurence of stall, either the operator or the control system must know where the fan is operating with respect to its stall curve. This can be accomplished in several ways, depending on the degree of protection desired.

1. By monitoring blade angle and head (ΔP compensated for temperature), the operator will be visually aware of his operating point. Acutal blade position can be transmitted electrically to the control room. Pressure increase across the fan is measured and also transmitted to the control room.

   These two inputs when compared to the fan's pre determined (by acutal site testing) performance field will enable the operator to judge his operating point relative to the stall line.

2. The next step is to monitor and transmit a flow signal to the control room, along with the pressure increase.

   This flow signal is readily available because the inlet nozzle of the fan is a calibrated Venturi (Figure G-1).

   These two signals can be manually compared by the performance field by the operator.

3. By putting the head and flow signals into an "X-Y" plotter with the fan performance field already engraved on the plotter (Figure G-2), the operator can scan his operating point with respect to the stall line.

4. A further refinement of Step 3 is to place an alarm mechanism on the "X-Y" plotter.

5. The final step is to utilize a function generator to place the stall curve into the control system. The generated curve is compared to the head and volume inputs. When the operating point approaches the stall line an alarm is sounded. If travel toward stall continued, the controls will automatically reduce the blade angle on both fans.

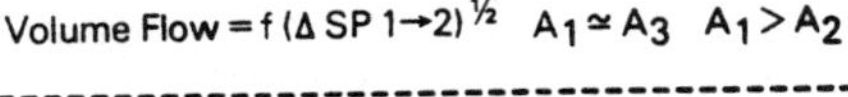

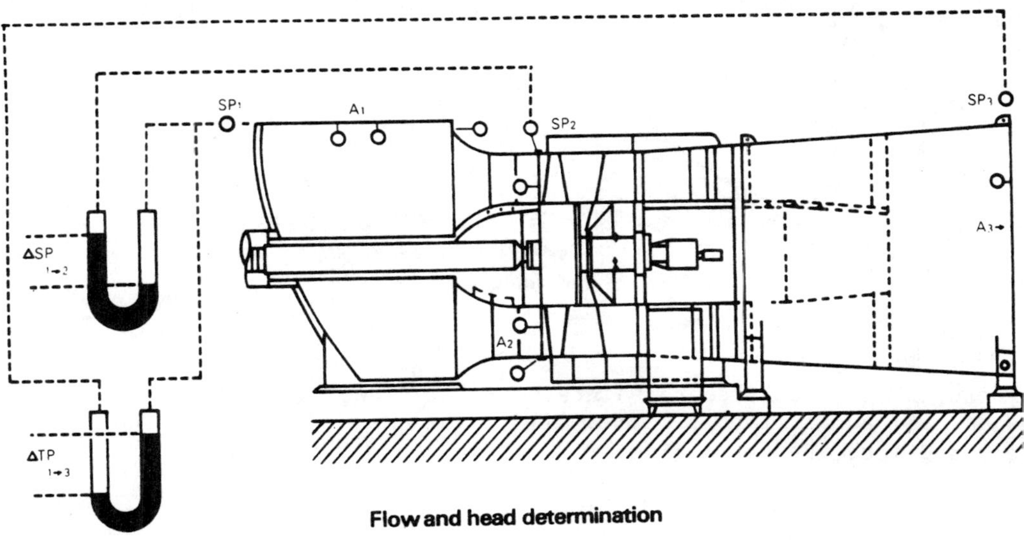

Flow and head determination

Figure G-1

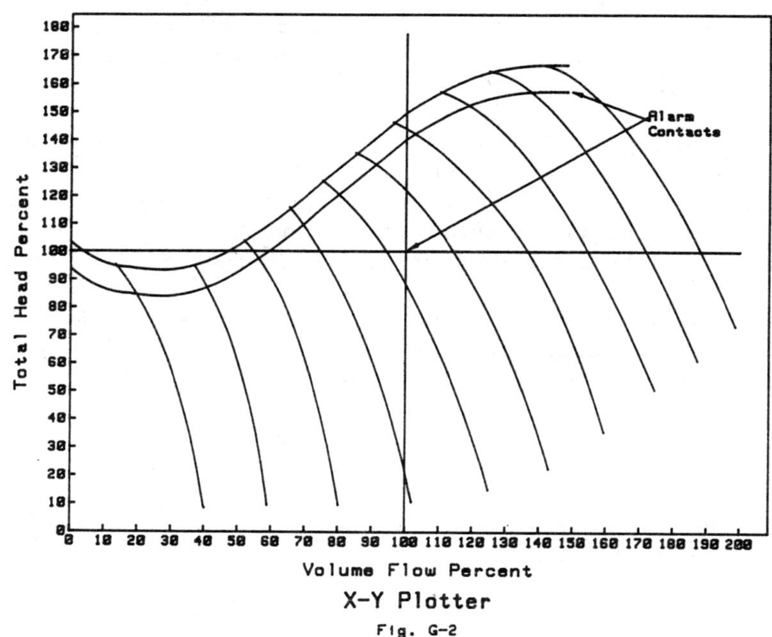

X-Y Plotter

Fig. G-2

It is necessary to close the blades of both fans to prevent fan "B" (Figure G-3) from maintaining the system flow while fan "A" is being run back.

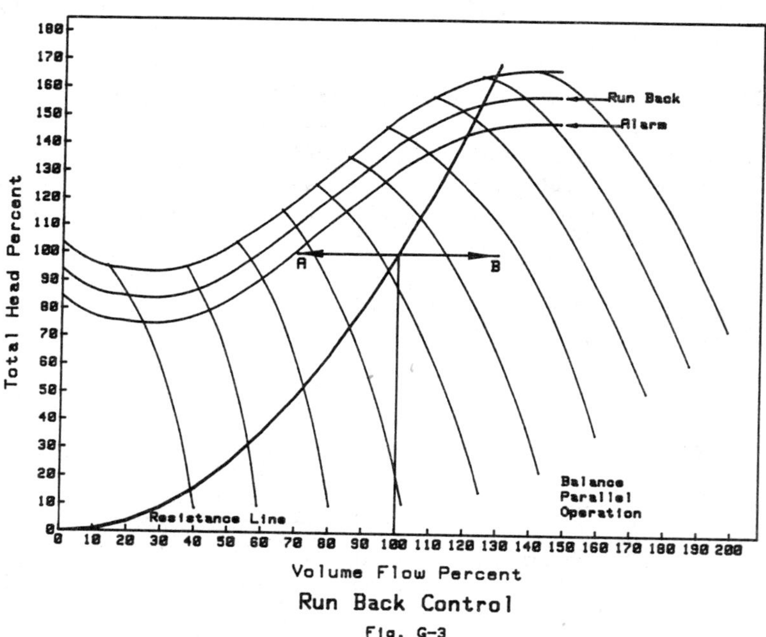

Run Back Control

Fig. G-3

If fan "B" is permitted to maintain system flow, system resistance will be maintained and fan "A" under run-back control will be forced to the left and into the stall line.

If the fan is forced into stall the change in air flow close to the blades can readily be detected by a simple pressure switch. This switch can be used to activate an alarm or as the input of a basic run-back control system. The disadvantage of this type of monitor is that it does not give an "early warning".

If a fan were to operate within the stall region for quite some time it could result in blade failure and mechanical components within the hub could be damaged. Therefore, to avoid such operating conditions the following obvious characteristics of a stalling fan should be noted.

- Increased sound transmission (distinctive)
- Pulsating air flow near the impeller blades
- Often higher vibration levels than normal

Vertical mounting of a single stage axial fan

Vertical mounted standby fan showing ease of interchangeability.

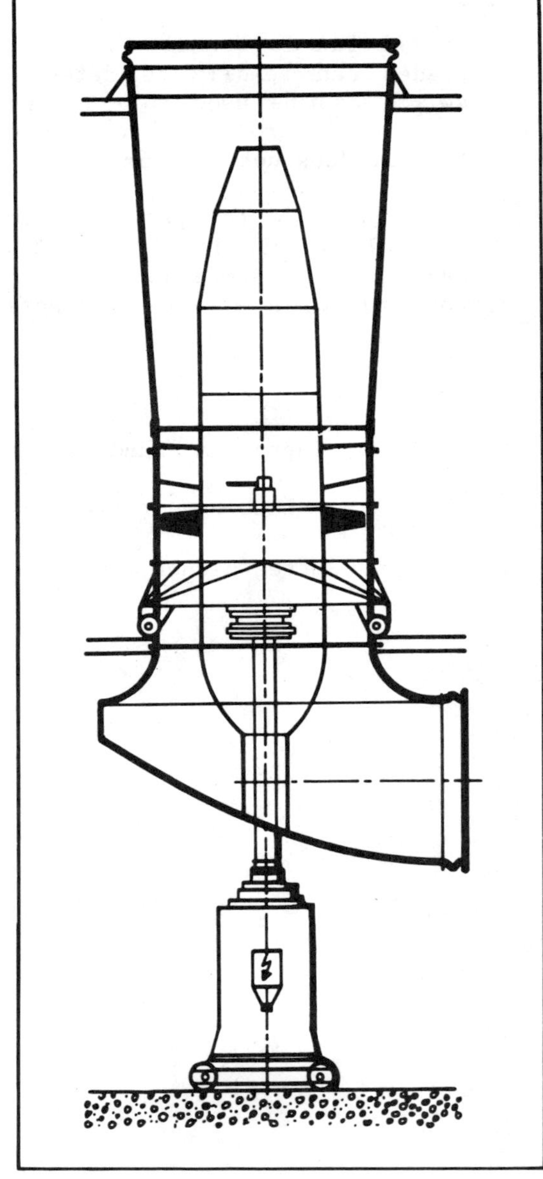

Horizontal mounting of a two stage axial fan

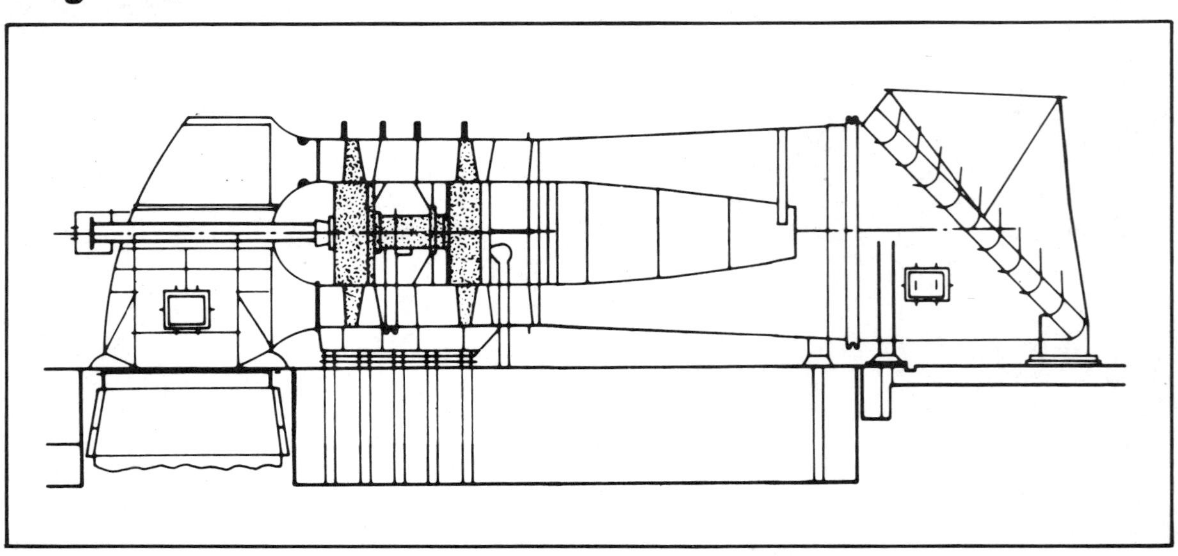

Appendix H – (i).

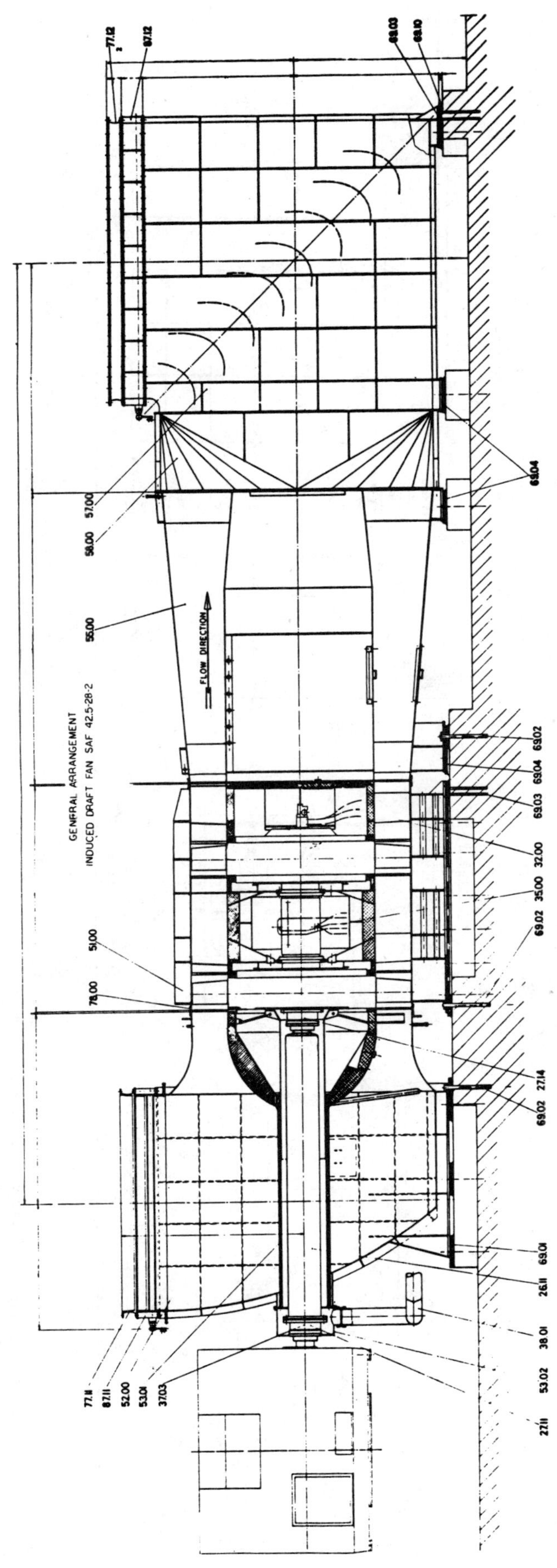

Appendix H - (ii)

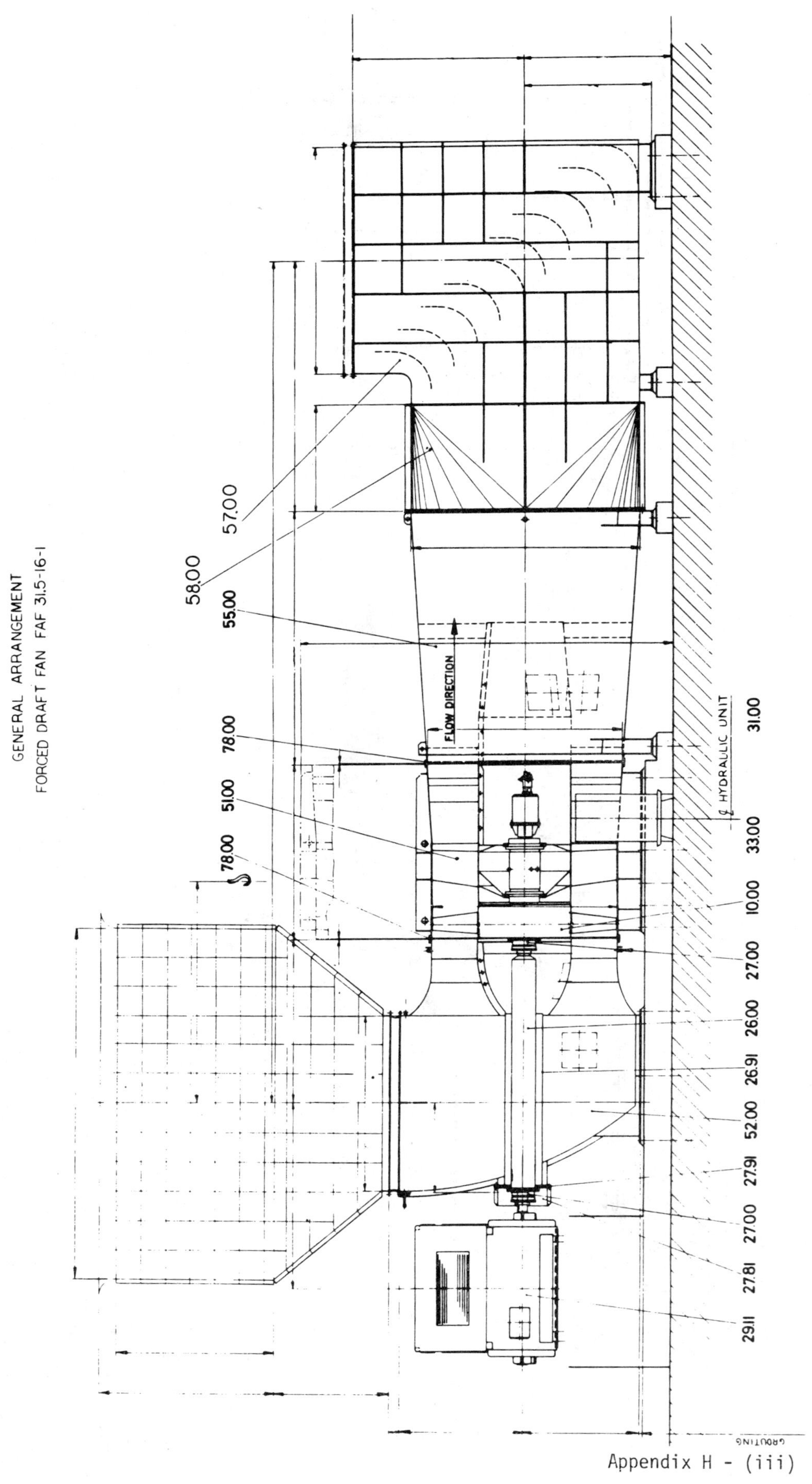

Appendix H - (iii)

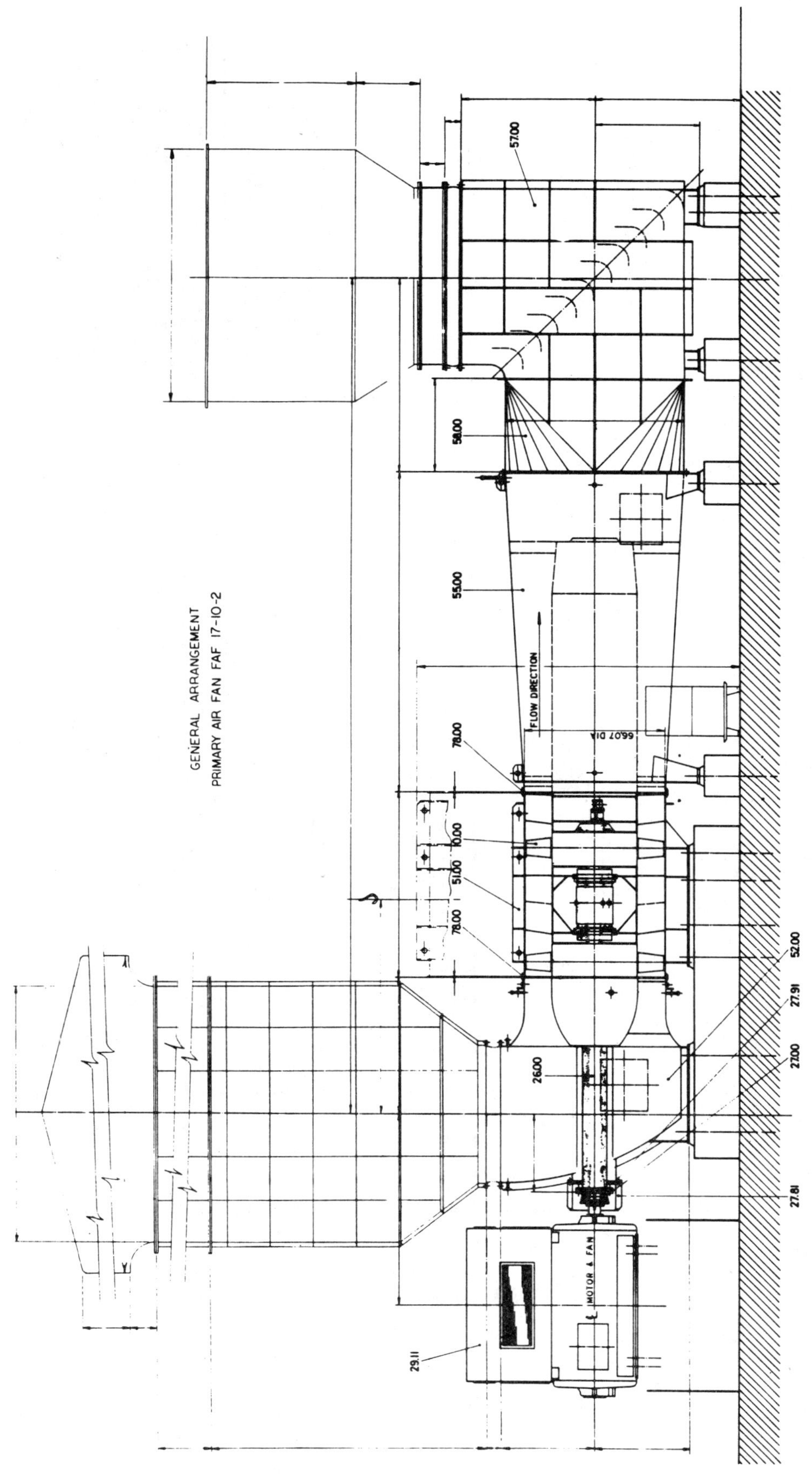

Appendix H - (iv)

GENERAL ARRANGEMENT
INDUCED DRAFT FAN SAF
BEAM FOR ROTOR
REMOVAL
TRACK GAUGE
SECTION A-A
PLATFORM
PLATFORM
EL.
GRADE

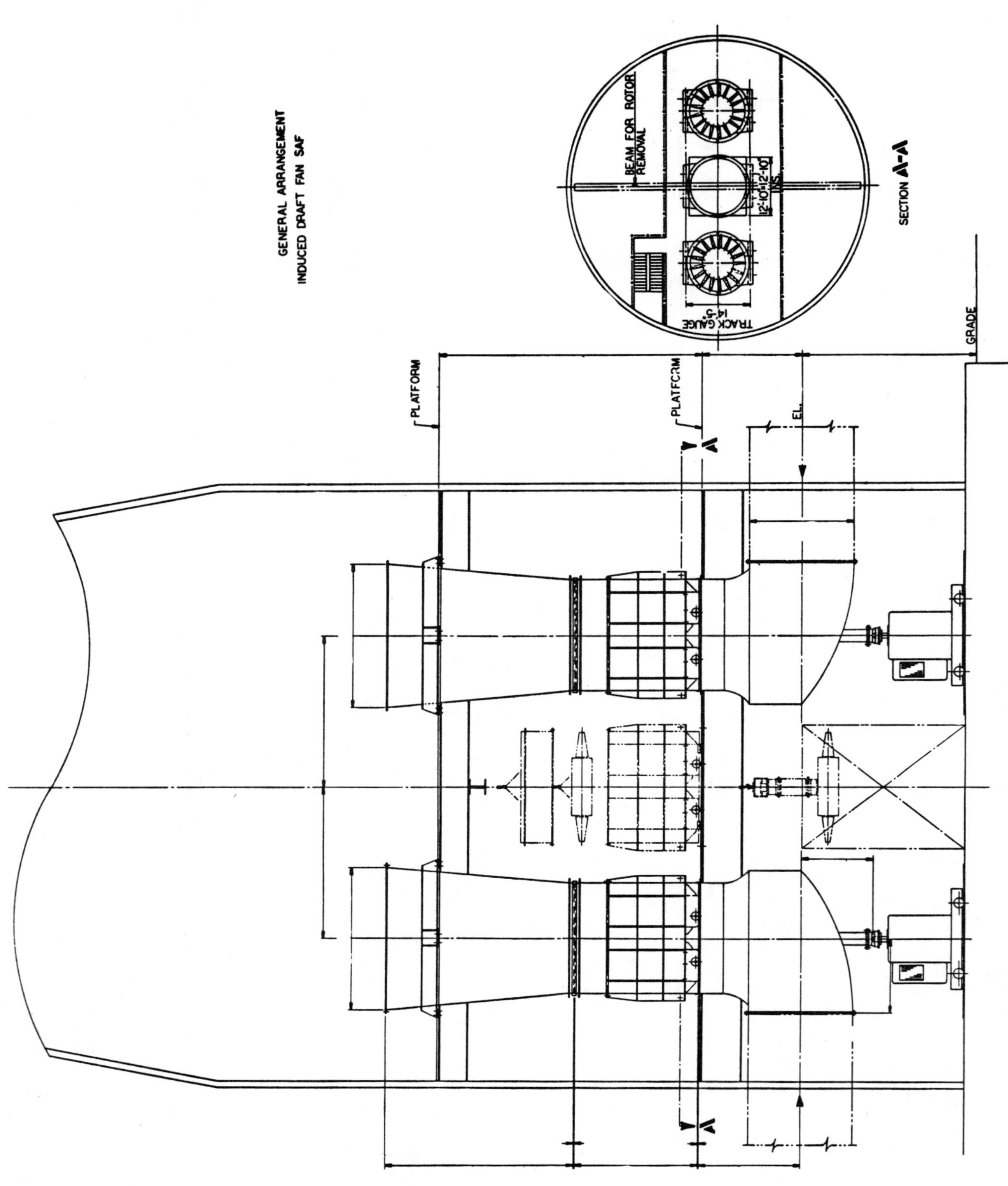

Appendix H - (v)

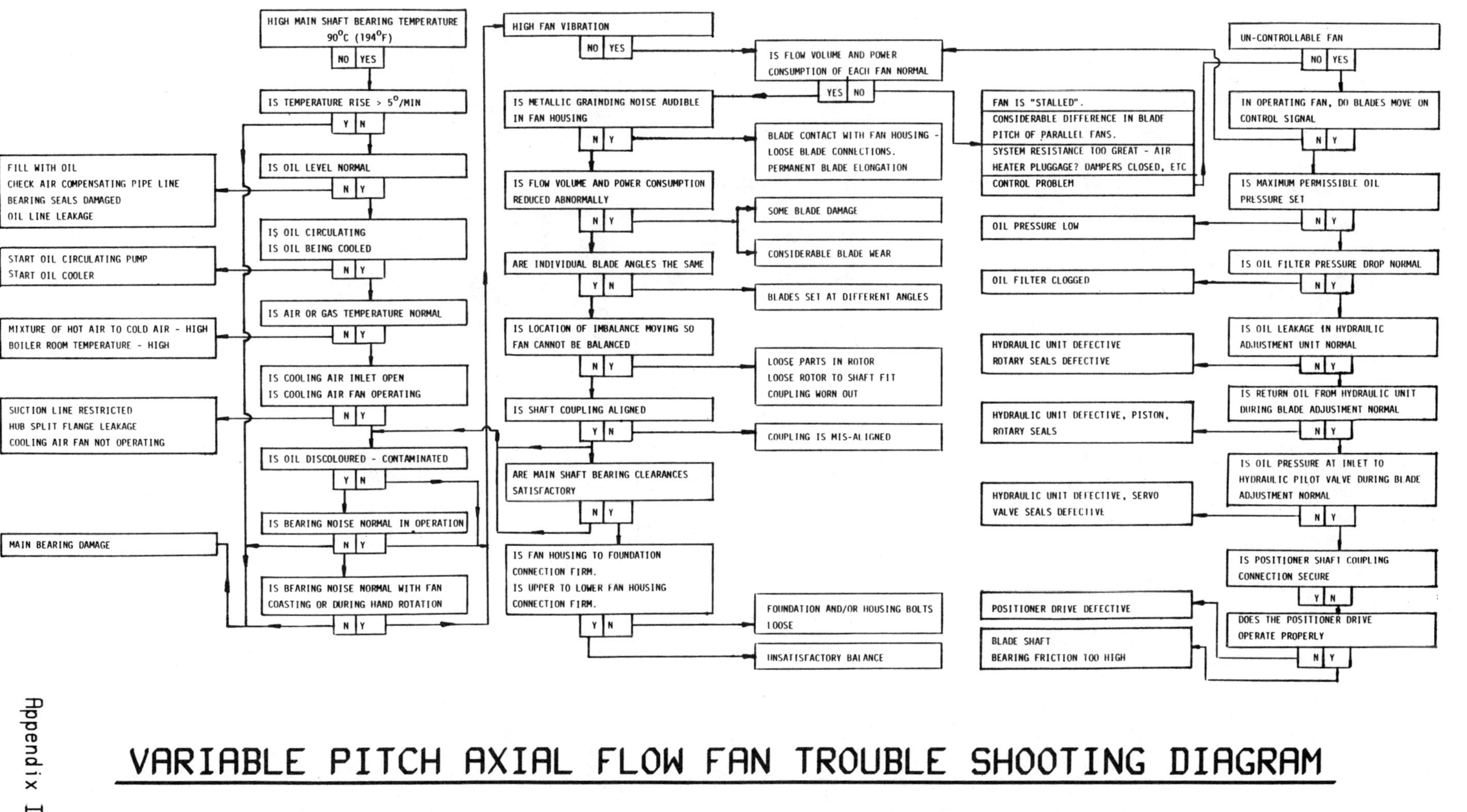

VARIABLE PITCH AXIAL FLOW FAN TROUBLE SHOOTING DIAGRAM

Appendix I

International Conference on

Fan Design & Applications

Guildford, England: September 7-9, 1982

PAPER K4

FLOW MODULATION WITH GOOD ENERGY SAVING ON CENTRIFUGAL FANS WITH INLET GUIDE VANES

C.W. Lack

Flakt Ltd., U.K.

Summary

The use of inlet guide vanes as a method of performance control on centrifugal fans is not new. However, in variable air volume (V.A.V.) systems, as applicable in air conditioning, their use can sometimes lead to problems in their operation.

New lightweight design, positive action inlet guide vane assemblies of the 'floating-ring' concept; overcome the previous problems associated with high actuation forces and consequent maintenance and un-reliability. With the addition of two speed fan operation an extremely energy conscious and stable system can result.

However as with most systems a complete understanding of the control network is necessary if the full energy saving potential is to be realised.

This paper indicates recent developments which have resulted in this reliable solution. With both first order and running costs being an ever increasing factor affecting selection of equipment today; this means of flow modulation is worthy of detailed consideration in the future.

Organised and sponsored by
BHRA Fluid Engineering, Cranfield, Bedford MK43 0AJ, England.

0263 - 421X/82/01 00 - 0001 $5.00
The entire volume can be purchased from
BHRA Fluid Engineering for $82.00

NOMENCLATURE

C	=	constant	-
J	=	moment of inertia	kgm^2
n	=	rotational speed	r/min
pd	=	dynamic pressure	Pa
Δpt	=	total pressure rise	Pa
P	=	fan absorbed power	kW
q	=	air flow	m^3 /sec
s	=	time	sec

1. INTRODUCTION

The aerodynamic principles of inlet guide vanes have been known for a long time, and there is generally no problem with their functioning provided that the centrifugal fan is suited for inlet vane control[1]. To be suitable, the fan must have a stable enough characteristic pressure curve which does not tend to surge because of the turbulance created by the vanes in the inlet eye when turning down.

This latter fact is of importance when fans incorporating inlet vanes are used in V.A.V. (Variable Air Volume) systems within the comfort field of air conditioning. Whereas, particularly in the Industrial field, the resulting operating system characteristic is substantially the basic square law relationship, In V.A.V. applications, the resulting system characteristic whilst having this basic square law relationship element, also has a high constant pressure one. This latter requirement needs to be considered carefully when proposing the use of I.G.V.'s and centrifugal impellers; however more about that aspect later.

It should also be realised that whilst the energy saving aspects of these comfort fans is important, initial prime cost comes into consideration. Hence, whilst the fans are built to high engineering standards to give satisfactory operating life, they are not subject to the severe loadings of their industrial counterparts. They can therefore be constructed in lighter gauge materials. Hence the incorporation of I.G.V.'s into these fan units needs to be compatible with the lighter gauge fan construction to be commercially viable, otherwise their cost, especially on double inlet fans will outweigh that of the basic fan unit.

2. HEAVY-DUTY GUIDE VANE DESIGN

In the past, the mechanical design of the vanes and particularly the mechanism has caused considerable problems due to the need of continuous maintenance and greasing. This was due mainly to the high friction and corresponding high operating torque required for the operation of the actuating mechanism.

The mechanism often involves an adjusting ring and a number of actuating levers, one lever for each vane (see Fig. 1). The adjusting ring is normally carried by a separate large hollow bearing at its centre so as to allow the fan shaft to pass through.

In the case that the adjusting ring is surrounding the vane section, it is normally supported by a number of rollers. The actuating levers are normally connected to the ring via double links, sliding pads in slots or some other more or less sophisticated arrangement so as to overcome the great differences between the paths of the levers and the actuating ring.

It is just these mechanical problems which have over the past years resulted in the doubt about their reliable operation for V.A.V. systems, especially as most fans are of double inlet design necessitating cross linkage between the two I.G.V. assemblies.

3. NEW LIGHTER WEIGHT DESIGN

However, there is a maintenance-free, low friction mechanism now available (see Figs. 2 and 3), which is denoted 'the floating-ring concept' applicable to I.G.V. design. As the name implies, the adjusting ring is floating directly on the inner portion of the vanes and consequently there is no bearing supporting the ring except for the ball joints between the ring and the vanes themselves.

Furthermore, there is no sliding movement between any parts in the mechanism. There is only a small turning movement in the ball-joints. Consequently the friction has been reduced to a minimum and by the selection of ideal material combinations in these components (stainless steel, nylon, ...) any lubrication can be omitted.

4. DESIGN GEOMETRY

This very attractive solution has become possible by inclining axially the radially disposed axes of the vanes (see Fig. 4). Thereby, the paths described by the ball joints at the lever-shaped portions of the vanes will be substantially concentric with the paths of the adjusting ring about the axis of the fan.

During the oscillation, there is substantially no sliding movement in the ball joint but free rotative and angular movement between the parts is permitted as well as lateral movement of the ring.

5. VARIABLE AIR VOLUME (VAV) SYSTEMS

It is necessary to understand the difference in operating characteristics of a VAV system to the many industrial systems whose fans incorporate IGV's. In a VAV system as applicable to building air conditioning the rate at which air is applied to each room/occupied space is varied individually and continuously to suit the instantaneous cooling demands of the room. The VAV system (see Fig. 5a) consists basically of a central unit (a), ducting (b), flow variator (c), and supply air terminal devices (d). The flow variator, which is controlled by the room thermostat (e), demands a constant pressure in the ducting. This pressure is maintained constant by the pressure transducer (f), controlling the inlet guide vanes of the fan.

The pressure drop in the system can be divided into three main parts as is shown in Fig. 5b. The sum of these parts result in an operating system curve far from the standard simple square law relationship.

Hence in selecting the suitability of a centrifugal fan with IGV's it is this resultant $\Delta pt \propto q^2 + C$ system that should be considered. Quite often C can be as great as one third of the overall pressure.

6. FAN INSTABILITY

It is sometimes not realised that many centrifugal impellers of backward bladed design, whilst represented on performance data as having a smooth continuous characteristic of pressure against flow; do in fact exhibit a small order discontinuity close to their peak pressure point. This discontinuity usually increases with impeller width. Many manufacturers attempt to obtain the maximum airflow from a given casing size by incorporating very wide impellers. This results in the maximum performance being obtained within the smallest space envelope.

However for a given resultant pressure rise there is a defined relationship between the blade inlet and outlet radii. The inlet cone throat diameter is dictated by the blade inlet diameter hence there is an optimum width of impeller so as to have the correct inlet throat area/impeller inlet blade area relationship. Any increase in this value will result in the impeller being very susceptable to inlet disturbance with the resultant discharge airflow containing pulsations of a disturbing order. These pulsations can be extremely difficult to deal with, as generally the downstream ducting becomes 'live' to these low frequency disturbances.

This area of secondary instability is shown in Fig. 6; which also indicates a typical VAV system curve. If speed control was used as a means of modulation then entry into this unstable area would be assured[2]. This fact is often overlooked of late with the availability of the low cost (but lower efficiency) prime movers.

7. INLET GUIDE VANE INSTABILITY

Similarly when considering the use of IGV's as a means of control, then the resultant area of instability should be looked at in detail. Generally on 'wider' impeller designs the area of instability can be extremely large, see Fig.7. This had led to the thought that one should not consider their use if a turndown to less than about 50% be required. Below this turndown, simple damper control would have to take over, with its resultant inefficiency.

However by correct impeller/IGV design relationship then the area of instability can be made very small with the modulation over the entire VAV system curve totally stable. The question then is how low can modulation be effected? Normally a turndown to 20% can be achieved with a single speed drive motor; and this is generally sufficient for VAV system use. However by using two speed fan operation turndown to 10-15% is realistic.

8. COMPARATIVE FAN ENERGY

If one compares the energy usage of a backward bladed centrifugal fan with IGV's to other forms of fan performance modulation, then the resultant power demand curves as shown in Fig. 8 result. Such a comparison must include all power losses in the drive train as it is input power to the drive motor that the operator pays for. The resulting curves should also reflect the minimum turndown achievable before the onset of instability. Curve (4b) reflects the power absorbed of a centrifugal fan + IGV's with single speed operation.

Should two speed operation be introduced then control would follow curve (4a) which can be seen to be more energy conscious. The use of two speed operation + IGV control is shown in more detail on Fig. 9. This figure highlights the further consideration to be taken account of; namely the relationship of the system line to the resultant lower speed area of instability as well.

In the past with the previously described doubt in stability on the wider impeller design then the added complication of two speed operation would have been considered to be compounding the problem. However completely safe operation can be assured with correct impeller/IGV design.

The two speed operation can be obtained by two means:-

a) a dual speed motor and belt drive,

or

b) a two motor drive arrangement with either varying motor speeds or differing belt drive ratios.

This latter alternative means that the two fan speeds, and hence the resultant 'step' in the power curve as shown in Fig. 9., can be positioned to be at a slightly greater airflow than the peak usage value. Thus the most beneficial power saving results.

9. CONTROL SYSTEM DESIGN

Any such control method is liable to incorrect setting-up and indeed operation. Hence it is important that the necessary control system including the actuator whether of electric or pneumatic form, be recommended by the fan supplier.

If this is not done then quite often a 'grey' area develops between the fan manufacturer and the equipment installer. Then the controls specialist, who has an overall responsibility for the total building installation, is expected to have detailed knowledge of this crucial control area.

Such typical information is shown in the Appendix which indicates the flow and circuitory recommended for a single fan with IGV's under two-speed operation and utilising electro-mechanical control. Not only should the necessary components and circuitory be clearly shown but the procedure for adjustment be given.

By covering the controls in this detail then the likelihood of achieving the possible energy saving is enhanced. There are at present too many installations that fail to achieve their expected energy levels due to incorrect adjustment of the necessary control systems.

By modern design techniques being applied to the manufacturing of both fans and their control devices then prime unit costs can be reduced. Together with more efficient operation of the IGV's and accurate matching of actuators and control networks then a more responsive and energy saving means of modulation results - to the satisfaction of the end user.

10. REFERENCES

1) ECK Dr Ing. B. Fans - Design and operation of centrifugal axial flow and cross flow fans. Pergaman Press.

2) Patterson N.R. Fan selection and control in high velocity VAV systesm. ASHRAE Journal April 1977.

APPENDIX. TWO SPEED OPERATION - WITH PRESSURE CONTROL OF ONE FAN

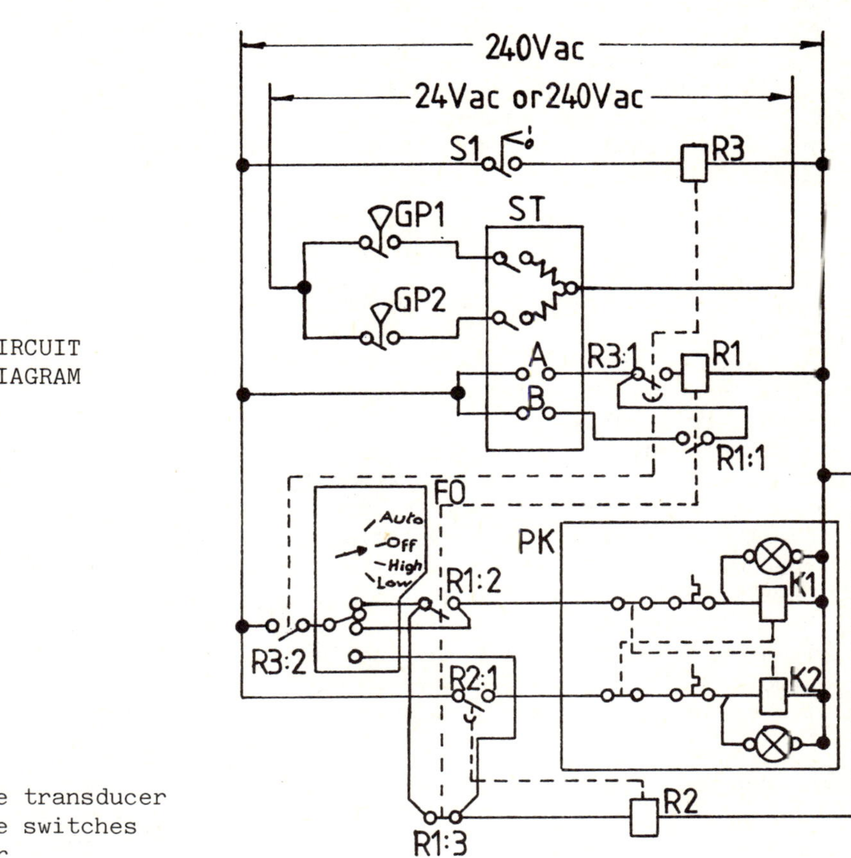

CIRCUIT DIAGRAM

DESIGNATIONS

GP	=	Pressure transducer
GP1, GP2	=	Pressure switches
ST	=	Actuator
RC	=	Control station
TO1, TO2	=	Voltage relays
PE1, PE2	=	Electro-pneumatic relays
R1	=	Auxiliary relay
R2	=	Time-delay relay (0-30 s) for delaying closure
R3	=	Time-delay relay with one pair of undelayed contacts and one pair of delayed contacts (0-120 s)
R4	=	Time-delay relay with one pair of contacts for delaying closure and two pairs of contacts for delaying opening (0-200 s)
S1	=	Starting switch or the like
PK	=	Pole-changing switch (drawn for separate windings) K1 = High-speed contactor K2 = Low-speed contactor
FO	=	Selector switch
F	=	Fan with variable inlet guide vanes

ST
R1
GP1
GP2
R2
F
PK

FLOW DIAGRAM

METHOD OF OPERATION ON ELECTRO-MECHANICAL CONTROL

The system is started by means of starting switch S1. Selector switch FO is used to select whether the fan is to be operated manually or automatically, at high or low speed. Manual operation is necessary for adjustment of auxiliary contacts A and B.

Automatic operation: Pressure switches GP1 and GP2 control actuator ST which controls the inlet guide vanes in fan F so that a constant static pressure will be maintained in the duct in which the pressure switches are fitted. GP1 is set to a higher pressure p_1 and GP2 is set to a lower pressure p_2. When the pressure has risen above p_1, ST will move the guide vanes in the closing direction. When the pressure has dropped below p_2, ST will move the guide vanes in the opening direction. In the deadband between p_1 and p_2, ST is inoperative. ST is equipped with two adjustable auxiliary contacts A and B which control change-over between high speed and low speed of the fan.

When the system is started, the inlet guide vanes are open and the auxiliary contacts A and B are closed. When S1 is closed, the non-delayed contacts R3:2 of time-delay relay R3 will close and time-delay relay R2 will then be energised. After about 10 s, contacts R2:1 will close and the fan will start at low speed. If the pressure in the duct is then too low, ST will move the inlet guide vanes in the closing direction until the correct operating point is reached. Contacts A will open as soon as the inlet guide vanes begin to close. On the other hand, if the pressure is too low, the inlet guide vanes will remain in the fully open position. After about 120 s, contacts R3:1 will close, thus causing contacts R1:1 and R1:2 of auxiliary relay R1 to close and R1:3 to open. The fan will then be switched over to high speed. Since the pressure will now presumably be too high, ST will move the guide vanes in the closing direction until the correct operating point is reached. When the guide vanes start to close, contacts A will open. Contacts B will remain closed and R1 will close its own holding circuit across contacts R1:1. If the guide vanes should close to a position at which the pressure is about 5% lower than that at which the fan is switched over from low speed to high speed, contacts B will open. The holding circuit of R1 will then be opened, and the fan will be switched back to low speed via R2 after a delay of about 10 s.

Contacts A and B are controlled by cams or the like in the actuator. The change-over points for A and B must be individually adjustable.

ADJUSTMENT

The end positions of the actuator and any linkage between the actuator and the control lever of the guide vanes are adjusted so that the guide vanes can be moved between fully open and fully closed positions without binding. This applies tc both fans if two are being controlled.

On parallel operation, the inlet guide vanes of both fans must always assume the same positions during the control process.

To avoid instability in the system on change-over between the two fan speeds, change-over from low speed to high speed must take place at a somewhat higher pressure than the change-over from high speed to low speed. This is achieved by means of auxiliary contacts A and B. A and B are both closed when the inlet guide vanes are fully open. As soon as the inlet guide vanes have started to close, A will open. B will remain in the close position and will not open before the inlet guide vanes have closed to the 'half-open' position, i.e. the change-over from high speed to low speed.

Setting of contacts B; Select low speed manually. Increase the set point of pressure so that the guide vanes open fully. Measure the static pressure in the duct in which the pressure switches (pressure transducer) are fitted. Select high speed manually. Reduce the set point of pressure so that the guide vanes close to a 'half-open' position which gives a pressure that is about 5% lower than that measured when the fan was running at low speed and the guide vanes were fully open. Contacts B should open when the inlet guide vanes are in this 'half-open' position.

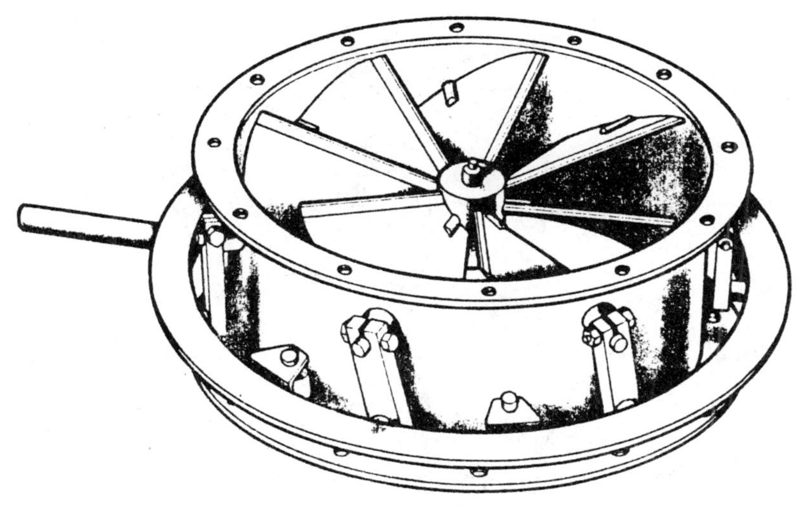

Fig. 1. HEAVY DUTY GUIDE VANE ASSEMBLY

Fig 2. LIGHTER WEIGHT DESIGN OF GUIDE VANE ASSEMBLY

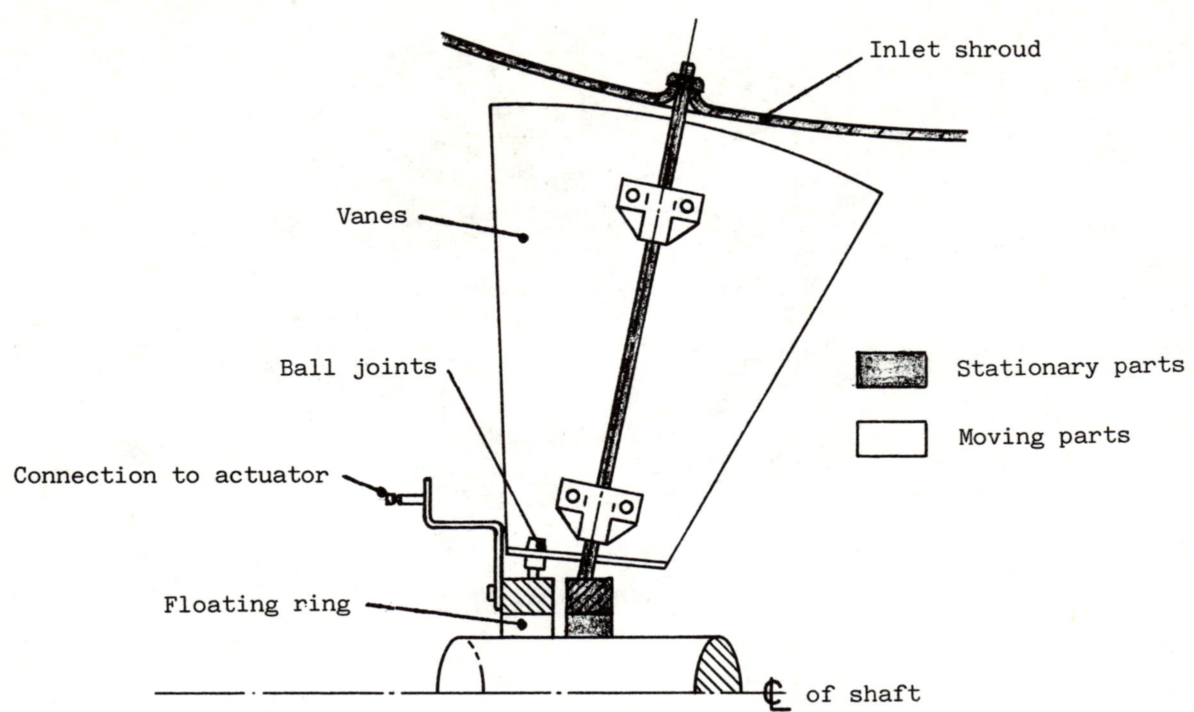

Fig. 3. 'FLOATING-RING' CONCEPT

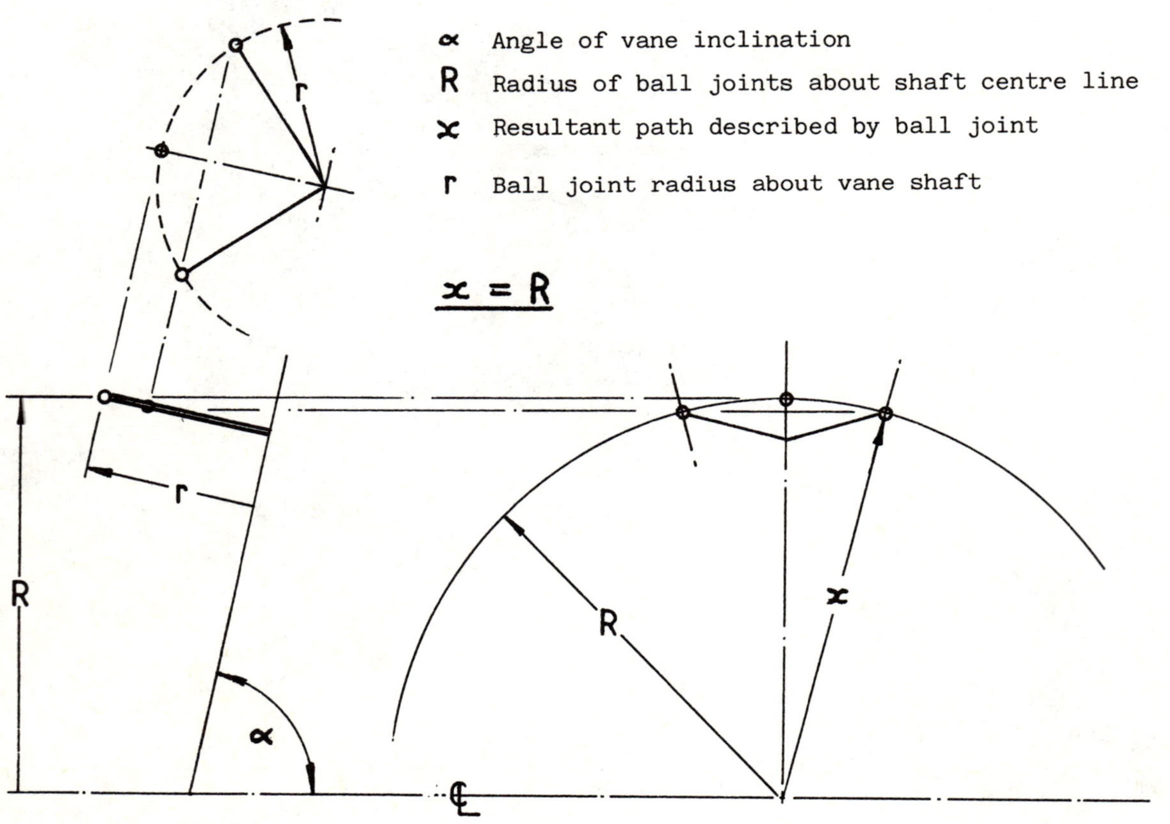

Fig. 4. BASIC DESIGN GEOMETRY

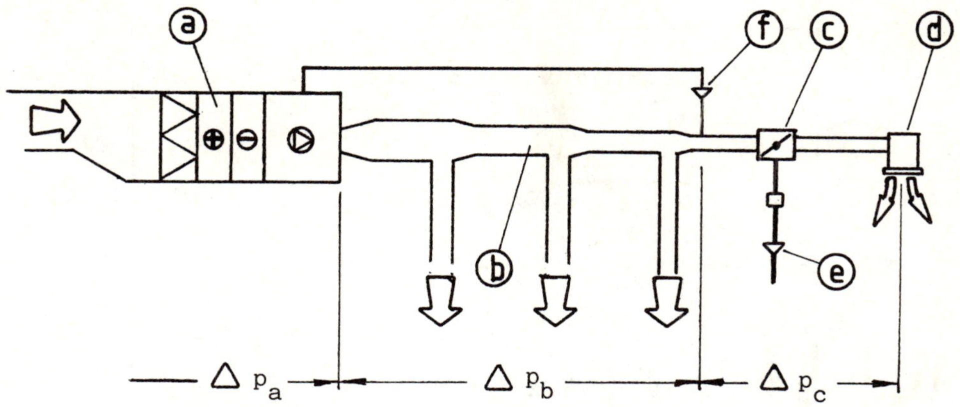

The pressure drop in the system can be divided into three parts:

Δp_a: Pressure drop in the air handling unit, which varies basically as the square of the air flow.

Δp_b: frictional pressure drop in the ducts, which varies as the square of the air flow.

Δp_c: constant pressure drop across the flow variator. This often amounts to between 25% and 50% of the total pressure drop in the system.

Fig. 5a. VARIABLE AIR VOLUME (VAV) SYSTEM

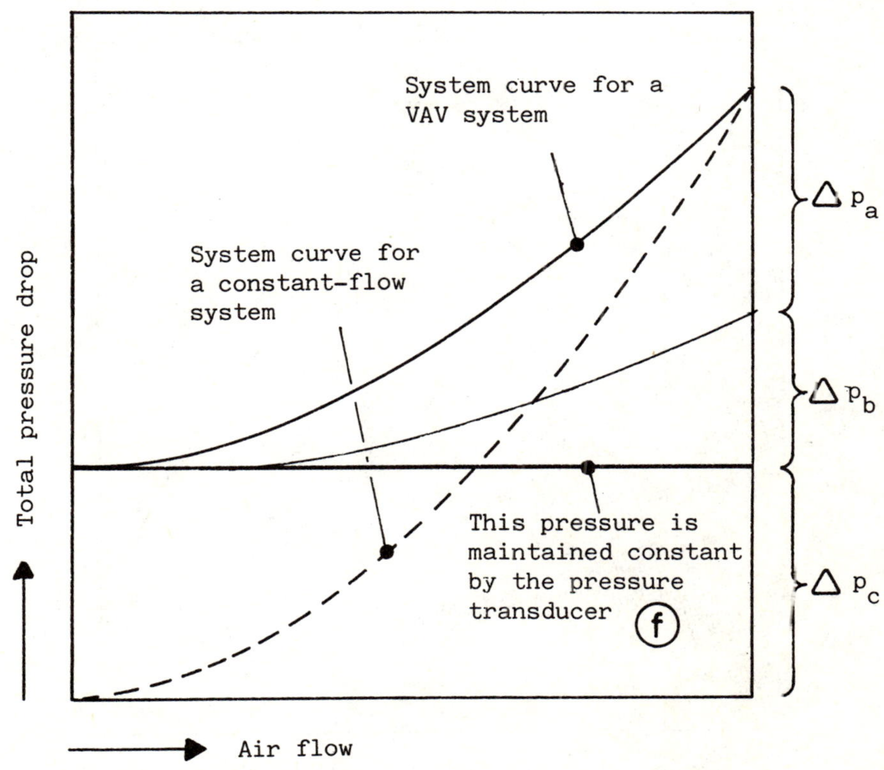

Fig. 5b. SYSTEM PRESSURE DROP IN A VAV SYSTEM

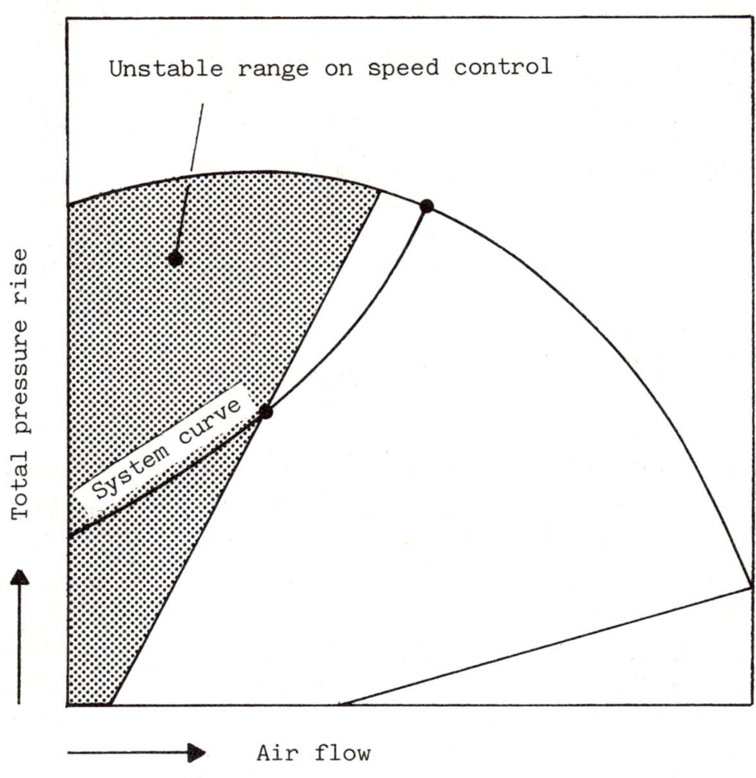

Fig. 6. UNSTABLE RANGE ON SPEED CONTROL

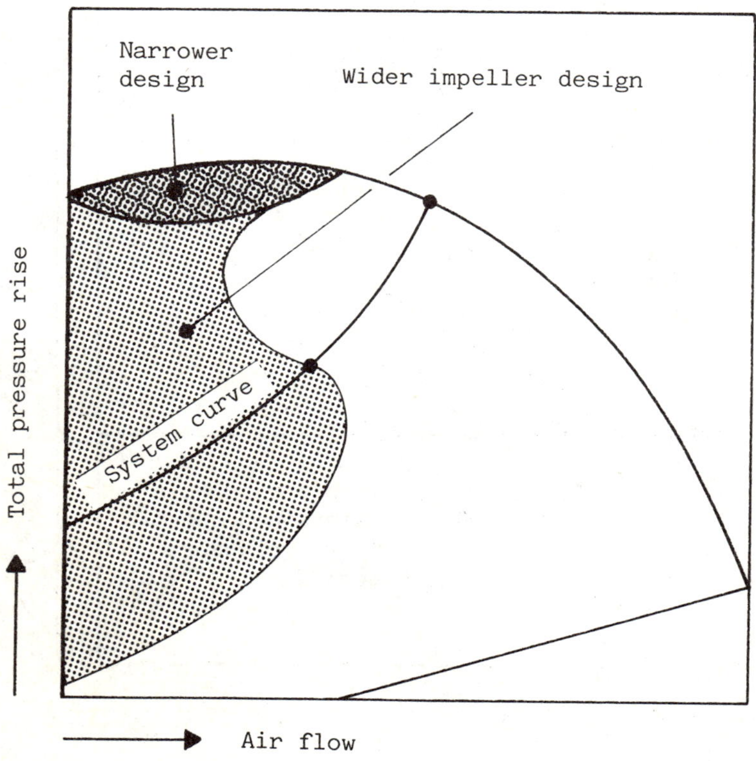

Fig. 7. UNSTABLE RANGE ON INLET GUIDE VANE CONTROL

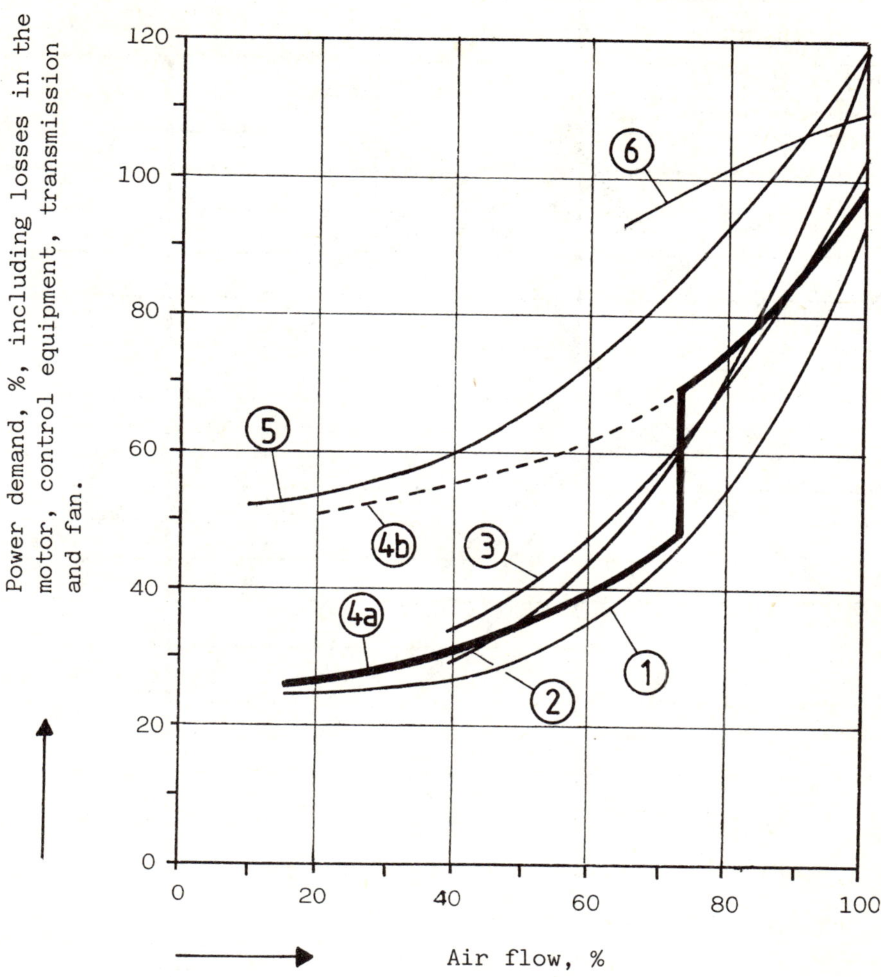

(1) Axial-flow fan with variable pitch blades.

(2) Speed control by frequency control.

(3) Speed control by eddy-current coupling.

(4a) Inlet guide vane control with two-speed fan.

(4b) Inlet guide vane control with constant speed of the fan.

(5) Damper control; fan with forward-curved blades.

(6) Damper control; fan with backward-curved blades.

Fig. 8. COMPARISON OF THE POWER DEMANDS AND CONTROL RANGES FOR VARIOUS METHODS OF CONTROL IN A VAV SYSTEM.

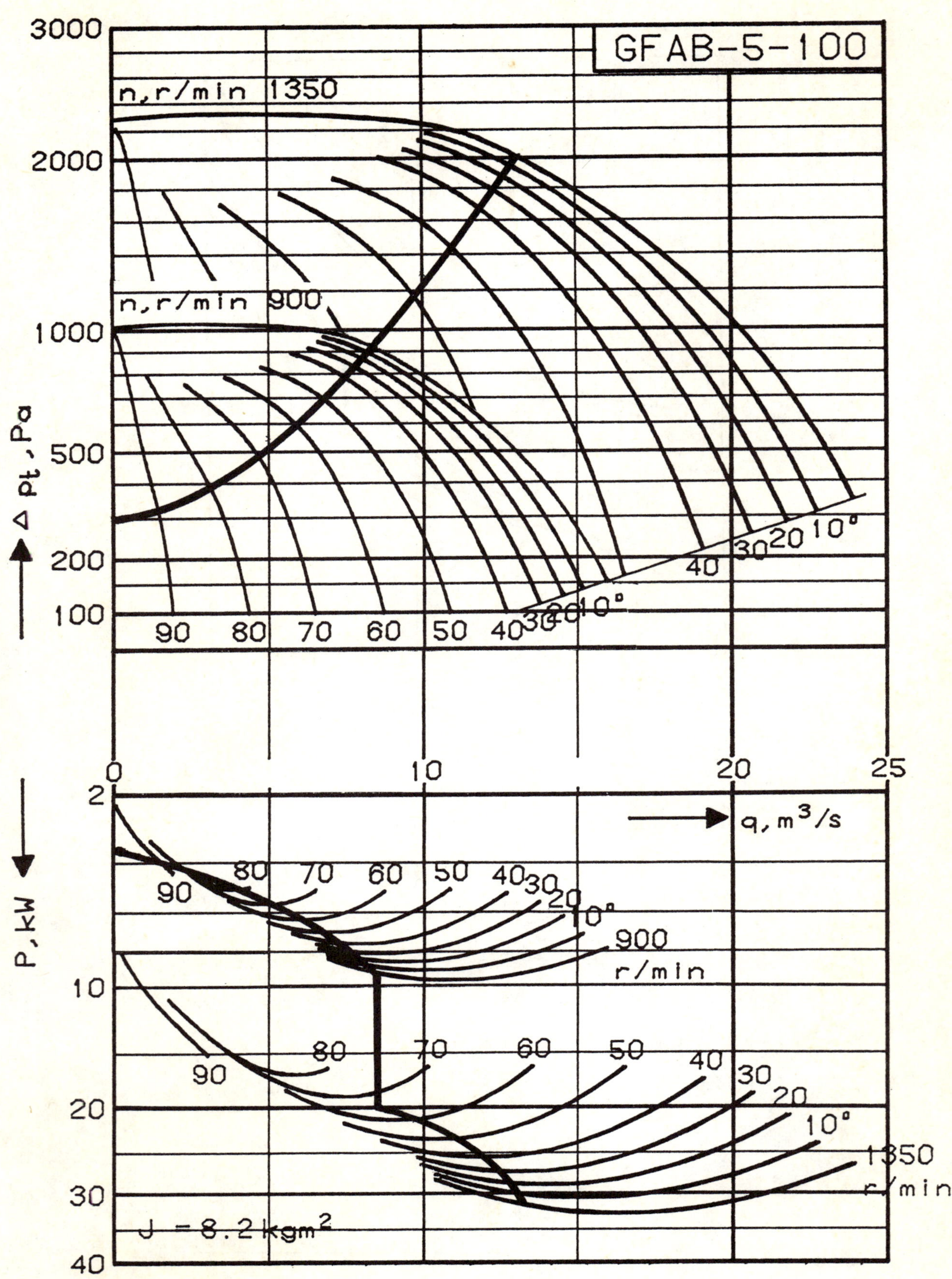

Fig. 9. TYPICAL PERFORMANCE CHART PLOTTED BY COMPUTER FOR 1000mm DIA. FAN UNIT AT 1350 r/min AND 900 r/min.